Plasmonic Nanosensors for Biological and Chemical Threats

Biological and chemical warfare agents, including viruses, bacteria, and explosive and radioactive compounds, can induce illness or death in humans, animals, and plants. Plasmonic nanosensors as detection tools of these agents offer significant advantages, including rapid detection, sensitivity, selectivity, and portability. This book explores novel and updated research on different types of plasmonic nanosensors for analysis of biological and chemical threat agents. It covers a brief theory of plasmonic nanosensors, summarizes the state-of-the-art molecular recognition of biological and chemical threat agents, and describes the application of various types of nanosensors in the detection of these threat agents.

This book

- Brings together recent academic research from an interdisciplinary approach, including chemistry, biology, and nanotechnology.
- Discusses current trends and developments.
- Describes applications of a variety of different types of plasmonic nanosensors.
- Explores outlooks and expectations for this technology.

Showcasing the latest achievements in plasmonic nanosensors, this book will appeal to researchers of materials, chemical and environmental engineering, and chemistry who are interested in exploring the application of sensors to support environmental monitoring and global health.

Plasmonic Nanosensors for Biological and Chemical Threats

Edited by
Adil Denizli

CRC Press
Taylor & Francis Group
Boca Raton London New York

CRC Press is an imprint of the
Taylor & Francis Group, an **informa** business

ISBN: 978-1-032-60474-9 (hbk)
ISBN: 978-1-032-60475-6 (pbk)
ISBN: 978-1-003-45931-6 (ebk)

DOI: 10.1201/9781003459316

Typeset in Times
by Deanta Global Publishing Services, Chennai, India

Contents

Chapter 4

Chapter 5

Preface

In an era marked by evolving biological and chemical threats, the imperative for advanced detection technologies has become increasingly evident. This book, *Plasmonic Nanosensors for Biological and Chemical Threats*, delves into the realm of cutting-edge research and innovation in the field of nanoscale sensing, specifically focusing on the utilization of plasmonic nanosensors.

The introductory chapter sets the stage by offering a comprehensive overview of biological and chemical threat agents, laying the foundation for understanding the urgency and significance of efficient detection mechanisms. As we navigate through the subsequent chapters, readers will be introduced to the key concepts of plasmonic nanosensors, exploring the principles that underpin their functionality and potential applications. Chapters 3 through 15 traverse a diverse landscape of topics, each dedicated to a specific aspect of plasmonic nanosensors. From the integration of synthetic biology in understanding biological and chemical threats to the development of microfluidic-based and surface-enhanced Raman spectroscopy-based plasmonic nanosensors, the book provides a detailed exploration of various methodologies. Readers will encounter specialized chapters addressing the detection of bacterial threats, nerve agents, viruses, toxins, and even the timely application of plasmonic nanosensors in the fight against the COVID-19 pandemic. The comprehensive coverage extends to molecularly imprinted polymers, explosive detection, and alternative analytical methods, ensuring a well-rounded understanding of the subject matter.

This compilation of chapters is not merely a collection of scientific findings; it is a testament to the collaborative efforts of researchers and experts who are dedicated to advancing the frontier of detection technologies. It is our hope that this book serves as a valuable resource for scientists, engineers, and students alike, inspiring further exploration and innovation in the critical domain of biosensing and threat detection. As the pages unfold, we invite you to embark on a journey through the fascinating world of plasmonic nanosensors and their pivotal role in fortifying our defenses against biological and chemical threats.

About the Editor

Prof. Dr. Adil Denizli was born in 1962 in Ankara. In 1985, he graduated from the Department of Chemical Engineering at Hacettepe University, Ankara, Turkey. He received his Master of Science (M.Sc.) and his Doctor of Philosophy (Ph.D.) degrees in 1986 and in 1992 at Hacettepe University, respectively. He became an Associate Professor of Chemical Technologies in 1994. Dr. Denizli has been a Professor in the Department of Chemistry, Hacettepe University since 2000. Dr. Denizli has written over 650 articles and 90 book chapters. He has published more than 200 popular science articles. He has 7 patents. He is one of the most-cited scientists in the chemistry and engineering field (over 30,000 citations, h-index of 88). He served as an associated member of the Turkish Academy of Sciences from 1999 to 2007, and as a full member since 2007. Dr. Denizli received "The Scientific and Technological Research Council of Turkey (TUBITAK) Encouragement Award" in 1998 for his studies on "Biomedicine, Biotechnology and Environmental Applications of Polymeric Matrices." He also received the TUBITAK Science Award and Basic Science Award of Popular Science Magazine in 2006. He worked as an editor for several national and international books. He is also a chief editor of 15 books of the 50th Year of Hacettepe University. He is an Editorial Board Member of 39 journals. He has supervised 69 M.Sc. and 51 Ph.D. theses, 10 post-doctoral and 5 visiting researchers to date. He organized various national and international congresses, summer schools, and graduate courses. His main research fields are molecular imprinting technologies, affinity chromatography, biosensors, polymers, and the application of these biosensors and polymers in medicine and biochemistry.

Contributors

Doğuş Akboğa
UNAM-Institute of Materials Science
and Nanotechnology
Bilkent University
Ankara, Turkey

Semra Akgönüllü
Department of Chemistry
Hacettepe University
Ankara, Turkey

Özge Altıntaş
Department of Chemistry
Hacettepe University
Ankara, Turkey

Ali Araz
Department of Chemistry
Dokuz Eylül University
Izmir, Turkey

Sevgi Aslıyüce
Department of Chemistry
Hacettepe University
Ankara, Turkey

Süleyman Aşır
Department of Materials Science and
Nanotechnology Engineering
Near East University
Nicosia, North Cyprus

Abdurahman Atilla
UNAM-Institute of Materials Science
and Nanotechnology
Bilkent University
Ankara, Turkey

Ece Avcı
UNAM-Institute of Materials Science
and Nanotechnology
Bilkent University
Ankara, Turkey

Abbas Aziz
National Center of Excellence in
Analytical Chemistry
University of Sindh
Jamshoro, Pakistan

Monireh Bakhshpour-Yucel
Department of Chemistry
Bursa Uludag University
Bursa, Turkey

Nilay Bereli
Department of Chemistry
Hacettepe University
Ankara, Turkey

Merve Çalışır
Department of Chemistry
Hacettepe University
Ankara, Turkey

Francesco Canfarotta
MIP Discovery
Sharnbrook, Bedfordshire, UK

Duygu Çimen
Department of Chemistry
Hacettepe University
Ankara, Turkey

Adil Denizli
Department of Chemistry
Hacettepe University
Ankara, Turkey

Sinem Diken-Gur
Department of Biology
Hacettepe Uludag University
Ankara, Turkey

Özgecan Erdem
UNAM-National Nanotechnology
    Research Center
Bilkent University
Ankara, Turkey

Muhammed Erkek
Department of Chemistry
Hacettepe University
Ankara, Turkey

Cem Esen
Department of Chemistry
Aydın Adnan Menderes University
Aydın, Turkey

Ilgım Göktürk
Department of Chemistry
Hacettepe University
Ankara, Turkey

Simge Balaban Hanoglu
Department of Biochemistry
Ege University
Izmir, Turkey

Neslihan İdil
Department of Chemistry
Hacettepe University
Ankara, Turkey

Fatih Inci
UNAM-National Nanotechnology
    Research Center & Institute of
    Materials Science and Nanotechnology
Bilkent University
Ankara, Turkey

Rüstem Keçili
Yunus Emre Vocational School of
    Health Services
Department of Medical Services and
    Techniques
Anadolu University
Eskişehir, Turkey

Seçkin Kılıç
Department of Chemistry
Hacettepe University
Ankara, Turkey

Beyza Nur Küçük
UNAM-National Nanotechnology
    Research Center & Institute
    of Materials Science and
    Nanotechnology
Bilkent University
Ankara, Turkey

Bo Mattiasson
Lund University
Lund, Sweden

Najma Memon
National Center of Excellence in
    Analytical Chemistry
University of Sindh
Jamshoro, Sindh, Pakistan

Bilgen Osman
Department of Chemistry
Bursa Uludag University
Bursa, Turkey

Mamajan Ovezova
Department of Chemistry
Hacettepe University
Ankara, Turkey

Merve Asena Özbek
Department of Chemistry
Hacettepe University
Ankara, Turkey

Yeşeren Saylan
Department of Chemistry
Hacettepe University
Ankara, Turkey

Urartu Özgür Şafak Şeker
UNAM-Institute of Materials Science
and Nanotechnology
Bilkent University
Ankara, Turkey

Huma Shaikh
National Center of Excellence in
Analytical Chemistry
University of Sindh
Jamshoro, Pakistan

Tufail Hussain Sherazi
National Center of Excellence in
Analytical Chemistry
University of Sindh
Jamshoro, Pakistan

Suna Timur
Department of Biochemistry &
Central Research Test and Analysis
Laboratory
Ege University
Izmir, Turkey

Aykut Arif Topçu
Medical Laboratory Program,
Vocational School of Health Services
Aksaray University
Aksaray, Turkey

Nazlıcan Tunç
UNAM-Institute of Materials Science
and Nanotechnology
Bilkent University
Ankara, Turkey

Deniz Türkmen
Department of Chemistry
Hacettepe University
Ankara, Turkey

Nursima Ucar
Department of Biochemistry
Ege University
Izmir, Turkey

Serhat Ünal
Department of Infectious Diseases and
Clinical Microbiology
Hacettepe University
Ankara, Turkey

Nese Lortlar Unlu
Department of Biomedical Engineering
Boston University
Boston, USA
and
Department of Histology and
Embryology
Atlas University
Istanbul, Turkey

Handan Yavuz
Department of Chemistry
Hacettepe University
Ankara, Turkey

Eylul Gulsen Yilmaz
UNAM-National Nanotechnology
Research Center & Institute
of Materials Science and
Nanotechnology
Bilkent University
Ankara, Turkey

Fatma Yılmaz
Department of Chemistry and Chemical
Processing Technologies
Bolu Abant Izzet Baysal University
Bolu, Turkey

Gaye Ezgi Yılmaz
Department of Chemistry
Hacettepe University
Ankara, Turkey

1 Introduction to Biological and Chemical Threat Agents

Semra Akgönüllü, Handan Yavuz, and Adil Denizli

1.1 INTRODUCTION

Recent events have demonstrated how chemical and biological warfare agents (CBWAs) present threats and risks (Russell et al. 2003; Gooding 2006; Eubanks, Dickerson, and Janda 2007; Seo et al. 2020). CBWAs can be pathogenic microorganisms, biological toxins, and toxic chemicals (Mayor 2019; Janik et al. 2019). They have been used in military operations to kill, mortally harm, and incapacitate those exposed to them by exerting physiological effects (Szinicz 2005). Despite the international prohibition established by the Chemical Weapons Convention (CWC) (Organisation for the Prohibition of Chemical Weapons (OPCW) 1997), in the last few years, the number of events involving chemical warfare agents (CWAs) has been increasing (Davidson et al. 2020; Black and Muir 2003; Picard et al. 2019). Toxic chemicals, including bad-smelling substances, irritants, poisonous plants and animals, and rotting corpses, have been employed in battle. It is possible to develop chemical agents that are only incapacitating or lethal. Moreover, it is also possible to develop chemical agents that cause physical damage but do not damage infrastructure, and as a result can still incapacitate or kill an enemy (Davidson et al. 2020; Diauudin et al. 2019; Sadik, Land, and Wang 2003; Seo et al. 2021). Biological weapons contain infectious agents and toxins. Toxins are poisons produced by living organisms (Anderson 2012). Biological warfare agents (BWAs) use disease-causing infectious agents to spread infection. The Biological Weapons Convention (BWC) globally bans their use in warfare (Jansen et al. 2014; Pohanka 2019).

Over the past two decades, CBWAs have been the subject of significant scientific research due to the increasing threat of terrorism and their use in war zones. A wide range of detection techniques have been advanced for the detection and determination of CBWAs (Akgönüllü et al. 2021; Ohrui et al. 2020). There are many existing determination analytical methods, such as ion mobility spectroscopy (Satoh et al. 2015), infrared spectroscopy (Ohrui et al. 2020), gas chromatography-mass spectroscopy (Valdez et al. 2018), capillary electrophoresis (Li, Hu, and Li 2012), and liquid chromatography-mass spectroscopy (Niemikoski, Söderström, and Vanninen 2017). In defense and homeland security applications of CBWAs, there is a need to develop

DOI: 10.1201/9781003459316-1

new sensing methods with sensitive and accurate detection capabilities (Saylan, Akgönüllü, and Denizli 2020). Sensors are a technology that uses mainly biochemical interactions to detect the level or presence of a target analyte. This target analyte or compound can be a small biological molecule, an inorganic compound, or a larger biological macromolecule component (Akgönüllü and Denizli 2022; Akgönüllü and Denizli 2023; Menger et al. 2016). Recent years have also witnessed a massive development in the field of sensors for the detection of analytes in many different fields because of the interaction of basic sciences and technology. The CBWA threat is an increasing worldwide concern. Toxic chemicals, biological toxins, radiological substances, nuclear bombs, and explosive agents pose significant threats in the 21st century, especially to first responders, including members of the armed forces, police, firefighters, and other emergency personnel (S. Wang et al. 2020). A suitable and convenient instrument for the accurate and fast detection of CBWAs must have qualities such as portability, ease of sample handling, selectivity, sensitivity, reusability, short measurement time, cost-effectiveness, low false response, and reliability. The fundamental challenge for commercially available instruments is to enhance the selectivity and sensitivity for the detection of CBWAs and to reduce the false response alert. This chapter critically examines the literature on the history of CWAs and BWAs, bioterrorism, and biocrimes.

1.2 BIOLOGICAL THREAT AGENTS

In 1346, the Mongol army besieging the city of Crimea threw the corpses of people with the plague over the castle walls. The disease that infected the city later spread across Europe as the "Black Death," killing 24 million people in 6 years. This is the elusive example of the force of rough biological warfare agents (Dembek 2008; Wheelis 2002). Pathogens such as bacteria, viruses, parasites, and fungi cause life-threatening infectious diseases, particularly through delayed diagnosis and treatment (Akgönüllü et al. 2020). It is not easy to pinpoint a specific time when the use of biological toxins as weapons began. There is ancient evidence reporting the use of dead animal carcasses to contaminate water wells (Rowland et al. 2016). While a basic understanding of the pathogenicity of most biothreat agents has been clarified and available treatments have increased significantly in recent years, (bio)terrorism still represents a major public health threat in an era of indiscriminate wars, pollution, climate change, uncontrolled population growth, and globalization (Walper et al. 2018; Sapsford et al. 2008; Georgiev 2009). Biological threat agents (biological threats or bioagents) are pathogens and/or their toxic products that pose a significant threat to human health. BWAs are a group that includes viruses, bacteria, toxins from biological resources, and other naturally existing pathogenic organisms (Shah and Wilkins 2003). This variety is reflected in the remarkable range of transmissibility, infectiousness, and lethality they demonstrate (Walper et al. 2018). The most well-known BWAs that cause diseases in humans, animals, or plants are shown in Table 1.1.

The most likely toxic agents on the BWA list are *Bacillus anthracis* (anthrax) and the variola major virus (smallpox). Extremely hazardous biological agents contain

TABLE 1.1

The Main Biological Warfare Agents and Caused Diseases (Pedrero, Campuzano, and Pingarrón 2012; Jansen et al. 2014; Franz et al. 1997; Pohanka 2019)

	Biological Agents	Diseases	Organism Persistence	Infective Dose	Symptoms	Mortality
Bacterial	*Bacillus anthracis*	Anthrax	In soil, spores with high stability can live for more than 40 years	8,000–50,000 spores	Tiredness, high fever, weakness, cough, mild chest distress, respiratory distress, shock	High
	Yersinia pestis	Plague	Can remain in soil for up to 1 year, but can only experience aerosol release in 1 hour	100–20,000 organisms	High fever, headache, malaise, chest pain, cough, hemoptysis, dyspnea, stridor, cyanosis	Very high if untreated, <10% with antibiotic
	Francisella tularensis	Tularemia (rabbit fever)	Weeks in water, soil, or carcasses and years in frozen meat	10–50 organisms	Fever, chills, myalgia, arthralgia, headache, nausea, vomiting, diarrhea, sore throat	4–50% mortality without treatment. With treatment, 1%
	Brucella melitensis	Brucellosis	Species *Brucella* (*B. melitensis, B. suis, B. abortus, B. canis*)	10–100 organisms	High fever, headache, malaise, chills, sweating, myalgia, arthralgia, depression	5% if untreated
	Coxiella burnetii	Q fever	Heat resistant, lasts for weeks or months	1–10 organisms	Fever, chills, headache, weakness, fatigue, anorexia, weight loss, endocarditis	1% untreated, chronic form 30–60%
	Salmonella species	Food poisoning	Resistant to heat up to 57–60°C	Unknown	Nausea, vomiting, mucopurulent or bloody diarrhea, abdominal cramps, headache, maculopapular exanthema	<1%
	Shigella	Food poisoning	Average survival time 2–3 days, up to 17 days in favorable conditions, several hours on infected hands	10–100 organisms	Fever, abdominal cramps, diarrhea, hemorrhagic colitis	<1%
	Burkholderia pseudomallei	Melioidosis	High, stable	Unknown	Pulmonary form: cough, chest pain, fever, rigors, sweating, pleuritis	Very high if untreated
	Vibrio cholerae	Cholera			Acute watery diarrhea and vomiting	

(Continued)

TABLE 1.1 (CONTINUED)

The Main Biological Warfare Agents and Caused Diseases (Pedrero, Campuzano, and Pingarrón 2012; Franz et al. 1997; Jansen et al. 2014; Pohanka 2019)

	Biological Agents	Diseases	Organism Persistence	Infective Dose	Symptoms	Mortality
	Escherichia coli O157:H7	Food poisoning			Vomiting, fever, hemorrhagic colitis, bloody diarrhea, thrombocytopenic purpura, hemolytic uremic syndrome	
Viral	Alphaviruses and flaviviruses	Viral encephalitis	Unstable in the environment	10–100 organisms	Malaise, spiking fevers, rigors, headache, myalgia, nausea, otophobia, vomiting, cough, encephalitis (4% children, <1% adults), diarrhea, sore throat	<1%
	Variola major virus	Smallpox	Highly stable for up to 1 year in dust and cloth	10–100 organisms	Severe headache, high fever, extreme prostration, backache, chest and joint pains, anxiety, exanthema, maculopapular rash that becomes vesicular	Ordinary-type smallpox: 30% if unvaccinated; 3% if vaccinated
Toxins	*Staphylococcal enterotoxin B*	Food poisoning	Resistant to freezing, inactivated at 100°C	0.03 μg/person	Fever, chills, dyspnea, un-productive, sough, headache, myalgia, retrosternal chest pain	<1%
	Ricinus communis (castor beans)	Ricin toxin poisoning	Stable until heated above 80°C	LD50 1 mg	Inhalation: fever, respiratory distress, cough. Ingestion: gastrointestinal hemorrhage. Both: multiorgan failure	High
	Clostridium botulinum toxin	Botulism	Weeks in unmoving food or water	LD50 is 0.001 μg/kg for type A (parenteral), 0.003 μg/kg (aerosol)	Acute, afebrile, symmetric paralysis descending from the head	In the absence of supportive care, mortality from respiratory failure is high
	Mycotoxins	Mycotoxicosis	Extremely stable, resistant to heat and ultraviolet light inactivation.	LD50 of T-2 toxin is approximately 1 mg/kg	Blistering, vomiting, and bloody diarrhea	

botulinum toxin, *Francisella tularensis*, *Salmonella typhimurium*, and *Yersinia pestis*. Terrorists tend to use biological agents because they are extremely difficult to detect, and detection is made more difficult, since infection and illness can occur anytime from a few hours to several days after exposure. Some bioterrorism agents can spread from person to person (e.g. smallpox) and some may not (e.g. anthrax). Other bioterrorism agents, such as Venezuelan equine encephalitis, Marburg, Ebola, and influenza viruses, have a lower position on the BWA list due to difficulties in their preparation, despite the fact that infections with these viruses are serious and the mortality rate is relatively high. In short, any pathogenic organism can be misused to make biological weapons, but the effect can be very different. Compared to CWAs, BWAs are much cheaper to produce and more effective in terms of the hazardous area and expected loss of life in a terrorist or military attack (Pohanka, Skládal, and Kroèa 2007).

1.3   CHEMICAL THREAT AGENTS

In the fight against terrorism using weapons of mass destruction, the identification and characterization of the hazardous substances involved is of paramount importance. Among hazardous substances, chemical warfare agents (CWAs) are of great concern because CWAs have been used by terrorists and in wars (Black et al. 1994). CWAs are made up of solid, liquid, and gaseous agents that are lethal, hazardous, and irritating, as well as inflammable and have an impact on humans, plants, and metals (Akgönüllü et al. 2023). CWAs in solid, liquid, gaseous, and aerosol form are dispersed and released by spray weapons and spray tanks carried by helicopters, or by artillery-rocket or missile shells, mines, grenades, hand and aircraft bombs, and aircraft (Akgönüllü et al. 2023). CWAs are quickly effective at low exposure levels. The treatment (if any) for CWA exposure should be rapid, considering the fast onset of symptoms.

Protection against CWAs can be achieved with large, durable, and comprehensive barrier suits, and caregivers should also be protected to prevent secondary CWA exposure to responding personnel. Rapid notification and identification of potential exposure means that the effectiveness of the countermeasures implemented is increased. The more robust a detector is against false alarms, the fewer resources need to be spent on protective measures such as the installation and removal of protective equipment and evacuation of personnel (Davidson et al. 2020).

CWAs are categorized according to their effects on human health. The main categories are nerve agents, those that cause vesicants/blisters, and blood/suffocation agents (Yao et al. 2020; Grissom et al. 2020). The chemical structures and volatility and inhalation toxicity data of CWAs examined are shown in Figure 1.1. Nerve gases are sarin (GB), soman (GD), tabun (GA), and VX. Blister agents are mustard gas (HD), and lewisite 1 (L1). Hydrogen cyanide (AC) and cyanogen chloride (CK) gases are blood agents (Kaledin et al. 2019). The main chemical warfare agents are summarized in Table 1.2.

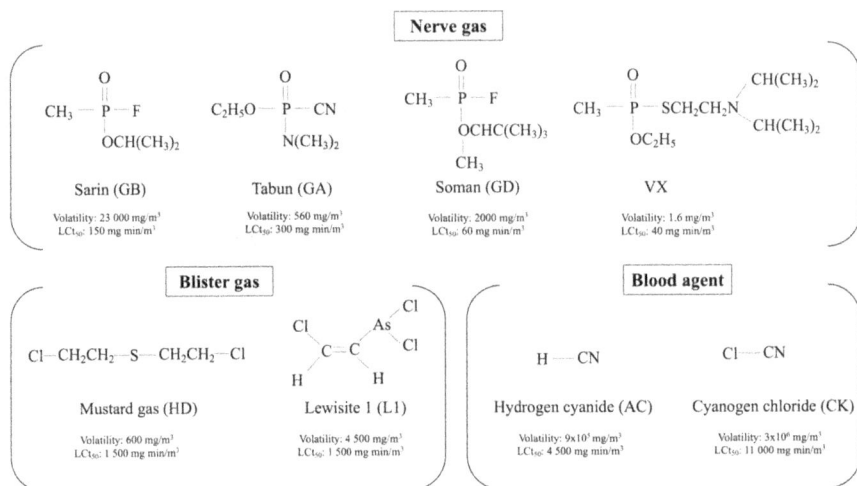

FIGURE 1.1 Chemical structures and volatility (at 25) and inhalation toxicity data of chemical warfare agents. (Reproduced from Seto et al. 2007.)

TABLE 1.2
Classes of Main and Common Chemical Warfare Agents (Yue et al. 2016)

Group	Agent	Action Mode
Nerve agents	G-series (Tabun, Sarin, Soman) V-series (VE, VX, VG, VM) A-series	It prevents the breakdown of the neurotransmitter acetylcholine (ACh) in the victim's synapses by inactivating the enzyme acetylcholinesterase (AChE), causing both muscarinic and nicotinic effects.
Vesicant/ blister	Sulfur mustard Nitrogen mustard Lewisite	They are acid forming compounds that damage the skin and respiratory system, causing burns and respiratory problems.
Blood/ asphyxiants	Cyanogen chloride Hydrogen cyanide	When cyanide is used directly, it prevents the cells from using oxygen. The cells then use anaerobic respiration to create excess lactic acid and metabolic acidosis.
Choking agent	Chlorine Chloropicrin Phosgene chloropicrin (PS) Phosgene (CG) Diphosgenenitrogen oxides	Hydrochloric and hypochlorous acid are produced when chlorine gas comes into contact with moist tissue, such as the eyes, throat, and lungs. Hypochlorous acid degenerates into hydrochloric acid and newly formed oxygen.

CWAs were used in World War I and World War II and during the Cold War and continue to be produced today, despite being prohibited as weapons (Gupta 2015; Seto 2020). Sarin and sulfur mustard were used by Iraq in the Iran-Iraq conflict in the 1980s (Haines and Fox 2014). The Chemical Weapons Convention, a treaty prohibiting the development, production, stockpiling, and use of chemical weapons and mandating their destruction, was approved in 1992 and entered into force in 1997 ("Chemical Weapons Convention" 2020). The 1995 sarin gas attack on the Tokyo subway (Nakamine et al. 2018), postal anthrax letter attacks in the US in 2001 (North et al. 2009), sarin and sulfur mustard used in the Syrian war (John et al. 2018), and Novichok, the nerve agent used in the UK assassination, in 2018 (Franca et al. 2019) are all examples of the use of CWAs. The use of CWAs is still ongoing following these events in the past. In the current conflict in Syria, CWAs are acknowledged to have been used in a number of incidents resulting in large numbers of deaths and long-term health impacts among affected populations (Diauudin et al. 2019). In recent years, the number of potential terrorist attacks and international conflicts has increased the need to protect against CWAs (H. Wang et al. 2019). CWAs are extremely toxic by inhalation and skin contact. These agents should be handled with special care in a fume hood using protective clothing and disposed of after inspection by treatment with sodium hypochlorite (Seto et al. 2007).

1.3.1 NERVE AGENTS

Nerve agents are a class of chemical agents that have been the most remarkable and effective. Substances belonging to the class of organophosphate compounds are all nerve agents (Dale and Rebek 2006; Zheng, Fu, and Xu 2010). They are persistent and easy to distribute. These substances are highly toxic. They show a very rapid effect when inhaled and in contact with the skin. Organophosphate (OP) compounds can be divided into two groups known as G series and V series agents (Chauhan et al. 2008). G-series nerve agents were synthesized by Dr. Gerhard Schrader and his team in Germany in the 1930s (Vásárhelyi and Földi 2007). G-series agents are tabun (GA) (Kim et al. 2011), sarin (GB) (Balali-Mood and Saber 2012), soman (GD), and cyclosarin (GF). All these compounds except tabun contain fluorine (Figure 1.1). The lethal concentration for G-series agents is typically an LC50 of 1 ppm over a 10-minute exposure time (McKenna et al. 2017) but there are some differences in both concentration and exposure time between the various members of the series. The odor of G-series agents is typically described as 'slightly fruity' or 'spicy' (McKenna et al. 2017).

1.3.2 BLISTER GAS

Blister agents or vesicants are a group of chemicals that cause severe blisters when they come into contact with the skin. These agents are not very lethal in terms of causing death, but they can incapacitate the enemy and overload already overburdened health services during war. These contain sulfur mustard, nitrogen mustard, and lewisite (Nair et al. 2021).

1.3.3   BLOOD AGENTS

Blood/asphyxiants are agents that are responsible for causing tissue hypoxia. They are categorized as simple agents or highly toxic chemicals. Simple asphyxiants such as methane and nitrogen physically replace oxygen in the inhaled air, causing oxygen deficiency and hypoxemia. Chemical asphyxiants such as cyanides inhibit oxygen transport at the cellular level, causing tissue hypoxia, anaerobic metabolism, and lactic acidosis. Important chemical asphyxiants used as CWAs include cyanogen chloride (CK), hydrogen cyanide (HCN), and arsine (SA) (Chauhan et al. 2008).

1.4   CONCLUSION AND FUTURE PERSPECTIVES

The manufacture of new chemical agents has become impossible to prevent with the advancement of science. The cheap and easy accessibility of these substances is a serious safety problem. The basic step toward almost complete prevention of biological and chemical agents, prevention strategies, post-exposure treatment, and mitigation is early monitoring and identification. Therefore, developing a strong capability to detect and correctly identify CBWA threats early is a high priority for responders. More rapid alert of potential exposure means increased effectiveness of implemented countermeasures. The more robust a detector is against false alarms, the fewer the resources that need to be spent on protective measures, such as the installation and removal of protective equipment and the evacuation of personnel. The use of CBWAs, from past to present, still has not been fully addressed by the widespread implementation of capable chemical and biological agent sensors in key locations around the world.

There is developing interest in sensing technologies for CBWA detection and quantification. Research is focusing on ways to fabricate portable tools that can allow fast, accurate, and in-situ detection. Research into the detection of CBWAs by plasmonic-based sensors is certainly of interest and well adapted for simultaneous monitoring. In particular, the focus is on the combination of polymer films and metal nanoparticles, which have been used for decades for more sensitive determinations. Sensitivity and selectivity are also promising. Meanwhile, these devices are affordable and portable. They are mechanically powerful to replace traditional methods. But in most cases, specialized training and high costs are still required to operate the devices and analyze the data. Real-time and in-situ sensing is in theory a critical point. For this reason, plasmonic based sensors are considered the most suitable sensors for application, providing a measurable response.

The use of plasmonic-based sensing systems for the detection of CBWAs is valuable for further studies. Based on the changes in signal transduction, due to the successful interaction between biological-based recognition elements and CBWAs, these have become vital in the construction of plasmonic-based biosensors capable of accurate response. The performance of sensing systems, their fast and accurate response, especially in terms of sensitivity and selectivity, can be improved through the improvement of novel biorecognition elements. As a result, new CBWAs can be produced and easily distributed. The intervention has led to the development of sensor technology with in-situ analysis. Nanomaterials can be designed and integrated into these systems for precise detection. Biological recognition elements, such as antibodies and aptamers, can be used for selectivity. Molecularly imprinted polymers can also be synthesized for the detection of these agents, as the recently preferred imprinting technique for target analysis is cheaper and reusable. The development of smart sensing systems is a priority for ensuring homeland security, reducing the risk to frontline security forces in the field, and protecting responders and the civilian population. To prevent the irreversible consequences and threats posed by highly dangerous CBWAs, plasmonic-based sensor systems have recently attracted attention.

REFERENCES

Akgönüllü, Semra, and Adil Denizli. 2022. "Nano-Sensors and Nano-Devices for Biological Disaster Monitoring (Virus/Disease Epidemics/Animal Plagues Detections)." In *Nanotechnology-Based Smart Remote Sensing Networks for Disaster Prevention*, 43–57. doi:10.1016/B978-0-323-91166-5.00006-9.

Akgönüllü, Semra, and Adil Denizli. 2023. "Recent Advances in Plasmonic Biosensors for the Detection of Food Allergens." In *Encyclopedia of Sensors and Biosensors* 4: 357–371. doi:10.1016/b978-0-12-822548-6.00143-6.

Akgönüllü, Semra, Duygu Çimen, Monireh Bakhshpour, Nilay Bereli, Handan Yavuz, and Adil Denizli. 2020. "Commercial Sensors for Pathogen Detection." In *Commercial Biosensors and Their Applications*, 89–106. Elsevier. doi:10.1016/B978-0-12-818592-6.00005-0.

Akgönüllü, Semra, Yeşeren Saylan, Nilay Bereli, Deniz Türkmen, Handan Yavuz, and Adil Denizli. 2021. "Plasmonic Sensors for Detection of Chemical and Biological Warfare Agents." In *Plasmonic Sensors and Their Applications*, edited by Adil Denizli, 71–85. John Wiley & Sons, Ltd. doi:10.1002/9783527830343.ch4

Akgönüllü, Semra, Merve Çalışır, Merve Asena Özbek, Muhammed Erkek, Nilay Bereli, and Adil Denizli. 2023. "Plasmonic Nanosensors for Chemical Warfare Agents." In *Sensing of Deadly Toxic Chemical Warfare Agents, Nerve Agent Simulants, and Their Toxicological Aspects*, 81–96. Elsevier. doi:10.1016/B978-0-323-90553-4.00016-0.

Anderson, Peter D. 2012. "Bioterrorism: Toxins as Weapons." *Journal of Pharmacy Practice* 25 (2): 121–129. doi:10.1177/0897190012442351.

Balali-Mood, Mahdi, and Hamidreza Saber. 2012. "Recent Advances in the Treatment of Organophosphorous Poisonings." *Iranian Journal of Medical Sciences* 37 (2): 74–91.

Black, Robin M., and Bob Muir. 2003. "Derivatisation Reactions in the Chromatographic Analysis of Chemical Warfare Agents and Their Degradation Products." *Journal of Chromatography A* 1000 (1–2): 253–281. doi:10.1016/S0021-9673(03)00183-3.

Black, Robin M., Raymond J. Clarke, Robert W. Read, and Michael T. J. Reid. 1994. "Application of Gas Chromatography-Mass Spectrometry and Gas Chromatography-Tandem Mass Spectrometry to the Analysis of Chemical Warfare Samples, Found to Contain Residues of the Nerve Agent Sarin, Sulphur Mustard and Their Degradation Products." *Journal of Chromatography A* 662 (2): 301–321. doi:10.1016/0021-9673(94)80518-0.

Chauhan, S., S. Chauhan, R. D'Cruz, S. Faruqi, K. K. Singh, S. Varma, M. Singh, and V. Karthik. 2008. "Chemical Warfare Agents." *Environmental Toxicology and Pharmacology* 26 (2): 113–122. doi:10.1016/j.etap.2008.03.003.

"Chemical Weapons Convention." 2020. OPCW.

Dale, Trevor J., and Julius Rebek. 2006. "Fluorescent Sensors for Organophosphorus Nerve Agent Mimics." *Journal of the American Chemical Society* 128 (14): 4500–4501. doi:10.1021/ja057449i.

Davidson, Charles E., Melissa M. Dixon, Barry R. Williams, Gary K. Kilper, Sung H. Lim, Raymond A. Martino, Paul Rhodes, et al. 2020. "Detection of Chemical Warfare Agents by Colorimetric Sensor Arrays." *ACS Sensors* 5 (4): 1102–1109. doi:10.1021/acssensors.0c00042.

Dembek, Zygmunt F. 2008. *Medical Aspects of Biological Warfare*. Edited by Zygmunt F. Dembek. US Government Printing Office.

Diauudin, Farah Nabila, Jahwarhar Izuan Abdul Rashid, Victor Feizal Knight, Wan Md Zin Wan Yunus, Keat Khim Ong, Noor Azilah Mohd Kasim, Norhana Abdul Halim, and Siti Aminah Mohd Noor. 2019. "A Review of Current Advances in the Detection of Organophosphorus Chemical Warfare Agents Based Biosensor Approaches." *Sensing and Bio-Sensing Research* 26: 100305. doi:10.1016/j.sbsr.2019.100305

Eubanks, Lisa M., Tobin J. Dickerson, and Kim D. Janda. 2007. "Technological Advancements for the Detection of and Protection against Biological and Chemical Warfare Agents." *Chemical Society Reviews* 36: 458–470. doi:10.1039/b615227a.

Franca, Tanos, Daniel Kitagawa, Samir Cavalcante, Jorge da Silva, Eugenie Nepovimova, and Kamil Kuca. 2019. "Novichoks: The Dangerous Fourth Generation of Chemical Weapons." *International Journal of Molecular Sciences* 20 (5): 1222. doi:10.3390/ijms20051222.

Franz, David R., P. B. Jahrling, A. M. Friedlander, A. M. McClain, D. J. Hoover, and D. L. Bryne. 1997. "Clinical Recognition and Management of Patients Exposed to Biological Warfare Agents." *JAMA: The Journal of the American Medical Association* 278 (5): 399. doi:10.1001/jama.1997.03550050061035.

Georgiev, Vassil St. 2009. "Defense Against Biological Weapons (Biodefense)." In *National Institute of Allergy and Infectious Diseases*. NIH. 221–305. doi:10.1007/978-1-60327-297-1_23.

Gooding, J. Justin. 2006. "Biosensor Technology for Detecting Biological Warfare Agents: Recent Progress and Future Trends." *Analytica Chimica Acta* 559 (2): 137–151. doi:10.1016/j.aca.2005.12.020.

Grissom, Tyler G., Anna M. Plonka, Conor H. Sharp, Amani M. Ebrahim, Yiyao Tian, Daniel L. Collins-Wildman, Alexey L. Kaledin, et al. 2020. "Metal–Organic Framework- and Polyoxometalate-Based Sorbents for the Uptake and Destruction of Chemical Warfare Agents." *ACS Applied Materials & Interfaces* 12 (13): 14641–14661. doi:10.1021/acsami.9b20833.

Gupta, Ramesh C. 2015. *Handbook of Toxicology of Chemical Warfare Agents: Second Edition*. doi:10.1016/C2013-0-15402-5.

Haines, D. D., and S. C. Fox. 2014. "Acute and Long-Term Impact of Chemical Weapons: Lessons from the Iran-Iraq War." *Forensic Science Review* 26(2):97-114.

Janik, Edyta, Michal Ceremuga, Joanna Saluk-Bijak, and Michal Bijak. 2019. "Biological Toxins as the Potential Tools for Bioterrorism." *International Journal of Molecular Sciences* 20 (5): 1181. doi:10.3390/ijms20051181.

Jansen, H. J., F. J. Breeveld, C. Stijnis, and M. P. Grobusch. 2014. "Biological Warfare, Bioterrorism, and Biocrime." *Clinical Microbiology and Infection* 20 (6). European Society of Clinical Infectious Diseases: 488–496. doi:10.1111/1469-0691.12699.

John, Harald, Marcel J. van der Schans, Marianne Koller, Helma E. T. Spruit, Franz Worek, Horst Thiermann, and Daan Noort. 2018. "Fatal Sarin Poisoning in Syria 2013: Forensic Verification within an International Laboratory Network." *Forensic Toxicology* 36 (1): 61–71. doi:10.1007/s11419-017-0376-7.

Kaledin, Alexey L., Diego Troya, Christopher J. Karwacki, Alex Balboa, Wesley O. Gordon, John R. Morris, Mark B. Mitchell, Anatoly I. Frenkel, Craig L. Hill, and Djamaladdin G. Musaev. 2019. "Key Mechanistic Details of Paraoxon Decomposition by Polyoxometalates: Critical Role of Para-Nitro Substitution." *Chemical Physics* 518 Elsevier: 30–37. doi:10.1016/j.chemphys.2018.11.013.

Kim, Kibong, Olga G. Tsay, David A. Atwood, and David G. Churchill. 2011. "Destruction and Detection of Chemical Warfare Agents." *Chemical Reviews* 111 (9): 5345–5403. doi:10.1021/cr100193y.

Li, Pingjing, Bin Hu, and Xiaoyong Li. 2012. "Zirconia Coated Stir Bar Sorptive Extraction Combined with Large Volume Sample Stacking Capillary Electrophoresis-Indirect Ultraviolet Detection for the Determination of Chemical Warfare Agent Degradation Products in Water Samples." *Journal of Chromatography A* 1247. 49–56. doi:10.1016/j.chroma.2012.05.059.

Mayor, Adrienne. 2019. "Chemical and Biological Warfare in Antiquity." In *Toxicology in Antiquity*, 243–255. Elsevier. doi:10.1016/B978-0-12-815339-0.00016-0.

McKenna, Josiah, Elizabeth S. Dhummakupt, Theresa Connell, Paul S. Demond, Dennis B. Miller, J. Michael Nilles, Nicholas E. Manicke, and Trevor Glaros. 2017. "Detection of Chemical Warfare Agent Simulants and Hydrolysis Products in Biological Samples by Paper Spray Mass Spectrometry." *Analyst* 142 (9). Royal Society of Chemistry: 1442–1451. doi:10.1039/c7an00144d.

Menger, Marcus, Aysu Yarman, Júlia Erdőssy, Huseyin Yildiz, Róbert Gyurcsányi, and Frieder Scheller. 2016. "MIPs and Aptamers for Recognition of Proteins in Biomimetic Sensing." *Biosensors* 6 (3): 35. doi:10.3390/bios6030035.

Nair, Ashrit, Pooja Yadav, Amanpreet Behl, Rakesh Kumar Sharma, Shweta Kulshrestha, Bhupendra Singh Butola, and Navneet Sharma. 2021. "Toxic Blister Agents: Chemistry, Mode of Their Action and Effective Treatment Strategies." *Chemico-Biological Interactions* 350 109654. doi:10.1016/j.cbi.2021.109654.

Nakamine, Shin, Maiko Kobayashi, Hiroyuki Fujita, Sachiko Takahashi, and Yutaka Matsui. 2018. "Posttraumatic Stress Symptoms in Victims of the Tokyo Subway Sarin Attack: Twenty Years Later." *Journal of Social and Clinical Psychology* 37 (10): 794–811. doi:10.1521/jscp.2018.37.10.794.

Niemikoski, Hanna, Martin Söderström, and Paula Vanninen. 2017. "Detection of Chemical Warfare Agent-Related Phenylarsenic Compounds in Marine Biota Samples by LC-HESI/MS/MS." *Analytical Chemistry* 89 (20): 11129–11134. doi:10.1021/acs.analchem.7b03429.

North, Carol S., Betty Pfefferbaum, Meena Vythilingam, Gregory J. Martin, John K. Schorr, Angela S. Boudreaux, Edward L. Spitznagel, and Barry A. Hong. 2009. "Exposure to Bioterrorism and Mental Health Response among Staff on Capitol Hill." *Biosecurity and Bioterrorism: Biodefense Strategy, Practice, and Science* 7 (4): 379–388. doi:10.1089/bsp.2009.0031.

Ohrui, Yasuhiko, Ryota Hashimoto, Takeshi Ohmori, Yasuo Seto, Hiroyuki Inoue, Hideki Nakagaki, Katsutoshi Yoshikawa, and Larry McDermott. 2020. "Continuous Monitoring of Chemical Warfare Agents in Vapor Using a Fourier Transform Infra-Red Spectroscopy Instrument with Multi Pass Gas Cell, Mercury Cadmium Telluride Detector and Rolling Background Algorithm." *Forensic Chemistry* 21 100292. doi:10.1016/j.forc.2020.100292.

Organisation for the Prohibition of Chemical Weapons (OPCW). 1997. "Convention on the Prohibition of the Development, Production, Stockpiling and Use of Chemical Weapons and on Their Destruction, Organisation for the Prohibition of Chemical Weapons (OPCW), the Hague, The Netherlands (1997)." https://www.opcw.org/chemical-weapons-convention.

Pedrero, María, Susana Campuzano, and José M. Pingarrón. 2012. "Magnetic Beads-Based Electrochemical Sensors Applied to the Detection and Quantification of Bioterrorism/ Biohazard Agents." *Electroanalysis* 24 (3): 470–482. doi:10.1002/elan.201100528.

Picard, Baptiste, Isabelle Chataigner, Jacques Maddaluno, and Julien Legros. 2019. "Introduction to Chemical Warfare Agents, Relevant Simulants and Modern Neutralisation Methods." *Organic and Biomolecular Chemistry* 17 (27). Royal Society of Chemistry: 6528–6537. doi:10.1039/c9ob00802k.

Pohanka, Miroslav. 2019. "Current Trends in the Biosensors for Biological Warfare Agents Assay." *Materials* 12 (14): 2303. doi:10.3390/ma12142303.

Pohanka, Miroslav, Petr Skládal, and Michal Kroèa. 2007. "Biosensors for Biological Warfare Agent Detection." *Defence Science Journal* 57 (3): 185–193. doi:10.14429/dsj.57.1760.

Rowland, Clare E., Carl W. Brown, James B. Delehanty, and Igor L. Medintz. 2016. "Nanomaterial-Based Sensors for the Detection of Biological Threat Agents." *Materials Today* 19 (8). 464–477. doi:10.1016/j.mattod.2016.02.018.

Russell, Alan J., Jason A. Berberich, Géraldine F. Drevon, and Richard R. Koepsel. 2003. "Biomaterials for Mediation of Chemical and Biological Warfare Agents." *Annual Review of Biomedical Engineering* 5 (1): 1–27. doi:10.1146/annurev.bioeng.5.121202. 125602.

Sadik, Omowunmi A., Walker H. Land, and Joseph Wang. 2003. "Targeting Chemical and Biological Warfare Agents at the Molecular Level." *Electroanalysis* 15 (14): 1149–1159. doi:10.1002/elan.200390140.

Sapsford, Kim E., Christopher Bradburne, James B. Delehanty, and Igor L. Medintz. 2008. "Sensors for Detecting Biological Agents." *Materials Today* 11 (3). 38–49. doi:10.1016/ S1369-7021(08)70018-X.

Satoh, Takafumi, Shintaro Kishi, Hisayuki Nagashima, Masumi Tachikawa, Mieko Kanamori-Kataoka, Takao Nakagawa, Nobuyoshi Kitagawa, Kenichi Tokita, Soichiro Yamamoto, and Yasuo Seto. 2015. "Ion Mobility Spectrometric Analysis of Vaporous Chemical Warfare Agents by the Instrument with Corona Discharge Ionization Ammonia Dopant Ambient Temperature Operation." *Analytica Chimica Acta* 865 (1). 39–52. doi:10.1016/j.aca.2015.02.004.

Saylan, Yeşeren, Semra Akgönüllü, and Adil Denizli. 2020. "Plasmonic Sensors for Monitoring Biological and Chemical Threat Agents." *Biosensors* 10 (10): 142. doi:10.3390/bios10100142.

Seo, Jin Young, Kie Yong Cho, Jung-Hyun Lee, Min Wook Lee, and Kyung-youl Baek. 2020. "Continuous Flow Composite Membrane Catalysts for Efficient Decomposition of Chemical Warfare Agent Simulants." *ACS Applied Materials & Interfaces* 12 (29): 32778–32787. doi:10.1021/acsami.0c08276.

Seo, Jin Young, Younghan Song, Jung-hyun Lee, Hyungsup Kim, Sangho Cho, and Kyung-youl Baek. 2021. "Robust Nanocellulose/Metal–Organic Framework Aerogel Composites: Superior Performance for Static and Continuous Disposal of Chemical Warfare Agent Simulants." *ACS Applied Materials & Interfaces* 13 (28): 33516–33523. doi:10.1021/acsami.1c08138.

Seto, Yasuo. 2020. *On-Site Detection of Chemical Warfare Agents. Handbook of Toxicology of Chemical Warfare Agents.* INC. doi:10.1016/b978-0-12-819090-6.00057-x.

Seto, Yasuo, Mieko Kanamori-Kataoka, Kouichiro Tsuge, Isaac Ohsawa, Hisashi Maruko, Hiroshi Sekiguchi, Yasuhiro Sano, et al. 2007. "Development of an On-Site Detection Method for Chemical and Biological Warfare Agents." *Toxin Reviews* 26 (3): 299–312. doi:10.1080/15569540701506756.

Shah, Jasmin, and Ebtisam Wilkins. 2003. "Electrochemical Biosensors for Detection of Biological Warfare Agents." *Electroanalysis* 15 (3): 157–167. doi:10.1002/ elan.200390019.

Szinicz, L. 2005. "History of Chemical and Biological Warfare Agents." *Toxicology* 214 (3): 167–181. doi:10.1016/j.tox.2005.06.011.

Valdez, Carlos A., Roald N. Leif, Saphon Hok, and Bradley R. Hart. 2018. "Analysis of Chemical Warfare Agents by Gas Chromatography-Mass Spectrometry: Methods for Their Direct Detection and Derivatization Approaches for the Analysis of Their Degradation Products." *Reviews in Analytical Chemistry* 37 (1): 1–27. doi:10.1515/revac-2017-0007.

Vásárhelyi, Györgyi, and László Földi. 2007. "History of Russia's Chemical Weapons." *Aarms History* 6 (1): 135–146.

Walper, Scott A., Guillermo Lasarte Aragonés, Kim E. Sapsford, Carl W. Brown, Clare E. Rowland, Joyce C. Breger, and Igor L. Medintz. 2018. "Detecting Biothreat Agents: From Current Diagnostics to Developing Sensor Technologies." *ACS Sensors* 3 (10): 1894–2024. doi:10.1021/acssensors.8b00420.

Wang, Hui, John J. Mahle, Trenton M. Tovar, Gregory W. Peterson, Morgan G. Hall, Jared B. DeCoste, James H. Buchanan, and Christopher J. Karwacki. 2019. "Solid-Phase Detoxification of Chemical Warfare Agents Using Zirconium-Based Metal Organic Frameworks and the Moisture Effects: Analyze via Digestion." *ACS Applied Materials & Interfaces* 11 (23): 21109–21116. doi:10.1021/acsami.9b04927.

Wang, S., N. L. Pomerantz, Z. Dai, W. Xie, E. E. Anderson, T. Miller, S. A. Khan, and G. N. Parsons. 2020. "Polymer of Intrinsic Microporosity (PIM) Based Fibrous Mat: Combining Particle Filtration and Rapid Catalytic Hydrolysis of Chemical Warfare Agent Simulants into a Highly Sorptive, Breathable, and Mechanically Robust Fiber Matrix." *Materials Today Advances* 8 100085. doi:10.1016/j.mtadv.2020.100085.

Wheelis, Mark. 2002. "Biological Warfare at the 1346 Siege of Caffa." *Emerging Infectious Diseases* 8 (9): 971–975. doi:10.3201/eid0809.010536.

Yao, Aonan, Xiuling Jiao, Dairong Chen, and Cheng Li. 2020. "Bio-Inspired Polydopamine-Mediated Zr-MOF Fabrics for Solar Photothermal-Driven Instantaneous Detoxification of Chemical Warfare Agent Simulants." *ACS Applied Materials & Interfaces* 12 (16): 18437–18445. doi:10.1021/acsami.9b22242.

Yue, Guozong, Song Su, Na Li, Maobing Shuai, Xinchun Lai, Didier Astruc, and Pengxiang Zhao. 2016. "Gold Nanoparticles as Sensors in the Colorimetric and Fluorescence Detection of Chemical Warfare Agents." *Coordination Chemistry Reviews* 311: 75–84. doi:10.1016/j.ccr.2015.11.009.

Zheng, Qi, Yong Chun Fu, and Jia Qiang Xu. 2010. "Advances in the Chemical Sensors for the Detection of DMMP A Simulant for Nerve Agent Sarin." *Procedia Engineering* 7. 179–184. doi:10.1016/j.proeng.2010.11.027.

2 The Key Concepts of Plasmonic Nanosensors

Aykut Arif Topçu, Merve Asena Özbek,
Gaye Ezgi Yılmaz, and Adil Denizli

2.1 INTRODUCTION

The scientific community, especially in the last decades, has made a transition to the multidisciplinary field with a dynamic and rapid acceleration. A number of admirable multidisciplinary studies are being published every few days, the scope of research is expanding, and new perspectives are coming to the fore.

Sensor systems are devices that bring many conveniences, which have been brought to us through the history of science. The main purpose of sensors is to detect events or changes in the environment and to send valuable informations to a processor unit (Khanna 2011). Thanks to sensor systems—one of the most outstanding examples of multidisciplinary studies—several phenomena that were previously put forward in theory have become tangible and reveal the great potential that the merging of different fields can have. However, with an increasing population and the resulting increasing to human needs, science, by its nature, always needs to be innovating more. Sensor systems have evolved over the years as a strong candidate to satisfy this gap and are divided into many sub-areas.

Sensors can be classified according to the way they process the received signal. Electrochemical sensors, piezo-electric sensors, and optical sensors are some of these subclasses (Wyszynski and Nakamoto 2015; Alegria 2021; MacCraith 1998). These classifications are also divided into branches among themselves. Sensor systems based on optical transducing take advantage of photon-atom interaction for molecule detection. Utilizing a light source that produces electromagnetic waves, the system causes a spectral shift in electromagnetic waves when interacting with the analyte desired to be determined and provides the opportunity to make quantitative and qualitative determinations based on the extent of this shift. Since the detection platform characterizes matter interaction, its design and engineering are crucial.

Nanotechnological discoveries have solidly proven that they can powerfully assist multidisciplinary work. One of the most robust areas in which nanotechnology has developed is nanomaterials (Baig et al. 2021; Bayda et al. 2020). Nanoparticles, fullerenes, carbon nanotubes, and nano-films are among the nanomaterials that have made revolutionary developments in many fields such as medicine, textile,

DOI: 10.1201/9781003459316-2

energy, and food (Jeevanandam et al. 2018). The main purpose of nanomaterials is to increase the quality of life of human beings, just like other technological developments. One of the biggest advantages of nanomaterials is the reduction in size of materials and devices (Law, Marsal, and Santos 2020). The setups where sensor devices and nanomaterials are combined and allowed much more effective analyses to be made. Recently, the development of nanotechnology has paved the way for optical sensing platforms with unique optical features (Munawar et al. 2019; Sadana and Sadana 2011). These nanoplatforms can be combined with various spectroscopic techniques, such as surface SPR, LSPR, PSPR, SERS (Tran et al. 2018; Yilmaz et al. 2015; Kim et al. 2017; Ryu et al. 2021).

In particular, SPR sensors have made significant advances in characterizing biomolecular interactions (Hotta, Yamaguchi, and Teramae 2012; Rosenstein et al. 2012). SPR is based on the excitation of a thin metal sheet with the help of light sent through a prism (Zalevsky and Abdulhalim 2014). The wavelength of the incident light is fixed and undergoes certain refraction as it passes through the prism. Surface changes occur as a result of modifications made on the thin metal film that are expected to interact with a particular molecule. This interaction changes the refractive index of the incident light and provides an opportunity for quantitative analysis thanks to the angle shift that occurs (Miyazaki, Shimizu, and Ferreira 2017; Khlebtsov 2008). This phenomenon, which makes it possible to perform highly sensitive and accurate analyses in many fields, especially in life sciences, has become more specialized in itself and is divided into two branches, LSPR and PSPR, where lower detection limits are targeted (McDonnell 2001).

The plasmonic concept lies in the working principles of optical-based sensing nanosensors. Plasmons are, in general terms, the oscillation of charged particles, and the most common type of plasmons are in conductive materials (metals) (Maier 2007). Free electrons in metals are where plasmons form, and these electrons are not localized. They can be symbolized as a sea of electrons. When these electrons are exposed to electromagnetic light, they create waves similar to the sea. These waves, which are formed by the interaction of some of the energy with free electrons, are plasmonic waves (Jiang, Zhuo, and Wang 2018). These plasmons are formed on the surface and in particles and also, they can be localized for further purposes.

Plasmonic nanosensors are devices based on plasmon logic using plasmonic materials. Detections that have used plasmonic nanoparticles and surface-based plasmons demonstrate the successful ongoing use of nanomaterials (Zhang and Hoshino 2019). Plasmonic nanosensors can be integrated into many fields, due to their potential. Plasmonic nanosensors are able to adapt to any sensing system that requires sensitive sensing, due to the versatile nature of nanomaterials. There are already many studies using plasmonic nanosensors for detection and sometimes prevention, especially in agriculture, medicine, and even warfare threats. In this chapter, the fundamental logic, types, and applications of plasmonic nanosensors are examined as key concepts and summarized with their potential for future studies using relevant articles (Anker et al. 2008).

2.2 THE GENERAL CONCEPT OF PLASMONICS

The term plasmonic is an optical phenomenon arising from the interaction of metal electrons with an electromagnetic field. Sub-diffraction field variation occurs after electron-photon interaction at insulating metal interfaces and metallic nanostructures. Competent plasmonic simulations are implemented; combining calculations based on Maxwell's equations and noted optical properties of metals. Plasmonic-based research explores the formation of surface plasmons or plasmonic nanoparticles (De Ceglia and Vincenti 2016).

2.2.1 SURFACE PLASMONS

The theory of surface plasmons formed between metal and a dielectric interface is frequently used in studies of plasmonic nanosensors. Surface plasmons have the advantage of being able to perform highly efficient analyses in a limited space and can be easily integrated into many other systems where nanomaterials are used (Curto et al. 2010; Akimov et al. 2007). Surface plasmon polaritons (SPPs) are electromagnetic waves propagating in a direction parallel to the metal/dielectric interface; they can also be called surface plasmon waves (SPWs) and are the basis of the surface plasmon principle. SPR occurs when the frequency of the excitation light matches the natural frequency of the emitted surface electrons. The plasmonic waves that occur are extremely sensitive to the smallest changes (molecule binding, adsorption, desorption) that may occur on the surface (Oldenburg et al. 1998).

2.2.2 PLASMONIC NANOPARTICLES

When nanomaterials are considered, nanoparticles come to the fore with their wide variety and numerous areas of application. The use of nanoparticles, which are very valuable due to their ability to be integrated into different systems including biosensors, has also taken the research one step further. The optical properties of metal nanoparticles (NPs) bring many advantages in plasmonic sensors. Their interaction with electromagnetic radiation is the fundamental phenomenon underlying plasmonic nanoparticles (Chang et al. 2021). Plasmonic nanoparticles induce a collective oscillation of surface conduction electrons, called surface plasmons, which are very strong in absorbing and scattering light. The movement of metal electrons upon excitation by light induces an additional electrical field near the surface (Krajczewski, Kołątaj, and Kudelski 2017). As mentioned, metal NPs have unique optical properties. The trademark of nanomaterials is that there can be changes in the characteristic chemical and physical properties of materials at the nanoscale. For example, the known bright yellow color of a gold (Au) ingot can have all the rainbow colors, due to excited plasmons, depending on its shape, when it is reduced to nanodimensions (Liz-Marzán 2020). When metals are symbolized as a sea of free electrons with conduction bands, they cause collective oscillation with the surface when excitation by light occurs. This synchronization makes many metal nanoparticles suitable for plasmonic nanosensors (de Aberasturi, Serrano-Montes, and Liz-Marzán 2015).

2.3 THE FABRICATION METHODS OF PLASMONIC NANOSTRUCTURES

Nanofabrication encompasses the processes and techniques used to create engineered nanostructures and devices smaller than 100 nm. This technology is prevalent in areas such as electronics, energy harvesting, storage, medicine, sensing, and human health. Some multidisciplinary fields, e.g., nanotechnology, engineering, and the material science branches, including biology, chemistry, and biotechnology, have facilitated the design of different types of sensors that convert biological or biochemical signals into electrical signals, through electrical, electrochemical, or optical transducers. Nanomaterial-based transducers are particularly interesting owing to the unique physicochemical, optical, and electrical properties that manifest at the nanoscale (Nocerino et al. 2022). Among the many nanotechnology-based transducers, optical nanobiosensors using noble NPs have seen substantial advancements. They have become a prominent technology, offering enhanced sensing performance and intelligent application (Sepúlveda et al. 2009). The plasmonic NPs can be implemented on plasmonic nanosensors with different methods, such as deposition, or embedding on or into polymeric materials; so, the choice of the implementation method on NPs as a transducer unit is a key issue for the success of the biosensor and boosting of the sensor signals.

Generally, plasmonic nanostructures have been prepared via two main approaches; bottom-up and top-down (Stewart et al. 2008). The bottom-up solution phase is a versatile method of fabricating NPs using metal salts. The advantages of the bottom-up method include better control on the shape, size, and composition of the obtained NPs by the optimization of the stabilizer and the reduction agents used. On the one hand, the stabilizer—such as ligands, surfactans, organic acid or polymers— is responsible for controlling the growth step of NPs and on the other hand, it prevents the agregation of the NPs. The reduction of the metal salts is accomplished with the use of chemical, photochemical, and sonochemical approaches or the use of chemical reductans; but the choice of stabilizer, reductant, or experimental conditions can affect the features of the NPs mentioned above. For instance, the spherical, nanoroad, nanocage, and nanobelt form of AuNPs can be fabricated using different methodologies from the Turkevitch method, from mediated syntheses to the polypol process (Stewart et al. 2008).

The other method, top-down, including some lithography techniques, e.g., electron beam (E-beam) lithography, scanning probe lithography, nanoimprint lithography, and optical litohography (Gates et al. 2005), allows the control of size, shape, and interparticle spacing of surface-bound metalic nanostructures (Stewart et al. 2008).

Of these methods, optical lithography is considered as an established nanofabrication tool, especially for chip fabrication. On the other hand, E-beam lithography is a very popular method with its ability to produce nanostructures with the desired shape, as well being a highly sensitive method and an effective nanofabrication tool for <20 nm. Soft nanoprinted lithography is a nanofabrication tool for producing simple, effective, ultra-small features based on model transfer. Block copolymer lithography is a high-throughput, low-cost method suitable for a variety of nano-structure shapes, including large-scale, densely packed nanostructures, cylinders, and spheres. Scanning probe lithography is capable of high-resolution chemical, molecular, and mechanical nanomodelling to manipulate large molecules and indi-vidual atoms (Biswas et al. 2012).

2.4   SURFACE PLASMON RESONANCE (SPR)-BASED SENSORS

Highly efficient biosensor and chemical sensor technologies are needed in many fields, such as life sciences, medical diagnostics, drug development, and food safety. Optical sensors based on SPR have been shown to have great potential for appli-cation in these fields for affinity biosensing and biomolecular interaction analysis (Piliarik, Vaisocherová, and Homola 2005). SPR sensors provide a label-free, real-time, fast, in-situ, high sensitivity, and selectivity analysis method for the study of molecular binding events (G. E. Yılmaz et al. 2022). These sensing platforms are used for the analysis of thermodynamic and kinetic parameters, quantification of biomolecule concentrations, and investigation of bio-affinity reactions due to their high sensitivity to the refractive index or thickness changes of the medium around the metal surface (Tabasi and Falamaki 2018).

Traditional SPR biosensors use the standard Kretschmann configuration for the SPR coupler (Jiri Homola, Yee, and Gauglitz 1999). The prism-based SPR method, known as the Kretschmann configuration with low-cost, and ease of use, is a commonly used method, which is based on the surface coating of the hypotenuse surface of a glass prism (typically right-angled) using gold or silver metals (Pandey, Raghuwanshi, and Kumar 2022). Owing to the prism, which allows light to pass through a material with a high refractive index, volatile waves are formed along the surface between metals and analytes. These waves are highly sensitive to changes in the refractive index in either the background or the analyte, enabling an examination of reactions on metal surfaces (Sumantri et al. 2020).

However, the key parameters for enhancing the sensing performance of SPR plat-forms are, on the one hand, the thickness and the type of the metallic film formed in the SPR platform with the Kretschmann configuration, and on the other hand, the optimization of incidence angle (Uddin, Chowdhury, and Kabir 2017). Moreover, the metallic nanostructures or grid structures could be used to improve the detection limit of SPR sensors (Wu and Wang 2008).

2.4.1   TYPES OF SPR-BASED SENSORS

2.4.1.1   Localized Surface Plasmon Resonance (LSPR) Sensors

The Kretschmann geometry, a universal scheme for SPR, and its details are mentioned above. Although high sensitivity is achieved in studies performed with SPR, extended smooth surfaces are needed (Nagpal et al. 2009; Jiří Homola 2003) and the SPR method has shortcomings in its application due to its intricate system and reduced spectral resolution. At this point, LSPR sensor systems supported by nanostructures are favorable platforms for in vivo or in situ studies (Brolo 2012; Jeong et al. 2016). LSPR, a connection between the electromagnetic field and spatially confined free electrons, offers the potential to overcome these challenges when detecting nanoscale biological interactions (Lee, Roh, and Park 2009).

Plasmonic NPs show high sensitivity to differences in the refractive index and therefore they are potential candidates for use in the design of label-free nanosensors. The approach of LSPR is based on the interactions between incident light and surface electrons in a conduction band. Here, the operating principle of the system can be summarized as follows: An intense oscillation takes place in the surface electrons of conductive NPs, which are smaller than the wavelength of the incident light. LSPR extinction is maximally achieved if the frequencies of the free electrons of the NPs and the incident light match. Besides, there is a significant increase in the scattered and absorbed light, as well as the electromagnetic fields that exist around and within the NPs (Vo-Dinh, Wang, and Scaffidi 2010; Brahmkhatri et al. 2021).

The interaction in question depends on the geometry, surface morphology, size, dielectric environment, and aggregation level of the nanoparticles. The basic principle of LSPR sensors is encapsulated in how adjustments to these parameters control the LSPR signal, affecting intensity, wavelength shift, or both, and is central to real-time monitoring applications (Gao, Li, and Huang 2019). Additionally, strategies like loading/unloading and binding/disconnection can be utilized to regulate the LSPR of plasmonic NPs.

The most preferred metals for nanoparticle fabrication in LSPR sensor systems are noble metals such as Au and silver (Ag). It is possible to functionalize these NPs to increase the performance of LSPR sensors.

In the comparison between Ag and Au metals used in nanoparticle production, it is stated that Ag has the strongest and sharpest bands. However, some properties such as biocompatibility, inertness of structure, and thiol-gold relationship for immobilization of biomolecules, make Au more suitable for biological application (Willets and Van Duyne 2007; Lu et al. 2009; Liz-Marzán 2006; Petryayeva and Krull 2011).

2.4.1.2 Propagating Surface Plasmon Resonance (PSPR) Sensors

The conformational changes of macromolecules and the binding analysis of biomolecules are topics of interest in fields such as pharmacy and medicine. The optical spectroscopy of metallic structures that present surface plasmon resonances with high sensitivity compared to other label-free methods is important in this context. Molecular binding takes place on the surface of metallic structures where light is captured in the form of SPPs. In excitations of SPPs, which are highly sensitive to surface molecules, circumstances such as incidence angles and illumination wavelengths are essential. Furthermore, sensor systems based on changes in plasmon excitation supply real-time data flow about the molecular adsorption process (Jatschka et al. 2016; Yonzon et al. 2004).

There are generally two main types of surface plasmonic modes: PSPR and LSPR (Live, Bolduc, and Masson 2010; Barnes, Dereux, and Ebbesen 2003). In the PSPR system, which offers low detection limits by using thin smooth metallic layers, plasmons propagate along the interface between the metal and the dielectric. SPPs are created at the interface between a metal and a dielectric, thanks to a grid structure or a high refractive index prism. They perform as a collective excitation of conduction electrons propagating in a wave-like way along this interface. In addition to the efficient conversion of the energy of the incident light into a surface wave, the dependence of emission and excitation on the angle of incident can be considered an important feature of PSPR systems. Greater bulk sensitivity is achieved with SPPs used in plasmonic sensing (X. Wang et al. 2013; Yang et al. 2021; Z. L. Zhang et al. 2016).

2.4.1.3 Surface-Enhanced Raman Scattering (SERS) Sensors

Raman spectroscopy is a technique used to clarify the surface interactions of molecules and determine molecular structures. Photons hit the molecule, producing inelastic scattered photons, and this method measures the frequency shifts of the scattered light. As photons of laser light are absorbed by the sample, the wave numbers of re-emitted photons are shifted down or up relative to the original monochrome waves. The resulting shift gives data about the properties of molecules such as rotation, wavelength transitions, and vibration (C. Wang et al. 2017). The applications realized by this method face some limitations due to the low efficiency of Raman scattering processes. But now, Raman scattering efficiency can be increased by adsorbing molecules to nanostructured metal surfaces with SERS (Vo-Dinh, Wang, and Scaffidi 2010).

In addition to the progress made by nanotechnology, detailed experimental and theoretical studies have greatly improved the SERS approach and highlighted its application in different fields such as physics, biomedicine, and chemistry. Some of these can be summarized as environmental analyses (Bai et al. 2018; De Zhang et al. 2021) and disease biomarkers (C. Li et al. 2020; Song et al. 2016; Pang et al. 2020; Huang et al. 2019). Exhibiting high sensitivity with the ability to detect multiple analytes simultaneously, SERS is a powerful technique that provides non-destructive detection that can detect down to the single molecule level.

Discovered by Fleischmann et al., SERS is used to detect analytes adsorbed on the surfaces of nano-sized metal particles by enhancing Raman scattering signals. The amplification mechanisms of Raman signals are considered electromagnetic and chemical enhancements. Electromagnetic enhancement is associated with plasmon excitation in metal particles acting as a SERS substrate, that is, with the phenomenon of LSPR. As a result of the electron transfer between the metal substrate and the target molecule of interest, the improvement of Raman signals is expressed as chemical enhancement (Zong et al. 2018; Laing et al. 2017).

The contribution of SERS to the tip enhanced Raman scattering (TERS) technique, which is the result of combining SERS with scanning tunneling microscopy (STM) or atomic force microscopy (AFM) techniques, should be particularly emphasized (Langer et al. 2020). However, efforts are still needed to increase the applicability and practicality of SERS, such as substrate stability, data reproducibility, and analyte-substrate interaction (Pérez-Jiménez et al. 2020).

2.5 APPLICATIONS OF PLASMONIC NANOSENSORS

In this section, the usability, design, and applicability of plasmonic nanosensor platforms, including SPR sensors, SERS, and NPs with localized plasmon properties in the fields of biomedical applications, agricultural studies, and warfare threats are discussed using relevant research articles.

2.5.1 BIOMEDICAL STUDIES

Creatinine, the waste product of creatine metabolism, is released by the kidneys at a constant rate and monitoring its level is a useful biomarker to investigate kidney function (Arif Topçu et al. 2019). The Jaffe reaction is the simplest and a colorimetric method for determining creatinine levels using picric acid in an alkaline medium; however, interference from other metabolites such as uric acid, glucose, and others can affect the accuracy of experimental findings. As such, the need for monitoring creatine levels with accuracy is a key issue for determining kidney function. To this aim, researchers prepared plastic creatinine receptors on the sensing surface via molecular imprinting technique (MIT) to enhance both the selectivity of the sensor and experimental accuracy. The sensor recognized this clinical analyte with higher selectivity than uric acid or creatine as competitor molecules in the aqueous solution and the urine mimic.

Urinary tract infection (UTI) is a common bacterial infection caused by *Escherichia coli* (*E. coli*) and if left untreated, UTI leads to serious health problems in human beings. Urine culture is an effortless and a cost-effective way to achieve diagnosis (Özgür et al. 2020); however, this technique is restricted by some factors, e.g., long incubation time, the need for a selective culture media, and labor-intensiveness. Thereby, new approaches including biosensors have paved the way for UTI diagnosis. In one such study, Özgür et al. (2020) constructed an SPR-based sensor with the assistance of AgNPs for the diagnosis of UTI and the researchers aimed to increase the sensitivity and selectivity of the SPR sensor and, with the implementation of AgNPs, created tailor-made *E. coli* receptors on the sensing surface. The limit of detection (LOD) of the proposed sensor was lower (0.57 CFU/mL) than its clinical range for this infection; furthermore, the whole analysis was accomplished in approximately 20 minutes.

Inherited diseases are generally caused by mutation of the genes, which results in dysfunction of proteins leading to a genetic disorder (Kaplanis et al. 2020; F. Zheng et al. 2022). The early diagnosis of inherited diseases is of paramount importance impacting the success of their treatment. Duchenne muscular dystrophy (DMD) is a hereditary disease caused by a mutation in the dystrophin gene (Verhaart and Aartsma-Rus 2019), which results in the loss of myofiber functions due to insufficient levels of the cytoskeletal protein. So, researchers (F. Zheng et al. 2022) employed the SPR sensor to sense DMD. Before the design of the SPR sensor, the

CRISPR-associated protein 9 (dCas9) was coupled with the single-guide RNA and were immobilized on the sensing surface. Following that, the clinical studies of determination of this inherited disease were evaluated using genomic DNA. The LOD of the prepared plasmonic nanosensor was calculated as 1.3 fM and during the clinical studies, the sensor detected DMD in real time within 5 minutes.

Corona disease, known as the ongoing global pandemic, has caused many deaths and serologic tests like enzyme-linked immunoassay sorbent (ELISA) are commonly used to diagnose COVID-19 (Qu et al. 2022). Qu and coworkers (Qu et al. 2022) designed a fiber-optic SPR (FO-SPR) sensor that not only examines antibody levels and their binding profiles in serum and whole blood samples but diagnoses COVID-19 as well. Before constructing the new sensing platforms, the researchers first prepared the SPR sensor using the His6-tagged receptor binding domain (RBD) of the SARS-CoV-2 viral protein as a bioreceptor and in the second approach, the sandwich bioassays were designed using goat anti-human antibodies [(with or without gold nanoparticles (AuNPs)] to improve the sensing capabilities. Finally, the sensing performances of the designed sensors were compared with in-house, and commercial ELISA kits. According to the experimental results, and the statistical analysis, the sandwich bioassay with the shorter analysis time showed a better performance than the ELISA.

Immunoglobulin M (IgM) was the first antibody produced to stop or fight infections. Therefore, measuring its levels in biological fluids is helpful to evaluate viral, bacterial infections, and some diseases such as dengue and rheumatoid arthritis (Bereli et al. 2021). With this in mind, the authors aimed to prepare an SPR-based immunosensor to measure the IgM levels in the aqueous solution and artificial plasma. For that purpose, the sensing surface of the plasmonic sensor was first modified for the attachment of the anti-human IgM and then, to enhance the sensing performance of the plasmonic nanosensor, the amine-functionalized AuNPs (N-AuNPs) were anchored onto the anti-human IgM molecules. Finally, the IgM levels were measured in the aqueous solution and artificial plasma. In light of the experimental results, the anti-IgM could successfully be immobilized on the sensing surface through the N-AuNPs and the implementation of N-AuNPs could enhance the sensing performance of the SPR-based immunosensor as compared with the plain SPR sensor.

Breast cancer is one of the most prevalent forms of cancer and the incidence rate is particularly high in women (Wong et al. 2021). Mammography is one of the most preferred approaches for a breast cancer diagnosis but the false positive results of mammography and the cost of treatment arising from false positive results are major limitations of this approach. So, the need for affordable, selective, and accurate detection methods has paved the way for biosensor-based detection for breast cancer diagnostics. Recently, miR-1249 has been a useful biomarker for the early detection of breast cancer diagnostics and some different types of sensors were employed in breast cancer diagnosis. For example, Wong and colleagues (Wong et al. 2021) designed an SPR-based sensor, which was used for the first time for early detection of breast cancer. To this aim, the authors immobilized the specific DNA probes to detect the miR-1249 breast cancer biomarker and the recognition capability of the sensor toward malignant and benign tumors was investigated. According to the experimental results, the resolution of the sensor was calculated as 63.5 nm and the sensor was capable of sensing malignant tumors during the clinical studies.

In the literature, most of the biomedical studies reported using SPR-based sensors and in Table 2.1, some relevant biomedical studies are summarized in relation to the modification of the sensing surface, the purpose of the sensor, and LOD.

TABLE 2.1
The Usability of SPR-Based Sensors for Biomedical Applications

Modification of Sensing Surface	Purpose of the Sensor	LOD	Reference
Procalcitonin	Sepsis	15–30 ng/mL	(Battaglia et al. 2021)
Graphene oxide	Pregnancy-associated plasma protein A2	0.12–0.13 pg/mL	(Chiu et al. 2019)
Anti-cardiac troponin I	Cardiac troponin I	0.00012 ng/mL	(Çimen et al. 2020)
Anti-leptin antibody	Leptin	0.07 ng /mL	(Sankiewicz et al. 2021)
DNA modified gold nanocube	Lung cancer	Up to 5 pm	(L. Zhang et al. 2017)
Star-shaped nanoparticle	Human genome DNA	6.1 nM–10 pM	(Mariani et al. 2015)
Fusion protein	Tuberculosis	-	(Sun et al. 2017)
Gold nanorods	DNA tumor point mutation	2 ng	(Tadimety et al. 2019)

As is well known, the transducer of a biosensor is one of the critical parts of the sensing system and the choice of transducer can affect the sensing performance of the constructed sensor toward the analyte of interest; so, the plasmonic NPs—particularly AuNPs and AgNPs due to their localized plasmon effects—are promising optical transducers for developing the biosensor platforms (Unser et al. 2015).

For example, Zhang and coworkers (Y. Zhang et al. 2018) employed SERS with the assistance of AgNPs to detect Lens culinaris agglutinin-reactive alpha-fetoprotein (AFP-L3) is a biomarker for the early diagnosis and used in monitoring of hepatocellular carcinoma disease. The authors prepared a polystyrene sphere @Ag/SiO$_2$/Ag SERS chip after which, the Raman reporter molecule, 5,5'-Dithiobis (succinimidyl-2-nitro-benzoate) (DSNB), was immobilized on the SERS chip and finally, the AFB-L3 anti-body was attached to the sensing surface before the Raman measurements were taken.

In addition, the SERS-based immunoassay was developed by researchers (Yu et al. 2021) for the early diagnosis of Alzheimer's disease and two biomarkers, amyloid β peptide (Aβ) and hyperphosphorylated tau protein (P-Tau-181), were chosen to design a sandwich SERS platform. Before the fabrication of the SERS platform, the prepared antibodies were immobilized on magnetic graphene oxide surfaces and were labelled with mercaptobenzoic acid (4-MBA). Following that, the sensing capability and the applicability of the sensor was investigated in the aqueous solution and in serum samples, respectively; moreover, their results were compared with ELISA.

TABLE 2.2

The Boost Effects of Plasmonic Nanoparticles on SERS Platforms for Biomedical Studies

Plasmonic Particles	Method	Purpose	LOD	Reference
AuNPs	SERS	cTnI	8.9 pg/mL	(Cheng et al. 2019)
		CK-MB	9.7 pg/mL	
Combination of Au and Ag nanostructures	SERS	CRP	1.14 pg/mL	(Hu et al. 2021)
AgNPs	SERS	CEA153	0.01 ng/mL	(Z. Zheng et al. 2018)
		CA125	0.01 ng/mL	
		CEA	0.001 ng/mL	
Au-Ag core-shell NPs	SERS	MUC1	0.1 U/mL	(N. Li et al. 2020)
Ag pyramids	SERS	PSA	0.96 aM	(Xu et al. 2015)
		Thrombin	85 aM	
		MUC1	9.2 aM	
AgNPs	SERS	mi-RNA	<1 pM	(Qi et al. 2017)
AuNPs	SERS	Haptoglobin	-	(Beffara et al. 2020)
AuNPs	SERS	HIV-1	0.24 pg/mL	(Fu et al. 2016)
AuNPs	SERS	HBV	0.1fM	(Zengin, Tamer, and Caykara 2017)

Abbreviations: cTnI (cardiac troponin I), CK-MB (creatine kinase-MB), CRP (C-reactive protein), CEA125 (carbohydrate antigen 15-3), CA125 (carbohydrate antigen 12-5), CEA (carcinoembryonic antigen), MUC1 (human mucin-1), PSA (prostate-specific antigen), ALP (alkaline phosphatase), HBsAg (Hepatitis B surface antigen), AFB (α-fetoprotein).

The results revealed that the P-Tau-181 biomarker is a better choice for the early diagnosis of Alzheimer's disease.

When researching SERS-based biosensors and how their signals are boosted with the help of plasmonic NPs for biomedical studies, it is possible to find various articles; so, in Table 2.2, some of the biomedical studies using SERS platforms are summarized with their purposes and LOD values.

2.5.2 AGRICULTURAL STUDIES

Aflatoxins are produced by the genus *Aspergillus*, and low concentrations of these secondary metabolites may exert mutagenic and teratogenic effects on humans and animals. In addition, aflatoxin-contaminated animal foods, such as milk, have adverse effects on human health (Akgönüllü, Yavuz, and Denizli 2021). Therefore, the use of biosensors to detect and monitor aflatoxin levels at low concentrations is of great importance for preventing these undesirable conditions. Akgönüllü et al.

(Akgönüllü, Yavuz, and Denizli 2021) prepared a molecularly imprinted polymer (MIP) on the sensing surface, which recognizes Aflatoxin M1 and utilized AuNPs to enhance the sensing capability of the SPR-based sensor to measure Aflatoxin M1 levels in dairy farmers' milk samples. Furthermore, the sensing and selectivity performance of this nanoplasmic sensor was also tested against other agricultural analog molecules, aflatoxin B1 (AFB1), citrinin (CIT), and ochratoxin A (OTA).

Additionally, another mycotoxin, zearalenone, is found in various grains and is an endocrine disruptor that causes dysfunction of the reproductive system in animals and humans (Choi et al. 2009). The authors (Choi et al. 2009) chose the same method, MIT, to prepare tailor-made zearalenone receptors by pyrrole electropolymerization on the sensing surface. After the characterization of the sensing surface, the LOD, and the recovery of the sensor were found to be 0.3 ng/g and 89%, respectively. Furthermore, the proposed sensor was more selective than the other zearalenone molecules, and results were comparable to those found using the ELISA results.

The next study reported by Yilmaz et al. (Yilmaz et al. 2017) proposed a new approach for the measurement of atrazine levels in aqueous solutions. The authors prepared atrazine imprinted nanoparticles using the N-methacryloyl-L-aspartic acid (MAAsp) as a functional monomer, which is a polymerizable form of L-aspartic acid, and then the atrazine imprinted NPs were physically attached to the sensing surface. After that, the sensing ability of the SPR-based sensor was characterized in the aqueous solution and the selectivity of the SPR sensor was determined against the competitor molecules. According to the experimental results, the recognition capability of the nanoparticle-attached SPR sensor was in accordance with the Langmuir isotherm model and the adsorption occurred in homogenous surfaces with identical binding sites. Furthermore, the LOD of the SPR sensor was calculated as 0.7134 ng/mL and the prepared sensor recognized atrazine with higher selectivity than simazine and amitrole.

Food contamination depending on microorganisms is another issue for public health and *Salmonella paratyphi (S. paratyphi)* is one of the pathogenic bacteria causing food-borne disease (Perçin et al. 2017); so, the choice of rapid, selective, and sensitive detection methods plays a key role in pathogen detection as compared to conventional techniques. The authors (Perçin et al. 2017) used the microcontact imprinted technique and aimed to develop the first SPR sensor to sense *S. paratyphi* in a real sample. During the microcontact approach, first, the glass slide was modified with 3-amino-propyltriethoxysilane (APTES) solution to generate the functional amine groups on the glass surface and then *S. paratyphi* was immobilized on the glutaraldehyde-activated glass surface containing the amine groups. Finally, the polymer solution was added to the chip surface and the glass slide containing the pathogenic bacteria was pressed onto the chip surface and UV polymerization occurred. Following the polymerization, the glass slide was removed from the chip surface to create the specific recognition cavities that recognize the whole bacterial cell. The experimental results revealed that the proposed selective and sensitive sensor with a low detection limit could be potentially used in sensing *S. paratyphi*.

In Table 2.3, we demonstrate some plasmonic platforms and their sensing capabilities for agricultural molecules.

TABLE 2.3

Some Other Agricultural Molecules and Their Sensing via Nanoplasmonic Platforms

Sensing Platform	Target Molecule	LOD	Reference
SPR	Imidacloprid	11.103 ppb	(S. H. Wang et al. 2020)
	Fipronil	9.862 ppb	
SPR	DON	26 µg/kg	(Joshi et al. 2016)
	ZEA	6 µg/kg	
	T-2	0.6 µg/kg	
	OTA	3 µg/kg	
	FB_1	2 µg/kg	
	AFB_1	0.6 µg/kg	
AuNPs-SPR	Atrazine	1.0 ng/mL	(Liu et al. 2015)

Abbreviations: DON (deoxynivalenol), ZEA (zearalenone), T-2 (T-2 toxin), OTA (ochratoxin A), FB1 (fumonisin B1), AFB1 (aflatoxin B1).

2.5.3 WARFARE THREATS

2,4,6-Trinitrotoluene (TNT) is an explosive and has risks to human health and environment even at low doses (J. Wang 2021); so, the sensing and monitoring of TNT levels are desirable for human health and environmental safety. Various sensing platforms that include nanoplasmonic sensors have been developed to measure its level. For instance, Wang (J. Wang 2021) prepared an SPR-based sensor to detect TNT levels using electron-rich amine groups of APTES and electron-poor of TNT via the Meisenheimer complex. Before the surface modification of the sensor chip, its surface was modified with a potassium hydroxide (KOH) solution to generate the OH groups on its surface and after that, the organic ligand, APTES, was attached to the sensing surface. Thereafter, the sensor was characterized with different TNT concentrations in real-time and label free. According to the experimental results, the TNT recognition by the SPR sensor modified with APTES was suitable with the Langmuir isotherm model, and the LOD of TNT was calculated as 134 ppb.

Additionally, the researchers (Liyanage et al. 2018) used gold triangular nanoprism (AuTNPs) as SERS substrates to enhance the prepared sensing capability against the three explosives, TNT, pentaerythritol tetranitrate (PETN), and cyclotrimethylene-trinitramine (RDX). During the design of the SERS-based sensor, the APTES functionalized glass surface was incubated with the AuTNPs to form a self-assembled layer and following that, the SERS substrate was removed from the surface using an adhesive tape and then transferred onto a flexible adhesive substrate to prepare the SERS sensor. Finally, the SERS measurements against the explosives were carried out using two approaches. In the first approach, each sample was added separately to the sensing surface, and in the second approach, the adhesive nanosensors were placed on the glass surface containing the explosives; but, in both cases, the sensing of explosives by the SERS-based sensor was dependent on physical adsorption via Au-N interactions. The LODs of the fabricated SERS nanosensor were found to be 900 ppq for TNT, 56 ppq for RDX, and 56 ppq for PETN and the fabricated sensor maintained its sensing performance toward these explosives after five months without any significant change in sensing capability.

In the next study, to sense TNT, peptides were used as a biorecognition unit of the nanoplasmonic sensor (Diming Zhang et al. 2016). The peptide was synthesized and the purity of the biorecognition unit was evaluated with the instrumental devices. After that, the peptide was immobilized on nanocup arrays (nanoCA) through Au-S covalent bonds without any chemicals, and dinitrotoluene (DNT) and 3-Nitrotoluene (3-NT) were chosen to test the selectivity of the fabricated sensor, due to their molecular structures being similar to TNT. The experimental studies showed that the LOD of the prepared sensor was calculated as 1.37 nm and the sensor, by using the specific peptide, became more selective for TNT than for the other explosives.

The vapor phase sensing platform reported by Aznar-Gadea et al. (2021) was based on sensing 3-NT as a taggant agent for TNT with the combination of Ag nanocomposites and MIT. In the design of this sensing platform, the in-situ formation of AgNPs and the specific nanocavities occurred in a single step using polyethyleneimine (PEI). During the design of the TNT sensor, PEI was used as a reducing agent of Ag (I) to AgNPs due to its functional primary and secondary amine groups, and the nanocavities were created inside PEI via the evaporation of the template molecule during the baking step. The LOD of the sensor was calculated as 10 nm and the sensor was more selective than the other competitive explosives.

Although the explosives are warfare agents, the biological and chemical molecules, such as botulinum, Tabun, VX, HD, etc., are also known as warfare agents; so, in Table 2.4, the plasmonic nanosensors used in the other warfare agents are illustrated.

TABLE 2.4
Other Warfare Agents and Their Sensing Methods

Sensing Method	Warfare Threat	Reference
SERS	VX and HD	(Heleg-Shabtai et al. 2020)
SERS	DMMP	(Lafuente, Pellejero, et al. 2018)
SERS	VX and tabun	(Hakonen et al. 2016)
SERS	VX, tabun, and cyclosarin	(Juhlin et al. 2020)
SERS	DMMP	(Lafuente, Berenschot, et al. 2018)

Abbreviations: DMMP (dimethyl methyl phosphonate), VX (O-ethyl S-[2-(diisopropylamine) ethyl] methyl phosphorothioate), HD (bis(2-chloroethyl) sulfide).

2.6 FUTURE ASPECTS AND CONCLUSION

The inspiration and observation of nature have paved the way for the new approaches to protecting human lives and one of these observations, biosensors, can make human life easier, as well as and bringing various disciplines together, such as engineering, nanotechonology, and natural sciences. Today, diagnostics and sensing are highly desirable concepts for protecting public health, monitoring the progression of diseases, and ensuring environmental safety.

In this section, we aimed to highlight state-of-the-art nanoplasmonic biosensor platforms, including SPR, SERS, LSPR, PSPRs, and plasmonic nanoparticles, all of which have use for biomedical purposes, agricultural studies, and warfare threats. In the first section, we gave a short overview of the key concepts of plasmonic nanosensors, with particular focus on their basic working principles and their configurations. In the next section, their use, applicability, and especially, their design and modification strategies were exemplified using relevant articles.

Consequently, the need for biosensor technology should not be underestimated and in the near future, the use of new sensor platforms will gradually increase and will offer new opportunities to improve the quality of human life.

REFERENCES

Aberasturi, Dorleta Jimenez de, Ana Belén Serrano-Montes, and Luis M. Liz-Marzán. 2015. "Modern Applications of Plasmonic Nanoparticles: From Energy to Health." *Advanced Optical Materials* 3 (5). https://doi.org/10.1002/adom.201500053.

Akgönüllü, Semra, Handan Yavuz, and Adil Denizli. 2021. "Development of Gold Nanoparticles Decorated Molecularly Imprinted–Based Plasmonic Sensor for the Detection of Aflatoxin M1 in Milk Samples." *Chemosensors* 9 (12): 1–15. https://doi.org/10.3390/chemosensors9120363.

Akimov, A. V., A. Mukherjee, C. L. Yu, D. E. Chang, A. S. Zibrov, P. R. Hemmer, H. Park, and M. D. Lukin. 2007. "Generation of Single Optical Plasmons in Metallic Nanowires Coupled to Quantum Dots." *Nature* 450 (7168): 402–6. https://doi.org/10.1038/nature06230.

Alegria, Francisco André Corrêa. 2021. *Sensors and Actuators*. https://doi.org/10.1142/12426.

Anker, Jeffrey N., W. Paige Hall, Olga Lyandres, Nilam C. Shah, Jing Zhao, and Richard P. Van Duyne. 2008. "2008 [Review] Biosensing with Plasmonic Nanosensors." *Nature Materials* 7 (June): 8–10.

Arif Topçu, Aykut, Erdoğan Özgür, Fatma Yılmaz, Nilay Bereli, and Adil Denizli. 2019. "Real Time Monitoring and Label Free Creatinine Detection with Artificial Receptors." *Materials Science and Engineering B: Solid-State Materials for Advanced Technology* 244 (April): 6–11. https://doi.org/10.1016/j.mseb.2019.04.018.

Aznar-Gadea, Eduardo, Pedro J. Rodríguez-Canto, Juan P. Martínez-Pastor, Andrii Lopatynskyi, Vladimir Chegel, and Rafael Abargues. 2021. "Molecularly Imprinted Silver Nanocomposites for Explosive Taggant Sensing." *ACS Applied Polymer Materials* 3 (6): 2960–70. https://doi.org/10.1021/acsapm.1c00116.

Bai, Shi, Daniela Serien, Anming Hu, and Koji Sugioka. 2018. "3D Microfluidic Surface-Enhanced Raman Spectroscopy (SERS) Chips Fabricated by All-Femtosecond-Laser-Processing for Real-Time Sensing of Toxic Substances." *Advanced Functional Materials* 28 (23): 1–10. https://doi.org/10.1002/adfm.201706262.

Baig, Nadeem, Irshad Kammakakam, Wail Falath, and Irshad Kammakakam. 2021. "Nanomaterials: A Review of Synthesis Methods, Properties, Recent Progress, and Challenges." *Materials Advances* 2 (6): 1821–71. https://doi.org/10.1039/d0ma00807a.

Barnes, William L., Alain Dereux, and Thomas W. Ebbesen. 2003. "Surface Plasmon Subwavelength Optics." *Nature* 424 (6950): 824–30. https://doi.org/10.1038/nature01937.

Battaglia, F., V. Baldoneschi, V. Meucci, L. Intorre, M. Minunni, and S. Scarano. 2021. "Detection of Canine and Equine Procalcitonin for Sepsis Diagnosis in Veterinary Clinic by the Development of Novel MIP-Based SPR Biosensors." *Talanta* 230 (January): 122347. https://doi.org/10.1016/j.talanta.2021.122347.

Bayda, Samer, Muhammad Adeel, Tiziano Tuccinardi, Marco Cordani, and Flavio Rizzolio. 2020. "The History of Nanoscience and Nanotechnology: From Chemical–Physical Applications to Nanomedicine." *Molecules* 25 (1): 1–15. https://doi.org/10.3390/molecules25010112.

Beffara, Flavien, Jayakumar Perumal, Aniza Puteri Mahyuddin, Mahesh Choolani, Saif A. Khan, Jean Louis Auguste, Sylvain Vedraine, Georges Humbert, U. S. Dinish, and Malini Olivo. 2020. "Development of Highly Reliable SERS-Active Photonic Crystal Fiber Probe and Its Application in the Detection of Ovarian Cancer Biomarker in Cyst Fluid." *Journal of Biophotonics* 13 (3): 1–11. https://doi.org/10.1002/jbio.201960120.

Bereli, Nilay, Monireh Bakhshpour, Aykut Arif Topçu, and Adil Denizli. 2021. "Surface Plasmon Resonance-Based Immunosensor for Igm Detection with Gold Nanoparticles." *Micromachines* 12 (9): 1–11. https://doi.org/10.3390/mi12091092.

Biswas, Abhijit, Ilker S. Bayer, Alexandru S. Biris, Tao Wang, Enkeleda Dervishi, and Franz Faupel. 2012. "Advances in Top-down and Bottom-up Surface Nanofabrication: Techniques, Applications & Future Prospects." *Advances in Colloid and Interface Science* 170 (1–2): 2–27. https://doi.org/10.1016/j.cis.2011.11.001.

Brahmkhatri, Varsha, Parimal Pandit, Pranita Rananaware, Aviva D'Souza, and Mahaveer D. Kurkuri. 2021. "Recent Progress in Detection of Chemical and Biological Toxins in Water Using Plasmonic Nanosensors." *Trends in Environmental Analytical Chemistry* 30. https://doi.org/10.1016/j.teac.2021.e00117.

Brolo, Alexandre G. 2012. "Plasmonics for Future Biosensors." *Nature Photonics* 6 (11): 709–13. https://doi.org/10.1038/nphoton.2012.266.

Ceglia, D. De, and M. A. Vincenti. 2016. "Plasmonics." *Fundamentals and Applications of Nanophotonics*, January, 233–52. https://doi.org/10.1016/B978-1-78242-464-2.00008-7.

Chang, Hyejin, Won Yeop Rho, Byung Sung Son, Jaehi Kim, Sang Hun Lee, Dae Hong Jeong, and Bong Hyun Jun. 2021. "Plasmonic Nanoparticles: Basics to Applications (I)." *Advances in Experimental Medicine and Biology* 1309: 133–59. Springer. https://doi.org/10.1007/978-981-33-6158-4_6.

Cheng, Ziyi, Rui Wang, Yanlong Xing, Linlu Zhao, Jaebum Choo, and Fabiao Yu. 2019. "SERS-Based Immunoassay Using Gold-Patterned Array Chips for Rapid and Sensitive Detection of Dual Cardiac Biomarkers." *Analyst* 144 (22): 6533–40. https://doi.org/10.1039/c9an01260e.

Chiu, Nan Fu, Ming Jung Tai, Hwai Ping Wu, Ting Li Lin, and Chen Yu Chen. 2019. "Development of a Bioaffinity SPR Immunosensor Based on Functionalized Graphene Oxide for the Detection of Pregnancy-Associated Plasma Protein A2 in Human Plasma." *International Journal of Nanomedicine* 14: 6735–48. https://doi.org/10.2147/IJN.S213653.

Choi, Sung Wook, Hyun Joo Chang, Nari Lee, Jae Ho Kim, and Hyang Sook Chun. 2009. "Detection of Mycoestrogen Zearalenone by a Molecularly Imprinted Polypyrrole-Based Surface Plasmon Resonance (SPR) Sensor." *Journal of Agricultural and Food Chemistry* 57 (4): 1113–18. https://doi.org/10.1021/jf804022p.

Çimen, Duygu, Nilay Bereli, Serdar Günaydın, and Adil Denizli. 2020. "Detection of Cardiac Troponin-I by Optic Biosensors with Immobilized Anti-Cardiac Troponin-I Monoclonal Antibody." *Talanta* 219 (June): 121259. https://doi.org/10.1016/j.talanta. 2020.121259.

Curto, Alberto G., Giorgio Volpe, Tim H. Taminiau, Mark P. Kreuzer, Romain Quidant, and Niek F. Van Hulst. 2010. "Unidirectional Emission of a Quantum Dot Coupled to a Nanoantenna." *Science* 329 (5994). https://doi.org/10.1126/science.1191922.

Fu, Xiuli, Ziyi Cheng, Jimin Yu, Priscilla Choo, Lingxin Chen, and Jaebum Choo. 2016. "A SERS-Based Lateral Flow Assay Biosensor for Highly Sensitive Detection of HIV-1 DNA." *Biosensors and Bioelectronics* 78: 530–37. https://doi.org/10.1016/j.bios.2015.11. 099.

Gao, Peng Fei, Yuan Fang Li, and Cheng Zhi Huang. 2019. "Localized Surface Plasmon Resonance Scattering Imaging and Spectroscopy for Real-Time Reaction Monitoring." *Applied Spectroscopy Reviews* 54 (3): 237–49. https://doi.org/10.1080/05704928.2018. 1554581.

Gates, Byron D., Qiaobing Xu, Michael Stewart, Declan Ryan, C. Grant Willson, and George M. Whitesides. 2005. "New Approaches to Nanofabrication: Molding, Printing, and Other Techniques." *Chemical Reviews* 105 (4): 1171–96. https://doi.org/10.1021/ cr030076o.

Hakonen, Aron, Tomas Rindzevicius, Michael Stenbæk Schmidt, Per Ola Andersson, Lars Juhlin, Mikael Svedendahl, Anja Boisen, and Mikael Käll. 2016. "Detection of Nerve Gases Using Surface-Enhanced Raman Scattering Substrates with High Droplet Adhesion." *Nanoscale* 8 (3): 1305–8. https://doi.org/10.1039/c5nr06524k.

Heleg-Shabtai, Vered, Hagai Sharabi, Amalia Zaltsman, Izhar Ron, and Alexander Pevzner. 2020. "Surface-Enhanced Raman Spectroscopy (SERS) for Detection of VX and HD in the Gas Phase Using a Hand-Held Raman Spectrometer." *Analyst* 145 (19): 6334–41. https://doi.org/10.1039/d0an01170c.

Homola, Jiří. 2003. "Present and Future of Surface Plasmon Resonance Biosensors." *Analytical and Bioanalytical Chemistry* 377 (3): 528–39. https://doi.org/10.1007/ s00216-003-2101-0.

Homola, Jiri, Sinclair S. Yee, and Gunter Gauglitz. 1999. "Surface Plasmon Resonance Sensors: Review." *Sensors and Actuators, B: Chemical* 54 (1): 3–15. https://doi.org/10. 1016/S0925-4005(98)00321-9.

Hotta, Kazuhiro, Akira Yamaguchi, and Norio Teramae. 2012. "Deposition of Polyelectrolyte Multilayer Film on a Nanoporous Alumina Membrane for Stable Label-Free Optical Biosensing." *Journal of Physical Chemistry C* 116 (44): 23533–39. https://doi.org/10. 1021/jp308724m.

Hu, Ziwei, Xia Zhou, Jun Duan, Xueqiang Wu, Jiamin Wu, Pengcheng Zhang, Wanzhen Liang, et al. 2021. "Aptamer-Based Novel Ag-Coated Magnetic Recognition and SERS Nanotags with Interior Nanogap Biosensor for Ultrasensitive Detection of Protein Biomarker." *Sensors and Actuators B: Chemical* 334 (November 2020): 129640. https://doi.org/10.1016/j.snb.2021.129640.

Huang, Zhipeng, Ren Zhang, Hui Chen, Wenhao Weng, Qiuyuan Lin, Di Deng, Zhi Li, and J. Kong. 2019. "Sensitive Polydopamine Bi-Functionalized SERS Immunoassay for Microalbuminuria Detection." *Biosensors and Bioelectronics* 142 (July): 111542. https://doi.org/10.1016/j.bios.2019.111542.

Jatschka, Jacqueline, André Dathe, Andrea Csáki, Wolfgang Fritzsche, and Ondrej Stranik. 2016. "Propagating and Localized Surface Plasmon Resonance Sensing – A Critical Comparison Based on Measurements and Theory." *Sensing and Bio-Sensing Research* 7: 62–70. https://doi.org/10.1016/j.sbsr.2016.01.003.

Jeevanandam, Jaison, Ahmed Barhoum, Yen S. Chan, Alain Dufresne, and Michael K. Danquah. 2018. "Review on Nanoparticles and Nanostructured Materials: History, Sources, Toxicity and Regulations." *Beilstein Journal of Nanotechnology* 9 (1): 1050–74. https://doi.org/10.3762/bjnano.9.98.

Jeong, Hyeon Ho, Andrew G. Mark, Mariana Alarcón-Correa, Insook Kim, Peter Oswald, Tung Chun Lee, and Peer Fischer. 2016. "Dispersion and Shape Engineered Plasmonic Nanosensors." *Nature Communications* 7. https://doi.org/10.1038/ncomms11331.

Jiang, Nina, Xiaolu Zhuo, and Jianfang Wang. 2018. "Active Plasmonics: Principles, Structures, and Applications." *Chemical Reviews* 118 (6): 3054–99. https://doi.org/10.1021/acs.chemrev.7b00252.

Joshi, Sweccha, Anna Segarra-Fas, Jeroen Peters, Han Zuilhof, Teris A. Van Beek, and Michel W. F. Nielen. 2016. "Multiplex Surface Plasmon Resonance Biosensing and Its Transferability towards Imaging Nanoplasmonics for Detection of Mycotoxins in Barley." *Analyst* 141 (4): 1307–18. https://doi.org/10.1039/c5an02512e.

Juhlin, Lars, Therese Mikaelsson, Aron Hakonen, Michael Stenbæk Schmidt, Tomas Rindzevicius, Anja Boisen, Mikael Käll, and Per Ola Andersson. 2020. "Selective Surface-Enhanced Raman Scattering Detection of Tabun, VX and Cyclosarin Nerve Agents Using 4-Pyridine Amide Oxime Functionalized Gold Nanopillars." *Talanta* 211 (August 2019): 120721. https://doi.org/10.1016/j.talanta.2020.120721.

Kaplanis, Joanna, Kaitlin E. Samocha, Laurens Wiel, Zhancheng Zhang, Kevin J. Arvai, Ruth Y. Eberhardt, Giuseppe Gallone, et al. 2020. "Evidence for 28 Genetic Disorders Discovered by Combining Healthcare and Research Data." *Nature* 586 (7831): 757–62. https://doi.org/10.1038/s41586-020-2832-5.

Khanna, Vinod Kumar. 2011. "Introduction to Nanosensors." *Nanosensors* (1): 29–112. https://doi.org/10.1201/b11289-5.

Khlebtsov, N. G. 2008. "Optics and Biophotonics of Nanoparticles with a Plasmon Resonance." *Quantum Electronics* 38 (6): 504–29. https://doi.org/10.1070/qe2008v038n06abeh013829.

Kim, Sae Wan, Jae Sung Lee, Sang Won Lee, Byoung Ho Kang, Jin Beom Kwon, Ok Sik Kim, Ju Seong Kim, Eung Soo Kim, Dae Hyuk Kwon, and Shin Won Kang. 2017. "Easy-to-Fabricate and High-Sensitivity LSPR Type Specific Protein Detection Sensor Using AAO Nano-Pore Size Control." *Sensors (Switzerland)* 17 (4). https://doi.org/10.3390/s17040856.

Krajczewski, Jan, Karol Kołątaj, and Andrzej Kudelski. 2017. "Plasmonic Nanoparticles in Chemical Analysis." *RSC Advances* 7 (28): 17559–76. https://doi.org/10.1039/c7ra01034f.

Lafuente, Marta, Erwin J. W. Berenschot, Roald M. Tiggelaar, Reyes Mallada, Niels R. Tas, and Maria P. Pina. 2018. "3D Fractals as SERS Active Platforms: Preparation and Evaluation for Gas Phase Detection of G-Nerve Agents." *Micromachines* 9 (2). https://doi.org/10.3390/mi9020060.

Lafuente, Marta, Ismael Pellejero, Víctor Sebastián, Miguel A. Urbiztondo, Reyes Mallada, M. Pilar Pina, and Jesús Santamaría. 2018. "Highly Sensitive SERS Quantification of Organophosphorous Chemical Warfare Agents: A Major Step towards the Real Time Sensing in the Gas Phase." *Sensors and Actuators, B: Chemical* 267: 457–66. https://doi.org/10.1016/j.snb.2018.04.058.

Laing, Stacey, Lauren E. Jamieson, Karen Faulds, and Duncan Graham. 2017. "Surface-Enhanced Raman Spectroscopy for in Vivo Biosensing." *Nature Reviews Chemistry* 1: 1–20. https://doi.org/10.1038/s41570-017-0060.

Langer, Judith, Dorleta Jimenez de Aberasturi, Javier Aizpurua, Ramon A. Alvarez-Puebla, Baptiste Auguié, Jeremy J. Baumberg, Guillermo C. Bazan, et al. 2020. "Present and Future of Surface-Enhanced Raman Scattering." *ACS Nano* 14 (1): 28–117. https://doi.org/10.1021/acsnano.9b04224.

Law, Cheryl Suwen, Lluís F. Marsal, and Abel Santos. 2020. "Electrochemically Engineered Nanoporous Photonic Crystal Structures for Optical Sensing and Biosensing." In *Handbook of Nanomaterials in Analytical Chemistry: Modern Trends in Analysis*, January, 201–26. https://doi.org/10.1016/B978-0-12-816699-4.00009-8.

Lee, Byoungho, Sookyoung Roh, and Junghyun Park. 2009. "Current Status of Micro- and Nano-Structured Optical Fiber Sensors." *Optical Fiber Technology* 15 (3): 209–21. https://doi.org/10.1016/j.yofte.2009.02.006.

Li, Chunxia, Yuan Liu, Xiaoyan Zhou, and Yuling Wang. 2020. "A Paper-Based SERS Assay for Sensitive Duplex Cytokine Detection towards the Atherosclerosis-Associated Disease Diagnosis." *Journal of Materials Chemistry B* 8 (16): 3582–89. https://doi.org/10.1039/c9tb02469g.

Li, Na, Shenfei Zong, Yizhi Zhang, Zhile Wang, Yujie Wang, Kai Zhu, Kuo Yang, Zhuyuan Wang, Baoan Chen, and Yiping Cui. 2020. "A SERS-Colorimetric Dual-Mode Aptasensor for the Detection of Cancer Biomarker MUC1." *Analytical and Bioanalytical Chemistry* 412 (23): 5707–18. https://doi.org/10.1007/s00216-020-02790-7.

Liu, Xia, Yang Yang, Lugang Mao, Zongjun Li, Chunjiao Zhou, Xianghua Liu, Shu Zheng, and Yuxin Hu. 2015. "SPR Quantitative Analysis of Direct Detection of Atrazine Traces on Au-Nanoparticles: Nanoparticles Size Effect." *Sensors and Actuators, B: Chemical* 218: 1–7. https://doi.org/10.1016/j.snb.2015.04.099.

Live, Ludovic S., Olivier R. Bolduc, and Jean François Masson. 2010. "Propagating Surface Plasmon Resonance on Microhole Arrays." *Analytical Chemistry* 82 (9): 3780–87. https://doi.org/10.1021/ac100177j.

Liyanage, Thakshila, Ashur Rael, Sidney Shaffer, Shozaf Zaidi, John V. Goodpaster, and Rajesh Sardar. 2018. "Fabrication of a Self-Assembled and Flexible SERS Nanosensor for Explosive Detection at Parts-per-Quadrillion Levels from Fingerprints." *Analyst* 143 (9): 2012–22. https://doi.org/10.1039/c8an00008e.

Liz-Marzán, Luis M. 2006. "Tailoring Surface Plasmons through the Morphology and Assembly of Metal Nanoparticles." *Langmuir* 22 (1): 32–41. https://doi.org/10.1021/la0513353.

Liz-Marzán, Luis M. 2020. "Nanometals: Formation and Color*." In *Colloidal Synthesis of Plasmonic Nanometals*, 1–13. https://doi.org/10.1201/9780429295188-1.

Lu, Xianmao, Matthew Rycenga, Sara E. Skrabalak, Benjamin Wiley, and Younan Xia. 2009. "Chemical Synthesis of Novel Plasmonic Nanoparticles." *Annual Review of Physical Chemistry* 60: 167–92. https://doi.org/10.1146/annurev.physchem.040808.090434.

MacCraith, Brian D. 1998. "Optical Chemical Sensors." *Chemical Analysis* 150: 195–233. https://doi.org/10.2320/materia.34.1233.

Maier, Stefan A. 2007. *Plasmonics: Fundamentals and Applications*. Springer US. https://doi.org/10.1007/0-387-37825-1.

Mariani, Stefano, Simona Scarano, Jolanda Spadavecchia, and Maria Minunni. 2015. "A Reusable Optical Biosensor for the Ultrasensitive and Selective Detection of Unamplified Human Genomic DNA with Gold Nanostars." *Biosensors and Bioelectronics* 74: 981–88. https://doi.org/10.1016/j.bios.2015.07.071.

McDonnell, James M. 2001. "Surface Plasmon Resonance: Towards an Understanding of the Mechanisms of Biological Molecular Recognition." *Current Opinion in Chemical Biology* 5 (5): 572–77. https://doi.org/10.1016/S1367-5931(00)00251-9.

Miyazaki, Celina M., Flávio M. Shimizu, and Marystela Ferreira. 2017. *Surface Plasmon Resonance (SPR) for Sensors and Biosensors. Nanocharacterization Techniques.* Elsevier Inc. https://doi.org/10.1016/B978-0-323-49778-7.00006-0.

Munawar, Anam, Yori Ong, Romana Schirhagl, Muhammad Ali Tahir, Waheed S. Khan, and Sadia Z. Bajwa. 2019. "Nanosensors for Diagnosis with Optical, Electric and Mechanical Transducers." *RSC Advances* 9 (12): 6793–803. https://doi.org/10.1039/c8ra10144b.

Nagpal, Prashant, Nathan C. Lindquist, Sang Hyun Oh, and David J. Norris. 2009. "Ultrasmooth Patterned Metals for Plasmonics and Metamaterials." *Science* 325 (5940): 594–97. https://doi.org/10.1126/science.1174655.

Nocerino, Valeria, Bruno Miranda, Chiara Tramontano, Giovanna Chianese, Principia Dardano, Ilaria Rea, and Luca De Stefano. 2022. "Plasmonic Nanosensors: Design, Fabrication, and Applications in Biomedicine." *Chemosensors* 10 (5). https://doi.org/10.3390/chemosensors10050150.

Oldenburg, S. J., R. D. Averitt, S. L. Westcott, and N. J. Halas. 1998. "Nanoengineering of Optical Resonances." *Chemical Physics Letters* 288 (2–4): 243–47. https://doi.org/10.1016/S0009-2614(98)00277-2.

Özgür, Erdoğan, Aykut Arif Topçu, Erkut Yılmaz, and Adil Denizli. 2020. "Surface Plasmon Resonance Based Biomimetic Sensor for Urinary Tract Infections." *Talanta* 212 (January): 120778. https://doi.org/10.1016/j.talanta.2020.120778.

Pandey, Purnendu Shekhar, Sanjeev Kumar Raghuwanshi, and Santosh Kumar. 2022. "Recent Advances in Two-Dimensional Materials-Based Kretschmann Configuration for SPR Sensors: A Review." *IEEE Sensors Journal* 22 (2): 1069–80. https://doi.org/10.1109/JSEN.2021.3133007.

Pang, Yuanfeng, Jinmaio Shi, Xingsheng Yang, Chongwen Wang, Zhiwei Sun, and Rui Xiao. 2020. "Personalized Detection of Circling Exosomal PD-L1 Based on Fe_3O_4@TiO_2 Isolation and SERS Immunoassay." *Biosensors and Bioelectronics* 148 (October 2019): 111800. https://doi.org/10.1016/j.bios.2019.111800.

Perçin, Isık, Neslihan Idil, Monireh Bakhshpour, Erkut Yılmaz, Bo Mattiasson, and Adil Denizli. 2017. "Microcontact Imprinted Plasmonic Nanosensors: Powerful Tools in the Detection of Salmonella Paratyphi." *Sensors (Switzerland)* 17 (6): 1375. https://doi.org/10.3390/s17061375.

Pérez-Jiménez, Ana Isabel, Danya Lyu, Zhixuan Lu, Guokun Liu, and Bin Ren. 2020. "Surface-Enhanced Raman Spectroscopy: Benefits, Trade-Offs and Future Developments." *Chemical Science* 11 (18): 4563–77. https://doi.org/10.1039/d0sc00809e.

Petryayeva, Eleonora, and Ulrich J. Krull. 2011. "Localized Surface Plasmon Resonance: Nanostructures, Bioassays and Biosensing-A Review." *Analytica Chimica Acta* 706 (1): 8–24. https://doi.org/10.1016/j.aca.2011.08.020.

Piliarik, Marek, Hana Vaisocherová, and Jiří Homola. 2005. "A New Surface Plasmon Resonance Sensor for High-Throughput Screening Applications." *Biosensors and Bioelectronics* 20 (10 SPEC. ISS.): 2104–10. https://doi.org/10.1016/j.bios.2004.09.025.

Qi, Lin, Mingshu Xiao, Xiwei Wang, Cheng Wang, Lihua Wang, Shiping Song, Xiangmeng Qu, Li Li, Jiye Shi, and Hao Pei. 2017. "DNA-Encoded Raman-Active Anisotropic Nanoparticles for MicroRNA Detection." *Analytical Chemistry* 89 (18): 9850–56. https://doi.org/10.1021/acs.analchem.7b01861.

Qu, Jia Huan, Karen Leirs, Wim Maes, Maya Imbrechts, Nico Callewaert, Katrien Lagrou, Nick Geukens, Jeroen Lammertyn, and Dragana Spasic. 2022. "Innovative FO-SPR Label-Free Strategy for Detecting Anti-RBD Antibodies in COVID-19 Patient Serum and Whole Blood." *ACS Sensors* 7 (2): 477–87. https://doi.org/10.1021/acssensors.1c02215.

Rosenstein, Jacob, Meni Wanunu, Marija Drndic, and Kenneth L. Shepard. 2012. "High-Bandwidth Solid-State Nanopore Sensors." *Biophysical Journal* 102 (3): 428a. https://doi.org/10.1016/J.BPJ.2011.11.2346.

Ryu, Jae Hoon, Ha Young Lee, Jeong Yeon Lee, Han Sol Kim, Sung Hyun Kim, Hyung Soo Ahn, Dong Han Ha, and Sam Nyung Yi. 2021. "Enhancing Sers Intensity by Coupling Pspr and Lspr in a Crater Structure with Ag Nanowires." *Applied Sciences (Switzerland)* 11 (24). https://doi.org/10.3390/app112411855.

Sadana, Ajit, and Neeti Sadana. 2011. "Nanobiosensors." *Handbook of Biosensors and Biosensor Kinetics*, January, 95–128. https://doi.org/10.1016/B978-0-444-53262-6.00005-X.

Sankiewicz, Anna, Adam Hermanowicz, Artur Grycz, Zenon Łukaszewski, and Ewa Gorodkiewicz. 2021. "An SPR Imaging Immunosensor for Leptin Determination in Blood Plasma." *Analytical Methods* 13 (5): 642–46. https://doi.org/10.1039/d0ay02047h.

Sepúlveda, Borja, Paula C. Angelomé, Laura M. Lechuga, and Luis M. Liz-Marzán. 2009. "LSPR-Based Nanobiosensors." *Nano Today* 4 (3): 244–51. https://doi.org/10.1016/j.nantod.2009.04.001.

Song, C. Y., Y. J. Yang, B. Y. Yang, Y. Z. Sun, Y. P. Zhao, and L. H. Wang. 2016. "An Ultrasensitive SERS Sensor for Simultaneous Detection of Multiple Cancer-Related MiRNAs." *Nanoscale* 8 (39): 17365–73. https://doi.org/10.1039/c6nr05504d.

Stewart, Matthew E., Christopher R. Anderton, Lucas B. Thompson, Joana Maria, Stephen K. Gray, John A. Rogers, and Ralph G. Nuzzo. 2008. "Nanostructured Plasmonic Sensors." *Chemical Reviews* 108 (2): 494–521. https://doi.org/10.1021/cr068126n.

Sumantri, Roni, Lilik Hasanah, Mohammad Arifin, Yuyu Rachmat Tayubi, Chandra Wulandari, Dadi Rusdiana, Cahya Julian, et al. 2020. "Simulation of Hemoglobin Detection Using Surface Plasmon Resonance Based on Kretschmann Configuration." *Journal of Engineering Science and Technology* 15 (4): 2239–47.

Sun, Wenxia, Shishan Yuan, Haowen Huang, Ning Liu, and Yunhong Tan. 2017. "A Label-Free Biosensor Based on Localized Surface Plasmon Resonance for Diagnosis of Tuberculosis." *Journal of Microbiological Methods* 142 (August): 41–45. https://doi.org/10.1016/j.mimet.2017.09.007.

Tabasi, Ozra, and Cavus Falamaki. 2018. "Recent Advancements in the Methodologies Applied for the Sensitivity Enhancement of Surface Plasmon Resonance Sensors." *Analytical Methods* 10 (32): 3906–25. https://doi.org/10.1039/c8ay00948a.

Tadimety, Amogha, Yichen Zhang, Kasia M. Kready, Timothy J. Palinski, Gregory J. Tsongalis, and John X. J. Zhang. 2019. "Design of Peptide Nucleic Acid Probes on Plasmonic Gold Nanorods for Detection of Circulating Tumor DNA Point Mutations." *Biosensors and Bioelectronics* 130 (January): 236–44. https://doi.org/10.1016/j.bios.2019.01.045.

Tran, B. M., N. N. Nam, S. J. Son, and N. Y. Lee. 2018. "Nanoporous Anodic Aluminum Oxide Internalized with Gold Nanoparticles for On-Chip PCR and Direct Detection by Surface-Enhanced Raman Scattering." *Analyst* 143 (4): 808–12. https://doi.org/10.1039/c7an01832k.

Uddin, Syed Mohammad Ashab, Sayeed Shafayet Chowdhury, and Ehsan Kabir. 2017. "A Theoretical Model for Determination of Optimum Metal Thickness in Kretschmann Configuration Based Surface Plasmon Resonance Biosensors." *ECCE 2017 – International Conference on Electrical, Computer and Communication Engineering*, 651–54. https://doi.org/10.1109/ECACE.2017.7912985.

Unser, Sarah, Ian Bruzas, Jie He, and Laura Sagle. 2015. "Localized Surface Plasmon Resonance Biosensing: Current Challenges and Approaches." *Sensors (Switzerland)* 15 (7): 15684–716. https://doi.org/10.3390/s150715684.

Verhaart, Ingrid E. C., and Annemieke Aartsma-Rus. 2019. "Therapeutic Developments for Duchenne Muscular Dystrophy." *Nature Reviews Neurology* 15 (7): 373–86. https://doi. org/10.1038/s41582-019-0203-3.

Vo-Dinh, Tuan, Hsin Neng Wang, and Jonathan Scaffidi. 2010. "Plasmonic Nanoprobes for SERS Biosensing and Bioimaging." *Journal of Biophotonics* 3 (1–2): 89–102. https:// doi.org/10.1002/jbio.200910015.

Wang, Cong, Lihua Zeng, Zhen Li, and Daoliang Li. 2017. "Review of Optical Fibre Probes for Enhanced Raman Sensing." *Journal of Raman Spectroscopy* 48 (8): 1040–55. https://doi.org/10.1002/jrs.5173.

Wang, Jin. 2021. "A Simple, Rapid and Low-Cost 3-Aminopropyltriethoxysilane (APTES)-Based Surface Plasmon Resonance Sensor for TNT Explosive Detection." *Analytical Sciences* 37 (7): 1029–32. https://doi.org/10.2116/analsci.20N028.

Wang, Sheng Hann, Shu Cheng Lo, Yung Ju Tung, Chia Wen Kuo, Yi Hsin Tai, Shu Yi Hsieh, Kuang Li Lee, et al. 2020. "Multichannel Nanoplasmonic Platform for Imidacloprid and Fipronil Residues Rapid Screen Detection." *Biosensors and Bioelectronics* 170 (June): 112677. https://doi.org/10.1016/j.bios.2020.112677.

Wang, Xinnan, Yuyang Wang, Ming Cong, Haibo Li, Yuejiao Gu, John R. Lombardi, Shuping Xu, and Weiqing Xu. 2013. "Propagating and Localized Surface Plasmons in Hierarchical Metallic Structures for Surface-Enhanced Raman Scattering." *Small* 9 (11): 1895–99. https://doi.org/10.1002/smll.201202424.

Willets, Katherine A., and Richard P. Van Duyne. 2007. "Localized Surface Plasmon Resonance Spectroscopy and Sensing." *Annual Review of Physical Chemistry* 58: 267–97. https://doi.org/10.1146/annurev.physchem.58.032806.104607.

Wong, Chi Lok, Sau Yeen Loke, Hann Qian Lim, Ghayathri Balasundaram, Patrick Chan, Bee Kiang Chong, Ern Yu Tan, Ann Siew Gek Lee, and Malini Olivo. 2021. "Circulating MicroRNA Breast Cancer Biomarker Detection in Patient Sera with Surface Plasmon Resonance Imaging Biosensor." *Journal of Biophotonics* 14 (11): 1–7. https://doi.org/10. 1002/jbio.202100153.

Wu, Bin, and Qing Kang Wang. 2008. "High Sensitivity Transmission-Type SPR Sensor by Using Metallic-Dielectric Mixed Gratings." *Chinese Physics Letters* 25 (5): 1668–71. https://doi.org/10.1088/0256-307X/25/5/040.

Wyszynski, B., and T. Nakamoto. 2015. "Chemical Sensors." *Flavour Development, Analysis and Perception in Food and Beverages*, January, 83–104. https://doi.org/10.1016/B978-1-78242-103-0.00005-9.

Xu, Liguang, Wenjing Yan, Wei Ma, Hua Kuang, Xiaoling Wu, Liqaing Liu, Yuan Zhao, Libing Wang, and Chuanlai Xu. 2015. "SERS Encoded Silver Pyramids for Attomolar Detection of Multiplexed Disease Biomarkers." *Advanced Materials* 27 (10): 1706–11. https://doi.org/10.1002/adma.201402244.

Yang, Kang, Xu Yao, Bowen Liu, and Bin Ren. 2021. "Metallic Plasmonic Array Structures: Principles, Fabrications, Properties, and Applications." *Advanced Materials* 33 (50): 1–21. https://doi.org/10.1002/adma.202007988.

Yilmaz, Erkut, Daryoush Majidi, Erdogan Ozgur, and Adil Denizli. 2015. "Whole Cell Imprinting Based Escherichia Coli Sensors: A Study for SPR and QCM." *Sensors and Actuators, B: Chemical* 209: 714–21. https://doi.org/10.1016/j.snb.2014.12.032.

Yılmaz, Erkut, Erdoğan Özgür, Nilay Bereli, Deniz Türkmen, and Adil Denizli. 2017. "Plastic Antibody Based Surface Plasmon Resonance Nanosensors for Selective Atrazine Detection." *Materials Science and Engineering C* 73: 603–10. https://doi.org/10.1016/ j.msec.2016.12.090.

Yılmaz, Gaye Ezgi, Yeşeren Saylan, Ilgım Göktürk, Fatma Yılmaz, and Adil Denizli. 2022. "Selective Amplification of Plasmonic Sensor Signal for Cortisol Detection Using Gold Nanoparticles." *Biosensors* 12 (7). https://doi.org/10.3390/bios12070482.

Yonzon, Chanda Ranjit, Eunhee Jeoung, Shengli Zou, George C. Schatz, Milan Mrksich, and Richard P. Van Duyne. 2004. "A Comparative Analysis of Localized and Propagating Surface Plasmon Resonance Sensors: The Binding of Concanavalin A to a Monosaccharide Functionalized Self-Assembled Monolayer." *Journal of the American Chemical Society* 126 (39): 12669–76. https://doi.org/10.1021/ja047118q.

Yu, Dan, Qilong Yin, Jiwei Wang, Jian Yang, Zimeng Chen, Zihan Gao, Qingli Huang, and Shibao Li. 2021. "Sers-Based Immunoassay Enhanced with Silver Probe for Selective Separation and Detection of Alzheimer's Disease Biomarkers." *International Journal of Nanomedicine* 16: 1901–11. https://doi.org/10.2147/IJN.S293042.

Zalevsky, Zeev, and Ibrahim Abdulhalim. 2014. "Plasmonics." *Integrated Nanophotonic Devices*, January, 179–245. https://doi.org/10.1016/B978-0-323-22862-6.00006-2.

Zengin, Adem, Ugur Tamer, and Tuncer Caykara. 2017. "SERS Detection of Hepatitis B Virus DNA in a Temperature-Responsive Sandwich-Hybridization Assay." *Journal of Raman Spectroscopy* 48 (5): 668–72. https://doi.org/10.1002/jrs.5109.

Zhang, De, Pei Liang, Wenwen Chen, Zhexiang Tang, Chen Li, Kunyue Xiao, Shangzhong Jin, Dejiang Ni, and Zhi Yu. 2021. "Rapid Field Trace Detection of Pesticide Residue in Food Based on Surface-Enhanced Raman Spectroscopy." *Microchimica Acta* 188 (11). https://doi.org/10.1007/s00604-021-05025-3.

Zhang, Diming, Qian Zhang, Yanli Lu, Yao Yao, Shuang Li, Jing Jiang, Gang Logan Liu, and Qingjun Liu. 2016. "Peptide Functionalized Nanoplasmonic Sensor for Explosive Detection." *Nano-Micro Letters* 8 (1): 36–43. https://doi.org/10.1007/s40820-015-0059-z.

Zhang, John X. J., and Kazunori Hoshino. 2019. "Optical Transducers: Optical Molecular Sensing and Spectroscopy." *Molecular Sensors and Nanodevices*, January, 231–309. https://doi.org/10.1016/B978-0-12-814862-4.00005-3.

Zhang, Lei, Jinghui Wang, Junxia Zhang, Yuqi Liu, Lingzhi Wu, Jingjing Shen, Ying Zhang, et al. 2017. "Individual Au-Nanocube Based Plasmonic Nanoprobe for Cancer Relevant MicroRNA Biomarker Detection." *ACS Sensors* 2 (10): 1435–40. https://doi.org/10.1021/acssensors.7b00322.

Zhang, Yongjun, Huanhuan Sun, Renxian Gao, Fan Zhang, Aonan Zhu, Lei Chen, and Yaxin Wang. 2018. "Facile SERS-Active Chip (PS@Ag/SiO2/Ag) for the Determination of HCC Biomarker." *Sensors and Actuators, B: Chemical* 272 (January): 34–42. https://doi.org/10.1016/j.snb.2018.05.139.

Zhang, Zheng Long, Yu Rui Fang, Wen Hui Wang, Li Chen, and Meng Tao Sun. 2016. "Propagating Surface Plasmon Polaritons: Towards Applications for Remote-Excitation Surface Catalytic Reactions." *Advanced Science* 3 (1): 1–14. https://doi.org/10.1002/advs.201500215.

Zheng, Fei, Zhi Chen, Jingfeng Li, Rui Wu, Bin Zhang, Guohui Nie, Zhongjian Xie, and Han Zhang. 2022. "A Highly Sensitive CRISPR-Empowered Surface Plasmon Resonance Sensor for Diagnosis of Inherited Diseases with Femtomolar-Level Real-Time Quantification." *Advanced Science* 9 (14): 1–14. https://doi.org/10.1002/advs.202105231.

Zheng, Zhihua, Lei Wu, Lang Li, Shenfei Zong, Zhuyuan Wang, and Yiping Cui. 2018. "Simultaneous and Highly Sensitive Detection of Multiple Breast Cancer Biomarkers in Real Samples Using a SERS Microfluidic Chip." *Talanta* 188 (December 2017): 507–15. https://doi.org/10.1016/j.talanta.2018.06.013.

Zong, Cheng, Mengxi Xu, Li Jia Xu, Ting Wei, Xin Ma, Xiao Shan Zheng, Ren Hu, and Bin Ren. 2018. "Surface-Enhanced Raman Spectroscopy for Bioanalysis: Reliability and Challenges." Review-article. *Chemical Reviews* 118 (10): 4946–80. https://doi.org/10.1021/acs.chemrev.7b00668.

3 Plasmonic Nanosensors for Biological and Chemical Threat Agent Detection

Yeşeren Saylan, Semra Akgönüllü, and Adil Denizli

3.1 INTRODUCTION

In an increasingly complex and interconnected world, the swift and accurate detection of biological and chemical threat agents is paramount to safeguarding public health, national security, and the environment (Saylan, Akgönüllü, and Denizli 2020; Schmid 2020). The emergence of plasmonic nanosensors has revolutionized the landscape of threat detection, offering an innovative and highly sensitive approach to identifying these agents with unprecedented precision (Ramalingam et al. 2023; Erdem et al. 2020; Lazarević-Pašti et al. 2023). Biological and chemical threat agents, which encompass a range of hazardous substances from lethal toxins and pathogens to noxious chemicals and explosives, pose significant risks when deployed intentionally or inadvertently (Clark and Pazdernik 2016; Hassard and Nieuwenhuizen 2022). The consequences of delayed or inaccurate detection can be dire, leading to casualties, environmental contamination, and widespread panic. Thus, the demand for rapid and reliable threat detection methodologies has never been more pressing (Li et al. 2022). Plasmonic nanosensors, grounded in the unique optical properties of plasmonic nanoparticles, have emerged as a transformative solution to address these challenges (Kumar et al. 2023; Inci et al. 2020). Their ability to harness localized surface plasmon resonance and amplify electromagnetic fields at their surfaces endows them with the capability to detect minuscule quantities of target molecules, even at trace levels (Smith and Gambhir 2017). This exceptional sensitivity and specificity make plasmonic nanosensors indispensable in various fields, including healthcare, environmental monitoring, and security (Javaid et al. 2021; Adam et al. 2023; Ha 2023; Saylan 2023).

In this context, this chapter explores the pivotal role played by plasmonic nanosensors in the swift and accurate detection of biological and chemical threat agents. It delves into the foundational principles governing these nanosensors, the diverse fabrication methods employed, and their multifaceted applications. Through a series of examples, we will illustrate how plasmonic nanosensors have revolutionized the landscape of threat detection, providing not only sensitivity but also rapid response capabilities. From the identification of hazardous pathogens and toxins to the detection of chemical threat agents and explosives, plasmonic nanosensors are paving the way for enhanced security, environmental protection, and public health. This chapter also underscores the pivotal role of plasmonic nanosensors as a transformative technology that bridges the gap between laboratory research and real-world threat detection, ensuring a safer and more secure future for us all.

3.2 PLASMONIC NANOSENSORS

The fundamental principles of plasmonic nanosensors are especially rooted in the unique optical properties of plasmonic nanoparticles (Guo et al. 2015; De and Kalita 2023). These properties enable nanosensors to detect changes in their local environment, including the presence of target molecules, with exceptional sensitivity (Jia et al. 2023). The key principles of plasmonic nanosensors include localized surface plasmon resonance (LSPR), enhanced electromagnetic fields, surface functionalization, changes in optical properties, spectral analysis, real-time monitoring, and multiplexing (Yang et al. 2021; Mauriz and Lechuga 2021). These principles collectively enable them to achieve highly sensitive and selective detection of target molecules, making the plasmonic nanosensors valuable tools in various applications, including threat detection, environmental monitoring, and medical diagnostics (Shrivastav, Cvelbar, and Abdulhalim 2021; Erdem et al. 2019; Wei, Abtahi, and Vikesland 2015; Spitzberg et al. 2019).

LSPR: Plasmonic nanoparticles, typically made of noble metals like gold or silver, exhibit LSPR, where free electrons at metal surface oscillate collectively in response to incident light. This resonance occurs at a specific wavelength and is sensitive to nanoparticles' shape, size, and the refractive index of the surrounding medium. Any alteration in local refractive index, created by the binding of target molecules, shifts the LSPR wavelength, enabling detection (Inci et al. 2020; Jin et al. 2022; Ha 2023).

Enhanced Electromagnetic Fields: Plasmonic nanoparticles concentrate electromagnetic fields at their surface. This leads to an enhancement of electromagnetic field strength in the vicinity of nanoparticles, known as "hot spots." Molecules situated within these hot spots experience significantly amplified Raman scattering and fluorescence signals, allowing for ultrasensitive detection (Kołątaj, Krajczewski, and Kudelski 2020; Krajczewski, Kołątaj, and Kudelski 2017; Linic et al. 2015).

Surface Functionalization: Plasmonic nanosensors are often functionalized with molecules that have a high affinity for the target molecules of interest. These functionalized surfaces can selectively capture and bind the target molecules, further enhancing the nanosensor's sensitivity and specificity (Tang and Li 2017; Ahijado-Guzmán et al. 2014; Swierczewska et al. 2012).

Changes in Optical Properties: When target molecules bind to the functionalized plasmonic nanoparticles, they induce changes in the nanoparticles' optical properties. These changes manifest as shifts in LSPR peak wavelength, alterations in the scattering and absorption spectra, and variations in intensity of surface-enhanced Raman scattering (SERS) signals (Mosquera et al. 2020; Chanda et al. 2011; Prakash, Harris, and Swart 2016).

Spectral Analysis: The detection of targets relies on analyzing the optical spectra of nanoparticles before and after exposure to the sample. Any shifts or changes in spectra supply information about the presence and concentration of target molecules (Im et al. 2014; Wang and Tao 2014).

Real-Time Monitoring: Plasmonic nanosensors offer real-time monitoring capabilities, making them suitable for continuous, on-site detection. Changes in the optical properties occur rapidly upon target molecule binding, allowing for immediate results (Brahmkhatri et al. 2021; Gerdan et al. 2022).

Multiplexing: Plasmonic nanosensors can be engineered to support multiplexed detection, where they simultaneously detect multiple target molecules by functionalizing different nanoparticles with specific receptors (Howes, Chandrawati, and Stevens 2014; Altug et al. 2022).

In addition, plasmonic nanosensors can be fabricated using various techniques, each offering advantages and limitations depending on the specific application and desired nanosensor properties (Manoj, Shanmugasundaram, and Anandharamakrishnan 2021). The choice of fabrication method depends on factors such as the desired nanoparticle size and shape, the substrate material, and the functionalization requirements. Each method offers a unique set of advantages and challenges, enabling the customization of plasmonic nanosensors for specific sensing applications (Harish et al. 2023).

Some common fabrication methods for plasmonic nanosensors include:

Chemical Synthesis:

(i) Chemical Reduction: Nanoparticles, often made of gold or silver, can be synthesized through chemical reduction methods. Reducing agents like sodium borohydride are used to reduce metal ions, forming nanoparticles with controlled sizes and shapes (Alaqad and Saleh 2016).

(ii) Seed-Mediated Growth: This method involves the controlled growth of nanoparticles from smaller seed particles. It allows for precise tuning of nanoparticle properties, such as size, shape, and composition (Ha 2023; He et al. 2023).

(iii) Galvanic Replacement: Plasmonic nanostructures can be formed by replacing one metal with another through a redox reaction. For instance, silver can be used to replace gold in gold nanorods, resulting in silver-coated nanorods (Wilson and Jain 2020).

(iv) Chemical Etching of Nanoparticles: Controlled etching processes can be applied to pre-synthesized nanoparticles to modify their shapes. For example, silver nanocubes can be etched to form nanorods with specific aspect ratios (Haase et al. 2020).

(v) Core-Shell Structures: Plasmonic nanosensors can be designed with a core-shell structure, where a plasmonic metal core is surrounded by a dielectric or another material. This configuration enhances plasmonic properties and allows for controlled surface functionalization (Huang et al. 2021).

Physical Methods:

(i) Sputtering and Evaporation: Thin films or nanoparticles can be deposited on substrates using techniques like sputtering or thermal evaporation. These methods are suitable for creating plasmonic structures on planar surfaces (Kibis et al. 2010).

(ii) Lithography: Photolithography or electron beam lithography can be employed to define patterns on substrates, creating nanoscale features with precise control over geometry and arrangement (Shu et al. 2023).

Bottom-Up Self-Assembly:

(i) Self-Assembled Monolayers (SAMs): Thiols or other molecules with affinity to the metal surface can be used to functionalize the metal nanoparticles' surfaces. SAMs enable the controlled assembly of nanoparticles into organized structures (Borah et al. 2021).

(ii) DNA-Based Assembly: DNA strands can be used to functionalize nanoparticles and guide their assembly into ordered structures through Watson-Crick base pairing (Dwivedi et al. 2022).

Plasmonic nanosensors find applications in various fields because of their unique capability to detect and analyze analytes with high sensitivity and specificity (Wujcik et al. 2014). Some of the prominent applications of plasmonic nanosensors include biological and medical sensing, environmental monitoring, food safety, chemical sensing, drug development, material characterization, point-of-care diagnostics, and chemical process monitoring (Özgür et al. 2020; Vo-Dinh et al. 2015; Razlansari et al. 2022).

Biological and Medical Sensing:

(i) Biomarker Detection: Plasmonic nanosensors can identify biomarkers related with diseases like cardiovascular disorders, cancer, and infectious diseases. They enable early diagnosis and monitoring of disease progression (Saylan, Akgönüllü, and Denizli 2022).

(ii) Protein Detection: These nanosensors are used in proteomics research to detect and quantify proteins, aiding drug discovery and the understanding of disease mechanisms (Saylan and Denizli 2018).

(iii) DNA/RNA Detection: Plasmonic nanosensors can detect and sequence DNA/RNA molecules, contributing to genomics research and personalized medicine (Xu et al. 2017).

Environmental Monitoring:

(i) Pollutant Detection: Plasmonic nanosensors are employed to detect pollutants, heavy metals, and chemical contaminants in air, water, and soil. They play a crucial role in environmental monitoring and assessment (Erdem et al. 2019).

(ii) Detection of Harmful Microorganisms: These nanosensors can identify harmful bacteria, viruses, and other microorganisms in environmental samples and water sources (Saylan et al. 2019).

Food Safety:

(i) Food Contaminant Detection: Plasmonic nanosensors are used to detect contaminants including pesticides, pathogens, toxins, and allergens in food products, ensuring food safety and quality (Çelik et al. 2023).

Chemical Sensing:

(i) Chemical Threat Agent Detection: Plasmonic nanosensors can identify chemical threat agents, explosives, and hazardous chemicals. They are vital for security and defense applications (Akgönüllü, Yavuz, and Denizli 2021).

(ii) Gas Sensing: These nanosensors are utilized for gas detection in industrial settings, including the monitoring of toxic gases and volatile organic compounds (El Kazzy et al. 2021).

Drug Development:

(i) Drug Screening: Plasmonic nanosensors are employed in the high-throughput screening of drug candidates and the study of drug-receptor interactions (Ahijado-Guzmán et al. 2017).

Point-of-Care Diagnostics:

(i) Portable Devices: Miniaturized plasmonic nanosensor platforms are being developed for use at the point of care, enabling rapid and on-site diagnostics in healthcare settings or remote locations (Li et al. 2019).

These diverse applications highlight the versatility and impact of plasmonic nanosensors across various fields, ranging from healthcare and environmental monitoring to materials science and security. Their sensitivity and specificity make them indispensable tools for research, diagnostics, and quality control (Zeng et al. 2014).

3.3 DETECTION OF THREATS WITH PLASMONIC NANOSENSORS

The ever-increasing danger to homeland public safety and security has caused the emergency request for advancement of high-throughput instruments that contribute to enhancing biodefense and helping to provide the survival of affected individuals. Development of modern, easy-to-use nanosensor technology for the quick recognition of threatening substances, including biological and chemical threat agents, radioactive and nuclear materials, and explosives is of high value to homeland security and defense.

3.3.1 DETECTION OF BIOLOGICAL THREATS

3.3.1.1 Detection of Plant-Based Biological Toxins

The detection of aflatoxin and ricin plant-based biotoxins is of crucial significance in defense applications. Ricin (64 kDa) is a highly lethal toxin reproduced from the castor bean plant (Lord, Roberts, and Robertus 1994; Tran et al. 2008). This toxic biomolecule has been recognized as a major biological threat agent because of its easy usage, lack of effective antidotes, and high stability after release. As a major, toxic bioterrorism threat, ricin is the only protein listed as a List 1 compound in the Chemical Weapons Convention (Tang et al. 2016). The lethal level of ricin in humans is approximately 5–10 µg/kg by inhalation or injection (Bradberry et al. 2003). Wang and co-workers (Wang et al. 2015) reported two different systems for the detection of the ricin molecule. The specific binding affinity with anti-ricin aptamer for ricin analytes was monitored using an SPR nanosensor. The atomic force microscopy (AFM) was also used to examine the surface conformations of the anti-ricin aptamer. The SPR nanosensor analysis of the detection of ricin showed a good linear concentration between 83.3 pM and 8.33 nM. The limit of detection (LOD) was found to be 1.5 ng/mL. They reported the AFM topography of the SPR chip

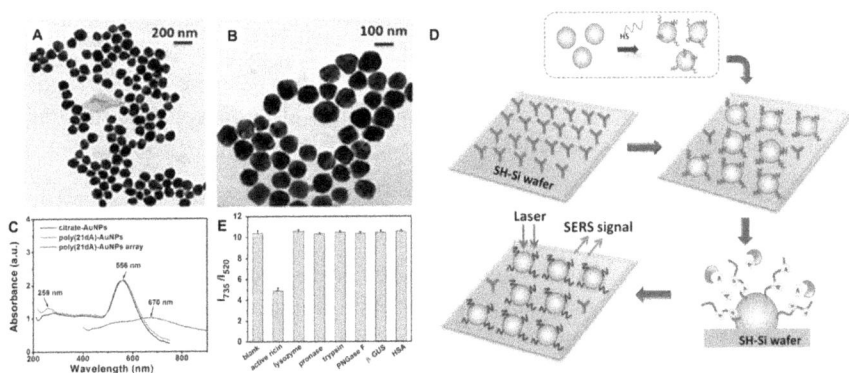

FIGURE 3.1 Transmission electron microscope (TEM) images of (A) citrate modified-AuNPs, (B) poly(21dA)-AuNPs, (C) UV-vis absorbance spectra, (D) SERS chip and active ricin detection, and (E) Selectivity of the SERS chip in the existence of various proteins (Tang et al. 2016).

surface and recognition images in the conformations of individual aptamer molecules on the Au(111) surface. Tang and co-workers (Tang et al. 2016) synthesized a single-stranded oligodeoxynucleotides (poly(21dA)) attached to gold nanoparticles (AuNPs) [poly(21dA)–AuNPs] for the detection of active ricin (Figure 3.1A–C). The poly(21dA)–AuNPs were assembled on a dihydrogen lipoic-acid-modified Si wafer (SH–Si) to design the SERS electrode [poly(21dA)–AuNPs@SH–Si] for the specific in-vitro ricin depurination (Figure 3.1D). The detection value limit of the developed SERS nanosensor was calculated as 8.9 ng/mL. They reported the selectivity of the plasmonic nanosensor in the presence of different proteins (Figure 3.1E).

The globular ricin protein consists of two different chains (ricin A and B chain) linked by a single disulfide-bond (Polito et al. 2019). Campos and co-workers (Campos et al. 2016) reported a sensitive detection of the ricin B chain (RBC) in human whole blood using an SERS nanosensor (Figure 3.2A). The aptamer-modified silver film-over-nanosphere (AgFON) substrate was used as a plasmonic surface for captured RBC (Figure 3.2B). The AgFON-aptamer complex showed good stability for ten days in whole human blood. The LOD value for direct ricin detection was 1.0 µg/mL. Luo and co-workers (Luo et al. 2022) described a label-free SPR nanosensor for the quantification and detection of ricin and abrin (Figure 3.2C). The protein G was a directed affinity molecule to fulfill multiple aims in a single SPR nanosensor. The limit of detection in the SPR nanosensor panel toward ricin, RCA120, abrin, and AAG was around 10–500 pM, depending on the affinity or construction format of different mAbs. Feltis and co-workers (Feltis et al. 2008) reported a hand-held SPR nanosensor for the detection of the biological toxin ricin (Figure 3.2D). The portable SPR nanosensor unit measured 15 × 8 cm, weighed 600 g, and was powered by a 9 V battery (Figure 3.2E). The gold film thickness was 200 nm. It could detect ricin at 200 ng/mL in 10 min.

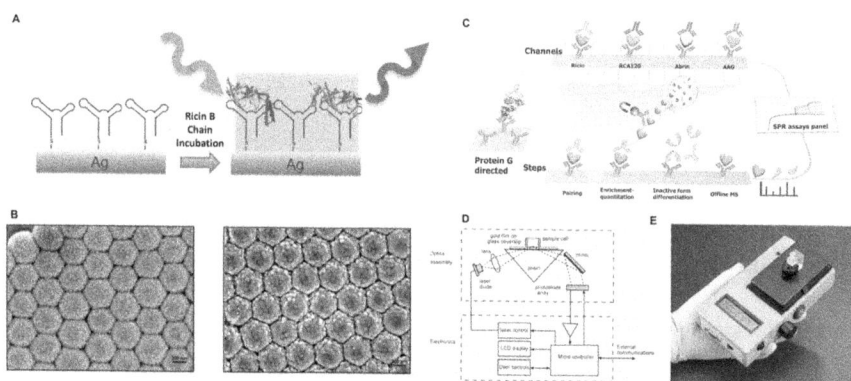

FIGURE 3.2 (A) The aptamer-modified SERS substrate, (B) SEM images of AgFON before (left) and after (right) thermal treatment (Campos et al. 2016), (C) the single protein G-based SPR nanosensor for multiple analysis (Luo et al. 2022), (D) diagram of the components of the sensor system, and (E) handheld SPR plasmonic sensor (Feltis et al. 2008).

Stern and co-workers (Stern et al. 2016) developed a real-time and sensitive SPR nanosensor for quick differentiation between ricin and agglutinin from different Ricinus communis cultivars. Ricin and agglutinin lectins were detected in a sandwich immunoassay-like setting by capturing with a cross-reactive antibody (R109). The limit of detection was found to be 3 ng/mL and 6 ng/mL for ricin and agglutinin, respectively. Fan and co-workers (Fan et al. 2017) developed a gold nanopin-based plasmonic sensor, which achieved a multicolor variation for the qualitative recognition and detection of ricin. The antibody-functionalized nanopin-cavity plasmonic nanosensor has shown high sensitivity and rapid response, enabling visual quantitative ricin detection in the range of 10–120 ng/mL. The LOD was found to be 10 ng/mL.

Aflatoxins (AFs) have posed a serious threat to food safety and human health. Therefore, it is important to detect aflatoxins in samples rapidly and accurately (Liu et al. 2023). Bhardwaj and co-workers (Bhardwaj, Sumana, and Marquette 2020) developed a gold nanoparticle (AuNPs) integrated microfluidic SPR nanosensor for the detection of Aflatoxin B1 (AFB1) in wheat samples. This multilayer functionalized AuNPs modified gold SPR nanosensor was successfully used for AFB1 detection ranging from 0.01 to 50 nM. The LOD was found to be 0.003 nM. In another study, Bhardwaj and co-workers (Bhardwaj, Sumana, and Marquette 2021) reported a gold nanobipyramids integrated plasmonic nanosensor for the detection of AFB1 in maize samples. The plasmonic-based SPR nanosensor showed a good linear range of 0.1–500 nM. The limit of detection was 0.4 nM. Akgönüllü and co-workers (Akgönüllü, Yavuz, and Denizli 2020) designed an AFB1-imprinted nanofilm coated SPR nanosensor. AuNPs were integrated into the MIP film for signal enrichment for the sensitive determination of small molecule AFB1. It showed a broad linear range from 0.0001 ng/mL to 10.0 ng/mL. The LOD for AFB1 was found to be 1.04 pg/mL.

3.3.1.2   Detection of Bacteria-Based Biological Toxins

B. anthracis is one of the most crucial biological threat agents (Goel 2015). Ghosh and co-workers (Ghosh et al. 2013) developed an SPR nanosensor for the detection of the protective antigen (PA) which is produced by all live B. anthracis bacteria. The anti-PA monoclonal antibody was immobilized on a carboxymethyldextran-modified SPR nanosensor surface. Different numbers of Bacillus spores were artificially added to the soil samples and detected. In another study, Ghosh and co-workers (Ghosh et al. 2013) sensitively detected an anthrax-specific toxin in human plasma using an SPR nanosensor. The protective antigen (PA) was immobilized on the carboxymethyldextran-modified gold chip. Human serum samples spiked with the PA toxin and concentrations between 1.0 pg/mL and 10 ng/mL were analyzed.

Botulinum neurotoxins (BoNTs) are the deadliest toxins. Tomar and co-workers (Tomar and Gupta 2016) developed a label-free plasmonic-based nanosensor for real-time detection and quantification of botulinum neurotoxin A (BoNT/A). They used an antibody fragment and synaptic vesicles immobilizing SPR nanosensor. The limit of detection was found to be 0.045 fM. Patel and co-workers (Patel et al. 2017) designed a novel Newton photonics SPR nanosensor for the sensitive and fast detection of botulinum neurotoxins. The limit of detection was 6.76 pg/mL for active BoNT/A light chains. Janardhanan and co-workers (Janardhanan et al. 2013) reported an RNA aptamer-based SPR nanosensor for type A botulinum neurotoxin detection in different samples. The LOD values were found to be 5.8 ng/mL, 20.3 ng/mL, 23.4 ng/mL, and 22.5 ng/mL in PBS buffer, carrot juice, milk, and human serum, respectively. Lévêque and co-workers (Lévêque et al. 2015) developed a target-specific SPR nanosensor for the detection of botulinum neurotoxins, types A and E (BoNT/A and E). The linear calibration curves were obtained at low and high concentrations (0.01–20 LD_{50}/mL). The LOD value was 0.01 LD_{50}/mL for BoNT/E.

3.3.2 DETECTION OF CHEMICAL THREATS

The G-series nerve agents are broadly recognized as one of the most toxic groups of chemical threat agents. Lafuente and co-workers (Lafuente et al. 2018) developed an SERS-based nanosensor for the label-free and real-time detection of gas phase dimethyl methylphosphonate (DMMP) which is a G-series nerve agent. The SERS citrate layer coated chip was designed using a self-assembly procedure with gold nanoparticles. Figure 3.3A shows the preparation of the SERS substrate.

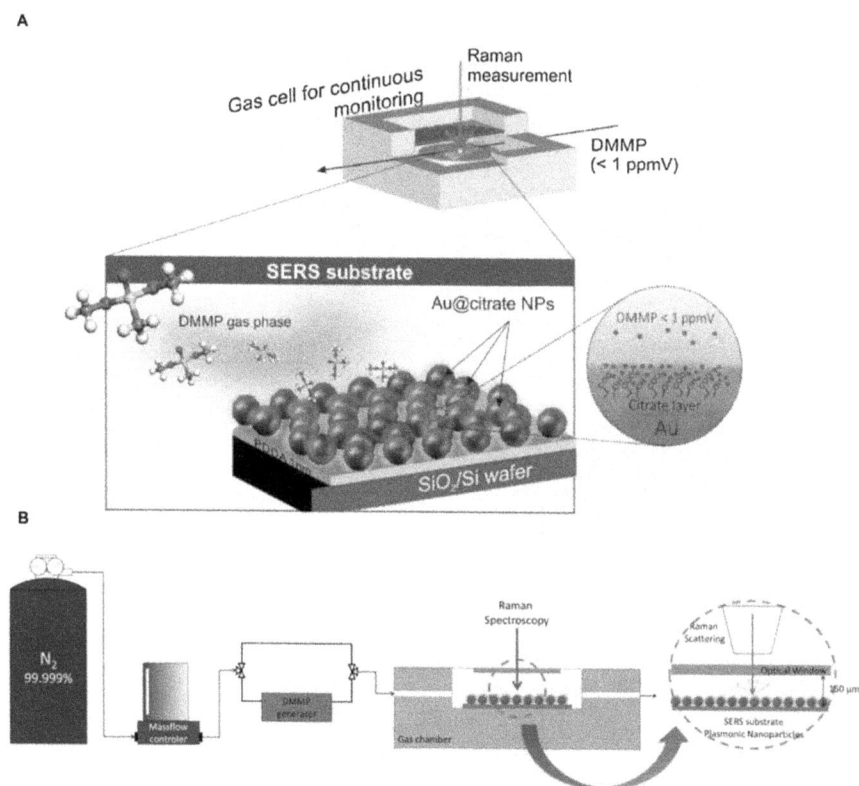

FIGURE 3.3 (A) SERS substrate and (B) experimental set-up for SERS analysis in gas phase (Lafuente et al. 2018).

Riskin and co-workers (Riskin, Tel-Vered, and Willner 2010) reported a highly sensitive MIP-based SPR nanosensor for the detection of hexahydro-1,3,5-trinitro-1,3,5-triazine (RDX), a hazardous explosive agent. They used bisaniline-cross-linked gold nanoparticles associated with the Au surface. The MIP surface was synthesized using the electropolymerization method. MIP-based plasmonic sensors provide a binding affinity for RDX with an LOD of 12 fM. In another study, they designed an MIP-based thioaniline-functionalized AuNPs composite for the detection of pentaerythritol tetranitrate (PETN), nitroglycerin (NG), and ethylene glycol dinitrate (Riskin et al. 2011). The limit of detection for the specific analysis of PETN and of NG was found to be 200 fM and 20 pM, respectively. The thioaniline monolayer-modified Au-coated surface was prepared using the electropolymerization method (Figure 3.4).

Zhao and co-workers (Zhao et al. 2017) reported a highly sensitive detection of the organophosphorus nerve agent, sarin, using an SERS sensing substrate. methanephosphonic acid (MPA) was selected as the sarin simulation agent. The Au-coated Si nanocone array is surface-modified with 2-aminoethanethiol and utilized as a SERS substrate for the detection of MPA. The LOD was low at ~1 ppb.

Kawaguchi and co-workers (Kawaguchi et al. 2007) developed a self-assembly chemistry-based SPR nanosensor for the detection of 2,4,6-trinitrotoluene (TNT). They presented an alternative method for the detection of TNT. The interaction between the self-assembly designed TNPh-β-alanine immobilized thiolate monolayer surface and the monoclonal anti-TNT Ab (M-TNT Ab) was monitored. TNT

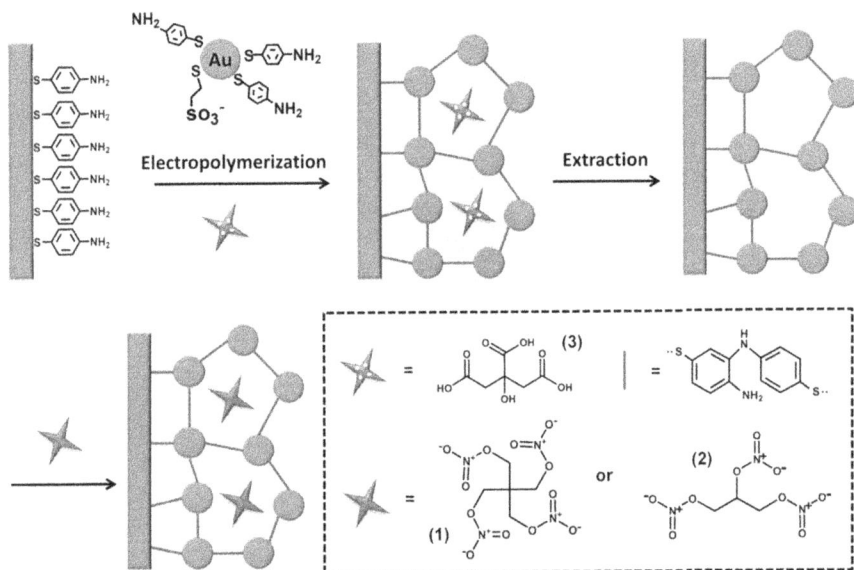

FIGURE 3.4 Electropolymerization procedure of MIPs with a composite of bisaniline-cross-linked AuNPs (Riskin et al. 2011).

was detected in the concentration range of 0.008 ng/mL to 30 ng/mL. The developed SPR nanosensor had a low LOD of 0.008 ng/mL for TNT.

3.4 CONCLUSION AND FUTURE PERSPECTIVES

Plasmonic nanosensors have emerged as a powerful and transformative technology in the realm of detecting biological and chemical threat agents. Their unique ability to exploit LSPR and amplify electromagnetic fields at the nanoscale has revolutionized the landscape of threat detection. Through this chapter, we have explored their fundamental principles, fabrication methods, and multifaceted applications, demonstrating their pivotal role in safeguarding public health, national security, and the environment.

The applications of plasmonic nanosensors in threat detection are both diverse and profound. From the rapid identification of hazardous pathogens and toxins to the detection of chemical threat agents, explosives, and environmental contaminants, these nanosensors have proven invaluable. Their sensitivity, specificity, and real-time monitoring capabilities make them indispensable tools for researchers, security professionals, and first responders alike.

The evolution of plasmonic nanosensors for threat detection holds exciting prospects:

- Enhanced Selectivity: Ongoing research aims to enhance the selectivity of plasmonic nanosensors by developing more advanced surface functionalization techniques and integrating artificial intelligence for real-time data analysis. This will minimize false positives and improve the accuracy of threat detection.
- Portable Devices: Efforts are underway to miniaturize and simplify plasmonic nanosensor platforms for use in handheld and field-deployable devices. These portable systems will empower first responders with immediate on-site threat assessment capabilities.
- Multiplexing: Advancements in multiplexing technology will enable the simultaneous detection of multiple threat agents, providing a comprehensive threat profile in a single analysis.
- Environmental Monitoring: Plasmonic nanosensors will play a growing role in environmental monitoring, aiding in the early detection of chemical spills, pollution, and the spread of infectious diseases in the environment.
- Integration with the Internet of Things (IoT): Plasmonic nanosensors can be integrated into IoT networks, allowing for remote monitoring and real-time data transmission. This will enable a more proactive and coordinated response to emerging threats.
- Customized Nanosensor Design: Tailoring plasmonic nanosensors to specific threat agents will become more common, allowing for optimized detection in different scenarios, from healthcare settings to critical infrastructure protection.

- Biosafety and Biosecurity: Plasmonic nanosensors will continue to contribute to biosafety and biosecurity efforts by enabling the rapid detection of biohazards in laboratory and clinical settings.

In conclusion, plasmonic nanosensors have ushered in a new era of threat detection, offering unparalleled sensitivity and rapid response capabilities. As research and technology development progress, these nanosensors will become increasingly integral to our collective security and preparedness efforts. Their role in safeguarding against biological and chemical threats is not only vital today but holds immense promise for a safer and more secure future.

REFERENCES

Adam, Hussaini, Subash C.B. Gopinath, M.K. Md Arshad, Tijjani Adam, Uda Hashim, Zaliman Sauli, Makram A. Fakhri, et al. 2023. "Integration of Microfluidic Channel on Electrochemical-Based Nanobiosensors for Monoplex and Multiplex Analyses: An Overview." *Journal of the Taiwan Institute of Chemical Engineers* 146 (March). Elsevier B.V.: 104814. doi:10.1016/j.jtice.2023.104814.

Ahijado-Guzmán, Rubén, Julia Menten, Janak Prasad, Christina Lambertz, Germán Rivas, and Carsten Sönnichsen. 2017. "Plasmonic Nanosensors for the Determination of Drug Effectiveness on Membrane Receptors." *ACS Applied Materials & Interfaces* 9 (1): 218–223. doi:10.1021/acsami.6b14013.

Ahijado-Guzmán, Rubén, Janak Prasad, Christina Rosman, Andreas Henkel, Lydia Tome, Dirk Schneider, Germán Rivas, and Carsten Sönnichsen. 2014. "Plasmonic Nanosensors for Simultaneous Quantification of Multiple Protein–Protein Binding Affinities." *Nano Letters* 14 (10): 5528–5532. doi:10.1021/nl501865p.

Akgönüllü, Semra, Handan Yavuz, and Adil Denizli. 2020. "SPR Nanosensor Based on Molecularly Imprinted Polymer Film with Gold Nanoparticles for Sensitive Detection of Aflatoxin B1." *Talanta* 219: 121219. doi:10.1016/j.talanta.2020.121219.

Akgönüllü, Semra, Handan Yavuz, and Adil Denizli. 2021. "Development of Gold Nanoparticles Decorated Molecularly Imprinted–Based Plasmonic Sensor for the Detection of Aflatoxin M1 in Milk Samples." *Chemosensors* 9 (12): 363. doi:10.3390/chemosensors9120363.

Alaqad, Khalid, and Tawfik A. Saleh. 2016. "Gold and Silver Nanoparticles: Synthesis Methods, Characterization Routes and Applications towards Drugs." *Journal of Environmental & Analytical Toxicology* 6 (4). doi:10.4172/2161-0525.1000384.

Altug, Hatice, Sang-Hyun Oh, Stefan A. Maier, and Jiří Homola. 2022. "Advances and Applications of Nanophotonic Biosensors." *Nature Nanotechnology* 17 (1). Springer US: 5–16. doi:10.1038/s41565-021-01045-5.

Bhardwaj, Hema, Gajjala Sumana, and Christophe A. Marquette. 2020. "A Label-Free Ultrasensitive Microfluidic Surface Plasmon Resonance Biosensor for Aflatoxin B1 Detection Using Nanoparticles Integrated Gold Chip." *Food Chemistry* 307: 125530. doi:10.1016/j.foodchem.2019.125530.

Bhardwaj, Hema, Gajjala Sumana, and Christophe A. Marquette. 2021. "Gold Nanobipyramids Integrated Ultrasensitive Optical and Electrochemical Biosensor for Aflatoxin B1 Detection." *Talanta* 222: 121578. doi:10.1016/j.talanta.2020.121578.

Borah, Rituraj, Rajeshreddy Ninakanti, Gert Nuyts, Hannelore Peeters, Adrián Pedrazo-Tardajos, Silvia Nuti, Christophe Vande Velde, et al. 2021. "Selectivity in the Ligand Functionalization of Photocatalytic Metal Oxide Nanoparticles for Phase Transfer and

Self-Assembly Applications." *Chemistry – A European Journal* 27 (35): 9011–9021. doi:10.1002/chem.202100029.

Bradberry, Sally M., Kirsten J. Dickers, Paul Rice, Gareth D. Griffiths, and J. Allister Vale. 2003. "Ricin Poisoning." *Toxicological Reviews* 22: 65–70. doi:10.2165/00139709-200322010-00007.

Brahmkhatri, Varsha, Parimal Pandit, Pranita Rananaware, Aviva D'Souza, and Mahaveer D. Kurkuri. 2021. "Recent Progress in Detection of Chemical and Biological Toxins in Water Using Plasmonic Nanosensors." *Trends in Environmental Analytical Chemistry* 30: e00117. doi:10.1016/j.teac.2021.e00117.

Campos, Antonio R., Zhe Gao, Martin G. Blaber, Rong Huang, George C. Schatz, Richard P. Van Duyne, and Christy L. Haynes. 2016. "Surface-Enhanced Raman Spectroscopy Detection of Ricin B Chain in Human Blood." *The Journal of Physical Chemistry C* 120 (37): 20961–20969. doi:10.1021/acs.jpcc.6b03027.

Çelik, Onur, Yeşeren Saylan, Ilgım Göktürk, Fatma Yılmaz, and Adil Denizli. 2023. "A Surface Plasmon Resonance Sensor with Synthetic Receptors Decorated on Graphene Oxide for Selective Detection of Benzylpenicillin." *Talanta* 253: 123939. doi: 10.1016/j.talanta.2022.123939.

Chanda, Debashis, Kazuki Shigeta, Tu Truong, Eric Lui, Agustin Mihi, Matthew Schulmerich, Paul V. Braun, Rohit Bhargava, and John A. Rogers. 2011. "Coupling of Plasmonic and Optical Cavity Modes in Quasi-Three-Dimensional Plasmonic Crystals." *Nature Communications* 2 (1): 479. doi:10.1038/ncomms1487.

Clark, David P., and Nanette J. Pazdernik. 2016. "Biological Warfare: Infectious Disease and Bioterrorism." In *Biotechnology*, 687–719. doi:10.1016/B978-0-12-385015-7.00022-3.

De, Anindita, and Dristie Kalita. 2023. "Bio-Fabricated Gold and Silver Nanoparticle Based Plasmonic Sensors for Detection of Environmental Pollutants: An Overview." *Critical Reviews in Analytical Chemistry* 53 (3): 672–688. doi:10.1080/10408347.2021.1970507.

Dwivedi, Manish, Swarn Lata Singh, Atul S. Bharadwaj, Vimal Kishore, and Ajay Vikram Singh. 2022. "Self-Assembly of DNA-Grafted Colloids: A Review of Challenges." *Micromachines* 13 (7): 1102. doi:10.3390/mi13071102.

El Kazzy, Marielle, Jonathan S. Weerakkody, Charlotte Hurot, Raphaël Mathey, Arnaud Buhot, Natale Scaramozzino, and Yanxia Hou. 2021. "An Overview of Artificial Olfaction Systems with a Focus on Surface Plasmon Resonance for the Analysis of Volatile Organic Compounds." *Biosensors* 11 (8): 244. doi:10.3390/bios11080244.

Erdem, Özgecan, Nilüfer Cihangir, Yeşeren Saylan, and Adil Denizli. 2020. "Comparison of Molecularly Imprinted Plasmonic Nanosensor Performances for Bacteriophage Detection." *New Journal of Chemistry* 44 (41): 17654–17663. doi: 10.1039/d0nj04053c.

Erdem, Özgecan, Yeşeren Saylan, Nilüfer Cihangir, and Adil Denizli. 2019. "Molecularly Imprinted Nanoparticles Based Plasmonic Sensors for Real-Time Enterococcus Faecalis Detection." *Biosensors and Bioelectronics* 126: 608–614. doi:10.1016/j.bios.2018.11.030.

Fan, Jiao-Rong, Jia Zhu, Wen-Gang Wu, and Yun Huang. 2017. "Plasmonic Metasurfaces Based on Nanopin-Cavity Resonator for Quantitative Colorimetric Ricin Sensing." *Small* 13 (1). John Wiley & Sons, Ltd: 1601710. doi:10.1002/smll.201601710.

Feltis, B.N., B.A. Sexton, F.L. Glenn, M.J. Best, M. Wilkins, and T.J. Davis. 2008. "A Hand-Held Surface Plasmon Resonance Biosensor for the Detection of Ricin and Other Biological Agents." *Biosensors and Bioelectronics* 23 (7): 1131–1136. doi:10.1016/j.bios.2007.11.005.

Gerdan, Zeynep, Yeşeren Saylan, Mukden Uğur, and Adil Denizli. 2022. "Ion-Imprinted Polymer-on-a-Sensor for Copper Detection." *Biosensors* 12 (2): 91. doi: 10.3390/bios12020091.

Ghosh, N., G. Gupta, M. Boopathi, V. Pal, A.K. Singh, N. Gopalan, and A.K. Goel. 2013. "Surface Plasmon Resonance Biosensor for Detection of Bacillus Anthracis, the Causative Agent of Anthrax from Soil Samples Targeting Protective Antigen." *Indian Journal of Microbiology* 53 (1): 48–55. doi:10.1007/s12088-012-0334-3.

Ghosh, Neha, Nidhi Gupta, Garima Gupta, Mannan Boopathi, Vijay Pal, and Ajay Kumar Goel. 2013. "Detection of Protective Antigen, an Anthrax Specific Toxin in Human Serum by Using Surface Plasmon Resonance." *Diagnostic Microbiology and Infectious Disease* 77 (1): 14–19. doi:10.1016/j.diagmicrobio.2013.05.006.

Goel, Ajay Kumar. 2015. "Anthrax: A Disease of Biowarfare and Public Health Importance." *World Journal of Clinical Cases* 3 (1): 20. doi:10.12998/wjcc.v3.i1.20.

Guo, Longhua, Joshua A. Jackman, Huang Hao Yang, Peng Chen, Nam Joon Cho, and Dong Hwan Kim. 2015. "Strategies for Enhancing the Sensitivity of Plasmonic Nanosensors." *Nano Today* 10 (2): 213–239. doi:10.1016/j.nantod.2015.02.007.

Ha, Ji Won. 2023. "Strategies for Sensitivity Improvement of Localized Surface Plasmon Resonance Sensors: Experimental and Mathematical Approaches in Plasmonic Gold Nanostructures." *Applied Spectroscopy Reviews* 58 (5): 346–365. doi:10.1080/057049 28.2022.2104864.

Haase, Frederik, Patrick Hirschle, Ralph Freund, Shuhei Furukawa, Zhe Ji, and Stefan Wuttke. 2020. "Beyond Frameworks: Structuring Reticular Materials across Nano-, Meso-, and Bulk Regimes." *Angewandte Chemie International Edition* 59 (50): 22350–22370. doi:10.1002/anie.201914461.

Harish, Vancha, M.M. Ansari, Devesh Tewari, Awadh Bihari Yadav, Neelesh Sharma, Sweta Bawarig, María-Luisa García-Betancourt, Ali Karatutlu, Mikhael Bechelany, and Ahmed Barhoum. 2023. "Cutting-Edge Advances in Tailoring Size, Shape, and Functionality of Nanoparticles and Nanostructures: A Review." *Journal of the Taiwan Institute of Chemical Engineers* 149: 105010. doi:10.1016/j.jtice.2023.105010.

Hassard, J.F., and M.S. Nieuwenhuizen. 2022. "Chemical Threats." In *Advanced Sciences and Technologies for Security Applications CBRNE: Challenges in the 21st Century*, edited by Peter D.E. Biggins and Deeph Chana, 17–46. Springer International Publishing.

He, Meng-Qi, Yongjian Ai, Wanting Hu, Liandi Guan, Mingyu Ding, and Qionglin Liang. 2023. "Recent Advances of Seed-Mediated Growth of Metal Nanoparticles: From Growth to Applications." *Advanced Materials*. 35(46): 2211915.

Howes, Philip D., Rona Chandrawati, and Molly M. Stevens. 2014. "Colloidal Nanoparticles as Advanced Biological Sensors." *Science* 346 (6205): 1–27. doi:10.1126/science.1247390.

Huang, Jijie, Xin Li Phuah, Luke Mitchell McClintock, Prashant Padmanabhan, K.S.N. Vikrant, Han Wang, Di Zhang, Haohan Wang, Ping Lu, and Xingyao Gao. 2021. "Core-Shell Metallic Alloy Nanopillars-in-Dielectric Hybrid Metamaterials with Magneto-Plasmonic Coupling." *Materials Today* 51: 39–47. doi:10.1016/j.mattod.2021.10.024.

Im, Hyungsoon, Huilin Shao, Yong Il Park, Vanessa M. Peterson, Cesar M. Castro, Ralph Weissleder, and Hakho Lee. 2014. "Label-Free Detection and Molecular Profiling of Exosomes with a Nano-Plasmonic Sensor." *Nature Biotechnology* 32 (5). Nature Publishing Group: 490–495. doi:10.1038/nbt.2886.

Inci, Fatih, Merve Goksin Karaaslan, Amideddin Mataji-Kojouri, Pir Ahmad Shah, Yeşeren Saylan, Yitian Zeng, Anirudh Avadhani, Robert Sinclair, Daryl T-Y Lau, and Utkan Demirci. 2020. "Enhancing the Nanoplasmonic Signal by a Nanoparticle Sandwiching Strategy to Detect Viruses." *Applied Materials Today* 20: 100709. doi: 10.1016/j. apmt.2020.100709.

Inci, Fatih, Yeşeren Saylan, Amideddin Mataji Kojouri, Mehmet Giray Ogut, Adil Denizli, and Utkan Demirci. 2020. "A Disposable Microfluidic-Integrated Hand-Held Plasmonic Platform for Protein Detection." *Applied Materials Today* 18: 100478. doi: 10.1016/j. apmt.2019.100478.

Janardhanan, Pavithra, Charlene M. Mello, Bal Ram Singh, Jianlong Lou, James D. Marks, and Shuowei Cai. 2013. "RNA Aptasensor for Rapid Detection of Natively Folded Type A Botulinum Neurotoxin." *Talanta* 117: 273–280. doi:10.1016/j.talanta.2013.09.012.

Javaid, Mohd, Abid Haleem, Ravi Pratap Singh, Shanay Rab, and Rajiv Suman. 2021. "Exploring the Potential of Nanosensors: A Brief Overview." *Sensors International* 2: 100130. doi:10.1016/j.sintl.2021.100130.

Jia, Zhixin, Ce Shi, Xinting Yang, Jiaran Zhang, Xia Sun, Yemin Guo, and Xiaoguo Ying. 2023. "QD-based Fluorescent Nanosensors: Production Methods, Optoelectronic Properties, and Recent Food Applications." *Comprehensive Reviews in Food Science and Food Safety*, 22(6):4644–4669. doi:10.1111/1541-4337.13236.

Jin, Congran, Ziqian Wu, John H. Molinski, Junhu Zhou, Yundong Ren, and John X.J. Zhang. 2022. "Plasmonic Nanosensors for Point-of-Care Biomarker Detection." *Materials Today Bio* 14: 100263.

Kawaguchi, T., D. Shankaran, S. Kim, K. Gobi, K. Matsumoto, K. Toko, and N. Miura. 2007. "Fabrication of a Novel Immunosensor Using Functionalized Self-Assembled Monolayer for Trace Level Detection of TNT by Surface Plasmon Resonance." *Talanta* 72 (2): 554–560. doi:10.1016/j.talanta.2006.11.020.

Kibis, L.S., A.I. Stadnichenko, E.M. Pajetnov, S.V. Koscheev, V.I. Zaykovskii, and A.I. Boronin. 2010. "The Investigation of Oxidized Silver Nanoparticles Prepared by Thermal Evaporation and Radio-Frequency Sputtering of Metallic Silver under Oxygen." *Applied Surface Science* 257 (2): 404–413. doi:10.1016/j.apsusc.2010.07.002.

Kołątaj, Karol, Jan Krajczewski, and Andrzej Kudelski. 2020. "Plasmonic Nanoparticles for Environmental Analysis." *Environmental Chemistry Letters* 18 (3): 529–542. doi:10.1007/s10311-019-00962-1.

Krajczewski, Jan, Karol Kołątaj, and Andrzej Kudelski. 2017. "Plasmonic Nanoparticles in Chemical Analysis." *RSC Advances* 7 (28): 17559–17576. doi:10.1039/C7RA01034F.

Kumar, Vinod, Heejeong Kim, Bipin Pandey, Tony D. James, Juyoung Yoon, and Eric V. Anslyn. 2023. "Recent Advances in Fluorescent and Colorimetric Chemosensors for the Detection of Chemical Warfare Agents: A Legacy of the 21st Century." *Chemical Society Reviews* 52: 663–704. doi:10.1039/D2CS00651K.

Lafuente, Marta, Ismael Pellejero, Víctor Sebastián, Miguel A. Urbiztondo, Reyes Mallada, M. Pilar Pina, and Jesús Santamaría. 2018. "Highly Sensitive SERS Quantification of Organophosphorous Chemical Warfare Agents: A Major Step towards the Real Time Sensing in the Gas Phase." *Sensors and Actuators B: Chemical* 267: 457–466. doi:10.1016/j.snb.2018.04.058.

Lazarević-Pašti, Tamara, Tamara Tasić, Vedran Milanković, and Nebojša Potkonjak. 2023. "Molecularly Imprinted Plasmonic-Based for Environmental Contaminants—Current State and Future Perspectives." *Chemosensors* 11 (1). doi:10.3390/chemosensors11010035.

Lévêque, Christian, Géraldine Ferracci, Yves Maulet, Christelle Mazuet, Michel R. Popoff, Marie-Pierre Blanchard, Michael Seagar, and Oussama El Far. 2015. "An Optical Biosensor Assay for Rapid Dual Detection of Botulinum Neurotoxins A and E." *Scientific Reports* 5 (1). Nature Publishing Group: 17953. doi:10.1038/srep17953.

Li, Chunsheng, Arlene Alves dos Reis, Armin Ansari, Luiz Bertelli, Zhanat Carr, Nicholas Dainiak, Marina Degteva, et al. 2022. "Public Health Response and Medical Management of Internal Contamination in Past Radiological or Nuclear Incidents: A Narrative Review." *Environment International* 163: 107222. doi:10.1016/j.envint.2022.107222.

Li, Zihan, Luca Leustean, Fatih Inci, Min Zheng, Utkan Demirci, and Shuqi Wang. 2019. "Plasmonic-Based Platforms for Diagnosis of Infectious Diseases at the Point-of-Care." *Biotechnology Advances* 37 (8): 107440.

Linic, Suljo, Umar Aslam, Calvin Boerigter, and Matthew Morabito. 2015. "Photochemical Transformations on Plasmonic Metal Nanoparticles." *Nature Materials* 14 (6): 567–576. doi:10.1038/nmat4281.

Liu, Shenqi, Shanxue Jiang, Zhiliang Yao, and Minhua Liu. 2023. "Aflatoxin Detection Technologies: Recent Advances and Future Prospects." *Environmental Science and Pollution Research* 30 (33): 79627–79653. doi:10.1007/s11356-023-28110-x.

Lord, J. Michael, Lynne M. Roberts, and Jon D. Robertus. 1994. "Ricin: Structure, Mode of Action, and Some Current Applications." *The FASEB Journal* 8 (2): 201–208. doi:10.1096/fasebj.8.2.8119491.

Luo, Li, Jiewei Yang, Zhi Li, Hua Xu, Lei Guo, Lili Wang, Yuxia Wang, et al. 2022. "Label-Free Differentiation and Quantification of Ricin, Abrin from Their Agglutinin Biotoxins by Surface Plasmon Resonance." *Talanta* 238 (P1): 122860. doi:10.1016/j.talanta.2021.122860.

Manoj, D., S. Shanmugasundaram, and C. Anandharamakrishnan. 2021. "Nanosensing and Nanobiosensing: Concepts, Methods, and Applications for Quality Evaluation of Liquid Foods." *Food Control* 126: 108017. doi:10.1016/j.foodcont.2021.108017.

Mauriz, Elba, and Laura M. Lechuga. 2021. "Plasmonic Biosensors for Single-Molecule Biomedical Analysis." *Biosensors* 11 (4). doi:10.3390/bios11040123.

Mosquera, Jesús, Yuan Zhao, Hee-Jeong Jang, Nuli Xie, Chuanlai Xu, Nicholas A. Kotov, and Luis M. Liz-Marzán. 2020. "Plasmonic Nanoparticles with Supramolecular Recognition." *Advanced Functional Materials* 30 (2). doi:10.1002/adfm.201902082.

Özgür, Erdoğan, Yeşeren Saylan, Nilay Bereli, Deniz Türkmen, and Adil Denizli. 2020. "Molecularly Imprinted Polymer Integrated Plasmonic Nanosensor for Cocaine Detection." *Journal of Biomaterials Science, Polymer Edition* 31 (9): 1211–1222. doi:10.1080/09205063.2020.1751524.

Patel, Kruti, Shmuel Halevi, Paul Melman, John Schwartz, Shuowei Cai, and Bal Singh. 2017. "A Novel Surface Plasmon Resonance Biosensor for the Rapid Detection of Botulinum Neurotoxins." *Biosensors* 7 (4): 32. doi:10.3390/bios7030032.

Polito, Letizia, Massimo Bortolotti, Maria Battelli, Giulia Calafato, and Andrea Bolognesi. 2019. "Ricin: An Ancient Story for a Timeless Plant Toxin." *Toxins* 11 (6): 324. doi:10.3390/toxins11060324.

Prakash, Jai, R.A. Harris, and H.C. Swart. 2016. "Embedded Plasmonic Nanostructures: Synthesis, Fundamental Aspects and Their Surface Enhanced Raman Scattering Applications." *International Reviews in Physical Chemistry* 35 (3). Taylor & Francis: 353–398. doi:10.1080/0144235X.2016.1187006.

Ramalingam, Murugan, Abinaya Jaisankar, Lijia Cheng, Sasirekha Krishnan, Liang Lan, Anwarul Hassan, Hilal Turkoglu Sasmazel, et al. 2023. "Impact of Nanotechnology on Conventional and Artificial Intelligence-Based Biosensing Strategies for the Detection of Viruses." *Discover Nano* 18 (1). Springer US: 58. doi:10.1186/s11671-023-03842-4.

Razlansari, Mahtab, Fulden Ulucan-Karnak, Masoud Kahrizi, Shekoufeh Mirinejad, Saman Sargazi, Sachin Mishra, Abbas Rahdar, and Ana M. Díez-Pascual. 2022. "Nanobiosensors for Detection of Opioids: A Review of Latest Advancements." *European Journal of Pharmaceutics and Biopharmaceutics* 179: 79–94. doi:10.1016/j.ejpb.2022.08.017.

Riskin, Michael, Yaniv Ben-Amram, Ran Tel-vered, Vladimir Chegel, Joseph Almog, and Itamar Willner. 2011. "Molecularly Imprinted Au Nanoparticles Composites on Au Surfaces." *Analytical Chemistry* 83: 3082–3088. doi:10.1021/ac1033424.

Riskin, Michael, Ran Tel-Vered, and Itamar Willner. 2010. "Imprinted Au-Nanoparticle Composites for the Ultrasensitive Surface Plasmon Resonance Detection of Hexahydro-1,3,5-Trinitro-1,3,5-Triazine (RDX)." *Advanced Materials* 22 (12): 1387–1391. doi:10.1002/adma.200903007.

Saylan, Y., O. Erdem, N. Cihangir, and A. Denizli. 2019. "Detecting Fingerprints of Waterborne Bacteria on a Sensor." *Chemosensors* 7 (3): 33. doi:10.3390/chemosensors7030033.

Saylan, Yeşeren. 2023. "Unveiling the Pollution of Bacteria in Water Samples through Optic Sensor." *Microchemical Journal* 193 (July): 109057. doi:10.1016/j.microc.2023.109057.

Saylan, Yeşeren, Semra Akgönüllü, and Adil Denizli. 2020. "Plasmonic Sensors for Monitoring Biological and Chemical Threat Agents." *Biosensors* 10 (10): 142. doi:10.3390/bios10100142.

Saylan, Yeşeren, Semra Akgönüllü, and Adil Denizli. 2022. "Preparation of Magnetic Nanoparticles-Assisted Plasmonic Biosensors with Metal Affinity for Interferon-α Detection." *Materials Science and Engineering: B* 280: 115687. doi:10.1016/j.mseb.2022.115687.

Saylan, Yeşeren, and Adil Denizli. 2018. "Molecular Fingerprints of Hemoglobin on a Nanofilm Chip." *Sensors* 18 (9). doi:10.3390/s18093016.

Schmid, Diana Coman. 2020. "Risks and Consequences of Hazard Agents to Human Health." In *NATO Science for Peace and Security Series A: Chemistry and Biology*, 129–141. doi:10.1007/978-94-024-2041-8_8.

Shrivastav, Anand M., Uroš Cvelbar, and Ibrahim Abdulhalim. 2021. "A Comprehensive Review on Plasmonic-Based Biosensors Used in Viral Diagnostics." *Communications Biology* 4 (1): 70. doi:10.1038/s42003-020-01615-8.

Shu, Zhiwen, Fuhua Ye, Peng Liu, Pei Zeng, Huikang Liang, Lei Chen, Xiaoqing Zhang, et al. 2023. "Programmable Nanoscale Crack Lithography for Multiscale PMMA Patterns." *ACS Applied Materials & Interfaces* 15 (23): 28349–28357. doi:10.1021/acsami.3c02625.

Smith, Bryan Ronain, and Sanjiv Sam Gambhir. 2017. "Nanomaterials for In Vivo Imaging." *Chemical Reviews* 117 (3): 901–986. doi:10.1021/acs.chemrev.6b00073.

Spitzberg, Joshua D., Adam Zrehen, Xander F. van Kooten, and Amit Meller. 2019. "Plasmonic-Nanopore Biosensors for Superior Single-Molecule Detection." *Advanced Materials* 31 (23): 1–18. doi:10.1002/adma.201900422.

Stern, Daniel, Diana Pauly, Martin Zydek, Christian Müller, Marc A. Avondet, Sylvia Worbs, Fred Lisdat, Martin B. Dorner, and Brigitte G. Dorner. 2016. "Simultaneous Differentiation and Quantification of Ricin and Agglutinin by an Antibody-Sandwich Surface Plasmon Resonance Sensor." *Biosensors and Bioelectronics* 78: 111–117. doi:10.1016/j.bios.2015.11.020.

Swierczewska, Magdalena, Gang Liu, Seulki Lee, and Xiaoyuan Chen. 2012. "High-Sensitivity Nanosensors for Biomarker Detection." *Chemical Society Reviews* 41 (7): 2641–2655. doi:10.1039/C1CS15238F.

Tang, Ji-jun, Jie-fang Sun, Rui Lui, Zong-mian Zhang, Jing-fu Liu, and Jian-wei Xie. 2016. "New Surface-Enhanced Raman Sensing Chip Designed for On-Site Detection of Active Ricin in Complex Matrices Based on Specific Depurination." *ACS Applied Materials & Interfaces* 8 (3): 2449–2455. doi:10.1021/acsami.5b12860.

Tang, Longhua, and Jinghong Li. 2017. "Plasmon-Based Colorimetric Nanosensors for Ultrasensitive Molecular Diagnostics." *ACS Sensors* 2 (7): 857–875. doi:10.1021/acssensors.7b00282.

Tomar, Arvind, and Garima Gupta. 2016. "Surface Plasmon Resonance Sensing of Biological Warfare Agent Botulinum Neurotoxin A." *Journal of Bioterrorism & Biodefense* 7 (2): 1000142. doi:10.4172/2157-2526.1000142.

Tran, Hung, Carol Leong, Weng Keong Loke, Con Dogovski, and Chun-Qiang Liu. 2008. "Surface Plasmon Resonance Detection of Ricin and Horticultural Ricin Variants in Environmental Samples." *Toxicon* 52 (4): 582–588. doi:10.1016/j.toxicon.2008.07.008.

Vo-Dinh, Tuan, Yang Liu, Andrew M. Fales, Hoan Ngo, Hsin-Neng Wang, Janna K. Register, Hsiangkuo Yuan, Stephen J. Norton, and Guy D. Griffin. 2015. "SERS Nanosensors and Nanoreporters: Golden Opportunities in Biomedical Applications." *WIREs Nanomedicine and Nanobiotechnology* 7 (1): 17–33. doi:10.1002/wnan.1283.

Wang, B., Z. Lou, B. Park, Y. Kwon, H. Zhang, and B. Xu. 2015. "Surface Conformations of an Anti-Ricin Aptamer and Its Affinity for Ricin Determined by Atomic Force Microscopy and Surface Plasmon Resonance." *Physical Chemistry Chemical Physics* 17 (1): 307–314. doi:10.1039/C4CP03190C.

Wang, Wei, and Nongjian Tao. 2014. "Detection, Counting and Imaging of Single Nanoparticles Wei." *Analytical Chemistry* 1 (86): 2–14. doi:10.1021/ac403890n.Detection.

Wei, Haoran, Seyyed M. Hossein Abtahi, and Peter J. Vikesland. 2015. "Plasmonic Colorimetric and SERS Sensors for Environmental Analysis." *Environmental Science: Nano* 2 (2): 120–135. doi:10.1039/C4EN00211C.

Wilson, Andrew J., and Prashant K. Jain. 2020. "Light-Induced Voltages in Catalysis by Plasmonic Nanostructures." *Accounts of Chemical Research* 53 (9): 1773–1781. doi:10.1021/acs.accounts.0c00378.

Wujcik, E.K., H. Wei, X. Zhang, J. Guo, X. Yan, N. Sutrave, S. Wei, and Z. Guo. 2014. "Antibody Nanosensors: A Detailed Review." *RSC Advances* 4 (82): 43725–43745. doi:10.1039/C4RA07119K.

Xu, Shicai, Jian Zhan, Baoyuan Man, Shouzhen Jiang, Weiwei Yue, Shoubao Gao, Chengang Guo, et al. 2017. "Real-Time Reliable Determination of Binding Kinetics of DNA Hybridization Using a Multi-Channel Graphene Biosensor." *Nature Communications* 8 (1): 14902. doi:10.1038/ncomms14902.

Yang, Kang, Xu Yao, Bowen Liu, and Bin Ren. 2021. "Metallic Plasmonic Array Structures: Principles, Fabrications, Properties, and Applications." *Advanced Materials* 33 (50): 1–21. doi:10.1002/adma.202007988.

Zeng, Shuwen, Dominique Baillargeat, Ho Pui Ho, and Ken Tye Yong. 2014. "Nanomaterials Enhanced Surface Plasmon Resonance for Biological and Chemical Sensing Applications." *Chemical Society Reviews* 43 (10): 3426–3452. doi:10.1039/c3cs60479a.

Zhao, Qian, Guangqiang Liu, Hongwen Zhang, Fei Zhou, Yue Li, and Weiping Cai. 2017. "SERS-Based Ultrasensitive Detection of Organophosphorus Nerve Agents via Substrate's Surface Modification." *Journal of Hazardous Materials* 324: 194–202. doi:10.1016/j.jhazmat.2016.10.049.

4 Synthetic Biology Applications for the Biosensing of Biological and Chemical Threats

Abdurahman Atilla, Ece Avcı, Nazlıcan Tunç, Doğuş Akboğa, and Urartu Özgür Şafak Şeker

4.1 INTRODUCTION

Biological threats encompass hazardous elements such as potent toxins, disease-causing bacteria, viruses, and fungi, including yeasts and molds (Banoub & Mikhael, 2020; Ellison, 2022). These agents embody a constant, irreversible danger to public health, potentially inflicting human sickness and mortality (CDC, Bioterrorism Agents/Diseases (by category), Emergency Preparedness & Response, 2019). Prominent examples of these threats are the incidents involving anthrax-tainted correspondences (Jernigan et al., n.d.), the recent emergence of the COVID-19 pandemic (SARS-CoV-2) (Y.-C.Wu, Chen, & Chan, 2020), and recurrent instances of maladies prompted by pathogens transmitted through food (Bintsis, 2017). The plausible deployment of biological threat agents in warfare scenarios poses a formidable obstacle to contemporary security measures. Pathogens or toxins from living organisms can be weaponized and disseminated to infect and incapacitate a large population (Frischknecht, 2003; Riedel, 2004). Historical precedents, such as the use of anthrax spores during World War II, underline the reality of this threat (Carus, 2015). Moreover, advancements in biotechnology have heightened concerns around the creation of genetically modified pathogens or the synthesis of known pathogens from scratch (van Aken & Hammond, 2003). Despite significant advances in comprehending the pathogenicity of numerous biothreat agents and the marked increase in available treatment options, these threats persist as a crucial public health concern in an era characterized by bioterrorism, unregulated warfare, environmental pollution, climate change, unrestricted population growth, and increasing globalization (Walper et al., 2018).

Biosensing technologies, which integrate a biological recognition element with a transducer, enable the detection and quantification of specific biological entities such as toxins, bacteria, viruses, or fungi (Turner, 2013). They utilize specialized

DOI: 10.1201/9781003459316-4

biosensors and have seen the inclusion of nanomaterials significantly augment their sensitivity and specificity (Malhotra & Ali, 2018). However, biosensing technologies face challenges, such as sensitivity and specificity issues, the impact of environmental factors on performance, and maintenance requirements (Bahadır & Sezginturk, 2015; Kahn & Plaxco, 2010; Quintela, Vasse, Lin, & Wu, 2022). The evolution of biosensors has moved towards miniaturized, portable devices that deliver rapid, real-time, and on-site detection (Kaushik et al., 2018). These include nano biosensors, which leverage nanomaterials for enhanced sensitivity and specificity (Arduini et al., 2016), and the adoption of electrochemical biosensors for their cost-effectiveness and swift response times (Pardee et al., 2016). Advances in genomics and proteomics have spurred the development of DNA and protein biosensors that can pinpoint specific sequences or markers linked to biothreat agents (J. Li & Macdonald, 2016). A significant growth has been the use of multiplexed lateral flow biosensors for simultaneous detection of multiple targets, significantly improving the diagnostic capacity of biosensors for biological warfare agents (Pohanka, 2019).

Synthetic biology, which encompasses the design and construction of new biological components and the refinement of existing systems, presents promising solutions to the existing limitations in biosensor technologies. Notably, it establishes the essential conditions, set by the evolution of biosensors, for sensitive and precise detection of biological threat agents. Engineered biological components could enhance the sensitivity and specificity of biosensors, and synthetic biology could aid in developing more robust biosensors that operate optimally under various environmental conditions (Saltepe et al., 2018; Slomovic, Pardee, & Collins, 2015; Way, Collins, Keasling, & Silver, 2014). Synthetic biology has the potential to revolutionize the field of biosensing and the detection and mitigation of biological threats. It promises to provide more reliable, efficient, and adaptable tools for the early detection of biothreat agents, significantly influencing public health in an era marked by biothreats and bioterrorism (Kumar & Rani, 2013; Pohanka, 2019).

4.2 GENETIC CIRCUIT-BASED BIOSENSORS FOR DETECTING BIOLOGICAL THREAT AGENTS

4.2.1 CELL-BASED BIOSENSORS

Cell-based biosensors, also known as Whole-Cell Biosensors (WCB), are sophisticated biosynthetic systems that utilize live cells as their sensing component to detect and quantify the presence of specific compounds. These biosensors harness the inherent abilities of cells to recognize, respond to, and generate signals in the presence of target analytes. They consist of three fundamental components: Sensing, processing, and actuating elements, as shown in Figure 4.1 (Saltepe et al., 2018). While it is difficult to pinpoint the date of the invention of the first WCB, an early notable example is the microbial glucose biosensor developed in the 1960s (Clark Jr. & Lyons, 1962). This pioneering work employed living microbial cells as the sensing element to detect the production of hydrogen peroxide upon the reaction of glucose with glucose oxidase. It laid the groundwork for subsequent successful

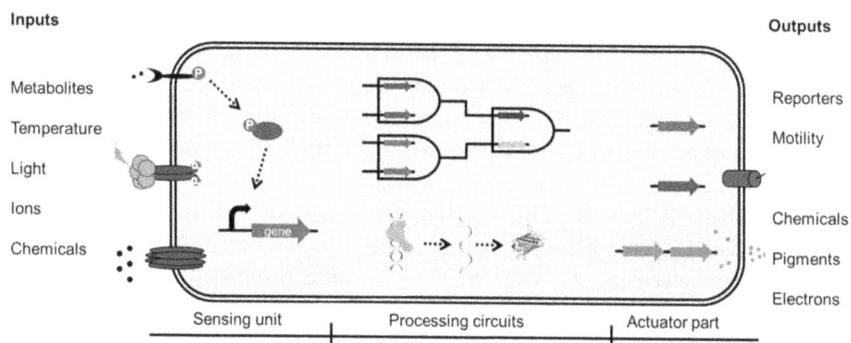

FIGURE 4.1 Components and the working principle of a WCB. The cell picks up cues from its surroundings, such as some metabolites, chemicals, ions, or changes in the temperature or light. These cues initiate internal cellular processes. The signal processing might occur through various methods, including DNA transcriptional control or artificial logic systems introduced into the cell. Depending on the internal processing, the cell reacts by secreting chemicals, altering its movement, or expressing a reporter. Reprinted with permission from Saltepe, B., Kehribar, E., Yirmibesoglu, S. S. S., & Seker, U. O. S. (2018, January 26). "Cellular biosensors with engineered genetic circuits", ACS publications. ACS SENSORS. https://pubs.acs.org/doi/pdfplus/10.1021/acssensors.7b00728. Copyright 2018 American Chemical Society.

advancements in this field. Numerous scientists have contributed to the progress of WCBs, resulting in various applications in environmental monitoring, medical diagnostics, and other areas.

Thanks to synthetic biology approaches, such as genetic engineering, genetic circuit design, promoter engineering, chassis selection and engineering, directed evolution, and standardization, the area of WCBs has been greatly transformed. Genetic engineering is the process of changing the genetic material of cells to create or improve specific sensory capabilities. This procedure permits the insertion of genes encoding sensing elements, such as receptors or enzymes, into the cellular genome, allowing cells to create the components required for detecting and reacting to target analytes. Synthetic biology makes genetic circuit design more straightforward, allowing for the creation of circuits that control gene expression in response to the presence of the target analyte, therefore maximizing the performance of the biosensor (Jaiswal & Shukla, 2020). When the target analyte is identified, promoter engineering focuses on changing or creating DNA sections to activate or repress gene expression, ensuring biosensors make detectable responses proportionate to the analyte concentration. Chassis selection and engineering include selecting and manipulating host organisms such as bacteria or yeast to improve biosensor sensitivity and resilience (S.-Y. Chen et al., 2019; H. J. Kim, Jeong, & Lee, 2018). Directed evolution uses evolutionary processes to improve biosensors' sensitivity, selectivity, and dynamic range. Standardization encourages the exchange and assembly of well-characterized genetic components, which ensures the reliability and scalability of WCB designs across laboratories and applications. WCBs with higher sensitivity, specificity, and performance may be built using synthetic biology principles, bringing up new possibilities for applications in environmental monitoring, healthcare diagnostics, and industrial processes (Saltepe et al., 2018).

WCBs may be categorized into many categories using a variety of factors. According to the origin of the cell type used in the WCB, five types can be listed:

1. Microbial Biosensors: Utilizing living microorganisms, such as bacteria and yeast, as the primary agents, these biosensors employ genetically engineered sensing elements to elicit specific responses towards target analytes, encompassing heavy metals, pollutants, metabolic intermediates, and toxins.

2. Plant Biosensors: With the employment of the tissues or cells of plants as the fundamental sensing elements, these biosensors capitalize on the innate ability of plants to react to environmental variations. Notably, plant WCBs are predominantly used to identify heavy metals and toxic compounds within pesticides, using alterations in gene expression or fluorescence as indicators.

   • Animal Cell Biosensors: These biosensors feature animal-originating cells, notably mammalian cells, as the core sensing component. In the realm of biomedical research, animal cell biosensors find extensive application in studies related to drug discovery. In this context, the target analytes of interest may encompass hormones, drugs, or neurotransmitters.

   • Enzyme-Based Biosensors: These biosensors use whole cells that contain specific enzymes capable of catalyzing reactions in the presence of the target analyte. The resulting enzymatic reaction subsequently generates a measurable signal, which can be quantified through current, fluorescence, or a discernible color change.

   • Hybrid Biosensors: By coupling transducing-surface immobilized whole cells with other complementary sensing elements, these biosensors achieve exceptional sensitivity and specificity. Prioritizing cellular responses, these biosensors seamlessly integrate electrochemical and optical methodologies to facilitate signal detection and analysis.

4.2.1.1 Biomedicine Applications of WCBs

WCBs play a significant role in biomedical research, providing valuable tools for studying cellular processes, drug development, and toxicity assessment. These biosensors can be specially designed to monitor specific biological responses induced by various stimuli or pharmacological compounds, including changes in gene expression, protein activity, or cell viability. The utilization of WCBs contributes significantly to advancing biomedical knowledge, enabling researchers to gain profound insights into fundamental biological systems and evaluate the efficacy and safety of therapeutic medications.

In medical diagnostics, WCBs find wide applications, particularly in detecting proteins that are biomarkers for various disorders. Traditionally, targeting proteins using WCBs has been challenging due to their inability to diffuse across the membrane. Consequently, immunoassays such as latex agglutination tests (LATs) have been commonly employed for protein detection. Nevertheless, Kylilis et al. (2019) developed an innovative biosensing platform based on cell agglutination, incorporating surface-displaying nanobodies capable of detecting target proteins (Kylilis et al., 2019). Their research presents a cost-effective point-of-care (PoC) device that aims for extracellular protein detection, potentially replacing immunoassays. This user-friendly and easily implementable gadget proves particularly advantageous in resource-constrained settings. Another remarkable application of WCBs in biomedical research is demonstrated by Woo and his colleagues, who designed a nitrate-sensing WCB using probiotics that utilized the Boolean AND logic gate to diagnose intestinal inflammation based on nitrate, the biomarker of gut inflammation (Woo et al., 2020). This approach offered a non-invasive means of fluorescence detection, holding great promise in practical diagnostics. Figure 4.2A explains the details of the study. The constitutively expressed NarXL proteins are phosphorylated in the presence of nitrate, which activates the expression of ThsS(L547T)R proteins through the yeaR promoter. When thiosulfate is present, expressed ThsS(L547T) R proteins are phosphorylated and activate the expression of the superfolder Green Fluorescent Protein (sfGFP) reporter protein via the phsA promoter.

Furthermore, the study conducted by Watstein and Styczynski showcases the use of WCBs in addressing micronutrient insufficiency, a global health concern (Watstein & Styczynski, 2018). Zinc (Zn) is an essential micronutrient for immune function, which provides a proper resistance against infectious diseases (Ackland & Michalczyk, 2016). Accurate assessment of micronutrient status is vital for maintaining health. Watstein and Styczynski engineered *Escherichia coli* cells to construct a three-color WCB to measure Zn levels in human serum with a broader range of Zn concentrations from 0μM to 20 μM. This new method for regulating violacein purple pigment (which behaves like an antibiotic) production is based on another Zn detection system driven by the PznuC promoter and the Xur repressor, which responds to Zn. This innovative approach offers a simple, rapid, on-site detection method for assessing micronutrient status. Moreover, when developing new biomaterials, detecting any potential cytotoxic elements they may contain is crucial before their clinical safety can be established. WCBs with genetically engineered components can assess materials' biocompatibility quickly. Saltepe et al. (2019) engineered promoters of the heat shock mechanism in *Mycobacterium tuberculosis* to detect nanomaterial-triggered toxicity, explicitly focusing on quantum dots (Saltepe et al., 2019). These examples show the versatility and great potential of WCBs in advancing biomedical research, from protein detection to non-invasive diagnostics and toxicity assessment of novel biomaterials.

4.2.1.2 Environmental Monitoring Applications of WCBs

WCBs play a vital role in monitoring and detecting environmental contaminants. They effectively detect pollutants such as heavy metals, organic compounds, and hazardous substances in water, soil, and air samples. These biosensors provide quick, sensitive, and cost-effective techniques for assessing environmental quality,

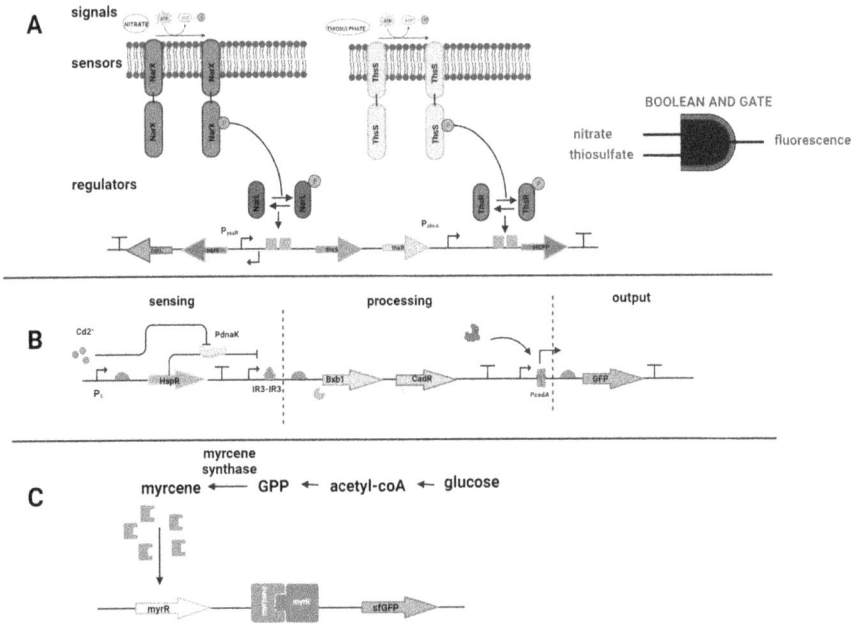

FIGURE 4.2 A summary of the applications of WCBs with an example for each application area. (A) Biomedicine. Here, constitutively expressed NarXL proteins are phosphorylated when nitrate, which is a biomarker of gut inflammation, is present, which leads to the expression of ThsS(L547T)R proteins through the yeaR promoter. In the presence of thiosulfate, ThsS(L547T)R proteins are phosphorylated and lead to the expression of the sfGFP reporter protein. This study harbors Boolean AND gate, which gives output only when both inputs are given. (B) Environmental monitoring. In the presence of cadmium ions, the stress-activating heat shock response causes HspR protein release. GFP is produced as a reporter upon CadR-Cadmium complex formation. The genetic circuit built in this study allows cadmium toxicity monitoring. (C) Other applications. Myrcene sensing promoter (Pmyr) and myrR gene cloned under the Ptrc promoter respond to the intracellular myrcene concentration. Myrcene binds with MyrR to induce a conformational change that is followed by the binding of MyrR to Pmyr, resulting in sfGFP fluorescence. This novel biosensor shows that the increase in the green fluorescence of sfGFP, is due to the higher myrcene production by myrcene synthase. (A) is adapted from Akboʾga et al. (2022); (B) is adapted from S. Chen et al. (2023); (C) is adapted from Woo et al. (2020). Created with BioRender.com.

facilitating environmental management, and supporting rehabilitation operations. Heavy metals such as arsenic (As), cadmium (Cd), mercury (Hg), lead (Pb), and chromium (Cr) have garnered significant attention among various pollutants since they can cause severe disorders even at low doses due to their high toxicity (H. Kim, Jang, & Yoon, 2020). As industrialization continues, these metals in environmental systems are expected to rise. Therefore, close monitoring of heavy metal concentrations in diverse environmental systems is paramount.

In their work published in 2022, Akboga et al. (2022) successfully developed a recombinase-based cadmium sensor by integrating two design approaches (Akboga et al., 2022). They combined a semi-specific transcriptional unit with a specific but cross-reactive one in a genetic circuit, enabling *Escherichia coli* cells to respond to toxic cadmium(II) (Figure 4.2B). To monitor cadmium toxicity, they built a biosensor that uses Green Fluorescent Protein (gfp) as the reporter. GFP expression is observed upon CadR-Cadmium complex formation that results from a stress-activating heat-shock response activated in the presence of cadmium(II). Moreover, Cayron et al. (2017) utilized engineered bioluminescent bacteria to create a Ni-specific sensor (Cayron et al., 2017). This sensor consists of a dual-acting protein-transcriptional regulator fused to reporter genes, allowing for the rapid detection of nickel ions in drinking water. The detection of toxic arsenic has been the subject of numerous studies. One particular study stood out by employing a positive feedback amplifier, which provided exceptional specificity to the sensor against As(II) ions (Jia, Bu, Zhao, & Wu, 2019). This finding highlights the significance of genetic circuit engineering in the design of WCBs.

Heavy metals pose a significant threat not only to human health but also to seawater and marine animals. F. Cui et al. (2020) successfully developed a bacterial WCB using salt-tolerant *Acinetobacter baylyi* to target heavy metal toxicity in seawater (F. Cui et al., 2020). Furthermore, heavy metals pose a contamination threat to cosmetics production. Guo et al. (2020) demonstrated the detection of toxic mercury using a test strip based on a bacterial WCB (Guo et al., 2020). This study introduced an instrument-independent method for sensing insoluble mercury in cosmetics, without the need for a predigestion step through inherently converting them into soluble forms. Finally, Elcin and Oktem concluded that biosensors are more prominent in the environmental monitoring of heavy metals than traditional methods by offering a fast, cost-effective, and high-throughput alternative (Elcin & Öktem, 2020). Their study focused on fluorescence-based in-situ cadmium WCBs. While monitoring heavy metals in various environments through biosensors has shown its importance and effectiveness, the field of biosensors has also seen innovations in addressing other challenges, notably leveraging the phenomenon of quorum sensing. Since its discovery, quorum sensing has generated significant interest in various fields. He et al. (2022) developed a sophisticated WCB that incorporated a quorum sensing module, focusing on the degradation of organophosphates. Organophosphates are commonly used as pesticides in agriculture and threaten public health, similar to other xenobiotics (He et al., 2022). Y. Wu et al. (2021) also utilized quorum sensing in their study (Y. Wu et al., 2021). As ensuring access to clean water is of the utmost importance, effective monitoring of bacterial contamination is crucial to prevent waterborne illnesses (Drinking-water, n.d.). Engineered WCBs hold great promise for detecting pathogenic bacteria. They designed a WCB targeting two pathogens: *Pseudomonas aeruginosa* and *Burkholderia pseudomallei*. This biosensor incorporates a quorum sensing module and employs synthetic biology approaches. The sensor was immobilized on paper, creating a portable device for PoC monitoring.

Another significant concern in environmental monitoring pertains to antibiotics. Antibiotics find their way into the environment through the urine and excreta of humans and animals. In addition to optimizing and limiting antibiotic usage, it is essential to detect antibiotics in samples. WCBs offer an easy means of detecting antibiotic concentrations in a sample (Parthasarathy, Monette, Bracero, & S. Saha, 2018). Bacteria inherently respond to the presence of antibiotics, so promoters sensitive to antibiotics can be fused to reporter elements to build cell-based systems for antibiotic detection.

Navigating from the concerns of antibiotic presence in the environment to the broader applications of WCBs highlights the versatility and importance of this technology. Beyond detecting antibiotics, these biosensors are poised to address numerous environmental issues, showcasing their adaptability and precision. One such area, which has recently gained traction in the field of environmental science, is the monitoring and treatment of wastewater. WCBs can be programmed to identify and monitor specific microbial activities or metabolic pathways associated with pollution breakdown (Jaiswal & Shukla, 2020). Monitoring oil and grease levels in wastewater is essential for environmental pollution control. While several WCBs for hydrocarbon detection have been previously developed, a review by Voon et al. (2022) concludes that there is still a need for a study that focuses on creating new devices specifically for measuring wastewater's oil and grease content (Voon et al., 2022). Despite the possible challenges posed by sample interferences, the stability of live cells, the presence of unwanted hydrocarbons in the mixture, and the specific growth requirements of bacteria in the sensor, Voon et al. (2022) lay out a promising approach for the field of biodegradation by bridging the knowledge gap between traditional methods and biosensor technology by introducing the idea of a bioluminescent bacteria-based WCB, providing a sensitive and fast analysis alternative.

Hydrogen peroxide is widely used in cosmetic products; however, it is a strong oxidizer and can be toxic when overexposed (Nelson & Porter, 2023). Wen et al. (2023) utilized electroactive bacteria, which naturally oxidize the colorless compound tetramethylbenzidine (TMB) only when hydrogen peroxide is present, resulting in color development (Wen et al., 2023). This method demonstrates the application of WCBs to detect hydrogen peroxide. In agriculture, WCBs are utilized to detect plant diseases, measure soil fertility, monitor pesticide residues, and analyze agricultural samples for quality control. These biosensors significantly contribute to agriculture and environmental management by improving agricultural techniques, reducing chemical inputs, and promoting sustainable farming practices.

4.2.1.3   Other Applications of WCBs

WCBs have emerged as powerful tools, revolutionizing diverse sectors from biomedicine, environmental monitoring, and food safety to sustainable energy production. In the food industry, WCBs can be applied to detect pathogens, toxins, allergens, and potential spoilage, thereby ensuring food safety. These biosensors also play a crucial role in developing and optimizing biofuel production. Moreover, they find widespread application in monitoring and regulating various industrial processes. Industries such as pharmaceuticals, chemicals, and biotechnology utilize WCBs to evaluate the synthesis of specific compounds. These biosensors contribute to increased yield, efficiency, and sustainability by monitoring critical enzymes involved in metabolic pathways.

One of the challenges natural enzymes face is their low activity, resulting in a low yield. C. Chen et al. (2023) developed a high-throughput screening (HTS) system based on a WCB to address this issue (C. Chen et al., 2023). The biosensor's ability to rapidly respond to environmental changes, diverse signal-transducing systems, and high sensitivity make it an excellent candidate for HTS applications. The study focused on myrcene synthase, an enzyme involved in the biosynthetic pathway of β-myrcene, an acyclic monoterpene used to produce elastic polymers and renewable surfactants. The novel biosensor constructed in this study facilitated the screening of enzymes with high catabolic activity, including enzyme variants by a genetic circuit that is based on the myrcene sensing promoter (Pmyr) and MyrR gene cloned under the Ptrc promoter, as shown in Figure 4.2C. Myrcene binds with MyrR to induce a conformational change, which triggers the MyrR binding to Pmyr, which leads to sfGFP fluorescence.

Another challenge in enzyme research is that many potential enzymes capable of catalyzing reactions in biorefineries still need to be discovered due to the difficulty in cultivating most microorganisms. Ho et al. (2018) designed a one-component biosensor that is a valuable tool for metagenomic screens (Ho et al., 2018). Their study aimed to identify candidate enzymes for lignin degradation through high-throughput screening based on a GFP-based WCB.

4.2.1.4 Challenges, Current State, and Future Perspectives

Effective WCBs need to exhibit selectivity and sensitivity. Addressing cross-reactivity and background noise and ensuring responsiveness to the target analyte is crucial. One significant challenge is converting biological responses into observable signals. While creating efficient and dependable signal transduction pathways is essential, it is intricate to maintain cellular integrity. Cells must be given specific growth conditions to thrive and perform in complex environments, ensuring biosensor stability (H. Kim et al., 2020). The variability in growing conditions, genetic changes, and detection procedures must be tackled using standardized processes and quality control mechanisms. Moreover, WCBs can be vulnerable to interference from non-target compounds in the sample, potentially leading to false results (Voon et al., 2022).

Another dimension of challenges pertains to regulatory hurdles. Biosensors employing genetically modified organisms (GMOs) or those used in specific sectors face regulatory constraints. Complying with safety, ethical, and approval laws is a time-intensive process (S. Chen et al., 2023; Jaiswal & Shukla, 2020; Saltepe et al., 2018). Addressing these multifaceted challenges demands expertise in molecular biology, genetics, biochemistry, engineering, and analytical sciences. Overcoming these barriers is essential to delivering robust, reliable, and cost-effective WCBs for diverse applications.

Different types of WCBs exist in many forms, each with unique advantages (S. Chen et al., 2023). A list of these forms is shown in Figure 4.3. Suspended cultures comprise cells that freely interact with their environment, enabling accurate response monitoring. In contrast, immobilized cells, anchored to solid supports, offer stability and ease of handling. Biofilms, complex colonies of microorganisms,

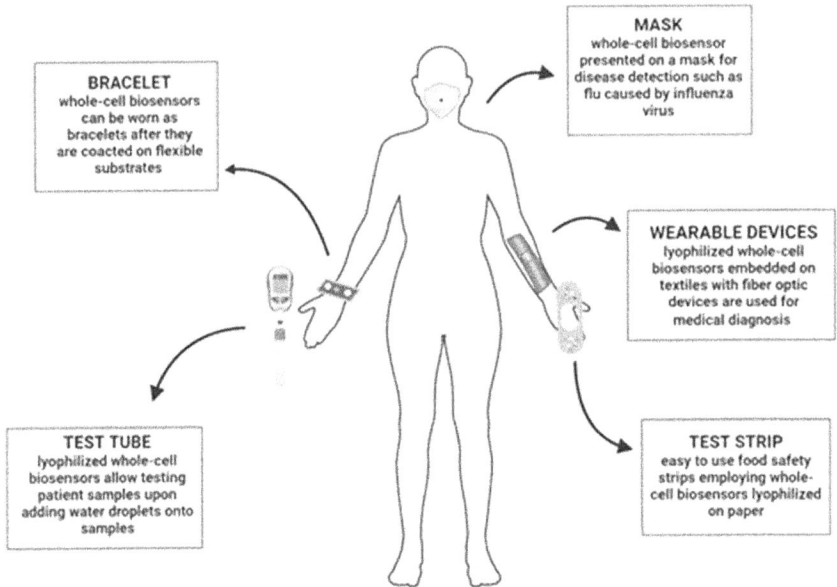

FIGURE 4.3   Presentation forms of WCBs, adapted from S. Chen et al. (2023). Created with BioRender.com.

provide increased stability and continuous sensing. Encapsulated cells, protected within semi-permeable barriers, maintain cellular response in controlled environments (Lu, Ye, Zhang, & Sun, 2021). Furthermore, certain organisms, like bioluminescent bacteria, can act as biosensors, giving quantitative indications (S. Chen et al., 2023). The target analyte, desired sensitivity, stability needs, and environmental contexts influence the choice of biosensor form.

4.2.2   CELL-FREE EXPRESSION SYSTEM-BASED BIOSENSORS

A cell-free protein synthesis (CFPS) system is used as an in vitro approach to study biological processes in a setting devoid of cellular components. This technique is frequently used to examine biological processes outside a fully developed cellular setting, simplifying complex relationships frequently found in a whole cell arrangement (Copeland, Langlois, Kim, & Kwon, 2021). The cell-free system can be categorized into two main types: The crude cell-extract system that utilizes extracts derived from whole cells by removing their internal components for external applications or the purified enzyme system that relies on purified enzymes and components known to participate in specific processes (Figure 4.4) (Dudley, Karim, & Jewett, 2015).

The crude cell lysate (extract) plays a crucial role in the system's functionality as it is the primary source of the cell-like environment in the microtube reaction and dictates the system's features. The removal of endogenous genetic material and

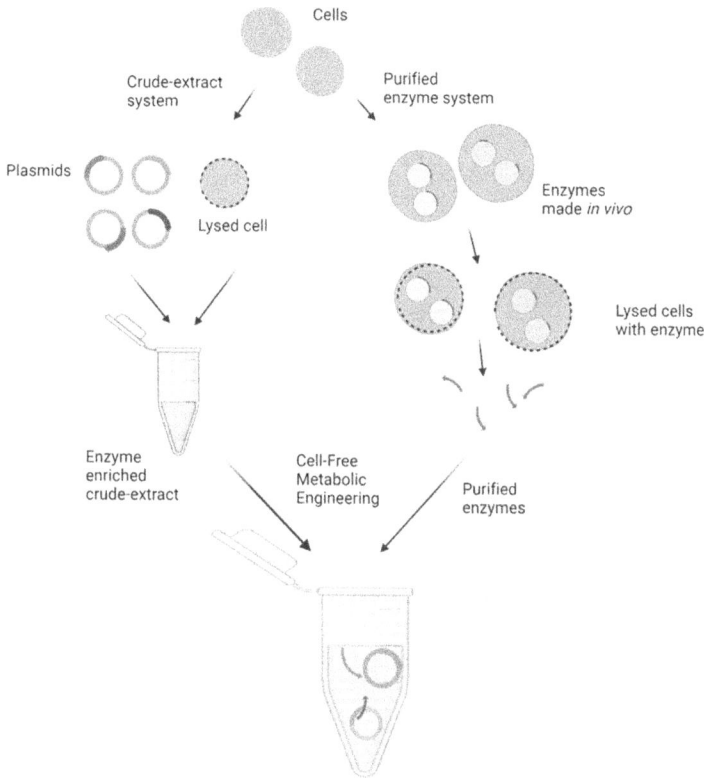

FIGURE 4.4 The crude extract and purified enzyme systems are two distinct cell-free metabolic engineering techniques. When cytotoxic chemical synthesis is the goal, dynamic process control is needed. When immobilization, encapsulation, or any type of carefully controlled spatial arrangement is of interest, purified cell-free systems are an especially effective strategy. Since membrane disintegration is the main cause of toxicity in bacteria, enzymes can tolerate far greater solvent titers than microbes can (Rollin et al., 2013; Schmidt-Dannert & Lopez-Gallego, 2016) Adapted from Copeland et al. (2021). Created with BioRender.com.

the cell wall, as well as the inclusion of supplements to enhance gene transcription and translation, are the fundamental components of the CFPS preparation, together with the cell extract. Thus, the burden of certain bacterial growth conditions, which restrict the platform expandability of live cells, may be avoided while synthesizing the protein (Copeland et al., 2021).

Synthetic biology has recently integrated cell-free protein synthesis (CFPS) systems because they can be used to prototype biological circuit architecture and create biosensors quickly. The basis for CFPS-based biosensors is to introduce more complicated genetic circuits laid through CFPS systems, carrying out biological transcription and translation (TX-TL) in an open environment, and demonstrating several unique characteristics superior to cellular systems. Their advantages over cell-based systems can be listed as (Brookwell, Oza, & Caschera, 2021; L. Zhang, Guo, & Lu, 2020):

- Biological noise-free: The absence of biological noise is a crucial benefit of cell-free systems over cellular ones, making them more efficient in controlling genetic circuits.
- Usefulness: Cell-free protein expression needs little laboratory space and equipment and is simple to utilize outside of a laboratory setting.
- Increased efficiency and adaptability: Cell-free systems are advantageous for biotechnology applications because of their higher efficiency and adaptability and lower cost when compared to in vivo systems.

Biosensors are analytical instruments for detecting several trace-level analytes that utilize biological components as their primary functional elements. As a basic approach, most cell-free biosensors are designed with the reporter gene and recognition mechanism chosen depending on the analytes. For cell-free expression, the encoded gene is then cloned into the vector. Additionally, mature cell-free biosensors were developed using the proper method for cell-free protein synthesis and the coding template DNA. Metal ions, pharmaceuticals, viruses, and other substances like amino acids can all be spotted using cell-free biosensors. (Voyvodic & Bonnet, 2020) (Figure 4.5).

The workflow of a cell-free biosensor may be separated into two parts: The production of readable signals via reporter gene expression and the recognition and

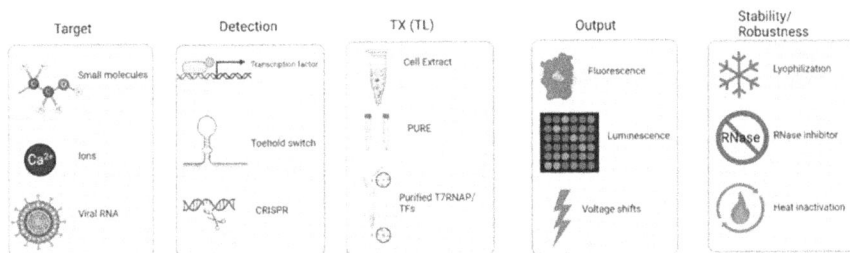

FIGURE 4.5 Cell-free biosensors use different techniques and technologies to make detections. Small compounds, ions, viral RNA, and bacterial genomic DNA may all be detected using sensors. Techniques for detection include CRISPR-based RNA recognition, RNA-based ribosome-binding site sequestration in toehold loops, and transcription factor activation/repression. Cell extract, the whole transcription and translation system, or selectively pure polymerases and transcription factors can all be used for transcription and/or translation [TX(TL)]. The signal output can then be seen as voltage shifts, observable color changes, fluorescence, or luminescence. Finally, before detection, clinical samples might be lyophilized, added with RNase inhibitors, or heated to boost the resilience of cell-free sensors and transcription factors. Adapted from Voyvodic and Bonnet (2020). Created with BioRender.com).

response of analytes by specific identification mechanisms. When introducing analytes, the cell-free system's recognition process is controlled by the particular chemical structure and binding. Next, the downstream reporter gene initiates the translation of reporter proteins. In addition to the often-employed optical signal, the detection findings can also be generated by electrochemical signals depending on the presentation method of various reporting proteins. For instance, Lopreside, Wan, Michelini, Roda, and Wang (2019) presented a design where particular DNA sequences may be recognized and electrochemical signals can be produced using nanoscale microelectrodes in cell-free fluids (Lopreside et al., 2019).

Many molecular structures have been created as recognition elements as a result of the constraints placed on the development of various analytes, such as the low selective ability of enzymes and antibodies. The main goal of the materials that follow is to provide an introduction to the recognition processes used by cell-free biosensors based on multiple molecular structures, including transcription factors (TFs), CRISPR-Cas components, toehold switches, and adaptors. For instance, the unique structure of a TF and its control and timing of gene expression often allow the integration of TF-based biosensing systems in cell-free platforms. As a simple example of a cell-free-based construction of a biosensing system, the TFs, MerR and BlcR were employed by Gräwe et al. (2019) to construct cell-free biosensors for the detection of mercury ions and gamma-hydroxybutyrate. They also created a 3D-printed box setup with filters compatible with mobile phone diagnostics to broaden the diagnostic platform for point-of-care applications (Gräwe et al., 2019). Another method involves using cell-free systems to examine how antibiotics that interfere with ribosome function affect the expression of a reporter that is consti-tutively expressed, in order to identify them. In addition to detecting paromomycin spiked into environmental water samples, Duyen et al. (2017) added paromomycin, tetracycline, chloramphenicol, and erythromycin to cell-free reactions in their puri-fied cell-free system, which they call PURE. Proteases and ribonucleases, which are frequently present as unwanted molecules in cell extracts, are, therefore, totally absent from the protein synthesis process of the PURE system. Additionally, the system eliminates membrane contamination of the reaction system. They detected each at concentrations between 0.5 and 1 g/mL at a reaction pH between 6 and 10 (Duyen et al., 2017).

4.2.2.1 Environmental Monitoring Applications of CFPS-Based Biosensor Systems

The most common harmful substances studied by CFPS-based biosensors are heavy metals, such as arsenic and mercury, as well as toxic small molecules and antibiotics. For instance, a paper-based mercury biosensor was constructed by the Beijing 2010 iGEM team (Jung et al., 2020). Their design improved the cell-free mercury sensor by modifying the plasmid concentrations and introducing fresh translation enhancer sequences, which raised the CFPS reaction yield and made it possible to quantify the analytes precisely. It can detect a mercury content of 6 g/L. The team extended the detection by creating a 3D-printed box that can be linked to a smartphone for real-time readings.

As an example of antibiotic detection from environmental sampling, a straight-forward biological sensor based on colorimetric paper was created by Duyen et al. (2017). The study aimed to identify the ranges of temperature and pH where a paper-based biosensor might perform well. At temperatures of 4°C, 15°C, 25°C, and 37°C, they tested in vitro transcription-translation (IVTT) processes using freeze-dried paper discs. At 25°C and 37°C, the color changed after two hours of incubation, but at 15°C, it took 12 hours and 24 hours, respectively, for the color to change entirely. Even after 24 hours at 4°C, the paper discs still had their original yellow tint. This implies that the paper-based biosensor may function between 15°C and 37°C, enabling usage in various environments without a constant-temperature incubator. The study also examined how pH affected the biosensor's reaction and discovered that water samples with pH levels between 6 and 10 showed a change in hue from yellow to purple. This shows that water samples with a pH range often encountered in practical applications may be utilized with the biosensor. The detection limits for paramomycin, tetracycline, erythromycin, and chloramphenicol were 0.5, 2.1, 0.8, and 6.1 g mL−11, respectively. However, this sensor lacks precision since most antibiotics are used directly to limit bacterial protein synthesis; hence, any variables impacting TX-TL will impact its accuracy.

A different cell-free biosensor for detecting tetracycline hydrochloride (Tc-HCl) was constituted by Pellinen, Huovinen, and Karp (2004). Plasmid pTetluc1, which contains the Tet operator/promoter upstream of the WreXy luciferase gene, was employed for Tc-induction assays. In the Tc tests, a reaction mixture consisting of S30 extract, reaction buffer, a combination of TetR and pTetluc1, and water samples that had been spiked with Tc was incubated for 30 minutes before the luciferase assay buffer was added, and bioluminescence was measured. Tc-HCl has been found in amounts lower than 10 ng mL−1. Tc-HCl has a maximum residue limit of 100 ng mL−1 in EU milk samples and a limit of 30 to 500 ng mL−1 in commercial microbial inhibition tests. Notably, the ppb threshold, which has become the environmental standard for detecting various contaminants, may be attained by the detection limit of cell-free biosensors. The test outcomes may be used to determine whether or not the tested samples adhere to the standards of unique PoC testing of CFPS-based systems. Despite the sensitivity of existing cell-free biosensors, which can reach a certain level, some compounds still exist or may develop with lower detection limits, necessitating ongoing optimization and improvement of the cell-free biosensors, as shown through the direct barrier of antibiotics detection.

Another advantage of CFPS-based biosensing is the detection of small molecules or indicators in a selective and concentration-dependent manner. In order to con-struct the nematode olfactory receptor ODR-10 for detecting diacetyl, to which long-term exposure by humans causes impaired lung function and airway disease, Chen et al. (2019) used an *Escherichia coli* extract cell-free system (Chen et al., 2019). They then functionalized it by attaching a His6-tag and anti-His6-tag aptamer to the surface of the electrolyte-insulator-semiconductor sensors. They demonstrated selectivity against 2,3-pentanedione, butanone, and isopentyl acetate and a concen-tration-dependent linear response to diacetyl from 0.01 nM to 1 nM.

Another chemical threat agent studied through the CFPS system is pesticides, which endanger the balance of the ecosystem and are harmful to human health if they come into the body from the outside. A cell-free biosensor was created by Silverman et al. (2020) to identify the herbicide atrazine. They paired a cell-free metabolic pathway that breaks down atrazine into CYA (cyanuric acid) over three stages in a one-pot reaction with a cell-free sensor for CYA, using a modular approach. A modular approach in cell-free systems refers to the design and construction of the system as a collection of distinct, independent modules or components, each with a particular purpose. Greater flexibility, scalability, and maintainability are all made possible by this method, along with the more straightforward incorporation of new parts or features. The total system may be developed, produced, and maintained more effectively and efficiently by dividing it into smaller, more manageable elements. The sensor can recognize large quantities of atrazine (10-100 μm) (Silverman et al., 2020). The quick cell-free detection of pesticides, other water pollutants, and environmental biomarkers became possible by combining metabolism and biosensing through a cell-free platform.

These illustrations show that using cell-free biosensors to detect environmentally hazardous compounds has become effective in recent years. Cell-free biosensors may identify pollutants, set the safety limit of analysis samples and keep an eye on the environment for timely prevention or repair due to their low detection limit and quick and handy qualities. The potential for cell-free biosensors in environmental monitoring is vast because of ongoing optimization.

4.2.2.2 Biomedicine Applications of CFPS-Based Biosensor Systems

The ability of cell-free biosensors to immediately determine the diagnosis outcome, which is crucial in biomedicine, is made possible by their sensitivity and quick reaction times. The current cell-free biosensors primarily concentrate on detecting bacteria, infections, and carcinogens that are therapeutically important.

Endocrine-disrupting chemicals (EDCs) are prevalent at high concentrations in the environment, food, and personal care items. Exposure to EDCs can potentially cause acute and chronic diseases, including cancer and diabetes. In order to detect endocrine xenoestrogens (XEs, one form of EDCs) in human blood and urine, Salehi et al. (2018) developed a human estrogen receptor (hER)-CFPS biosensor. These test samples (blood and urine) may partially interact with the cell-free system due to their complexity. They could alter the precise calibration of protein production processes, rendering the test method ineffective. By doing precise sample measurements before analysis, this problem may be solved. Inhibitors of the RNA enzyme, which can diminish the impact of urine or blood samples on the CFPS system, were added to the samples. By simplifying the detection process, these changes might increase the utility of cell-free biosensors and be used with more intricate analytes.

The one class of pathogens that requires early and quick detection are viruses that spread quickly, and certain strains are highly contagious. Some cell-free biosensors in this situation have viral detection capabilities. A rapid diagnostic biosensor based on a toehold switch that is cell-free was created by Pardee et al. (2016) to identify the Zika virus in a viremic macaque's plasma. By using isothermal amplification, they increased the sensor's sensitivity even further. Using the NASBA (nucleic acid sequence-based amplification)-CRISPR cleavage assay, the sensor can discriminate between several strains (Zika and dengue). Other viruses, including the Norovirus, SARS-CoV-2, and the Ebola virus, may all be detected using cell-free biosensors. In other words, cell-free biosensors may identify numerous viral kinds in addition to the presence or absence of a specific virus. Also, Koksaldi et al. (2021) have created a novel riboregulator system to detect specific genomic regions of the SARS-CoV-2 virus. This system, developed from scratch, utilizes in vitro synthetic biology tools and has been optimized to detect the presence of the virus in patient samples. The successful development of this diagnostic platform shows promise for quick and portable detection without the need for expensive equipment or specialized laboratory settings. This diagnostic method may use a very sensitive microplate reader to identify particular viral genome regions within 60 minutes, according to their kinetic analysis of the cell-free systems. However, visual identification takes around two hours to be made with the unaided eye under blue light illumination and via a clear orange filter. This shows that the technology may detect viral genetic material quickly and accurately, providing a viable method for diagnostic applications.

The benefits of cell-free biosensors for clinical sample analysis and identifying illnesses are clearly demonstrated by these examples. They can be employed for the early diagnosis of disease. For the purpose of observing the spread of disease, they may also respond rapidly to viral outbreaks or the appearance of genetic variants and mutations. Despite the fact that these technologies have made tremendous progress in environmental detection and medical diagnosis, there is still potential for development in commercial uses of cell-free biosensors, which should be the primary objective of current research.

4.2.2.3 Challenges and Future Trends

Cell-free biosensors have demonstrated considerable benefits in various domains, including environmental detection and medical diagnostics. Cell-free biosensors will take some time to become widely used in the industry, for worldwide monitoring, and used independently by the general population.

Strong chemical tolerance and rapid reaction times are benefits of an entirely open cell-free environment. However, it also has the drawback of being susceptible to interference from other compounds in the sample. In contrast to a cell-based biosensor, a cell-free biosensor loses a natural physical barrier due to the absence of a cell membrane. Salehi et al. (2017) tested a cell-free transcription-translation system in various samples of untreated water, untreated sewage, and human body fluids. Their findings demonstrated that, in comparison to high-purity laboratory samples, CFPS activity was reduced in all samples. Because human urine contains a large amount of protein denaturing urea (approximately 280 mM), the activity of CFPS was dramatically reduced (Salehi et al., 2017). Therefore, one of the main objectives of future developments is to avoid possible inhibitory effects and increase the robustness of cell-free biosensors in complex samples.

For transcription and translation, the majority of cell-free biosensors depend on cellular extract systems, which include template DNA, cellular crude extract, and different inorganic ions. Trehalose was used by Karig, Bessling, Thielen, Zhang, and Wolfe (2017) to keep cell-free components during high-temperature drying in an oven. As an outcome, they were capable of enduring temperatures as high as 37°C for several months without compromising sensor functionality. Additional tries have been made to categorize cell-free components into hydrogels, protein-based structures, and polymer substrates, in order to assure the stability of sensor systems. The complete enhancement of the efficiency and stability of cell-free biosensors remains one of the most significant directions for future development.

4.3 SYNTHETIC BIOLOGY-BASED IN VITRO METHODS FOR DETECTING BIOLOGICAL THREAT AGENTS

4.3.1 TOEHOLD SWITCH-BASED BIOSENSORS

The RNA-based regulatory system, particularly riboswitches and toehold switches, has attracted research attention, leading to the development of platforms for biological detection applications, including biosensing and molecular diagnostics. A riboswitch is a transcriptional or translational level cis-regulatory part of gene expression. The non-coding section of mRNA can attach to ligands and change structure, which in turn regulates how genes are expressed. By base-pairing with target RNA sequences, riboregulators, frequently referred to as toehold switches, control gene expression in translation and transcription. The original Isaacs et al. (2004) riboregulator design has been utilized to detect RNAs regulating gene translation. Afterwards, in 2014, Green, Silver, Collins, and Yin identified a new riboregulator called a toehold switch, opening the way for its application in molecular diagnostics. The research team demonstrated with outstanding accuracy and precision how toehold switches may be employed to identify a number of lethal viruses, including Ebola and Zika, from their RNAs. (Pardee et al., 2014, 2016).

When typical riboregulators lock up the RBS sequence, translation halts. The target mRNA's 5′ end cis-repressor complementarily binds to its RBS region to halt translation. In order to allow for translation, the cis-repressor detaches from the trans-activator RNA. Because of the inhibition caused by RBS binding, only a limited number of trigger sequences may be used (Geraldi, Puspaningsih, & Khairunnisa, 2022).

The toehold switch was created by Green et al. (2014) as a potential improvement. The specific "trigger RNA" is detected using the toehold switch (Figure 4.6). The RBS and start codon (AUG) in the toehold switch are located in a loop and a bulge, respectively, of an RNA hairpin. The RNA hairpin has a single-stranded toehold domain that is complementary to the trigger RNA upstream and a distinct protein-coding sequence downstream. Once the switch's toehold domain has made connection with the trigger RNAs, the strand displacement process will get going. This enables translation by releasing the RBS from the loop and opening the hairpin. Target trigger RNAs can be optically recognized in a similar way to riboswitch-based

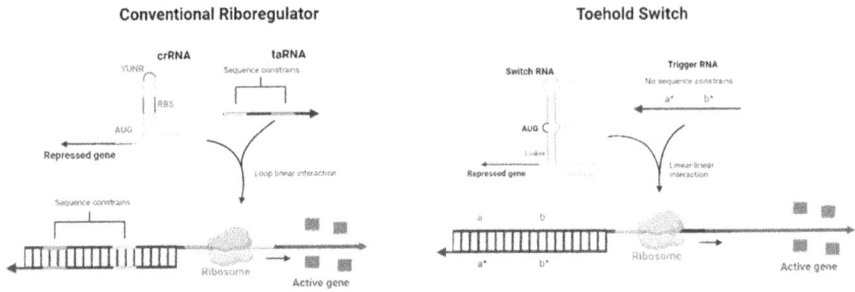

FIGURE 4.6 Typical riboregulator and toehold switch design schematics. Toehold switch systems are made up of two RNA strands called the switch and the trigger. The switch RNA contains the gene's coding region, which is under regulatory control. A typical 21 nt linker sequence that codes for low-molecular-weight amino acids that have been added to the target gene's N terminus follows this coding sequence. A hairpin-based processing module upstream of this coding region has a start codon and a strong RBS. At position 50 of the hairpin module, the trigger RNA strand first binds to a single-stranded toehold sequence. When connected to the hairpin, the trigger molecule's expanded single-stranded section completes a branch migration process, revealing the target gene's RBS and start codon and initiating translation. Adapted from Green et al. [18]. Created with BioRender.com.

biosensors by adding a reporter gene (such as GFP) in the downstream region of the toehold switch hairpin. The toehold switch has features that might be used as molecular diagnostics equipment. With its capabilities, the toehold switch might be used as a molecular diagnostics tool. The toehold switch isolates the area near the start codon, in contrast to the usual riboswitch. Since the RBS and start codon are not linked, there are several potential toehold and trigger sequences. (Green et al., 2014).

There are several advantages of integrating toeholds into biosensing applications. Firstly, toehold switches have a high sensitivity and specificity, making them practical for detecting various targets, such as nucleic acids and proteins. Toehold switches are also less expensive than other biosensors, making them appropriate for application in environments with limited resources. Next, toehold switch-based diagnostic tools can be used in regions with few medical resources since they are stable for long-term preservation at room temperature (Geraldi et al., 2022; Green et al., 2014). The main drawbacks of prior riboregulator designs, such as poor dynamic range, orthogonality, and programmability, are addressed via toehold switches. Toehold switches are more advantageous from a kinetic and thermodynamic standpoint than conventional riboregulators because they use linear-linear interactions rather than loop-loop and loop-linear interactions. Toehold switches have typically demonstrated high protein expression regulation with a dynamic range of nearly two orders of magnitude (Hoang Trung Chau et al., 2020). Toehold switch trigger RNA sequences may be freely created and modified for various purposes. The use of toehold switches in biosensing and molecular diagnostics benefits from their versatility. For instance, toehold switches have been utilized to detect microRNA in mammalian cells and several pathogenic viruses (Ma, Shen, Wu, Diehnelt, & Green, 2018;

Meagher, Negrete, & Van Rompay, 2016). The conventional architecture of a synthetic diagnostic system is made up of three modules: A sensor module, a processing module, and a reporting module. The trigger RNAs (sensing module) and the gene of interest (reporting module) in the toehold switch may be freely changed and included as parts of systems with greater complexity.

Computational techniques are utilized to build and validate the trigger RNA sequence collection, which then serves as a model for toehold domain design, after defining the fundamental toehold switch secondary structure, preserved sequences, and interaction domain sizes. In general, the design of a suitable toehold switch should consider the following elements for any particular trigger RNA sequence. To avoid leaky expression in the absence of the trigger RNA, the switch must, first and foremost, have a stable conformation (hairpin loop). Second, it should be useful to unwind the hairpin loop for activation, expression, and translation of the downstream gene of interest upon exposure to the energy state of the toehold switch-trigger RNA duplex. Third, for complete translation of the downstream gene of interest, the sequences used to generate the toehold switch should not have in-frame stop codons. Finally, to avoid unwanted off-target interactions, the sequence of the trigger RNA sequence recognized by the RNA toehold switch should be specific to the RNA toehold switch (Auslander & Fussenegger, 2014; Liang, Bloom, & Smolke, 2011).

The Nucleic Acid Package (NUPACK) program, an application of software suitable for the construction and analysis of nucleic acid systems, has been used to create a number of toehold switch-based devices. The tool is used to calculate the predicted minimum free energy (MFE) secondary structures of the screened toehold switches and the switch-trigger RNA complex structures, as well as their structure/sequence features. These energy functions were described by Serra and Turner (1995) and Mathews, Sabina, Zuker, and Turner (1999). The toehold switches are identified using the calculated ensemble defect values, the free energy of MFE secondary structures of the particular RNA sequences, and the switch-trigger RNA complexes. NUPACK was effectively employed by Green et al. (2014) to design and screen the original toehold switches. They then used the design constraints to construct libraries of potential riboswitches, and then used the Monte Carlo simulation to produce a subset of riboswitches with strong orthogonality. The study team carried out a number of tests to evaluate the toehold switch activity critically and examine elements connected to the design that could influence toehold switch activities. (Green et al., 2014).

4.3.1.1 Toehold-Based Biosensing Applications

When it first expanded over West Africa between 2013 and 2016, the Ebola pandemic posed a major threat to human life. Pardee et al. (2014) created a portable biosensor that could swiftly detect the RNA of the Ebola virus in vitro through the use of a cell-free method on paper, based on an earlier biosensor that could detect a specific RNA in Ebola. The created toehold switches showed efficient Ebola RNA detection, and after 60 to 120 minutes of incubation, the maximum fold change in absorbance was reached. The Ebola toehold switches' high level of strain discrimination is further demonstrated by their ability to distinguish among the Sudan and Zaire strains, which only differ in size by three nucleotides. The toehold switch-based tool also showed a 30 nM threshold of detection for trigger RNA (Pardee et al., 2014).

For paper-based synthetic gene networks, toehold switches were also employed. These switches were employed to activate previously inactive GFP constructs with T3 or T7 promoters and activate the expression of a toehold switch and trigger pair controlled by T7 and T3, respectively. Paper-based sensors that recognize full-length active mRNA targets were also developed using toehold switches. The sensors produced fluorescence induction in the presence of GFP and mCherry mRNAs because they were made to target sequences that were probably accessible for binding. A year after the Zika virus significantly damaged Brazil at the start of 2015, an identical approach was used in genetic testing. A toehold switch-based sensor was developed using the same cell-free paper-based substrate as this device. The CRISPR-Cas9 system's specificity and sensitivity are improved using the NASBA isothermal amplification technique. The technique can distinguish between the RNA sequences of Zika with dengue and two forms of the Zika virus, the African and American Zika strains, thanks to CRISPR's selective capacity to cleave. Without the addition of thermocyclers, the technique can also detect the Zika virus in plasma samples from macaque monkeys at a concentration of 2.8 fM (Pardee et al., 2016).

Additionally, for sequence-specific Zika virus detection, the scientists created a collection of 48 toehold switch sensors. An algorithm for in silico design was tweaked to create these sensors. The toehold switch constructions from DNA templates were amplified and joined to the lacZ reporter gene to create the toehold switch sensors. The amplified sensor DNA and trigger RNA, which code for a complementary section of the Zika virus genome, were combined with freeze-dried, paper-based processes to verify the sensors. A portable electronic reader was used to keep track of the sensors' activation. The efficiency of the authors' technique was shown by the successful detection of sensor 32B activation from a diagnostic workflow started with live Zika virus. Additionally, using the freeze-drying technique and embedding the synthetic system into the paper improves its mobility (Pardee et al., 2016).

Integrating toehold switches into traditional methods can be performed with their inclusion into surface plasmon resonance (SPR) technology. SPR technology is suggested as a biosensing technique for sensitive and label-free DNA detection. Ding et al. (2017) use a nonlinear hybridization chain reaction (HCR) technology, which amplifies signals without the need for enzymes or labels. The technique entails attaching capture probes to a gold sensing chip. When the target DNA partially combines with the capture probes, the unpaired fragment of the target DNA triggers the nonlinear HCR. High molecular weight dendritic DNA nanostructures are produced due to the HCR system. Upon introducing the nonlinear HCR system, the SPR signal is shown to be amplified. The toehold number of the substrates and the nonlinear HCR reaction time are a few of the experimental parameters that the researchers improve. They create a calibration curve by identifying the target DNA at various concentrations and assessing the analytical performance of the biosensing technique. The findings show a strong linear correlation between target DNA concentrations in the 1 pM–1 nM range and SPR signals (Ding et al., 2017). The predicted detection threshold is 0.85 pM. By recognizing diverse DNA sequences, including target DNA, single-base mismatch oligonucleotides, double-base mismatch oligonucleotides, and non-complementary oligonucleotides, the specificity of the biosensing approach is evaluated. The method shows significant potential for single nucleotide polymorphism analysis and exhibits excellent specificity in determining target DNA. Investigations are being made on the biosensing strategy's repeatability and reusability. The technique has high repeatability, and the analytical performance may be maintained after at least 50 reuses of the sensor chip (Ding et al., 2017).

4.3.1.2   Future Perspectives

The toehold switches are still in their infancy as a commercial product. Their extensive development has made them useful in several domains, including molecular diagnostics, synthetic biology, and metabolic engineering. Toehold switches, however, still need a fully established system that may be used in large-scale applications, which opens up various opportunities for future research and development.

A significant turning point in the evolution of the riboregulator, the toehold switch opened the door for more extensive and in-depth implementations. Toehold switches are flexible with diverse uses due to their capacity to bind to any RNA sequence, as current research shows. In order to satisfy different system requirements, the toehold switch system may be used with a range of sensor outputs, including electrochemical, fluorescent, and colorimetric outputs. Toehold switches in conjunction with the paper-based platform and the cell-free system provide a straightforward, transportable, and inexpensive technique for molecular diagnostics. These outcomes have an encouraging future for the toehold switch system, particularly the synthetic biological system.

4.3.2   CRISPR-Cas-Based Biosensors for Biological Threat Agents

Fast and apprehensive molecular testing is important for the early identification of infectious diseases. While polymerase chain reaction (PCR) has found widespread use in this regard, it is often limited by the requirement for well-equipped laboratories and skilled technicians. Consequently, the gene-editing capabilities of CRISPR technology make it a promising alternative for PoC detection. The CRISPR-Cas9 technique has been regarded as a revolutionary method for gene editing. This perception could be attributed to the emphasis on the CRISPR/Cas9 toolkit. Our knowledge of the mechanisms behind various CRISPR-Cas systems is expanded by pioneering studies on bacterial adaptive immune systems, which also find new Cas genes with different enzymatic activity. Following those developments, a brand-new area of study has been established that makes use of CRISPR-Cas systems as unique biosensing platforms with unmatched sensitivity and specificity for identifying nucleic acids (Ganbaatar & Liu, 2022).

Pre-CRISPR RNA (crRNA), the guide for the Cas effectors within a CRISPR-Cas system, undergoes transcription from the CRISPR array and subsequent processing to generate mature crRNA. The enzymatic entities endowed with target-dependent cleavage activity are referred to as Cas effectors, which can manifest as either singular

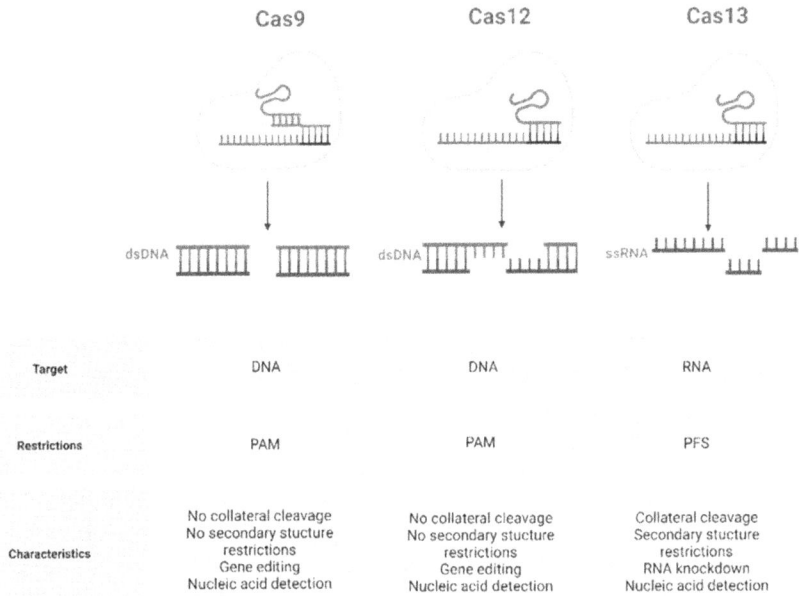

FIGURE 4.7 Comparison of CRISPR-Cas systems. A brief trinucleotide protospacer adjacent motif (PAM) is also required for the first DNA binding, and both target and non-target strands of DNA can be cut by Cas9. Under the control of gRNA, Cas12a is capable of cleaving dsDNA. By binding to crRNA, Cas13 can start cleaving single-stranded RNA (ssRNA). In addition, its second cleavage activity is induced by the target RNA (J. Li et al., 2022)). Adapted from Puig-Serra et al. (2022), Created with BioRender.com.

proteins or intricate protein complexes. While various Cas effectors (Cas9, Cas13, and Cas12) have been harnessed in current CRISPR-Cas-based systems, notable distinctions persist among them (Y. Li, Li, Wang, & Liu, 2019) (Figure 4.7). This technology presents a sensitive nucleic acid detection method with improved accuracy and speed, potentially impacting biosensor performance. Additionally, the development of deployable PoC paper devices utilizing CRISPR-Cas seems plausible.

4.3.2.1 CRISPR-Cas9 Biosensing Systems

The single guide RNA (sgRNA) and the Cas9 endonuclease are the two basic parts of the CRISPR-Cas9 system. The sgRNA is a combination of two RNA molecules, namely the trans-activating crRNA (tracrRNA) and the crRNA (Y. Cui, Xu, Cheng, Liao, & Peng, 2018). When double-stranded DNA (dsDNA) molecules interact with the guide RNA of Cas9, they activate Cas9's cis-cleavage activity, resulting in the cleavage of the targeted dsDNA. It is noteworthy that Cas9 generates blunt ends upon cleavage (S.-Y. Wang et al., 2021).

The CRISPR-Cas9 system can be combined with traditional methods for specific gene detection. In order to discover particular nucleic acid sequences for pathogen genotyping, Zhou's research group has made substantial progress in combining Cas9 cleavage with nucleic acid amplification (Xiong et al., 2021). They have developed the CRISPR-Cas9-mediated lateral flow assay (CASLFA), a technique that enables simple and precise gene detection with remarkable sensitivity. In their design, the Cas9 protein and guide RNA (sgRNA) facilitate the interaction between gold nanoparticles (AuNPs) conjugated with DNA probes and the target DNA. The scaffold sequence that is found in the sgRNA's loop region has been designed to provide a binding site for the AuNP-DNA probes. The AuNP-DNA probes bind to the sgRNA, and they accumulate near the test line on a lateral flow strip. Visual detection is possible due to the accumulation being visible to the naked eye. Initially, the CASLFA version focused on single-gene analysis. However, the research team created a triple-line lateral flow assay (TL-LFA) by combining multiplex reverse transcription-recombinase polymerase amplification (RT-RPA) with CRISPR-Cas9 technology in order to increase its capabilities. This combination enabled the simultaneous analysis of multiple genes, enhancing the assay's utility for multiplex detection.

The enzymatically deactivated Cas9 effector (dCas9) has been used to construct a number of novel CRISPR-Cas-based biosensing devices besides the effector's enzymatic cleavage function. Zhang and co-workers proposed and demonstrated the potential of using dCas9 (catalytically inactive Cas9) as an effective in vitro platform for detecting Mycobacterium tuberculosis (M. tb) DNA (Y. Zhang et al., 2017). The NFluc and CFluc are the N- and C-terminal parts of the firefly luciferase enzyme and they are linked with dCas9 in the Paired dCas9 (PC) reporter system. The NFluc and CFluc halves of luciferase are in close proximity in the presence of the target DNA sequence and identified by a pair of dCas9 that is being directed by the appropriate sgRNAs. This spatial proximity allows for the reconstitution of the complete catalytic activity of luciferase, leading to the generation of luminescence (Y. Zhang et al., 2017). These studies show that the introduction of CRISPR-/Cas9 can potentially overcome the technological constraints of conventional methods, enabling the fast and precise detection of infections caused by pathogens and viruses (H. Wang et al., 2022).

4.3.2.2 CRISPR-Cas12 Biosensing Systems

Cas12 (Cas12a), a member of the class 2 type V CRISPR effector proteins, has a prominent role in this group. A single crRNA molecule controls its activity. Contrary to type II Cas9, Cas12a needs a TTTN PAM. It creates sticky-ended dsDNA cleavage products.

In 2018, a pivotal advancement in comprehending Cas12a's cleavage activity was reported by S.-Y. Li et al. (2018). They unveiled that the CRISPR-Cas12a system possesses the capacity for both cis- and trans-cleavage. J. S. Chen et al. (2018) proposed a DNA endonuclease targeted CRISPR trans reporter (DETECTR), a CRISPR-Cas diagnostic tool, as proof of the idea. DETECTR also has enriched target sequences with an isothermal preamplification step (Figure 4.8). RPA helps minimize the need for complicated and expensive equipment by enhancing the analytical sensitivity of the diagnostic test. Another team (Tsou, Leng, & Jiang, 2019) employed DETECTR for human papillomavirus (HPV) detection, and it was successful in distinguishing between HPV16 and HPV18, the two HPV strains that are the most

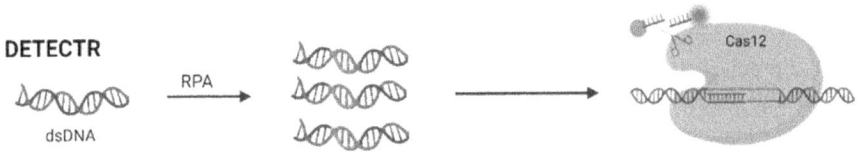

FIGURE 4.8 The general principle of the DETECTR system. DNA is amplified with RPA in the DNA endonuclease-targeted CRISPR trans reporter (DETECTR). The Cas12 system's DNase activity starts when the target single-stranded DNA (ssDNA) pairs with the Cas12 system (Hillary et al., 2021). Adapted from Mustafa & Makhawi (2021). Created with BioRender.com.

pro-oncogenic. Broughton et al. (2020) also use this method to detect SARS-COV-2. Reverse transcription and isothermal amplification utilizing loop-mediated amplification (RT-LAMP) are included in the DETECTR assay. Following the amplification, Cas12 detects a predefined coronavirus. The E (envelope) and N (nucleoprotein) genes from SARS-CoV-2 are the targets of their primer design. Using FAM-biotin reporter in lateral flow assay, the Cas12 detection reaction is visualized within five minutes. Additionally, they used swab samples from the typical human coronavirus and the influenza virus to assess the assay's cross-reactivity. They found no cross-reactivity with other respiratory viruses and then it compares SARS-COV-2. The US Centers for Disease Control and Prevention (CDC) assay and other qRT-PCR assays use similar techniques for sample collecting and RNA extraction, and the DETECTR assay is vulnerable to the same possible drawbacks as these common standard assays. However, DETECTR uses simple reporting formats, like lateral flow strips, and does not necessitate a sophisticated laboratory infrastructure (Broughton et al., 2020).

Researchers have also looked at how extending the 5' end of RNA affects CRISPR detection, which has led to the creation of a CRISPR-Cas12a enzyme system that has been improved. By fine-tuning the crRNA/CRISPR enzyme ratio and buffer conditions, sensitivity was markedly improved, culminating in the development of an exact and sensitive nucleic acid detection technique dubbed "5' eNd EXTension CRISPR (NEXT CRISPR)." To meet the requirements of clinical diagnostics, the NEXT CRISPR method was further integrated with RPA. By fusing RPA/NEXT CRISPR with a lateral flow assay, a sophisticated biosensor that can precisely detect the human papillomavirus (HPV) 16 strain was created (Ganbaatar & Liu, 2022). When the HPV 16 target is present, active Cas12a cleaves the ssDNA-FB probe, resulting in a purple test band. Conversely, when the HPV 16 target is absent, Cas12a cannot cleave the ssDNA-FB probe, resulting in a signal in the control band.

In another study conducted by Gong et al. (2021), they ingeniously employed the RPA-Cas12a system, this time proposing a non-genotyping technique to detect 13 high-risk HR-HPV types simultaneously. It is the first study using the RPA-Cas method to realize multiple HPV testing in a single response. Their innovative test incorporated RPA amplification employing a primer pool followed by a discerning detection process utilizing CRISPR-Cas12a technology. Their findings elegantly emphasize the indispensability of integrating RPA amplification with Cas12a technology. While conventional PCR-based HPV testing may only discern a solitary HPV form within a single reaction, this novel approach successfully identified an impressive repertoire of 13 HR-HPV strains, marking a significant stride forward in HPV detection capabilities.

The CRISPR-Cas12 system is not only used for HPV or other viruses. It is also used for parasitic organisms. Toxoplasma gondii is a parasite protozoan that may cause potentially deadly illnesses such as toxoplasmic encephalitis and extracerebral toxoplasmosis in practically all warm-blooded species, including humans. This parasitic organism has been found in various environments, including water, soil, animals, and clinical specimens. Lei et al. (2022) presented a cutting-edge, portable, and user-friendly method for detecting T. gondii in environmental samples: A one-pot paper-based RPA/CRISPR-Cas12a system. Also, they demonstrate remarkable specificity by effectively distinguishing *T. gondii* from 14 other parasitic species. To fulfill the urgent need for on-site detection of T. gondii, they have devised a combination method that leverages the RPA/CRISPR-Cas12a approach in conjunction with lateral flow strips (Lei et al., 2022).

4.3.2.3 CRISPR-Cas13 Biosensing Systems

The Cas13 protein represents a type VI class 2 CRISPR effector. It recognizes and cleaves RNA targets by recognizing the protospacer flanking site (PFS). Recent research has shown that Cas13a has both cis- and trans-cleavage activity and it is activated by RNA targets, as opposed to Cas12a, which is activated by DNA targets and degrades single-stranded DNA (ssDNA). It has trans-cleavage action against single-stranded RNA (ssRNA) molecules with arbitrary sequences. This unique attribute renders Cas13a more adept than Cas12a at discerning RNA targets. Furthermore, the combination of Cas13a with various signal amplification techniques enables the construction of highly sensitive biosensors, offering enhanced detection capabilities (S.-Y. Wang et al., 2021).

Specific high-sensitivity enzymatic reporter unlocking (SHERLOCK) and DETECTR have both specificity and sensitivity. Also, they do not necessitate costly equipment. Gootenberg et al. (2017) introduced SHERLOCK, a new diagnostic tool built upon the foundation of DETECTR, harnessing the remarkable capabilities of Leptotrichia wadei's Cas13 nuclease. Distinguishing itself from Cas12a, which predominantly recognizes and cleaves DNA, Cas13 exhibits an exclusive affinity for RNA, presenting a distinct detection avenue. Target molecules can be enhanced and sensitivity raised by RPA and if they are present in the sample, the fluorophore and the quencher link is broken because Cas13 identifies them via crRNA and cleaves fluorescent RNA probes indiscriminately (by collateral action). The researchers demonstrated that SHERLOCK has attomolar sensitivity for detecting single nucleotide polymorphisms (SNPs) in DNA and harmful microorganisms such as Zika and dengue viruses (Mustafa & Makhawi, 2021).

To advance the SHERLOCK, SHERLOCKv2 was built on the principle of multiplexed quantitative and highly sensitive detection (Gootenberg et al., 2018) (Figure 4.9). The scientists studied 3 CRISPR-Cas13a and 14 CRISPR-Cas13b enzymes to find ones that could work together on many things simultaneously. They

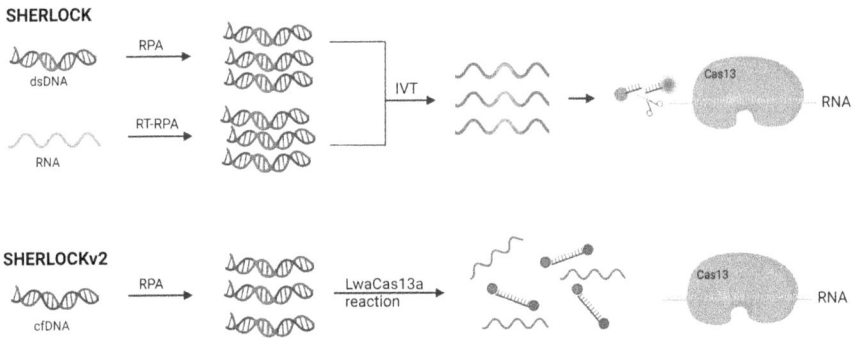

FIGURE 4.9 General schematic for comparison of SHERLOCK and SHERLOCKv2 methods. SHERLOCK targets ssRNA with Cas13 that has been coded with crRNA. RPA and RT-RPA are used respectively to amplify DNA and RNA. DNA is transformed into RNA by T7 transcription, which Cas13 will then process. Cell-free DNA (cfDNA), Cas13, Cas12a, and Csm6 are used in SHERLOCKv2. Naked-eye detection can be achieved by performing a lateral flow experiment on strips (Alamri et al., 2022). Adapted from Mustafa and Makhawi (2021). Created with BioRender.com.

demonstrated their abilities by detecting synthetic dengue virus (DENV) ssRNA and synthetic Zika virus (ZIKV) ssRNA at the same time (Gootenberg et al., 2018). Notably, Gootenberg and colleagues integrated the CRISPR type III effector nuclease Csm6 to enhance the detection signal. SHERLOCKv2 achieved a stunning 3.5-fold improvement in sensitivity by combining Cas13a with Csm6. Furthermore, for quantifiable outcomes, diluted isothermal amplification primers were employed. The true strength of SHERLOCKv2 lies in its capacity to screen multiple sequences, making its use a possibility in clinical or field settings where differential diagnosis of viruses with comparable symptoms is crucial. This achievement was made possible by the discovery that various species' Cas13 exhibited nonspecific transcleavage activities with distinct and skewed preferences for specific sequence motifs. The enzymes were also equipped with orthogonal guide RNA sequences, which efficiently distinguished the multiplex target locations. The SHERLOCKv2 was, unsurprisingly, designed to provide a visual analysis on commercially available lateral flow strips, eliminating the requirement for special equipment. In addition, the superiority of SHERLOCKv2 over its predecessor is highlighted by the fact that the entire reaction may be carried out in a single step without the time-consuming requirement of purifying and isolating nucleic acid (Mustafa & Makhawi, 2021).

Ultimately, Ackerman et al. (2020) invented the Combinatorial Arrayed Reactions for Multiplexed Evaluation of Nucleic Acids (CARMEN) technique, which has revolutionized scalable and multiplexed pathogen detection. Each sample may be examined using this approach against duplicate CRISPR RNA (crRNA) sequences. Leveraging the power of CARMEN-Cas13, they have successfully crafted a multiplexed test capable of rapidly identifying SARS-CoV-2 and all 169 human-associated viruses, with at least 10 published genome sequences identified. The versatility of CARMEN-Cas13 extends even further, allowing for the identification of hundreds of drug-resistance mutations in HIV and facilitating the extensive subtyping of influenza A strains. Utilization of the collateral cleavage activity enables CARMEN-Cas13 to identify Zika sequences with increased sensitivity, reaching attomolar levels. Notably, this sensitivity aligns with that of SHERLOCK and PCR-based assays, underscoring the performance of CARMEN-Cas13. Furthermore, CARMEN-Cas13 benefits from the specificity inherited from SHERLOCK. Through the precise binding and recognition of Cas13 with crRNA, CARMEN-Cas13 achieves sequence-specific identification, allaying concerns regarding off-target amplification that often plague other nucleic acid detection techniques (Ackerman et al., 2020).

4.3.2.4 Future Perspectives of CRISPR-Cas-Based Biosensing Detection of Biological Threat Agents

The SARS-CoV-2 epidemic underlined the importance of the development of nucleic acid tests that are precise, fast, and inexpensive, allowing for the detection of infectious viruses. This testing approach not only proves effective for identifying infections like Mtb, ZIKV, and HPV, but also demonstrates its versatility for diverse infections. CRISPR-Cas-based techniques have emerged as superior alternatives to previous technologies. Their heightened sensitivity, specificity, efficiency, cost-effectiveness, portability, and user-friendliness position them as a remarkable advancement over conventional methods (Huang, Zhang, & Li, 2023).

4.4 CONCLUSION

In the landscape of biosensor development, synthetic biology has been a game-changer in recent years. This discipline has provided powerful tools and methodologies for designing these biosensors with precision. Innovations like standardized genetic components, advanced gene editing with CRISPR-Cas systems, sophisticated promoter engineering, and a variety of platforms with optimized conditions have bolstered the efficiency of biosensor creation. Coupled with this, the integration of high-throughput screening methods and automation has quickened the pace of biosensor discovery, simultaneously facilitating the screening of numerous variants.

Multiplexed biosensors, capable of detecting multiple analytes concurrently, have become a focal point of interest. They offer efficiencies in time, resources, and data coverage. Notably, advancements in signal transduction and reporting have expanded these biosensors' capabilities. These developments encompass improved fluorescent proteins, bioluminescent reporters, and genetically encoded biosensors that cover a more comprehensive dynamic range. Some innovative transduction mechanisms like electrical or acoustic signals pave the way for real-time and non-invasive cellular response monitoring.

Marrying artificial intelligence and machine learning with WCBs has ushered in a new era of performance enhancements and data analysis (Cui et al., 2020). AI algorithms analyze complex data, identify patterns, and offer actionable insights. Artificial neural networks are non-linear signal processing algorithms specializing in processing data beyond what can be done analytically. These networks use various learning methods to assess the data and then train their models to represent it accurately. Recurrent neural networks (RNNs) are a well-known type of such networks. In RNN architectures, the network receives feedback from earlier inputs as a new input. Thanks to this looping process, the network can recall its state, which enables it to fine-tune its weights during training for a more precise data representation. For instance, an innovative approach accompanying this network architecture with biosensing performance analysis has been achieved by our group. The study integrated such a deep neural network with a complex WCB design to predict the final state of the WCB's performance to detect gold ions much more earlier than its standard incubation time which decreased the sensor's response time significantly (Saltepe et al., 2021). Such combination-based works will ensure efficient data processing, real-time decision-making, and predictive analytics under varied conditions in the future.

These recent strides in research enhance performance and broaden application horizons, promising a brighter future for affecting biosensing of biological and chemical threats in multiple domains. Therefore, the integration of synthetic biology with new technologies not only transforms the creation of novel biosensors but also establishes an optimistic path for advanced diagnostic approaches, signaling a new era in environmental surveillance, medical care, and biological safety.

ABBREVIATIONS (IN ALPHABETICAL ORDER)

As: Arsenic, As(II): Arsenic ion, AuNPs: Gold nanoparticles, Bxb1: Bxb1 serine recombinase, CadR: Cadmium-binding transcription factor of CadA, CARMEN: Combinatorial Arrayed Reactions for Multiplexed Evaluation of Nucleic Acids, CASLFA: CRISPR-Cas9-mediated lateral flow assay, Cd: Cadmium, Cd(II): Cadmium ion, CDC: The US Centers for Disease Control and Prevention, cfDNA: Cell-free DNA, CFPS: Cell-free protein synthesis, CFluc: C-terminal half of the firefly luciferase enzyme, CRISPR: Clustered regularly interspaced short palindromic repeats, Cr: Chromium, crRNA: Pre-CRISPR RNA, CYA: Cyanuric acid, dCas9: Deactivated Cas9 effector, DENV: Dengue virus, DETECTR: DNA endonuclease-targeted CRISPR trans reporter, DNA: Deoxyribonucleic acid, dnaK: Promoter of dnaK, dsDNA: Double-stranded DNA, EDCs: Endocrine-disrupting chemicals, EU: European Union, GFP: Green fluorescent protein, GMO: Genetically modified organism, gRNA: Guide RNA, HCR: Hybridization chain reaction, Hg: Mercury, hER: Human estrogen receptor, HPV: Human papillomavirus (HPV), HspR: Heat shock protein transcriptional repressor, HTS: High-throughput screening, iGEM: International Genetically Engineered Machine competition, IR3: Heat shock protein binding motif, LAT: Latex agglutination test, M. tb: *Mycobacterium tuberculosis*, MFE: Minimum free energy, mRNA: Messenger RNA, MyrR: Myrcene regulator, NASBA: Nucleic acid sequence-based amplification, NEXT CRISPR: 5′ eNd EXTension CRISPR (NEXT CRISPR), NFluc: N-terminal half of the firefly luciferase enzyme, NUPACK: Nucleic Acid Package, NarL: Nitrate reductase X, NarX: Nitrate reductase X, NarXL: Nitrate reductase X-nitrate reductase L, Pb: Lead, PC: Paired dCas9, PCadA: Promoter of CadA, PCR: Polymerase chain reaction, PFS:

Protospacer flanking site, PAM: Protospacer adjacent motif, PoC: Point-of-care, Pc: Constitutively active promoter, Pc: Constitutive promoter, Ppb: Parts per billion, PyeaR: Promoter of yeaR repressor, Pmyr: Myrcene sensing promoter, RBS: Ribosome binding site, RNA: Ribonucleic acid, RNN: Recurrent neural network, RT-LAMP: Reverse transcription and isothermal amplification using loop-mediated amplification, RT-RPA: Transcription-recombinase polymerase amplification (RT-RPA), qRT-PCR: Quantitative reverse transcriptase PCR, SARS-CoV-2: Severe acute respiratory syndrome coronavirus 2, sfGFP: Super folder green fluorescent protein, sgRNA: Single guide RNA, SHERLOCK: Specific high-sensitivity enzymatic reporter unlocking, SNP: Single nucleotide polymorphism, SPR: Surface plasmon resonance, ssDNA: Single-stranded DNA, Tc-HCl: Tetracycline hydrochloride (Tc-HCl), TMB: Tetramethylbenzidine, TL-LFA: Triple-line lateral flow assay, TF: Transcription factor, ThsR: Thiosulfate response regulator, ThsS: Thiosulfate sensor kinase, tracrRNA: Trans-activating crRNA. TX-TL: Transcription and translation, WCB: Whole-cell biosensor, XE: Xenoestrogen, ZIKV: Zika virus, Zn: Zinc.

REFERENCES

Ackerman, C. M., Myhrvold, C., Thakku, S. G., Freije, C. A., Metsky, H. C., Yang, D. K., Ye, S. H., Boehm, C. K., Kosoko-Thoroddsen, T.-S. F., Kehe, J., Nguyen, T. G., Carter, A., Kulesa, A., Barnes, J. R., Dugan, V. G., Hung, D. T., Blainey, P. C., & Sabeti, P. C. (2020). Massively multiplexed nucleic acid detection with Cas13. *Nature*, *582*(7811), 277–282. https://doi.org/10.1038/s41586-020-2279-8

Ackland, M. L., & Michalczyk, A. A. (2016). Zinc and infant nutrition. *Archives of Biochemistry and Biophysics*, *611*, 51–57. https://doi.org/10.1016/j.abb.2016.06.011

Akboğa, D., Saltepe, B., Bozkurt, E. U., & Şeker, U. Ö. Ş. (2022). A recombinase-based genetic circuit for heavy metal monitoring. *Biosensors*, *12*(2), 122. https://doi.org/10.3390/bios12020122

Alamri, A. M., Alkhilaiwi, F. A., & Ullah Khan, N. (2022). Era of molecular diagnostics techniques before and after the COVID-19 pandemic. *Current Issues in Molecular Biology*, *44*(10), 4769–4789. https://doi.org/10.3390/cimb44100325

Arduini, F., Micheli, L., Moscone, D., Palleschi, G., Piermarini, S., Ricci, F., & Volpe, G. (2016). Electrochemical biosensors based on nanomodified screen-printed electrodes: Recent applications in clinical analysis. *TrAC Trends in Analytical Chemistry*, *79*, 114–126. https://doi.org/10.1016/j.trac.2016.01.032

Ausländer, S., & Fussenegger, M. (2014). Synthetic biology: Toehold gene switches make big footprints. *Nature*, *516*(7531), 333–334. https://doi.org/10.1038/516333a

Bahadır, E. B., & Sezgintürk, M. K. (2015). Applications of commercial biosensors in clinical, food, environmental, and biothreat/biowarfare analyses. *Analytical Biochemistry*, *478*, 107–120. https://doi.org/10.1016/j.ab.2015.03.011

Banoub, J. H., & Mikhael, A. (2020). Detection of biological warfare agents using biosensors. In *Toxic chemical and biological agents* (pp. 11–46). Dordrecht: Springer. https://doi.org/10.1007/978-94-024-2041-8_2

Bintsis, T. (2017). Foodborne pathogens. *AIMS Microbiology*, *3*(3), 529–563. https://doi.org/10.3934/microbiol.2017.3.529

Brookwell, A., Oza, J. P., & Caschera, F. (2021). Biotechnology applications of cell-free expression systems. *Life*, *11*(12), 1367. https://doi.org/10.3390/life11121367

Broughton, J. P., Deng, X., Yu, G., Fasching, C. L., Servellita, V., Singh, J., Miao, X., Streithorst, J. A., Granados, A., Sotomayor-Gonzalez, A., Zorn, K., Gopez, A., Hsu, E., Gu, W., Miller, S., Pan, C.-Y., Guevara, H., Wadford, D. A., Chen, J. S., & Chiu, C. Y. (2020). CRISPR–Cas12-based detection of SARS-CoV-2. *Nature Biotechnology*, *38*(7), 870–874. https://doi.org/10.1038/s41587-020-0513-4

Carus, W. S. (2015). The history of biological weapons use: What we know and what we don't. *Health Security*, *13*(4), 219–255. https://doi.org/10.1089/hs.2014.0092

Cayron, J., Prudent, E., Escoffier, C., Gueguen, E., Mandrand-Berthelot, M.-A., Pignol, D., Garcia, D., & Rodrigue, A. (2017). Pushing the limits of nickel detection to nanomolar range using a set of engineered bioluminescent *Escherichia coli*. *Environmental Science and Pollution Research International*, *24*(1), 4–14. https://doi.org/10.1007/s11356-015-5580-6

CDC, *Bioterrorism agents/Diseases (by category), Emergency preparedness & response*. (2019, May 15). https://emergency.cdc.gov/agent/agentlist-category.asp

Chen, C., Liu, J., Yao, G., Bao, S., Wan, X., Wang, F., Wang, K., Song, T., Han, P., Liu, T., & Jiang, H. (2023). A novel, genetically encoded whole-cell biosensor for directed evolution of myrcene synthase in *Escherichia coli*. *Biosensors and Bioelectronics*, *228*, 115176. https://doi.org/10.1016/j.bios.2023.115176

Chen, F., Wang, J., Du, L., Zhang, X., Zhang, F., Chen, W., Cai, W., Wu, C., & Wang, P. (2019). Functional expression of olfactory receptors using cell-free expression system for biomimetic sensors towards odorant detection. *Biosensors and Bioelectronics*, *130*, 382–388. https://doi.org/10.1016/j.bios.2018.09.032

Chen, J. S., Ma, E., Harrington, L. B., Da Costa, M., Tian, X., Palefsky, J. M., & Doudna, J. A. (2018). CRISPR-Cas12a target binding unleashes indiscriminate single-stranded DNase activity. *Science*, *360*(6387), 436–439. https://doi.org/10.1126/science.aar6245

Chen, S., Chen, X., Su, H., Guo, M., & Liu, H. (2023). Advances in synthetic-biology-based whole-cell biosensors: Principles, genetic modules, and applications in food safety. *International Journal of Molecular Sciences*, *24*(9), 7989. https://doi.org/10.3390/ijms24097989

Chen, S.-Y., Wei, W., Yin, B.-C., Tong, Y., Lu, J., & Ye, B.-C. (2019). Development of a highly sensitive whole-cell biosensor for arsenite detection through engineered promoter modifications. *ACS Synthetic Biology*, *8*(10), 2295–2302. https://doi.org/10.1021/acssynbio.9b00093

Clark Jr., L. C., & Lyons, C. (1962). Electrode systems for continuous monitoring in cardiovascular surgery. *Annals of the New York Academy of Sciences*, *102*(1), 29–45. https://doi.org/10.1111/j.1749-6632.1962.tb13623.x

Copeland, C. E., Langlois, A., Kim, J., & Kwon, Y.-C. (2021). The cell-free system: A new apparatus for affordable, sensitive, and portable healthcare. *Biochemical Engineering Journal*, *175*, 108124. https://doi.org/10.1016/j.bej.2021.108124

Cui, F., Yue, Y., Zhang, Y., Zhang, Z., & Zhou, H. S. (2020). Advancing biosensors with machine learning. *ACS Sensors*, *5*(11), 3346–3364. https://doi.org/10.1021/acssensors.0c01424

Cui, Y., Xu, J., Cheng, M., Liao, X., & Peng, S. (2018). Review of CRISPR/Cas9 sgRNA design tools. *Interdisciplinary Sciences: Computational Life Sciences*, *10*(2), 455–465. https://doi.org/10.1007/s12539-018-0298-z

Ding, X., Cheng, W., Li, Y., Wu, J., Li, X., Cheng, Q., & Ding, S. (2017). An enzyme-free surface plasmon resonance biosensing strategy for detection of DNA and small molecule based on nonlinear hybridization chain reaction. *Biosensors and Bioelectronics*, *87*, 345–351. https://doi.org/10.1016/j.bios.2016.08.077

Drinking-water. (n.d.). Retrieved September 26, 2023, from https://www.who.int/news-room/fact-sheets/detail/drinking-water

Dudley, Q. M., Karim, A. S., & Jewett, M. C. (2015). Cell-free metabolic engineering: Biomanufacturing beyond the cell. *Biotechnology Journal*, *10*(1), 69–82. https://doi. org/10.1002/biot.201400330

Duyen, T. T. M., Matsuura, H., Ujiie, K., Muraoka, M., Harada, K., & Hirata, K. (2017). Paper-based colorimetric biosensor for antibiotics inhibiting bacterial protein synthesis. *Journal of Bioscience and Bioengineering*, *123*(1), 96–100. https://doi.org/10.1016/j. jbiosc.2016.07.015

Elcin, E., & Öktem, H. A. (2020). Inorganic cadmium detection using a fluorescent whole-cell bacterial bioreporter. *Analytical Letters*, *53*(17), 2715–2733. https://doi.org/10.1080/ 00032719.2020.1755867

Ellison, D. H. (2022). *Handbook of chemical and biological warfare agents: Nonlethal chemical agents and biological warfare agents* (3rd ed., Vol. 2). CRC Press. https://doi. org/10.4324/9781003230564

Frischknecht, F. (2003). The history of biological warfare. *EMBO Reports*, *4*(Suppl 1), S47– S52. https://doi.org/10.1038/sj.embor.embor849

Ganbaatar, U., & Liu, C. (2022). NEXT CRISPR: An enhanced CRISPR-based nucleic acid biosensing platform using extended crRNA. *Sensors and Actuators B: Chemical*, *369*, 132296. https://doi.org/10.1016/j.snb.2022.132296

Geraldi, A., Puspaningsih, N. N. T., & Khairunnisa, F. (2022). Update on the development of toehold switch-based approach for molecular diagnostic tests of COVID-19. *Journal of Nucleic Acids*, *2022*, 7130061. https://doi.org/10.1155/2022/7130061

Gong, J., Zhang, G., Wang, W., Liang, L., Li, Q., Liu, M., Xue, L., & Tang, G. (2021). A simple and rapid diagnostic method for 13 types of high-risk human papillomavirus (HR-HPV) detection using CRISPR-Cas12a technology. *Scientific Reports*, *11*(1), 12800. https://doi.org/10.1038/s41598-021-92329-2

Gootenberg, J. S., Abudayyeh, O. O., Kellner, M. J., Joung, J., Collins, J. J., & Zhang, F. (2018). Multiplexed and portable nucleic acid detection platform with Cas13, Cas12a, and Csm6. *Science*, *360*(6387), 439–444. https://doi.org/10.1126/science.aaq0179

Gootenberg, J. S., Abudayyeh, O. O., Lee, J. W., Essletzbichler, P., Dy, A. J., Joung, J., Verdine, V., Donghia, N., Daringer, N. M., Freije, C. A., Myhrvold, C., Bhattacharyya, R. P., Livny, J., Regev, A., Koonin, E. V., Hung, D. T., Sabeti, P. C., Collins, J. J., & Zhang, F. (2017). Nucleic acid detection with CRISPR-Cas13a/C2c2. *Science*, *356*(6336), 438– 442. https://doi.org/10.1126/science.aam9321

Gräwe, A., Dreyer, A., Vornholt, T., Barteczko, U., Buchholz, L., Drews, G., Ho, U. L., Jackowski, M. E., Kracht, M., Lüders, J., Bleckwehl, T., Rositzka, L., Ruwe, M., Wittchen, M., Lutter, P., Müller, K., & Kalinowski, J. (2019). A paper-based, cell-free biosensor system for the detection of heavy metals and date rape drugs. *PLOS ONE*, *14*(3), e0210940. https://doi.org/10.1371/journal.pone.0210940

Green, A. A., Silver, P. A., Collins, J. J., & Yin, P. (2014). Toehold switches: De-novo-designed regulators of gene expression. *Cell*, *159*(4), 925–939. https://doi.org/10.1016/j.cell.2014. 10.002

Guo, M., Wang, J., Du, R., Liu, Y., Chi, J., He, X., Huang, K., Luo, Y., & Xu, W. (2020). A test strip platform based on a whole-cell microbial biosensor for simultaneous on-site detection of total inorganic mercury pollutants in cosmetics without the need for predigestion. *Biosensors and Bioelectronics*, *150*, 111899. https://doi.org/10.1016/j.bios.2019. 111899

He, J., Zhang, X., Qian, Y., Wang, Q., & Bai, Y. (2022). An engineered quorum-sensing-based whole-cell biosensor for active degradation of organophosphates. *Biosensors & Bioelectronics*, *206*, 114085. https://doi.org/10.1016/j.bios.2022.114085

Hillary, V. E., Ignacimuthu, S., & Ceasar, S. A. (2021). Potential of CRISPR/Cas system in the diagnosis of COVID-19 infection. *Expert Review of Molecular Diagnostics, 21.* https://doi.org/10.1080/14737159.2021.1970535

Ho, J. C. H., Pawar, S. V., Hallam, S. J., & Yadav, V. G. (2018). An improved whole-cell biosensor for the discovery of lignin-transforming enzymes in functional metagenomic screens. *ACS Synthetic Biology, 7*(2), 392–398. https://doi.org/10.1021/acssynbio. 7b00412

Hoang Trung Chau, T., Hoang Anh Mai, D., Ngoc Pham, D., Thi Quynh Le, H., & Yeol Lee, E. (2020). Developments of riboswitches and toehold switches for molecular detection—biosensing and molecular diagnostics. *International Journal of Molecular Sciences, 21*(9), 3192. https://doi.org/10.3390/ijms21093192

Huang, T., Zhang, R., & Li, J. (2023). CRISPR-Cas-based techniques for pathogen detection: Retrospect, recent advances, and future perspectives. *Journal of Advanced Research, 50,* 69–82. https://doi.org/10.1016/j.jare.2022.10.011

Isaacs, F. J., Dwyer, D. J., Ding, C., Pervouchine, D. D., Cantor, C. R., & Collins, J. J. (2004). Engineered riboregulators enable post-transcriptional control of gene expression. *Nature Biotechnology, 22*(7), 841–847. https://doi.org/10.1038/nbt986

Jaiswal, S., & Shukla, P. (2020). Alternative strategies for microbial remediation of pollutants via synthetic biology. *Frontiers in Microbiology, 11,* 808. https://doi.org/10.3389/fmicb. 2020.00808

Jernigan, D. B., Raghunathan, P. L., Bell, B. P., Brechner, R., Bresnitz, E. A., Butler, J. C., Cetron, M., Cohen, M., Doyle, T., Fischer, M., Greene, C. M., Griffith, K. S., Guarner, J., Hadler, J. L., Hayslett, J. A., Meyer, R., Petersen, L. R., Phillips, M., Pinner, R. W., ... Gerberding, J. L. (n.d.). *Investigation of bioterrorism-related anthrax, United States, 2001: Epidemiologic findings – volume 8, number 10—October 2002 – Emerging Infectious Diseases Journal – CDC.* https://doi.org/10.3201/eid0810.020353

Jia, X., Bu, R., Zhao, T., & Wu, K. (2019). Sensitive and specific whole-cell biosensor for arsenic detection. *Applied and Environmental Microbiology, 85*(11), e00694–19. https://doi.org/10.1128/AEM.00694-19

Jung, J. K., Alam, K. K., Verosloff, M. S., Capdevila, D. A., Desmau, M., Clauer, P. R., Lee, J. W., Nguyen, P. Q., Pastén, P. A., Matiasek, S. J., Gaillard, J.-F., Giedroc, D. P., Collins, J. J., & Lucks, J. B. (2020). Cell-free biosensors for rapid detection of water contaminants. *Nature Biotechnology, 38*(12), 1451–1459. https://doi.org/10.1038/s41587-020-0571-7

Kahn, K., & Plaxco, K. W. (2010). Principles of biomolecular recognition. In M. Zourob (Ed.), *Recognition receptors in biosensors* (pp. 3–45). Springer. https://doi.org/10.1007/978-1-4419-0919-0_1

Karig, D. K., Bessling, S., Thielen, P., Zhang, S., & Wolfe, J. (2017). Preservation of protein expression systems at elevated temperatures for portable therapeutic production. *Journal of The Royal Society Interface, 14*(129), 20161039. https://doi.org/10.1098/rsif. 2016.1039

Kaushik, A., Yndart, A., Kumar, S., Jayant, R. D., Vashist, A., Brown, A. N., Li, C.-Z., & Nair, M. (2018). A sensitive electrochemical immunosensor for label-free detection of Zika-virus protein. *Scientific Reports, 8*(1), 9700. https://doi.org/10.1038/s41598-018-28035-3

Kim, H. J., Jeong, H., & Lee, S. J. (2018). Synthetic biology for microbial heavy metal biosensors. *Analytical and Bioanalytical Chemistry, 410*(4), 1191–1203. https://doi.org/10.1007/s00216-017-0751-6

Kim, H., Jang, G., & Yoon, Y. (2020). Specific heavy metal/metalloid sensors: Current state and perspectives. *Applied Microbiology and Biotechnology, 104*(3), 907–914. https://doi.org/10.1007/s00253-019-10261-y

Köksaldı, İ. Ç., Köse, S., Ahan, R. E., Hacıosmanoğlu, N., Şahin Kehribar, E., Güngen, M. A., Baştuğ, A., Dinç, B., Bodur, H., Özkul, A., & Şeker, U. Ö. Ş. (2021). SARS-CoV-2 detection with de novo-designed synthetic riboregulators. *Analytical Chemistry*, *93*(28), 9719–9727. https://doi.org/10.1021/acs.analchem.1c00886

Kumar, H., & Rani, R. (2013). Development of biosensors for the detection of biological warfare agents: Its issues and challenges. *Science Progress*, *96*(3), 294–308.

Kylilis, N., Riangrungroj, P., Lai, H.-E., Salema, V., Fernández, L. Á., Stan, G.-B. V., Freemont, P. S., & Polizzi, K. M. (2019). Whole-cell biosensor with tunable limit of detection enables low-cost agglutination assays for medical diagnostic applications. *ACS Sensors*, *4*(2), 370–378. https://doi.org/10.1021/acssensors.8b01163

Lei, R., Li, L., Wu, P., Fei, X., Zhang, Y., Wang, J., Zhang, D., Zhang, Q., Yang, N., & Wang, X. (2022). RPA/CRISPR/Cas12a-Based on-site and rapid nucleic acid detection of toxoplasma gondii in the environment. *ACS Synthetic Biology*, *11*(5), 1772–1781. https://doi.org/10.1021/acssynbio.1c00620

Li, J., & Macdonald, J. (2016). Multiplexed lateral flow biosensors: Technological advances for radically improving point-of-care diagnoses. *Biosensors & Bioelectronics*, *83*, 177–192. https://doi.org/10.1016/j.bios.2016.04.021

Li, J., Wang, Y., Wang, B., Lou, J., Ni, P., Jin, Y., Chen, S., Duan, G., & Zhang, R. (2022). Application of CRISPR/Cas systems in the nucleic acid detection of infectious diseases. *Diagnostics*, *12*(10), 2455. https://doi.org/10.3390/diagnostics12102455

Li, S.-Y., Cheng, Q.-X., Liu, J.-K., Nie, X.-Q., Zhao, G.-P., & Wang, J. (2018). CRISPR-Cas12a has both cis- and trans-cleavage activities on single-stranded DNA. *Cell Research*, *28*(4), 491–493. https://doi.org/10.1038/s41422-018-0022-x

Li, Y., Li, S., Wang, J., & Liu, G. (2019). CRISPR/Cas systems towards next-generation biosensing. *Trends in Biotechnology*, *37*(7), 730–743. https://doi.org/10.1016/j.tibtech.2018.12.005

Liang, J. C., Bloom, R. J., & Smolke, C. D. (2011). Engineering biological systems with synthetic RNA molecules. *Molecular Cell*, *43*(6), 915–926. https://doi.org/10.1016/j.molcel.2011.08.023

Lopreside, A., Wan, X., Michelini, E., Roda, A., & Wang, B. (2019). Comprehensive profiling of diverse genetic reporters with application to whole-cell and cell-free biosensors. *Analytical Chemistry*, *91*(23), 15284–15292. https://doi.org/10.1021/acs.analchem.9b04444

Lu, X., Ye, Y., Zhang, Y., & Sun, X. (2021). Current research progress of mammalian cell-based biosensors on the detection of foodborne pathogens and toxins. *Critical Reviews in Food Science and Nutrition*, *61*(22), 3819–3835. https://doi.org/10.1080/10408398.2020.1809341

Ma, D., Shen, L., Wu, K., Diehnelt, C. W., & Green, A. A. (2018). Low-cost detection of norovirus using paper-based cell-free systems and synbody-based viral enrichment. *Synthetic Biology (Oxford, England)*, *3*(1), ysy018. https://doi.org/10.1093/synbio/ysy018

Malhotra, B. D., & Ali, Md. A. (2018). Chapter 1 – Nanomaterials in biosensors: Fundamentals and applications. In B. D. Malhotra & Md. A. Ali (Eds.), *Nanomaterials for biosensors* (pp. 1–74). William Andrew Publishing. https://doi.org/10.1016/B978-0-323-44923-6.00001-7

Mathews, D. H., Sabina, J., Zuker, M., & Turner, D. H. (1999). Expanded sequence dependence of thermodynamic parameters improves prediction of RNA secondary structure. *Journal of Molecular Biology*, *288*(5), 911–940. https://doi.org/10.1006/jmbi.1999.2700

Meagher, R. J., Negrete, O. A., & Van Rompay, K. K. (2016). Engineering paper-based sensors for Zika virus. *Trends in Molecular Medicine*, *22*(7), 529–530. https://doi.org/10.1016/j.molmed.2016.05.009

Mustafa, M. I., & Makhawi, A. M. (2021). SHERLOCK and DETECTR: CRISPR-Cas systems as potential rapid diagnostic tools for emerging infectious diseases. *Journal of Clinical Microbiology*, *59*(3), e00745–20. https://doi.org/10.1128/JCM.00745-20

Nelson, A. L., & Porter, L. (2023). Hydrogen peroxide toxicity. In *StatPearls*. StatPearls Publishing. http://www.ncbi.nlm.nih.gov/books/NBK585102/

Pardee, K., Green, A. A., Ferrante, T., Cameron, D. E., DaleyKeyser, A., Yin, P., & Collins, J. J. (2014). Paper-based synthetic gene networks. *Cell*, *159*(4), 940–954. https://doi.org/10.1016/j.cell.2014.10.004

Pardee, K., Green, A. A., Takahashi, M. K., Braff, D., Lambert, G., Lee, J. W., Ferrante, T., Ma, D., Donghia, N., Fan, M., Daringer, N. M., Bosch, I., Dudley, D. M., O'Connor, D. H., Gehrke, L., & Collins, J. J. (2016). Rapid, low-cost detection of Zika virus using programmable biomolecular components. *Cell*, *165*(5), 1255–1266. https://doi.org/10.1016/j.cell.2016.04.059

Parthasarathy, R., Monette, C. E., Bracero, S., & Saha, M. S. (2018). Methods for field measurement of antibiotic concentrations: Limitations and outlook. *FEMS Microbiology Ecology*, *94*(8). https://doi.org/10.1093/femsec/fiy105

Pellinen, T., Huovinen, T., & Karp, M. (2004). A cell-free biosensor for the detection of transcriptional inducers using firefly luciferase as a reporter. *Analytical Biochemistry*, *330*(1), 52–57. https://doi.org/10.1016/j.ab.2004.03.064

Pohanka, M. (2019). Current trends in the biosensors for biological warfare agents assay. *Materials*, *12*(14), 2303. https://doi.org/10.3390/ma12142303

Puig-Serra, P., Casado-Rosas, M. C., Martinez-Lage, M., Olalla-Sastre, B., Alonso-Yanez, A., Torres-Ruiz, R., & Rodriguez-Perales, S. (2022). CRISPR approaches for the diagnosis of human diseases. *International Journal of Molecular Sciences*, *23*(3), 1757. https://doi.org/10.3390/ijms23031757

Quintela, I. A., Vasse, T., Lin, C.-S., & Wu, V. C. H. (2022). Advances, applications, and limitations of portable and rapid detection technologies for routinely encountered foodborne pathogens. *Frontiers in Microbiology*, *13*, 1054782. https://doi.org/10.3389/fmicb.2022.1054782

Riedel, S. (2004). Biological warfare and bioterrorism: A historical review. *Proceedings (Baylor University. Medical Center)*, *17*(4), 400–406.

Rollin, J. A., Tam, T. K., & Zhang, Y.-H. P. (2013). New biotechnology paradigm: Cell-free biosystems for biomanufacturing. *Green Chemistry*, *15*(7), 1708–1719. https://doi.org/10.1039/C3GC40625C

Salehi, A. S. M., Shakalli Tang, M. J., Smith, M. T., Hunt, J. M., Law, R. A., Wood, D. W., & Bundy, B. C. (2017). Cell-free protein synthesis approach to biosensing hTRβ-specific endocrine disruptors. *Analytical Chemistry*, *89*(6), 3395–3401. https://doi.org/10.1021/acs.analchem.6b04034

Salehi, A. S. M., Yang, S. O., Earl, C. C., Shakalli Tang, M. J., Porter Hunt, J., Smith, M. T., Wood, D. W., & Bundy, B. C. (2018). Biosensing estrogenic endocrine disruptors in human blood and urine: A RAPID cell-free protein synthesis approach. *Toxicology and Applied Pharmacology*, *345*, 19–25. https://doi.org/10.1016/j.taap.2018.02.016

Saltepe, B., Bozkurt, E. U., Güngen, M. A., Çiçek, A. E., & Şeker, U. Ö. Ş. (2021). Genetic circuits combined with machine learning provides fast responding living sensors. *Biosensors and Bioelectronics*, *178*, 113028. https://doi.org/10.1016/j.bios.2021.113028

Saltepe, B., Bozkurt, E. U., Hacıosmanoğlu, N., & Şeker, U. Ö. Ş. (2019). Genetic circuits to detect nanomaterial triggered toxicity through engineered heat shock response mechanism. *ACS Synthetic Biology*, *8*(10), 2404–2417. https://doi.org/10.1021/acssynbio.9b00291

Saltepe, B., Kehribar, E. Ş., Yirmibeşoğlu, S. S. S., & Şeker, U. Ö. Ş. (2018). Cellular biosensors with engineered genetic circuits. *ACS Sensors*, *3*(1), 13–26. https://doi.org/10.1021/ACSSENSORS.7B00728

Schmidt-Dannert, C., & Lopez-Gallego, F. (2016). A roadmap for biocatalysis—Functional and spatial orchestration of enzyme cascades. *Microbial Biotechnology*, *9*(5), 601–609. https://doi.org/10.1111/1751-7915.12386

Serra, M. J., & Turner, D. H. (1995). Predicting thermodynamic properties of RNA. *Methods in Enzymology*, *259*, 242–261. https://doi.org/10.1016/0076-6879(95)59047-1

Silverman, A. D., Akova, U., Alam, K. K., Jewett, M. C., & Lucks, J. B. (2020). Design and optimization of a cell-free atrazine biosensor. *ACS Synthetic Biology*, *9*(3), 671–677. https://doi.org/10.1021/acssynbio.9b00388

Slomovic, S., Pardee, K., & Collins, J. J. (2015). Synthetic biology devices for in vitro and in vivo diagnostics. *Proceedings of the National Academy of Sciences*, *112*(47), 14429–14435. https://doi.org/10.1073/pnas.1508521112

Tsou, J.-H., Leng, Q., & Jiang, F. (2019). A CRISPR test for detection of circulating nuclei acids. *Translational Oncology*, *12*(12), 1566–1573. https://doi.org/10.1016/j.tranon.2019.08.011

Turner, A. P. F. (2013). Biosensors: Sense and sensibility. *Chemical Society Reviews*, *42*(8), 3184–3196. https://doi.org/10.1039/C3CS35528D

van Aken, J., & Hammond, E. (2003). Genetic engineering and biological weapons. *EMBO Reports*, *4*(Suppl 1), S57–S60. https://doi.org/10.1038/sj.embor.embor860

Voon, C. H., Yusop, N. M., & Khor, S. M. (2022). The state-of-the-art in bioluminescent whole-cell biosensor technology for detecting various organic compounds in oil and grease content in wastewater: From the lab to the field. *Talanta*, *241*, 123271. https://doi.org/10.1016/j.talanta.2022.123271

Voyvodic, P. L., & Bonnet, J. (2020). Cell-free biosensors for biomedical applications. *Current Opinion in Biomedical Engineering*, *13*, 9–15. https://doi.org/10.1016/j.cobme.2019.08.005

Walper, S. A., Lasarte Aragonés, G., Sapsford, K. E., Brown, C. W., Rowland, C. E., Breger, J. C., & Medintz, I. L. (2018). Detecting biothreat agents: From current diagnostics to developing sensor technologies. *ACS Sensors*, *3*(10), 1894–2024. https://doi.org/10.1021/acssensors.8b00420

Wang, H., Wu, Q., Yan, C., Xu, J., Qin, X., Wang, J., Chen, W., Yao, L., Huang, L., & Qin, P. (2022). CRISPR/Cas9 bridged recombinase polymerase amplification with lateral flow biosensor removing potential primer-dimer interference for robust Staphylococcus aureus assay. *Sensors and Actuators B: Chemical*, *369*, 132293. https://doi.org/10.1016/j.snb.2022.132293

Wang, S.-Y., Du, Y.-C., Wang, D.-X., Ma, J.-Y., Tang, A.-N., & Kong, D.-M. (2021). Signal amplification and output of CRISPR/Cas-based biosensing systems: A review. *Analytica Chimica Acta*, *1185*, 338882. https://doi.org/10.1016/j.aca.2021.338882

Watstein, D. M., & Styczynski, M. P. (2018). Development of a pigment-based whole-cell zinc biosensor for human serum. *ACS Synthetic Biology*, *7*(1), 267–275. https://doi.org/10.1021/acssynbio.7b00292

Way, J. C., Collins, J. J., Keasling, J. D., & Silver, P. A. (2014). Integrating biological redesign: Where synthetic biology came from and where it needs to go. *Cell*, *157*(1), 151–161. https://doi.org/10.1016/j.cell.2014.02.039

Wen, J., He, D., Zhuo, T., Li, W., Liu, J., Wu, J., Zhao, Y., & Yuan, Y. (2023). Electroactive bacteria as whole-cell biosensor for sensitive detection of hydrogen peroxide. *Journal of Environmental Chemical Engineering*, *11*(2), 109333. https://doi.org/10.1016/j.jece.2023.109333

Woo, S.-G., Moon, S.-J., Kim, S. K., Kim, T. H., Lim, H. S., Yeon, G.-H., Sung, B. H., Lee, C.-H., Lee, S.-G., Hwang, J. H., & Lee, D.-H. (2020). A designed whole-cell biosensor for live diagnosis of gut inflammation through nitrate sensing. *Biosensors and Bioelectronics*, *168*, 112523. https://doi.org/10.1016/j.bios.2020.112523

Wu, Y., Wang, C.-W., Wang, D., & Wei, N. (2021). A whole-cell biosensor for point-of-care detection of waterborne bacterial pathogens. *ACS Synthetic Biology, 10*(2), 333–344. https://doi.org/10.1021/acssynbio.0c00491

Wu, Y.-C., Chen, C.-S., & Chan, Y.-J. (2020). The outbreak of COVID-19: An overview. *Journal of the Chinese Medical Association, 83*(3), 217. https://doi.org/10.1097/JCMA. 0000000000000270

Xiong, E., Jiang, L., Tian, T., Hu, M., Yue, H., Huang, M., Lin, W., Jiang, Y., Zhu, D., & Zhou, X. (2021). Simultaneous dual-gene diagnosis of SARS-CoV-2 based on CRISPR/ Cas9-mediated lateral flow assay. *Angewandte Chemie (International Ed. in English), 60*(10), 5307–5315. https://doi.org/10.1002/anie.202014506

Zhang, C., Zhang, P., Ren, H., Jia, P., Ji, J., Cao, L., Yang, P., Li, Y., Liu, J., Li, Z., You, M., Duan, X., Hu, J., & Xu, F. (2022). Synthetic biology-powered biosensors based on CRISPR/Cas mediated cascade signal amplification for precise RNA detection. *Chemical Engineering Journal, 446*, 136864. https://doi.org/10.1016/j.cej.2022.136864

Zhang, L., Guo, W., & Lu, Y. (2020). Advances in cell-free biosensors: Principle, mechanism, and applications. *Biotechnology Journal, 15*(9), 2000187. https://doi.org/10.1002/biot. 202000187

Zhang, Y., Qian, L., Wei, W., Wang, Y., Wang, B., Lin, P., Liu, W., Xu, L., Li, X., Liu, D., Cheng, S., Li, J., Ye, Y., Li, H., Zhang, X., Dong, Y., Zhao, X., Liu, C., Zhang, H. M., … Lou, C. (2017). Paired design of dCas9 as a systematic platform for the detection of featured nucleic acid sequences in pathogenic strains. *ACS Synthetic Biology, 6*(2), 211–216. https://doi.org/10.1021/acssynbio.6b00215

5 Microfluidic-Based Plasmonic Nanosensors for Biological and Chemical Threats

Özgecan Erdem, Eylul Gulsen Yilmaz,
Beyza Nur Küçük, Fatih Inci, and Yeşeren Saylan

5.1 INTRODUCTION

The potential for diseases or harmful chemicals to spread quickly is greater than ever in the modern world due to expanding trade, quick international travel, and the increased rate of communication. If hazardous chemical and biological threats cannot be detected early and accurately, they can cause severe public health problems and large-scale catastrophes. Recently, the world has faced one of the most urgent pandemics due to the severe coronavirus disease 2019 (COVID-19), which caused approximately 7 million deaths worldwide. Since early diagnosis is crucial to initiate treatment and prevent the spread of the virus, during that period many researchers gave priority to developing sensing devices (Ji et al., 2023; Torres et al., 2021). In 1995, as a result of a bioterrorism attack, thousands of people were affected in the Tokyo subway Sarin attack which was caused by a colorless and toxic nerve agent that can cause death in minutes. After such an attack, the detection of toxic chemicals, which can create residues and contaminate water and soil, also became important for protecting public health in mass living areas (Tan et al., 2006). In addition, research into biological pathogen detection, including *Salmonella* and *Escherichia coli* in food products, has been ongoing for decades (Wang et al., 2019). In the developing world and contemporary analytical science—in light of newly emerging infectious diseases and pollution owing to released hazardous substances—the requirement for enhanced, robust, and rapid sensing methods able to detect biological and chemical threats with high sensitivity and specificity has reached extraordinary levels. Since conventional detection methods, characterized by requiring high sample volume and prolonged analysis time, are not sufficient, researchers are trying to develop new sensor platforms and adapt conventional techniques to them (Figure 5.1).

One of the biggest contributions to the development and progress of biotechnological systems has been the adaptation of microfluidic platforms (Sackmann et

DOI: 10.1201/9781003459316-5

91

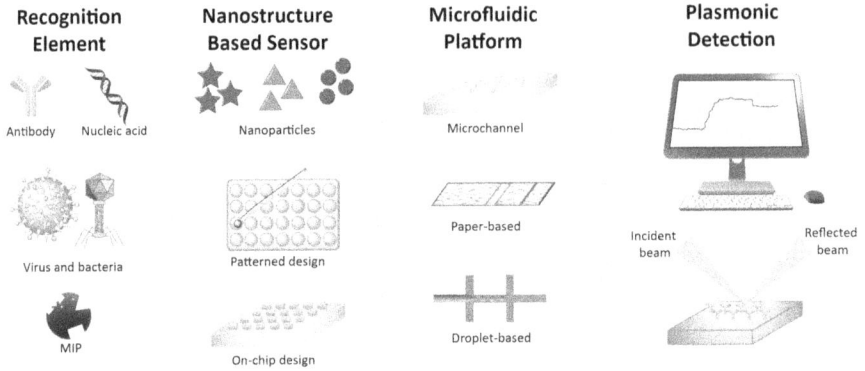

Recognition Element	Nanostructure Based Sensor	Microfluidic Platform	Plasmonic Detection

Antibody   Nucleic acid

Nanoparticles

Microchannel

Virus and bacteria

Patterned design

Paper-based

MIP

On-chip design

Droplet-based

Incident beam   Reflected beam

FIGURE 5.1   Scheme of microfluidic integrated nanosensors via various recognition elements. Readapted from C. Jin et al. (2022) with the permission of Elsevier.

al., 2014; Whitesides, 2006). Microfluidics can be defined as a micro total analysis system (µTAS) that allows the manipulation of micro or less volume of liquid (10^{-9} to 10^{-18} L) in submillimeter-sized channels. By applying a drawing or repulsing force in a controlled manner, sample liquid and analyte molecules in the sample such as cells, micro and nanoparticles, protein, or nucleic acid residues, etc. can be directed and detected with high sensitivity from a small amount of sample (Oh, 2020). Considering the parallelized structure of the microfluidics, more than one transaction can take place, which provides high throughput processing and miniaturized sensor units (Lakhera et al., 2022). Microfluidics can be integrated with different types of nanosensor platforms, such as electrochemical (Schmidt-Speicher and Länge, 2021), mechanical (Thies et al., 2017), or optical (Tokel et al., 2015). These small-scale functional devices can be defined as lab-on-a-chip (LOC) platforms. After their surface is modified with antibody, nucleic acid, or molecularly imprinted polymer (MIP) as target-specific recognition elements, they can handle multi-step processes including sample loading, washing, signal amplification, and conversion. Among the optical nanosensors, plasmonic nanosensors such as surface plasmon resonance (SPR), localized surface plasmon resonance (LSPR), and nanostructured photocrystals are well-established systems in terms of monitoring molecular interactions without the need for educated or expert people (Estevez et al., 2012). These methods are thorough-paced compared to regular methods in terms of label-free detection, real-time monitoring, short response time, and reusability (Shrivastav et al., 2021). Plasmonic nanosensors typically consist of a plasmonic material (generally a thin metallic film) and a dielectric material (such as liquid solution). When exposed to light, the free electrons in these continuous metallic structures collectively oscillate at the boundary of the plasmonic surface, creating what is known as SPR (Masson, 2020). These resonances are extremely responsive to alterations in the immediate surroundings, such as fluctuations in the refractive index, brought on by the presence of analytes on or close to the nanosensor's surface, such as cells, pathogens, or proteins. With the development of nanotechnology in recent years, the utilization of especially gold (Au) and silver (Ag) nanostructures (particles, rods, pillars, etc.) have gained importance for enhancing the plasmonic properties of nanosensor platforms with their improved electrical and optical features (Anker et al., 2008; Inci et al., 2020a). The utilization of nanostructures in nanosensor technologies was initiated with the use of metallic nanoparticles for LSPR. Unlike surface plasmons (SP), localized surface plasmons (LSP) are collective electron oscillations confined to nanoparticles or nanostructures which are significantly smaller than the wavelength of incident light (Hutter and Fendler, 2004). Both material type (Au, Ag, titanium, etc.) and morphological features of the nanostructures (size, size distribution, shape, etc.) can significantly affect the absorption bands of the plasmonic materials (Brahmkhatri et al., 2021).

In this chapter, microfluidic integrated plasmonic nanosensor applications for biological and chemical threat detection will be covered in detail. The fabrication methods of microfluidics and commonly used microfluidic nanosensor types, the basics of plasmonics, the detection of biological and chemical targets via these systems, advantages and disadvantages, and future expectations are discussed in the next sections of the chapter.

5.2 MICROFLUIDICS

Microfluidic technology, a fast-developing branch of science distinguished by microscale effects, has become more important in a variety of applications, from micro-arrays to cellular biophysics. The term "lab-on-a-chip" was initially used by Tabeling in 2005 to describe these devices, which typically range in size from 10 to 100 micrometers (Tabeling, 2023). With this cutting-edge technology, numerous processes—including sample processing, reactions, separation, and detection—are combined onto a little chip that is only a few square centimeters in size. The main goal is to carry out analyses that are quick, effective, and resource-efficient. The creation of tiny devices that painstakingly control physicochemical processes inside fluid containers is where microfluidic technology thrives. These microstructures support a variety of microfluidic phenomena, including diffusion, laminar flow, surface tension, capillary effects, and rapid heat conduction, providing fine fluid control and prompt responses (Mou et al., 2016). Fluids have particular analytical capabilities because of the peculiar features they exhibit due to their micron-scale structures that are not visible at macroscopic scales. These chips' diminutive size and larger surface area improve analytical throughput and mass transfer (Zhang et al., 2022). The need for significant sample and reagent volumes is greatly reduced by this breakthrough, which also makes multiplexing and high-throughput screening possible (Cui and Wang, 2019). In addition to its broader applications, microfluidics is increasingly playing a pivotal role in several specialized fields. For instance, microfluidics empowers the creation of highly precise microenvironments for regulating cell growth and differentiation in tissue engineering, thus enabling the generation of functional tissue for applications in regenerative medicine (Gharib et al., 2022). In the realm of drug discovery, microfluidic systems enhance high-throughput screening by facilitating the rapid evaluation of multiple drug candidates within controlled settings. These systems emulate natural physiological processes and contribute to the assessment of drug effectiveness and safety. Conversely, microfluidic-based sensing devices offer a level of accuracy and sensitivity in the detection of a wide array of biological substances, a capability of paramount importance for tasks such as environmental monitoring, point-of-care diagnostics, and biomedical research.

5.2.1 MATERIALS FOR MICROFLUIDICS

Making a careful choice of the best material for creating devices is a crucial step in microfluidic applications. This decision is crucial because materials' features are greatly enhanced at the microscale, where they may have an impact on the properties of newly created nanomaterials (Song et al., 2008). Shorter retention durations, laminar flows, increased mass, and heat transfer rates, as well as significant surface-to-volume ratios, especially in capillary microfluidics, bring out different phenomena (Ren et al., 2013). Contrary to macroscale vessels, aqueous solution wetting behavior and contact angles on chip materials are crucial. Material selection involves considering several crucial factors beyond simple composition. These include but are not limited to resilience, fabrication simplicity, translucency, suitability for biological applications, compatibility with planned reagents, the ability to withstand required temperature and pressure conditions during reactions, and the potential for surface customization (Shakeri et al., 2019).

There are several materials competing to fulfill these strict requirements and be acceptable for producing microfluidic devices. Glass, silicon, polymers, and ceramics are common substrates, while new materials of variable quality are constantly being developed. Based on the intended use, each material has a unique set of benefits and disadvantages. Due to its chemical inertness, thermostability, electrical insulation, rigidity, biocompatibility, and ease of surface functionalization, glass is an extremely suitable material for microreactors (Shakeri et al., 2019).

Glass microreactors are the best choice for carrying out chemical reactions when pressures, temperatures, and solvents are high due to these characteristics (Hwang et al., 2019). In comparison to other materials, glass microcapillary reactors have finer resolution at the micrometer scale, allowing for more exact control over the creation of polymeric nanoparticles and emulsions (Ofner et al., 2017). Excellent optical transparency, cost-effectiveness, and the ability to integrate active components are additional benefits of glass over silicon. Despite difficulties in production and clean room preparation, glass is still one of the most generally utilized materials for microfluidic chips due to its compatibility with biological samples, which makes it ideal for biochemical studies (Nielsen et al., 2019).

Silicon is a popular choice for microfluidic chips owing to its chemical compatibility, availability, and thermostability (Campbell et al., 2020). It offers easy fabrication, flexibility of design, semiconductor features, and surface modification potential, making it the dominant material in microfluidics for years. However, it has limitations, notably its opacity, which is unsuitable for visible and ultraviolet optical detection (Nielsen et al., 2019). In-situ imaging may require non-silicon components, and silicon's fragility and high modulus pose challenges for integrating active elements like valves and pumps. Silicon is also relatively expensive. Despite this, it finds use in medical diagnostics and drug screening (Jiang et al., 2022).

Polymers have gained traction in microfluidic chip manufacturing, moving away from silicon and glass, due to their cost-effectiveness and ease of production. These versatile platforms find applications in nanoparticle synthesis, fluid manipulation, and more, accommodating temperatures up to 200°C, ideal for large-scale production. These transparent polymeric materials offer optical access for monitoring reactions, crucial for processes like nanocrystallization. Common polymers used include polydimethylsiloxane (PDMS), poly (methyl methacrylate) (PMMA), cyclo-olefin polymers, fluoropolymers, and copolymers (COPs/COCs) and hydrogels. PDMS, a representative polymer, stands out for its affordability, ease of molding, optical transparency, and gas permeability (Syed et al., 2014). Valued for prototyping, it finds use in bio-related research, disease diagnosis, and cellular assays. However, its porosity limits compatibility with certain solvents and leads to water evaporation issues. PMMA, another choice, offers good solvent compatibility, optical transparency, and mechanical properties, suitable for research settings, organ-on-a-chip platforms, and micro-physiological systems (Kotz et al., 2020; Nielsen et al., 2019). Perfluorinated polymers, like Teflon, excel with thermal processability, chemical inertness, and antifouling properties. Polytetrafluoroethylene(PTFE), in particular, tolerates a broad range of chemicals and high temperatures, making it useful for synthesis devices (Liao et al., 2019). COPs/COCs have gained interest for their optical transparency, chemical resistance, and minimal water absorption, ideal for aggressive solvents (Bruijns et al., 2019). Hydrogels, though less common as the primary material, offer biocompatibility, low cytotoxicity, and tunable pore sizes, suitable to encapsulate cells in tissue engineering and substance delivery (Yilmaz et al., 2023). These materials can enhance microfluidic component functionality, including semipermeable barriers and smart valves, with chips primarily composed of rigid materials (Goy et al., 2019).

5.2.2 Fabrication Methods for Microfluidics

The fabrication of microfluidic chips demands a meticulous approach rooted in biological principles. Various methodologies exist for crafting these chips, each offering distinct advantages and limitations (Akceoglu et al., 2021). As shown in Figure 5.2, soft lithography, for instance, stands as one of the most prevalent techniques, employing flexible materials like polymers to generate microfluidic channels and structures (Scott and Ali, 2021). This process involves molding or patterning the polymer, followed by curing it through heat or ultraviolet radiation. It excels in crafting intricate three-dimensional structures, proving cost-effective and user-friendly. Photolithography represents another avenue for microfluidic chip fabrication. Here, light is harnessed to pattern a photosensitive material, like SU-8 photoresist, on a substrate (Lee et al., 2015). The developed photoresist yields a patterned mask, guiding the etching of desired channels and structures in the substrate (Ma et al., 2010). While photolithography boasts high resolution and the ability to craft complex

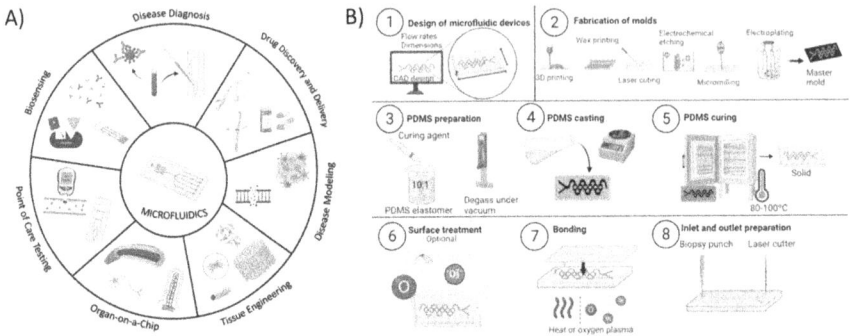

FIGURE 5.2 Common applications (A) and fabrication methods of microfluidics (B). Reproduced from Gharib et al. (2022) with the permission of MDPI and reproduced from Rodríguez et al. (2023) with the permission of Frontiers.

structures, it can be costly, time-consuming, and less suited for high-volume production (Nguyen et al., 2022).

Micromachining emerges as a third fabrication technology, involving cutting, drilling, or etching microfluidic channels and structures into a substrate through various tools and techniques (Juang and Chiu, 2022). For instance, concentrated ion beams may be employed to shape channels in silicon wafers or drill holes in glass substrates. While this method enables the creation of exceedingly small and intricate structures, it can be costly and challenging to control. Micro-molding, another prevalent approach, utilizes a mold to craft desired structures in a polymer material, such as PDMS (Scott and Ali, 2021). Typically, molds are fashioned using photolithography or other methods and are employed to create channels and structures in polymeric materials. Micro-molding offers advantages like affordability and high throughput and can be readily scaled up for extensive production. However, it may lack the precision of photolithography and can pose challenges in crafting complex structures. Lastly, microfluidic systems can also be manufactured using 3D printing techniques, necessitating specialized printers capable of fine-resolution material deposition, including polymers and metals. Typically guided by computer control, these printers allow for precise construction of intricate 3D structures and enable the customization of designs (Jin et al., 2022).

5.2.3 TYPES OF MICROFLUIDICS

Droplet microfluidic technology harnesses tiny liquid droplets enclosed within an immiscible fluid, serving as minuscule independent reaction chambers, whether in continuous or segmented flow (Cui and Wang, 2019). These methods bring forth notable benefits, including minimized sample usage, accelerated reaction rates, and heightened reliability and consistency, all while enabling precise control over composition within these extremely minute volumes. Also, these methods have a variety of uses, including chemical synthesis, drug screening, and biochemical analysis.

In contrast, "Digital Microfluidics" elevates droplet manipulation to an art form by using electrodes and automation to precisely direct discrete droplets over open surfaces (Zhai et al., 2020). This digital precision streamlines laboratory processes, making it essential for uses like single-cell analysis and digital polymerase chain reaction (PCR). Contrarily, "Paper-Based Microfluidics" values affordability and simplicity, relying on capillary action to support point-of-care diagnostics even in environments with low resources. This method offers quick and affordable testing solutions that are essential for both environmental monitoring and medical diagnosis (Nishat et al., 2021). "Lab-on-a-chip Systems" are microchip-sized wonders that can house full laboratories. These systems combine tasks ranging from sample preparation to analysis, revolutionizing analytical chemistry, DNA sequencing, and medical diagnostics (Francesko et al., 2019). "Organ-on-a-chip Technology," however, transports us to a world where physiological organ conditions are precisely duplicated in tiny settings (Bhatia and Ingber, 2014). By bridging the gap between controlled lab trials and real-world circumstances, this breakthrough redefines drug testing, disease modeling, and organ function studies. In summary, the field of microfluidics is a multidimensional setting where accuracy and miniaturization intersect. These various microfluidic technologies, each equipped with special capacities, demonstrate the enormous potential of microscale activities. They enable scientists to rethink possibilities and shape the future of science and technology, from high-throughput chemical synthesis to mimicking human organ functioning.

5.3 MICROFLUIDIC-BASED PLASMONIC NANOSENSORS

5.3.1 PRINCIPLES OF PLASMONIC NANOSENSORS

A surface plasmon is created as photons interact when the surface of a metal dielectric with SPPs is struck, and the wavelengths in the SP are sensitive to the refractive index of the medium (Akgönüllü et al., 2023) (Figure 5.3A). Because of these properties, there are several different types of plasmons that can be generated from SPP and used in optical-based measurements, including SPR and LSPR (Duan et al., 2021). Both these nanosensors mentioned are sensitive to local refractive index changes that happen when the target is bound to the surface. Therefore, these nanosensors can generate a response without labeling (via chromophore and fluorophore, etc.), unlike conventional optical nanosensors (Choi and Choi, 2012). In SPR, the interacting electromagnetic field at the metal and dielectric interface initially excites collective coherent oscillations of free electrons in the metal's conduction band, and SPPs are generated with subsequent charge density oscillations as shown in Figure 5.3B. After penetrating a few nanometers into the surrounding matrix, SPPs generate an electric field that decays exponentially (Srivastava et al., 2018). The capacity of nanoparticles to resonantly couple incoming photons to collective oscillations of free electrons on their surface is defined as LSPR, as indicated in Figure 5.3C (Yildirim et al., 2021). Optical nanosensors based on metallic nanoparticles' LSPR have gained growing popularity for detecting various chemical and biological species (Taghavi et al., 2020; Tian et al., 2021; Versiani et al., 2020).

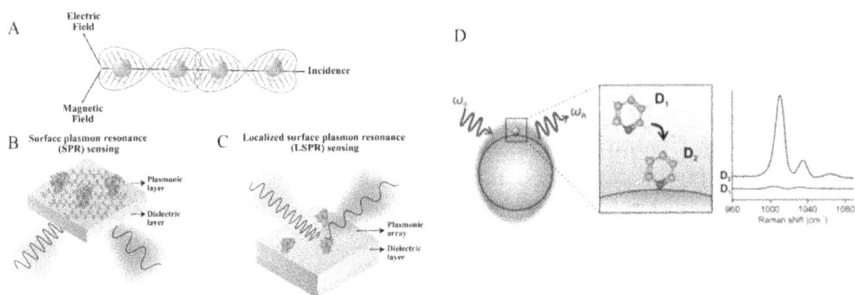

FIGURE 5.3 Illustration of plasmonic effects which are created by collective oscillation of free electrons in response to a component of the incident light (A), geometry-dependent properties of plasmonic materials of SPR (B) and LSPR (C), and conduction band electron resonance oscillation and the generation of a localized strong electromagnetic field zone on the surface of materials related to SERS enhancement (D). Reproduced from Amrollahi et al. (2021) with the permission of ACS Publications and reproduced from Leong et al. (2022) with the permission of Elsevier.

Because the stimulation of surface plasmons is deeply dependent on the dielectric medium's refractive index, every contact between the analyte and the plasmonic nanoparticle modifies the resonance state (Amirjani et al., 2022). Gold nanoparticles (AuNPs), in particular, when paired with a thin metal surface, improve the SPR response in nanosensors, whereas silver nanoparticles create a robust and acute plasmon resonance (Saylan, 2023). When exposed to incoming light, these resonances are caused by the collective oscillation of free electrons on the nanoparticle surface. This interaction with light produces stronger electromagnetic fields surrounding nanoparticles, which can be exploited for sensing purposes (Liu et al., 2018).

Surface-enhanced Raman scattering (SERS), in addition to the previously described SPR and LPSR nanosensors, is employed for the quantitative measurement and accurate determination of (bio)chemical compounds. Raman spectroscopy is fundamentally based on the inelastic scattering of a photon by molecules. The vibrational or rotational modes of the molecules are relaxed or excited during this scattering process, creating a shift in the wavelength of the photons as depicted in Figure 5.3D. Because each chemical species has unique vibrational energy levels, a fingerprint Raman spectrum for each molecule is theoretically possible (Demirel et al., 2018). SERS has evolved as a strong, label-free sensing approach, capable of capturing "fingerprint" data from biological cells, organelles, and molecules down to the single-molecule scale (Bai et al., 2019; Lv et al., 2016; Nikelshparg et al., 2022). Due to local field enhancement, surface plasmons are essential in increasing Raman signals for target molecules near plasmonic nanostructures (Peng et al., 2021).

Overall, in the context of nanosensors, the interaction between plasmonic nanomaterials and target analytes can be understood in several key steps: (i) Surface functionalization of plasmonic nanomaterials with antibodies (Panchal et al., 2022; Pereira et al., 2021), aptamers (Shi et al., 2013; Xiao et al., 2016), or MIPs

(Ahmad et al., 2015; Erdem et al., 2023; Yılmaz et al., 2022) which enable the selective binding of the target analyte; (ii) Local environment changes after binding, i.e., changes in the refractive index (Omrani et al., 2021), electric field distribution (Ndukaife et al., 2014), and dielectric constant (Farmani et al., 2020) of the environment of the nanomaterials; (iii) Shifts in plasmon resonance, which is related to the presence and concentration of the target analyte (Li et al., 2019; Ye et al., 2022).

5.3.2 INTEGRATION OF MICROFLUIDICS AND PLASMONIC NANOSENSORS

Microfluidic lab-on-a-chip systems, which are miniaturized and require only small volumes of liquid, have recently been created and are suited for a variety of applications such as infectious illness (Kanitthamniyom et al., 2021; Saylan and Denizli, 2019) and cancer diagnostics (De Oliveira et al., 2018; Zilberman and Sonkusale, 2015), as well as environmental monitoring (Li et al., 2022; Tavakoli et al., 2022). Integrating microfluidic platforms and optical imaging systems joins the benefits of LOC systems and optical technology. Photonics and plasmonics (i.e., SPR, LSPR, nanostructured photonic crystals) are at crossroads of optics and nanotechnology, assisting in developing reliable, accurate, and simple-to-use sensor platforms. Plasmonic LOC systems may be created as low-cost platforms for point-of-care (POC) testing by utilizing disposable chips (Aslan et al., 2023).

Integrating microfluidics with plasmonic nanosensors offers many advantages, creating a powerful platform for varied sensing and analytical applications. Microfluidic devices have the capability to handle minimal sample volumes while providing a controlled flow of liquids at the micro and nanoscale (Gimondi et al., 2023). Reduced diffusion times enhance interactions between plasmonic nanomaterials and analytes, resulting in faster and more reliable detection (Srivastava et al., 2018). Multiple parallel channels make the simultaneous analysis of numerous analytes possible (Qu et al., 2020). By combining multiple channels and different channel plans, the platform can handle more than one sample/chip, and it can be changeable for detecting multiple targets on the same chip (Inci et al., 2020b). The combination of microfluidics with plasmonic nanosensors offers a lot of opportunities for POC applications. Microfluidic systems' real-time monitoring and continuous measuring capabilities make it simple to keep track of the quick reactions brought on by the analyte's binding in plasmonic nanosensors (Zhou et al., 2016). Additionally, the mobility and compactness of microfluidic tools, and the receptiveness of plasmonic nanosensors (Duan et al., 2021), can provide rapid and precise diagnosis for bedside applications (Zhou et al., 2019).

In summary, the combination of microfluidics with plasmonic nanosensors affords an adaptable platform for a variety of applications, including healthcare (Ameen et al., 2015; Escobedo et al., 2013; Tian et al., 2022) and environmental monitoring (Mishra et al., 2022; Pilát et al., 2018; Rasheed et al., 2023). This integration improves sensitivity, selectivity, speed, and automation, making it a significant tool in enhancing sensing technology (Luka et al., 2015).

5.3.3   Detection of Biological Threats

Pathogenic bacteria and viruses pose a significant threat to security through food contamination, bioterrorism, or biological threat (Walper et al., 2018). Most food-borne diseases are caused by a pathogenic bacteria/virus or toxins released from microorganisms. For bioterrorism/biological threats, examining the relationship between aerosol contagion and toxicity and the amount of the substance reduces the number of active substances that can cause the highest number of casualties (Kirsch et al., 2013). Any contagious organism is likely to be misused for criminal or threat approaches. Some biological agents pose a more significant threat than others due to factors such as their virulence or toxicity, as well as the agent's stability in the environment, capacity to infiltrate the host organism, and so on (Pohanka, 2019).

The threat from biological agents is scaled via the Centers for Disease Control and Prevention (CDC) using the letters A, B, and C, from highest to lowest, to identify them according to their hazard level (Centers for Disease Control and Prevention, 2018). Table 5.1 provides an overview of significant biological threat agents.

Determining infectious disease threats is a fundamental function of public health. The early detection of a bioterrorism event or the danger of a naturally occurring disease depends on clinicians diagnosing the first few cases or the

TABLE 5.1
Bioterrorism Agents and Diseases (Centers for Disease Control and Prevention, 2018)

Category	Biological Agent	Disease/Toxins
A	*Bacillus anthracis*	Anthrax
	Clostridium botulinum	Botulism
	Yersinia pestis	Plague
	Variola major	Smallpox
	Francisella tularensis	Tularemia
	Filoviruses (Ebola, Marburg)	Viral hemorrhagic fevers
	Arenaviruses (Lassa, Machupo)	
B	*Brucella* sp.	Brucellosis
	Clostridum perfringens	Epsilon toxin
	Salmonella sp., *E. coli* O157:H7, *Shigella* sp.	Food safety threats
	Burkholderia mallei	Glanders
	Chlamydia psittaci	Psittacosis
	Coxiella burnetii	Q fever
	Ricinus communis	Ricin toxin
	Staphylococcus aureus	Staphylococcal enterotoxin B
	Rickettsia prowazekii	Typhus fever
	Alphaviruses	Viral encephalitis
	Vibrio cholerae, Cryptosporidium parvum	Water safety threats
C	Nipah virus, Hantavirus	Emerging infectious diseases

identification of suspicious clinical appearances by infectious disease specialists. However, recognizing a rare illness that may be seen for the first time is a challenging prospect (National Research Council, 2011). Traditional detection methods, such as biochemical and immunological recognition-based assays and biomolecular techniques such as PCR, or cell culture, are used to detect pathogenic microorganisms or toxins, but detection times are lengthy and necessitate the use of specialized personnel. These approaches must be faster and more selective for environmental monitoring and initiating treatment of exposed persons (Rowland et al., 2016).

The detection platform for POC applications must be simple to apply and understand, inexpensive, durable, stable across a broad variety of operating conditions, and preferably disposable and portable (Wang et al., 2021) One of the other significant requirements for pathogen detection equipment is the capacity to run multiplex testing. To address the present constraints of traditional approaches, LOC and microfluidic devices are gaining traction (Mondal et al., 2020). Owing to the rising security concerns and ongoing war on terrorism, the requirement for fast detection nanosensors against threat agents for civil and military defense applications has risen. However, the availability of rapid, low-cost, accurate diagnostic tests is the primary requirement for the monitoring and controlling of illnesses. Microfluidic systems are a significant option with the attributes discussed in this context (Srinivasan and Tung, 2015). This section briefly summarizes microfluidic integrated nanosensors developed for the fast and accurate detection of biological threat agents.

Bacillus anthracis (*B. anthracis*), which causes anthrax, is a zoonotic disease primarily transmitted from animals and contaminated soil and has also been determined by the CDC as a Category A potential threat agent. The bacterium was utilized as a weapon in the First and Second World Wars and was the cause of deaths in the Soviet Union in 1979, Japan in 1995, and the United States in 2001. *B. anthracis* has aerobic/facultative anaerobic, Gram-positive properties and dormant spores that can survive for decades in soil and are very resistant to extreme environmental conditions. These spores can be transmitted easily from one host to another (Savransky et al., 2020). In a study, a fully automated, SERS-based microfluidic chip was developed for poly-γ-D-glutamic acid (PGA) detection, which is an anthrax biomarker (Gao et al., 2015) (Figure 5.4A–C). In the system developed utilizing anti-PGA-immobilized magnetic beads and AuNPs with PGA in a microfluidic environment, magnetic immune complexes were analyzed by directly measuring their SERS signals after being captured by solenoids placed inside the chip. To increase mixing effectiveness, a mixer was integrated to the channel as shown in Figure 5.4D–E. The limit of detection (LOD) value of PGA in serum in the microfluidic nanosensor was determined as 100 pg/mL (Figure 5.4F).

Humans and other warm-blooded animals' intestines contain the naturally occurring bacteria *E. coli*, which aids in the body's synthesis of vital vitamins. Some strains, nevertheless, are pathogenic. For instance, the strain *E. coli O157:H7* can create toxins that can damage the intestinal lining and result in illness (Dhull et al., 2019). With a Gram-negative rod shape and all associated characteristics, *E. coli O157:H7* belongs to the Enterobacteriaceae family. *E. coli O157:H7*'s primary

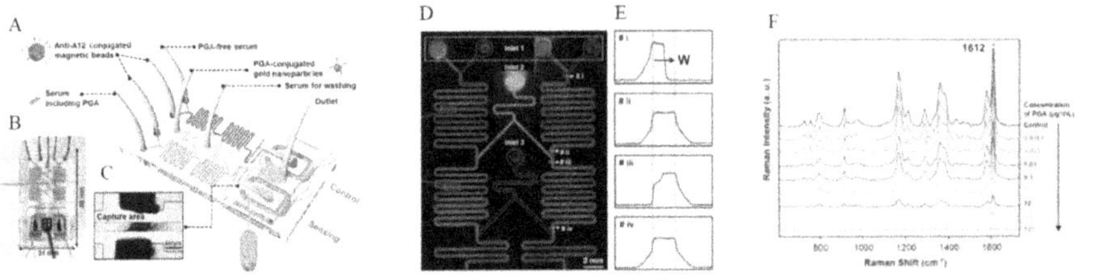

FIGURE 5.4 Scheme of dual-channel of SERS-based microfluidic nanosensor for PGA and control detection (A), photo of chip filled with ink and capture area for magnetic immune complexes (B), process of mixing efficiency analysis using Rhodamine B and Rhodamine 6G solutions (C), photo of fluorescent microfluidic channels (D), fluorescence intensity patterns corresponding to the places indicated via dashed lines (E), and concentration-dependent SERS spectra of PGA detection (F). Reproduced from Gao et al. (2015) with the permission of Elsevier.

reservoir is in the intestines of ruminant animals, which include cattle, sheep, deer, and goats. Animal feces are the pathogen's principal route of environmental release (Sharma et al., 2011). The World Health Organization (WHO) has recently stated that significant *E. coli O157:H7* outbreaks in the world's highest-income nations, including the UK, Germany, and Canada, have been caused by contaminated food or water (Caprioli et al., 2005). *Salmonella sp.* is another noteworthy pathogen causing gastrointestinal problems in Enterobacteriaceae with Gram-negative, non-spore forming and rod-shaped characteristics. Ingesting *Salmonella*, which is one of the most common causes of diarrhea, by drinking infected water or food, may also result in additional symptoms (Ehuwa et al., 2021). *Salmonella* sp. and *E. coli* O157:H7 have been determined by CDC as a Category B potential agent for biothreat. In a study for the rapid and reliable screening of food-borne bacterial pathogens, an SPR nanosensor depended on very low contamination and functionalizable poly (carboxybetaine acrylamide) (pCBAA) brushes were developed for the detection of bacterial pathogens in raw food samples (Vaisocherová-Lísalová et al., 2016). In this study, raw food samples (cucumber and hamburger) were used. Firstly, the bacteria in the raw food samples were detected by antibodies immobilized to pCBAA. In the second step, the secondary biotinylated antibody (Ab2) was attached to the pre-captured bacteria. Finally, the streptavidin-modified AuNPs were bound to the biotinylated Ab2, increasing the nanosensor response. Also, an SPR nanosensor depended on a pCBAA brush with a thickness as low as 20 nm. In detection studies performed on complex hamburger and cucumber samples, LOD values were 57 cfu/mL and 17 cfu/mL for *E. coli* O157:H7, while these values were 7.4×10^3 cfu/mL and 11.7×10^3 cfu/mL for *Salmonella* sp. A hybrid microfluidic chip was developed in different research to determine *E. coli* O157:H7 using an array of SPR and fluorescence imaging (Zordan et al., 2009). A PDMS microfluidic flow chamber surrounds a sequence of gold dots; each modified with a capture molecule targeting a particular pathogen and supplying a magnetically concentrated sample for testing. To functionalize the gold spots, three biomolecules were used: (i) An antibody that interacts with *E. coli*

O157:H7 specifically; (ii) Rabbit pre-immune serum as a negative control; and (iii) 0.5% bovine serum albumin solution in water as a second negative control. The chip's SPR at the bottom and epifluorescence at the top provides imaging. According to research, the chip's capacity to successfully catch *E. coli* O157:H7 was 97.7% effective. In a different study, vertically aligned carbon nanofibers embedded at the base of a microfluidic chip were used to combine nanostructured dielectrophoresis with nanotag-labeled SERS to monitor pathogen concentration, detection, and kinetics (Madiyar et al., 2015). Nanotag-labeled *E. coli* DH5α cells were successfully collected by a nanoelectrode array. The core-shell nano-ovals in SERS nanotags are around 50 nm in size, covered with a QSY21 Raman reporter, and attached to *E. coli* by particular immunochemistry. At a concentration as low as 210 cfu/mL and in less than 50 seconds, SERS tests are sensitive enough to identify a single bacterial cell.

An SPR nanosensor was developed for *Brucella melitensis* (*B. melitensis*) detection in milk samples, addressing the global concern of brucellosis, a zoonotic disease often transmitted through contaminated milk products. Through a whole bacteria-SELEX procedure, specific aptamers recognizing *B. melitensis* were selected. Two aptamers, chosen for their remarkable specificity and affinity, played distinct roles in the detection process. The high-affinity B70 aptamer was modified on magnetic silica core-shell nanoparticles to initially purify *B. melitensis* cells from milk matrices. Meanwhile, a B46 aptamer, known for its exceptional specificity to *B. melitensis* cells, was employed to fabricate an SPR nanosensor. These nanosensor chips enabled highly sensitive *Brucella* detection in eluted samples obtained post-magnetic purification. This integrated approach combines efficient magnetic isolation with instant SPR-based detection, achieving an impressively low detection limit of 27 ± 11 cells (Dursun et al., 2022).

A low-noise and affordable SPR nanosensor, utilizing extraordinary optical transmission (EOT) in nanoholes, has shown quantification of a wide range of antibody-ligand binding kinetics. These kinetics encompass equilibrium dissociation constants spanning from 200 pM to 40 nM. This SPR nanosensor is constructed around a standard microscope and a portable fiber-optic spectrometer, making it straightforward to assemble and operate. Remarkably, it achieves a refractive index resolution of 3×10^{-6}, even without on-chip cooling, representing a notable achievement in EOT-based SPR nanosensors. This impressive performance is made possible by the rapid acquisition of full-spectrum data in just 10 ms, after the frame averaging of EOT spectra. This efficiency is facilitated via producing template-stripped gold nanoholes characterized by optical properties across extensive surfaces. To extend its utility, the study conducted sequential SPR measurements employing a 12-channel microfluidic flow cell. These experiments involved meticulous optimization of surface modification procedures and antibody injection protocols to minimize mass-transport artifacts. In the real-time monitoring of the immobilization process, the protective antigen of anthrax, on the gold surface, yielded a remarkable signal-to-noise ratio of approximately 860. After that, the research ventured into quantitatively measuring real-time binding kinetics between antigen and a selection of compact, 25 kDa single-chain antibodies, even at concentrations as low as 1 nM. These findings underscore the vast potential of nanohole-based SPR nanosensors, not only for quantitative antibody screening but also as a versatile platform, seamlessly integrating SPR nanosensors with a wide array of bioanalytical tools (Im et al., 2012).

The necessity to battle upcoming epidemics and bioterrorism risks cannot be stressed enough, hence the search for speedy and sensitive virus detection technologies that are adaptable across multiple regions. The novel optofluidic label-free nanosensor described in this chapter may detect intact viruses in biological samples at clinically relevant concentrations without sample pretreatment. The nanosensor platform's extraordinary light transmission phenomenon within plasmonic nanoholes serves as its core. Additionally, it uses a large range of rapidly evolving viral strains to target group-specific antibodies. The similar size of the pathogens and the depth of surface plasmon polaritons' penetration, however, raised concerns about the method's possible limitations for detecting viruses. They showed the nanosensor's ability to detect and distinguish between small enveloped RNA viruses, such vesicular stomatitis virus and pseudotyped Ebola, and large enveloped DNA viruses, like vaccinia virus, in order to allay these worries. Their platform has an impressive dynamic detection range of three orders of magnitude. Additionally, it produces data with a high signal-to-noise ratio without the need for mechanical or optical separation, opening up new possibilities for pathogen identification in conventional biology laboratory settings (Yanik et al., 2010).

There has been a continual requirement for quick and visible detection, which has made it a difficult task thus far, given the terrifying potential of ricin-based bioweapon threats. This study presents a method for qualitative ricin analysis and recognition utilizing a gold nanopin-based colorimetric nanosensor that supports multicolor variants. The unique reflected hues produced by the activation of narrow-bandwidth standing-wave resonances in the visible spectrum are attributable to the plasmonic metasurfaces constructed around nanopin-cavity resonators, which are the basis for this nanosensor. As a result of the reflected color amalgamation resulting from various resonant wavelengths, a wide spectrum of colors can be seen. Additionally, these colored metasurfaces display clear fluctuations over a constrained refractive index range, making them particularly well-suited for a variety of applications. As a result, this biosensor, which has nanopin-cavities that have been functionalized with antibodies, has extraordinary sensitivity and quick response times, making it possible to detect ricin visually and quantitatively in a concentration range of 10–120 ng/mL. This method has a lot of potential thanks to its outstanding detection limit of 10 ng/mL and typical measurement duration of under 10 minutes. Furthermore, the on-chip integration of these nanopin metasurfaces into transportable colorimetric microfluidic devices opens doors for quantitative studies spanning a wide range of biological compounds (Fan et al., 2017).

It is extremely difficult to quantify botulinum neurotoxin A (BoNT/A), a bacterial toxin with extraordinary toxicity and medicinal potential in neurological therapies and cosmetics. Accurate measurement is challenging due to its great physiological potency and the requirement for three essential functionalities: Receptor binding, catalytic activity, and internalization-translocation. The mouse bioassay, a long-standing industrial standard that was formerly universally recognized, has become obsolete because of ethical considerations. As a potential solution to this conundrum, cell-based tests were made available. Their labor-intensive nature, however, has prevented their broad use. This study proposes a novel in vitro method that combines microfluidic technology with a nanosensor made with nerve cell-inspired nanoreactors. This nanosensor can assess all three essential functions, making it a useful instrument for estimating the amount of physiologically active BoNT/A. This technique detects and quantifies BoNT/A levels quickly when integrated into a microfluidic device, taking much less time than the mouse bioassay (less than 10 hours versus 2–4 days). Additionally, this approach shows its effectiveness in accurately measuring physiologically active BoNT/A in a straightforward pharmaceutical formulation. This strategy represents a significant ethical advance over in vivo methods by providing a thorough in vitro testing platform that combines a highly sensitive nanosensor with microfluidic technology. It also offers a workable substitute for time-consuming cell-based in vitro detection methods (Weingart et al., 2019).

These state-of-the-art sensing technologies provide a number of critical issues that call for careful evaluation. Making sure that these nanosensors operate dependably and consistently in a variety of environmental scenarios, especially field settings with limited resources, is one of the biggest problems. Assuring that these nanosensors are capable of detecting a variety of biological threat substances with the highest sensitivity and accuracy is another crucial component. Continuous research and development efforts are devoted to improving these technologies to handle these difficulties. The nanosensors' sensitivity and specificity are now being improved by scientists, making them better able to respond to the constantly changing landscape of biological threats. Efforts are also being made to make these technologies easier to use, with the aim of making it possible for anyone to use them effectively in emergency response situations. Wearable technology and artificial intelligence (AI) integration are two notable trends in this area. By aiding in data processing, pattern identification, and real-time decision-making, artificial intelligence, in particular, has the potential to significantly improve the capabilities of these nanosensors. The benefit of continuous monitoring provided by wearables with these nanosensors allows for the early identification of biological risks and prompt notification of users. It is imperative to remain aware of these technologies' ethical implications as they develop. To guarantee responsible implementation, considerations for privacy, data security, and potential for abuse must be taken into account. Furthermore, regulatory frameworks and policies are crucial in regulating the creation and application of these technologies, ensuring that they are ethically acceptable, accurate, and safe. In conclusion, there is constant development taking place in the field of biological hazard detecting technology. Conquering obstacles, enhancing capabilities, and making these nanosensors more useful for actual applications are all ongoing goals. The integration of AI and wearable technology offers the potential for sensors that are more sensitive, accurate, and user-friendly in the future. But it also calls for a watchful attitude to moral considerations and legal compliance.

5.3.4 Detection of Chemical Threats

Chemical threat agents are substances that can cause harm to humans, animals, and the environment when intentionally released and/or deployed with harmful intent (Saylan et al., 2020). These agents are typically toxic chemicals that can have a wide range of effects, from causing injury or illness to death, based on the type and concentration of chemical and the exposure duration (Kumar et al., 2023). Chemical threat agents can be categorized into several classes, including blister agents, nerve agents, choking agents, blood agents, vesicant agents, and toxic industrial chemicals (Chen et al., 2018). Chemical threat agents are prohibited by international conventions, including the Chemical Weapons Convention, which aims to eliminate the production, stockpiling, and usage of chemical weapons. The use of chemical threat agents is considered a violation of international law and is widely condemned (Gupta, 2015). Preparedness, detection, and response measures are in place in many countries to mitigate the impact of potential chemical threats, whether they result from intentional acts or accidental releases. The detection of chemical threat agents is of paramount importance for several critical reasons, such as protection of public health, prevention of mass casualties, environmental protection, public safety and security, international security, emergency response planning, public awareness and preparedness, prevention of accidental releases, supporting law enforcement and investigations, and scientific research and monitoring (Black, 2016).

There are some recent studies about microfluidic-based plasmonic nanosensors for chemical threat agents. For instance, Wang et al. introduced an approach to material fabrication, centered around a fluid-based construction method to produce three-dimensional (3D) nanostructures with microchannels (Wang et al., 2016). These nanostructures are precisely positioned to form an integrated microfluidic system designed for the SERS method (Figure 5.5A). This fluid-based construction method

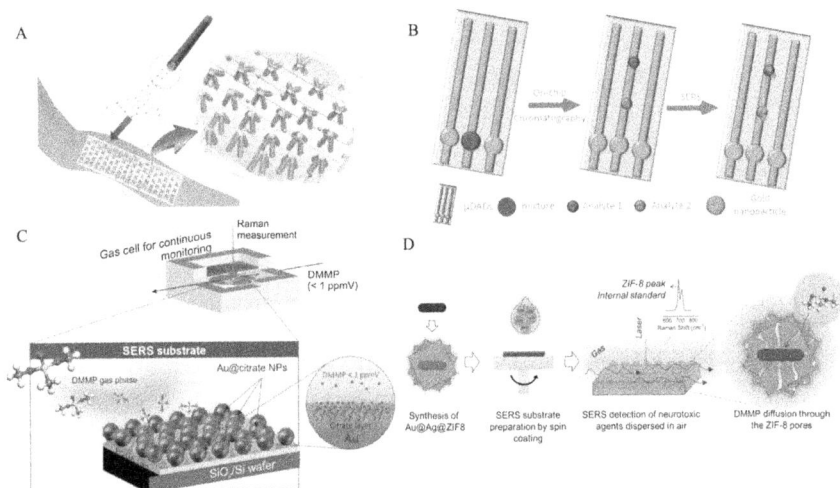

FIGURE 5.5   Scheme of 3D clustered nanostructures-based microfluidic system (A), on-chip chromatography-SERS sensing with DADs (B), SERS substrate and experimental set-up (C), and fabrication of SERS substrates for neurotoxic agent detection (D). Reproduced from Wang et al. (2016) with the permission of ACS. Reproduced from Kong et al. (2018) with the permission of Elsevier. Reproduced from Lafuente et al. (2018) with the permission of Elsevier. Reproduced from Lafuente et al. (2021) with the permission of ACS.

simplifies the creation of nanostructured substrates, which have been demonstrated to incorporate a significant number of hot spots essential for supporting an SERS response. Finite-difference time-domain experiments have indicated that these 3D clustered arrangements of the nanostructures actively promote hot spots formation. The performance of the SERS microfluidic device was assessed and revealed high sensitivity and efficient recyclability when tested with 4-mercaptobenzoic acid and Rhodamine 6G and 4-mercaptobenzoic acid mixture, both serving as target organic pollutants. Furthermore, these 3D clustered nanostructures exhibited effectiveness in detecting a representative nerve agent. The outcomes of this study present a materials and manufacturing approach for creating the necessary nanostructures, paving the way for the detection of organic pollutants, real-time monitoring of environmental threats, and personal health screening devices.

Kong et al. exemplified microfluidic paper-based plasmonic nanosensor-based diatomite analytical devices (DADs) for illicit drugs detection (Kong et al., 2018). The DADs are characterized by highly porous photonic crystal biosilica channels. The manufacturing process of these DADs involves the application of diatomaceous earth onto standard glass slides using spin-coating and tape-stripping techniques, resulting in microfluidic channels. A distinctive attribute of the DADs is their ability to simultaneously execute on-chip chromatography, effectively separating small molecules from complex samples. Additionally, they enable the acquisition of SERS with exceptional specificity for the target chemicals (Figure 5.5B). Leveraging the extremely small dimensions of microfluidic channels and harnessing the photonic crystal effects, they showcased an unprecedented level of sensitivity, determining pyrene (1 ppb) in a mixture of Raman dye and cocaine (10 ppb) in plasma. This research underscores the distinctive advantages of DADs as an emerging microfluidic platform for sensing, particularly in the context of drug screening.

Lafuente et al. crafted a tailor-made SERS-based nanosensor for the detection of dimethyl methylphosphonate (DMMP) (Lafuente et al., 2018). DMMP serves as a substitute molecule for nerve agents, a group of chemicals of significant concern because of their exceptionally high toxicity, historical deployment, and persistence. The platform was meticulously designed utilizing straightforward components, primarily employing AuNPs coated with a citrate layer (Figure 5.5C). The citrate coating plays a pivotal role by effectively capturing the targets in the immediate surroundings of the AuNPs surface through mutual hydrogen bonding effects. These marks detected DMMP in the gas phase at ppm concentrations, specifically as low as 130 ppb. Such levels of sensitivity are indicative of the potential for real-world deployment, particularly in emergency scenarios. Moreover, the SERS platforms developed in this study offer advantages in terms of ease of preparation and reusability, presenting a promising avenue for the urgent detection of chemical threat agents in practical and realistic settings.

Lafuente et al. also presented a plasmonic-sorbent based thin-film nanosensor, incorporating a Raman, in the detection of chemical threat agents (Lafuente et al., 2021). This film is composed of densely packed core-shell Ag and Au nanorods, each meticulously enclosed with a ZIF-8 framework, in trapping 2-chloroethyl ethyl sulfide (CEES) of DMMP from the gas phase (Figure 5.5D). They delved into the underlying adsorption mechanisms responsible for capturing molecules with the framework, as well as interactions between DMMP and the Ag surface, through computational calculations. The thin films exhibit remarkable SERS sensing capabilities. For DMMP, they reported a limit of detection as low as 0.2 ppbV. Furthermore, experiments conducted with the Raman system successfully detected 2.5 ppmV for DMMP in air and 76 ppbV for CEES in a nitrogen atmosphere, with reply times of 21 and 54 seconds, respectively. This system heralds the advent of handheld SERS-based sensing at extraordinarily low concentrations, with vast applications encompassing home security, safeguarding infrastructure, chemical process monitoring, and personalized medicine.

Saito et al. presented an advanced self-governing air sampling and detection system designed to assess the existence of chemical threat agents (Saito et al., 2018). This developed system incorporates a mist generator-assisted air collection mechanism, capable of sampling at a rate of 338 L/min. It harnesses cutting-edge sensing technologies, including electrochemical measurement, AuNPs-based localized surface plasmon resonance, and microfluidic platform PCR. These technologies enable the system to detect minute concentrations of hazardous substances, even below the mean lethal dose (LD_{50}), encompassing nerve gases like sarin and VX, toxic proteins such as BTX/A/Hc and ricin, and pathogens like an anthrax simulant. One of the key advantages of this system is its remarkable operational efficiency, with a collection and detection process requiring just 5 to 15 minutes. Additionally, the system integrates sample preparation seamlessly, eliminating the requirement for direct human mediation. Beyond its sensitivity and user-friendliness, the system's portability stands out as a significant asset, particularly for first responders. It facilitates immediate risk assessment and on-site event mitigation, underscoring its potential to be a valuable tool in critical situations.

5.4 CONCLUSION AND FUTURE PERSPECTIVES

Microfluidic-based plasmonic nanosensors have emerged as a pioneering platform at the crossroads of microfluidics and plasmonic nanotechnology, holding immense promise for the rapid and precise detection of biological and chemical threat agents. This chapter has illuminated the synergistic potential of integrating these two technologies, showcasing their ability to address critical challenges in threat detection. The case studies provided demonstrate the wide applicability of microfluidic-based plasmonic nanosensors, not only in terms of the threats they can detect but also in the versatility of their sensing mechanisms. Looking ahead, the future of microfluidic-based plasmonic nanosensors in threat detection is both exciting and challenging. As researchers continue to innovate, the integration of microfluidic-based plasmonic nanosensors with miniaturized and portable devices is expected to enable field-deployable sensing solutions. Advances in nanofabrication techniques and material science will further enhance the sensitivity, selectivity, and stability of plasmonic nanosensors. The incorporation of artificial intelligence and machine learning algorithms will contribute to real-time data analysis and decision-making, expediting the identification of threats. However, the journey is not without obstacles. The translation of microfluidic-based plasmonic nanosensors from the laboratory to practical applications demands addressing issues related to sample preparation, long-term stability, and cost-effectiveness. Standardization protocols and regulatory considerations will play a pivotal role in establishing the credibility and widespread adoption of microfluidic-based plasmonic nanosensors-based detection systems. In conclusion, the marriage of microfluidics and plasmonic nanosensors has opened new frontiers in the realm of threat detection. Microfluidic-based plasmonic nanosensors offer a transformative approach that combines precision, speed, and accuracy, paving the way for advancements in security, healthcare, and environmental monitoring. As this technology matures, interdisciplinary collaboration, innovative research, and a keen awareness of practical implementation will be crucial in harnessing the full potential of microfluidic-based plasmonic nanosensors for the betterment of society.

The future of microfluidic-based plasmonic nanosensors holds tremendous potential for revolutionizing threat detection and sensing technologies across various domains. As researchers continue to push the boundaries of knowledge and innovation, several exciting future perspectives emerge:

- Multimodal Sensing: Integrating multiple sensing modalities, such as electrical, optical, and mechanical, could enable microfluidic-based plasmonic nanosensors to provide more comprehensive and accurate threat detection. Combining different sensing mechanisms can enhance sensitivity and reduce false positives.
- Real-Time Monitoring: The development of real-time monitoring systems that can continuously track changes in complex samples will be crucial for early threat detection. This could be achieved through the integration of microfluidic-based plasmonic nanosensors with automated sampling, data analysis, and decision-making algorithms.
- Nanomaterial Innovations: Advances in nanomaterials, including novel plasmonic materials and nanostructures, will enhance the performance of microfluidic-based plasmonic nanosensors. Tailoring the properties of these nanomaterials can lead to improvements in sensitivity, specificity, and stability.
- Point-of-Care Diagnostics: Microfluidic-based plasmonic nanosensors integrated into portable and handheld devices could bring threat detection capabilities to the point of need, whether in the field, clinical settings, or remote locations. This would democratize access to advanced detection technologies.
- Environmental Monitoring: Beyond security, microfluidic-based plasmonic nanosensors have significant potential for monitoring environmental pollutants, water quality, and air contaminants. Their ability to operate in complex samples makes them valuable tools for safeguarding ecosystems and public health.
- Bioinformatics Integration: Incorporating bioinformatics and machine learning algorithms will enable the interpretation of complex sensor data.

This integration can improve the accuracy of threat identification and facilitate data-driven decision-making.

- Global Collaborations: Collaboration between researchers, industries, and policymakers will accelerate the development, validation, and deployment of microfluidic-based plasmonic nanosensors-depended technologies. Open dialogue and partnerships are essential to overcoming technical and regulatory challenges.

In conclusion, the future of microfluidic-based plasmonic nanosensors is marked by groundbreaking advancements that will reshape threat detection and sensing paradigms. With sustained innovation, interdisciplinary collaboration, and a commitment to addressing real-world challenges, microfluidic-based plasmonic nanosensors have the potential to become pivotal tools in ensuring global security, health, and sustainability.

REFERENCES

Ahmad, R., Félidj, N., Boubekeur-Lecaque, L., Lau-Truong, S., Gam-Derouich, S., Decorse, P., Lamouri, A., Mangeney, C., 2015. Water-soluble plasmonic nanosensors with synthetic receptors for label-free detection of folic acid. *Chem. Commun.* 51, 9678–9681.

Akceoglu, G.A., Saylan, Y., Inci, F., 2021. A snapshot of microfluidics in point-of-care diagnostics: Multifaceted integrity with materials and sensors. *Adv. Mater. Technol.* 6, 2100049.

Akgönüllü, S., Çalışır, M., Özbek, M.A., Erkek, M., Bereli, N., Denizli, A., 2023. Chapter 3 – Plasmonic nanosensors for chemical warfare agents, in: Das, S., Thomas, S. (Eds.), *Sensing of Deadly Toxic Chemical Warfare Agents, Nerve Agent Simulants, and their Toxicological Aspects*, Elsevier, pp. 81–96.

Ameen, A., Gartia, M.R., Hsiao, A., Chang, T.W., Xu, Z., Liu, G.L., 2015. Ultra-sensitive colorimetric plasmonic sensing and microfluidics for biofluid diagnostics using nanohole array. *J. Nanomater.* 2015, 460895.

Amirjani, A., Kamani, P., Hosseini, H.R.M., Sadrnezhaad, S.K., 2022. SPR-based assay kit for rapid determination of Pb^{2+}. *Anal. Chim. Acta* 1220, 340030.

Amrollahi, P., Zheng, W., Monk, C., Li, C.-Z., Hu, T.Y., 2021. Nanoplasmonic sensor approaches for sensitive detection of disease-associated exosomes. *ACS Appl. Bio Mater.* 4, 6589–6603.

Anker, J.N., Hall, W.P., Lyandres, O., Shah, N.C., Zhao, J., Van Duyne, R.P., 2008. Biosensing with plasmonic nanosensors. *Nat. Mater.* 7, 442–453.

Aslan, Y., Atabay, M., Chowdhury, H.K., Göktürk, I., Saylan, Y., Inci, F., 2023. Aptamer-based point-of-care devices: Emerging technologies and integration of computational methods. *Biosensors* 13, 569.

Bai, L., Wang, X., Zhang, K., Tan, X., Zhang, Y., Xie, W., 2019. Etchable SERS nanosensor for accurate pH and hydrogen peroxide sensing in living cells. *Chem. Commun.* 55, 12996–12999.

Bhatia, S.N., Ingber, D.E., 2014. Microfluidic organs-on-chips. *Nat. Biotechnol.* 32, 760–772.

Black, R., 2016. Development, historical use and properties of chemical warfare agents, in: Worek, F., Jenner, J. H. Thiermann, H. (Eds.), *Chemical Warfare Toxicology*, The Royal Society of Chemistry, pp. 1–28.

Brahmkhatri, V., Pandit, P., Rananaware, P., D'Souza, A., Kurkuri, M.D., 2021. Recent progress in detection of chemical and biological toxins in water using plasmonic nanosensors. *Trends Environ. Anal. Chem.* 30, e00117.

Bruijns, B., Veciana, A., Tiggelaar, R., Gardeniers, H., 2019. Cyclic olefin copolymer microfluidic devices for forensic applications. *Biosensors* 9, 85.

Campbell, S.B., Wu, Q., Yazbeck, J., Liu, C., Okhovatian, S., Radisic, M., 2020. Beyond polydimethylsiloxane: Alternative materials for fabrication of organ-on-a-chip devices and microphysiological systems. *ACS Biomater. Sci. Eng.* 7, 2880–2899.

Caprioli, A., Morabito, S., Brugère, H., Oswald, E., 2005. Enterohaemorrhagic *Escherichia coli*: Emerging issues on virulence and modes of transmission. *Vet. Res.* 36, 289–311.

Centers for Disease Control and Prevention, 2018. *CDC/Bioterrorism Agents/Diseases (by category)* [WWW Document]. URL https://emergency.cdc.gov/agent/agentlist-category.asp

Chen, L., Wu, D., Yoon, J., 2018. Recent advances in the development of chromophore-based chemosensors for nerve agents and phosgene. *ACS Sensors* 3, 27–43.

Choi, I., Choi, Y., 2012. Plasmonic nanosensors: Review and prospect. *IEEE J. Sel. Top. Quantum Electron.* 18, 1110–1121.

Cui, P., Wang, S., 2019. Application of microfluidic chip technology in pharmaceutical analysis: A review. *J. Pharm. Anal.* 9, 238–247.

De Oliveira, R.A.G., Nicoliche, C.Y.N., Pasqualeti, A.M., Shimizu, F.M., Ribeiro, I.R., Melendez, M.E., Carvalho, A.L., Gobbi, A.L., Faria, R.C., Lima, R.S., 2018. Low-cost and rapid-production microfluidic electrochemical double-layer capacitors for fast and sensitive breast cancer diagnosis. *Anal. Chem.* 90, 12377–12384.

Demirel, G., Usta, H., Yilmaz, M., Celik, M., Alidagi, H.A., Buyukserin, F., 2018. Surface-enhanced Raman spectroscopy (SERS): An adventure from plasmonic metals to organic semiconductors as SERS platforms. *J. Mater. Chem. C* 6, 5314–5335.

Dhull, N., Kaur, G., Jain, P., Mishra, P., Singh, D., Ganju, L., Gupta, V., Tomar, M., 2019. Label-free amperometric biosensor for *Escherichia coli* O157:H7 detection. *Appl. Surf. Sci.* 495, 143548.

Duan, Q., Liu, Y., Chang, S., Chen, H., Chen, J., 2021. Surface plasmonic sensors: Sensing mechanism and recent applications. *Sensors* 21, 5262.

Dursun, A.D., Borsa, B.A., Bayramoglu, G., Arica, M.Y., Ozalp, V.C., 2022. Surface plasmon resonance aptasensor for *Brucella* detection in milk. *Talanta* 239, 123074.

Ehuwa, O., Jaiswal, A.K., Jaiswal, S., 2021. *Salmonella*, food safety and food handling practices. *Foods* 10, 907.

Erdem, Ö., Eş, I., Saylan, Y., Atabay, M., Gungen, M.A., Ölmez, K., Denizli, A., Inci, F., 2023. In situ synthesis and dynamic simulation of molecularly imprinted polymeric nanoparticles on a micro-reactor system. *Nat. Commun.* 14, 4840.

Escobedo, C., Chou, Y.W., Rahman, M., Duan, X., Gordon, R., Sinton, D., Brolo, A.G., Ferreira, J., 2013. Quantification of ovarian cancer markers with integrated microfluidic concentration gradient and imaging nanohole surface plasmon resonance. *Analyst* 138, 1450–1458.

Estevez, M.C., Alvarez, M., Lechuga, L.M., 2012. Integrated optical devices for lab-on-a-chip biosensing applications. *Laser Photon. Rev.* 6, 463–487.

Fan, J.R., Zhu, J., Wu, W.G., Huang, Y., 2017. Plasmonic metasurfaces based on nanopincavity resonator for quantitative colorimetric ricin sensing. *Small* 13, 1601710.

Farmani, H., Farmani, A., Biglari, Z., 2020. A label-free graphene-based nanosensor using surface plasmon resonance for biomaterials detection. *Phys. E Low-dimensional Syst. Nanostructures* 116, 113730.

Francesko, A., Cardoso, V.F., Lanceros-Méndez, S., 2019. Lab-on-a-chip technology and microfluidics, in: Santos, H.A., Liu, D., Zhang, H. (Eds.), *Microfluidics for Pharmaceutical Applications* William Andrew Publishing, New York, US, pp. 3–36.

Gao, R., Ko, J., Cha, K., Ho Jeon, J., Rhie, G., Choi, J., DeMello, A.J., Choo, J., 2015. Fast and sensitive detection of an anthrax biomarker using SERS-based solenoid microfluidic sensor. *Biosens. Bioelectron.* 72, 230–236.

Gharib, G., Bütün, İ., Muganlı, Z., Kozalak, G., Namlı, İ., Sarraf, S.S., Ahmadi, V.E., Toyran, E., Van Wijnen, A.J., Koşar, A., 2022. Biomedical applications of microfluidic devices: A review. *Biosensors* 12, 1023.

Gimondi, S., Ferreira, H., Reis, R.L., Neves, N.M., 2023. Microfluidic devices: A tool for nanoparticle synthesis and performance evaluation. *ACS Nano* 17, 14205–14228.

Goy, C.B., Chaile, R.E., Madrid, R.E., 2019. Microfluidics and hydrogel: A powerful combination. *React. Funct. Polym.* 145, 104314.

Gupta, R.C., 2015. *Handbook of Toxicology of Chemical Warfare Agents.* Academic Press.

Hutter, E., Fendler, J.H., 2004. Exploitation of localized surface plasmon resonance. *Adv. Mater.* 16, 1685–1706.

Hwang, J., Cho, Y.H., Park, M.S., Kim, B.H., 2019. Microchannel fabrication on glass materials for microfluidic devices. *Int. J. Precis. Eng. Manuf.* 20, 479–495.

Im, H., Sutherland, J.N., Maynard, J.A., Oh, S.H., 2012. Nanohole-based surface plasmon resonance instruments with improved spectral resolution quantify a broad range of antibody-ligand binding kinetics. *Anal. Chem.* 84, 1941–1947.

Inci, F., Karaaslan, M.G., Mataji-Kojouri, A., Shah, P.A., Saylan, Y., Zeng, Y., Avadhani, A., Sinclair, R., Lau, D.T.Y., Demirci, U., 2020a. Enhancing the nanoplasmonic signal by a nanoparticle sandwiching strategy to detect viruses. *Appl. Mater. Today* 20, 100709.

Inci, F., Saylan, Y., Kojouri, A.M., Ogut, M.G., Denizli, A., Demirci, U., 2020b. A disposable microfluidic-integrated hand-held plasmonic platform for protein detection. *Appl. Mater. Today* 18, 100478.

Ji, C., Zhou, L., Chen, Y., Fang, X., Liu, Y., Du, M., Lu, X., Li, Q., Wang, H., Sun, Y., Lan, T., Ma, J., 2023. Microfluidic-LAMP chip for the point-of-care detection of gene-deleted and wild-type African swine fever viruses and other four swine pathogens. *Front. Vet. Sci.* 10, 1116352.

Jiang, L., Li, Q., Liang, W., Du, X., Yang, Y., Zhang, Z., Xu, L., Zhang, J., Li, J., Chen, Z., 2022. Organ-on-a-chip database revealed—Achieving the human avatar in silicon. *Bioengineering* 9, 685.

Jin, C., Wu, Z., Molinski, J.H., Zhou, J., Ren, Y., Zhang, J.X.J., 2022. Plasmonic nanosensors for point-of-care biomarker detection. *Mater. Today Bio* 14, 100263.

Jin, Y., Xiong, P., Xu, T., Wang, J., 2022. Time-efficient fabrication method for 3D-printed microfluidic devices. *Sci. Rep.* 12, 1233.

Juang, Y.J., Chiu, Y.J., 2022. Fabrication of polymer microfluidics: An overview. *Polymers.* 14, 2028.

Kanitthamniyom, P., Hon, P.Y., Zhou, A., Abdad, M.Y., Leow, Z.Y., Yazid, N.B.M., Xun, V.L.W., Vasoo, S., Zhang, Y., 2021. A 3D-printed magnetic digital microfluidic diagnostic platform for rapid colorimetric sensing of carbapenemase-producing Enterobacteriaceae. *Microsystems Nanoeng.* 7, 47.

Kirsch, J., Siltanen, C., Zhou, Q., Revzin, A., Simonian, A., 2013. Biosensor technology: Recent advances in threat agent detection and medicine. *Chem. Soc. Rev.* 42, 8733–8768.

Kong, X., Chong, X., Squire, K., Wang, A.X., 2018. Microfluidic diatomite analytical devices for illicit drug sensing with ppb-level sensitivity. *Sensors Actuators B Chem.* 259, 587–595.

Kotz, F., Mader, M., Dellen, N., Risch, P., Kick, A., Helmer, D., Rapp, B.E., 2020. Fused deposition modeling of microfluidic chips in polymethylmethacrylate. *Micromachines* 11, 873.

Kumar, V., Kim, H., Pandey, B., James, T.D., Yoon, J., Anslyn, E.V., 2023. Recent advances in fluorescent and colorimetric chemosensors for the detection of chemical warfare agents: A legacy of the 21st century. *Chem. Soc. Rev.* 52, 663–704.

Lafuente, M., De Marchi, S., Urbiztondo, M., Pastoriza-Santos, I., Pérez-Juste, I., Santamaria, J., Mallada, R., Pina, M., 2021. Plasmonic MOF thin films with Raman internal standard for fast and ultrasensitive SERS detection of chemical warfare agents in ambient air. *ACS Sensors* 6, 2241–2251.

Lafuente, M., Pellejero, I., Sebastián, V., Urbiztondo, M.A., Mallada, R., Pina, M.P., Santamaría, J., 2018. Highly sensitive SERS quantification of organophosphorous chemical warfare agents: A major step towards the real time sensing in the gas phase. *Sensors Actuators B Chem.* 267, 457–466.

Lakhera, P., Chaudhary, V., Bhardwaj, B., Kumar, P., Kumar, S., 2022. Development and recent advancement in microfluidics for point of care biosensor applications: A review. *Biosens. Bioelectron.* X11, 100218.

Lee, J.B., Choi, K.-H., Yoo, K., 2015. Innovative SU-8 lithography techniques and their applications. *Micromachines* 6, 1–18.

Leong, S.X., Leong, Y.X., Koh, C.S.L., Chen, J.R.T., Ling, X.Y., 2022. Chapter 2 – Nanoplasmonic materials for surface-enhanced Raman scattering, in: Wang, Y.B. (Ed.), in *Principles and Clinical Diagnostic Applications of Surface-Enhanced Raman Spectroscopy*. Elsevier, pp. 33–79.

Li, Z., Leustean, L., Inci, F., Zheng, M., Demirci, U., Wang, S., 2019. Plasmonic-based platforms for diagnosis of infectious diseases at the point-of-care. *Biotechnol. Adv.* 37, 107440.

Li, Z., Liu, H., Wang, D., Zhang, M., Yang, Y., Ren, T., 2022. Recent advances in microfluidic sensors for nutrients detection in water. *TrAC Trends Anal. Chem.* 158, 116790.

Liao, S., He, Y., Chu, Y., Liao, H., Wang, Y., 2019. Solvent-resistant and fully recyclable perfluoropolyether-based elastomer for microfluidic chip fabrication. *J. Mater. Chem. A* 7, 16249–16256.

Liu, J., He, H., Xiao, D., Yin, S., Ji, W., Jiang, S., Luo, D., Wang, B., Liu, Y., 2018. Recent advances of plasmonic nanoparticles and their applications. *Materials.* 11, 1833.

Luka, G., Ahmadi, A., Najjaran, H., Alocilja, E., DeRosa, M., Wolthers, K., Malki, A., Aziz, H., Althani, A., Hoorfar, M., 2015. Microfluidics integrated biosensors: A leading technology towards lab-on-a-chip and sensing applications. *Sensors* 15, 30011–30031.

Lv, Y., Qin, Y., Svec, F., Tan, T., 2016. Molecularly imprinted plasmonic nanosensor for selective SERS detection of protein biomarkers. *Biosens. Bioelectron.* 80, 433–441.

Ma, J., Jiang, L., Pan, X., Ma, H., Lin, B., Qin, J., 2010. A simple photolithography method for microfluidic device fabrication using sunlight as UV source. *Microfluid. Nanofluidics* 9, 1247–1252.

Madiyar, F.R., Bhana, S., Swisher, L.Z., Culbertson, C.T., Huang, X., Li, J., 2015. Integration of a nanostructured dielectrophoretic device and a surface-enhanced Raman probe for highly sensitive rapid bacteria detection. *Nanoscale* 7, 3726–3736.

Masson, J.F., 2020. Portable and field-deployed surface plasmon resonance and plasmonic sensors. *Analyst* 145, 3776–3800.

Mishra, N., Dhwaj, A., Verma, D., Prabhakar, A., 2022. Cost-effective microabsorbance detection based nanoparticle immobilized microfluidic system for potential investigation of diverse chemical contaminants present in drinking water. *Anal. Chim. Acta* 1205, 339734.

Mondal, B., Bhavanashri, N., Mounika, S.P., Tuteja, D., Tandi, K., Soniya, H., 2020. Microfluidics application for detection of biological warfare agents, in: Flora, S.J.S,Pachauri, V., (Eds.), *Handbook on Biological Warfare Preparedness*. Academic Press, Cambridge, US, pp. 103–131.

Mou, X.B., Ali, Z., Li, B., Li, T.T., Yi, H., Dong, H.M., He, N.Y., Deng, Y., Zeng, X., 2016. Multiple genotyping based on multiplex PCR and microarray. *Chinese Chem. Lett.* 27, 1661–1665.

National Research Council, 2011. *BioWatch and Public Health Surveillance: Evaluating Systems for the Early Detection of Biological Threats: Abbreviated Version*. National Academies Press.

Ndukaife, J.C., Mishra, A., Guler, U., Nnanna, A.G.A., Wereley, S.T., Boltasseva, A., 2014. Photothermal heating enabled by plasmonic nanostructures for electrokinetic manipulation and sorting of particles. *ACS Nano* 8, 9035–9043.

Nguyen, T., Sarkar, T., Tran, T., Moinuddin, S.M., Saha, D., Ahsan, F., 2022. Multilayer soft photolithography fabrication of microfluidic devices using a custom-built wafer-scale PDMS slab aligner and cost-efficient equipment. *Micromachines* 13, 1357.

Nielsen, J.B., Hanson, R.L., Almughamsi, H.M., Pang, C., Fish, T.R., Woolley, A.T., 2019. Microfluidics: Innovations in materials and their fabrication and functionalization. *Anal. Chem.* 92, 150–168.

Nikelshparg, E.I., Baizhumanov, A.A., Bochkova, Z. V, Novikov, S.M., Yakubovsky, D.I., Arsenin, A. V, Volkov, V.S., Goodilin, E.A., Semenova, A.A., Sosnovtseva, O., 2022. Detection of hypertension-induced changes in erythrocytes by SERS nanosensors. *Biosensors* 12, 32.

Nishat, S., Jafry, A.T., Martinez, A.W., Awan, F.R., 2021. Paper-based microfluidics: Simplified fabrication and assay methods. *Sensors Actuators B Chem.* 336, 129681.

Ofner, A., Moore, D.G., Rühs, P.A., Schwendimann, P., Eggersdorfer, M., Amstad, E., Weitz, D.A., Studart, A.R., 2017. High-throughput step emulsification for the production of functional materials using a glass microfluidic device. *Macromol. Chem. Phys.* 218, 1600472.

Oh, K.W., 2020. Microfluidic devices for biomedical applications: Biomedical microfluidic devices 2019. *Micromachines* 11, 370.

Omrani, M., Mohammadi, H., Fallah, H., 2021. Ultrahigh sensitive refractive index nanosensors based on nanoshells, nanocages and nanoframes: Effects of plasmon hybridization and restoring force. *Sci. Rep.* 11, 2065.

Panchal, N., Jain, V., Elliott, R., Flint, Z., Worsley, P., Duran, C., Banerjee, T., Santra, S., 2022. Plasmon-enhanced bimodal nanosensors: An enzyme-free signal amplification strategy for ultrasensitive detection of pathogens. *Anal. Chem.* 94, 13968–13977.

Peng, X., Kotnala, A., Rajeeva, B.B., Wang, M., Yao, K., Bhatt, N., Penley, D., Zheng, Y., 2021. Plasmonic nanotweezers and nanosensors for point-of-care applications. *Adv. Opt. Mater.* 9, 2100050.

Pereira, R.H.A., Keijok, W.J., Prado, A.R., de Oliveira, J.P., Guimarães, M.C.C., 2021. Rapid and sensitive detection of ochratoxin A using antibody-conjugated gold nanoparticles based on localized surface plasmon resonance. *Toxicon* 199, 139–144.

Pilát, Z., Kizovský, M., Ježek, J., Krátký, S., Sobota, J., Šiler, M., Samek, O., Buryška, T., Vaňáček, P., Damborský, J., 2018. Detection of chloroalkanes by surface-enhanced raman spectroscopy in microfluidic chips. *Sensors* 18, 3212.

Pohanka, M., 2019. Current trends in the biosensors for biological warfare agents assay. *Materials.* 12, 2303.

Qu, J.H., Dillen, A., Saeys, W., Lammertyn, J., Spasic, D., 2020. Advancements in SPR biosensing technology: An overview of recent trends in smart layers design, multiplexing concepts, continuous monitoring and in vivo sensing. *Anal. Chim. Acta* 1104, 10–27.

Rasheed, S., ul Haq, M.A., Ahmad, N., Hussain, D., 2023. Smartphone-integrated colorimetric and microfluidic paper-based analytical devices for the trace-level detection of permethrin. *Food Chem.* 429, 136925.

Ren, K., Zhou, J., Wu, H., 2013. Materials for microfluidic chip fabrication. *Acc. Chem. Res.* 46, 2396–2406.

Rodríguez, C.F., Andrade-Pérez, V., Vargas, M.C., Mantilla-Orozco, A., Osma, J.F., Reyes, L.H., Cruz, J.C., 2023. Breaking the clean room barrier: Exploring low-cost alternatives for microfluidic devices. *Front. Bioeng. Biotechnol.* 11, 1176557.

Rowland, C.E., Brown, C.W., Delehanty, J.B., Medintz, I.L., 2016. Nanomaterial-based sensors for the detection of biological threat agents. *Mater. Today* 19, 464–477.

Sackmann, E.K., Fulton, A.L., Beebe, D.J., 2014. The present and future role of microfluidics in biomedical research. *Nature* 507, 181–189.

Saito, M., Uchida, N., Furutani, S., Murahashi, M., Espulgar, W., Nagatani, N., Nagai, H., Inoue, Y., Ikeuchi, T., Kondo, S., Uzawa, H., Seto, Y., Tamiya, E., 2018. Field-deployable rapid multiple biosensing system for detection of chemical and biological warfare agents. *Microsystems Nanoeng.* 4, 17083.

Savransky, V., Ionin, B., Reece, J., 2020. Current status and trends in prophylaxis and management of anthrax disease. *Pathogens* 9, 370.

Saylan, Y., 2023. Unveiling the pollution of bacteria in water samples through optic sensor. *Microchem. J.* 193, 109057.

Saylan, Y., Akgönüllü, S., Denizli, A., 2020. Plasmonic sensors for monitoring biological and chemical threat agents. *Biosensors* 10, 142.

Saylan, Y., Denizli, A., 2019. Molecularly imprinted polymer-based microfluidic systems for point-of-care applications. *Micromachines* 10, 766.

Schmidt-Speicher, L.M., Länge, K., 2021. Microfluidic integration for electrochemical biosensor applications. *Curr. Opin. Electrochem.* 29, 100755.

Scott, S., Ali, Z., 2021. Fabrication methods for microfluidic devices: An overview. *Micromachines* 12, 319.

Shakeri, A., Jarad, N.A., Leung, A., Soleymani, L., Didar, T.F., 2019. Biofunctionalization of glass-and paper-based microfluidic devices: A review. *Adv. Mater. Interfaces* 6, 1900940.

Sharma, M., Luo, Y., Buchanan, R., 2011. 8 – Microbial safety of tropical and subtropical fruits, in: Yahia, E.M. (Ed.), *Woodhead Publishing Series in Food Science, Technology and Nutrition*. Woodhead Publishing, pp. 288–314.

Shi, H., Zhao, G., Liu, M., Fan, L., Cao, T., 2013. Aptamer-based colorimetric sensing of acetamiprid in soil samples: Sensitivity, selectivity and mechanism. *J. Hazard. Mater.* 260, 754–761.

Shrivastav, A.M., Cvelbar, U., Abdulhalim, I., 2021. A comprehensive review on plasmonic-based biosensors used in viral diagnostics. *Commun. Biol.* 4, 70.

Song, Y., Hormes, J., Kumar, C.S.S.R., 2008. Microfluidic synthesis of nanomaterials. *Small* 4, 698–711.

Srinivasan, B., Tung, S., 2015. Development and applications of portable biosensors. *J. Lab. Autom.* 20, 365–389.

Srivastava, A.K., Dev, A., Karmakar, S., 2018. Nanosensors and nanobiosensors in food and agriculture. *Environ. Chem. Lett.* 16, 161–182.

Syed, A., Mangano, L., Mao, P., Han, J., Song, Y.-A., 2014. Creating sub-50 nm nanofluidic junctions in a PDMS microchip via self-assembly process of colloidal silica beads for electrokinetic concentration of biomolecules. *Lab Chip* 14, 4455–4460.

Tabeling, P., 2023. *Introduction to Microfluidics*. Oxford University Press.

Taghavi, A., Rahbarizadeh, F., Abbasian, S., Moshaii, A., 2020. Label-free LSPR prostate-specific antigen immune-sensor based on GLAD-fabricated silver nano-columns. *Plasmonics* 15, 753–760.

Tan, H.Y., Nguyen, N.-T., Loke, W.K., Tan, Y.T., 2006. Microfluidic chip with optical sensor for rapid detection of nerve agent Sarin in water samples, in: *Proc.SPIE*, p. 64160M.

Tavakoli, H., Mohammadi, S., Li, X., Fu, G., Li, X., 2022. Microfluidic platforms integrated with nano-sensors for point-of-care bioanalysis. *TrAC Trends Anal. Chem.* 157, 116806.

Thies, J.-W., Kuhn, P., Thürmann, B., Dübel, S., Dietzel, A., 2017. Microfluidic quartz-crystal-microbalance (QCM) sensors with specialized immunoassays for extended measurement range and improved reusability. *Microelectron. Eng.* 179, 25–30.

Tian, F., Liu, C., Deng, J., Sun, J., 2022. Microfluidic separation, detection, and engineering of extracellular vesicles for cancer diagnostics and drug delivery. *Accounts Mater. Res.* 3, 498–510.

Tian, Y., Chen, Y., Chen, M., Song, Z.-L., Xiong, B., Zhang, X.-B., 2021. Peroxidase-like Au@ Pt nanozyme as an integrated nanosensor for Ag^+ detection by LSPR spectroscopy. *Talanta* 221, 121627.

Tokel, O., Yildiz, U.H., Inci, F., Durmus, N.G., Ekiz, O.O., Turker, B., Cetin, C., Rao, S., Sridhar, K., Natarajan, N., 2015. Portable microfluidic integrated plasmonic platform for pathogen detection. *Sci. Rep.* 5, 9152.

Torres, M.D.T., de Araujo, W.R., de Lima, L.F., Ferreira, A.L., de la Fuente-Nunez, C., 2021. Low-cost biosensor for rapid detection of SARS-CoV-2 at the point of care. *Matter* 4, 2403–2416.

Vaisocherová-Lísalová, H., Víšová, I., Ermini, M.L., Špringer, T., Song, X.C., Mrázek, J., Lamačová, J., Scott Lynn, N., Šedivák, P., Homola, J., 2016. Low-fouling surface plasmon resonance biosensor for multi-step detection of foodborne bacterial pathogens in complex food samples. *Biosens. Bioelectron.* 80, 84–90.

Versiani, A.F., Martins, E.M.N., Andrade, L.M., Cox, L., Pereira, G.C., Barbosa-Stancioli, E.F., Nogueira, M.L., Ladeira, L.O., da Fonseca, F.G., 2020. Nanosensors based on LSPR are able to serologically differentiate dengue from Zika infections. *Sci. Rep.* 10, 11302.

Walper, S.A., Lasarte Aragonés, G., Sapsford, K.E., Brown, C.W.I.I.I., Rowland, C.E., Breger, J.C., Medintz, I.L., 2018. Detecting biothreat agents: From current diagnostics to developing sensor technologies. *ACS Sensors* 3, 1894–2024.

Wang, G., Li, K., Purcell, F.J., Zhao, D., Zhang, W., He, Z., Tan, S., Tang, Z., Wang, H., Reichmanis, E., 2016. Three-dimensional clustered nanostructures for microfluidic surface-enhanced Raman detection. *ACS Appl. Mater. Interfaces* 8, 24974–24981.

Wang, S., Zheng, L., Cai, G., Liu, N., Liao, M., Li, Y., Zhang, X., Lin, J., 2019. A microfluidic biosensor for online and sensitive detection of *Salmonella typhimurium* using fluorescence labeling and smartphone video processing. *Biosens. Bioelectron.* 140, 111333.

Wang, C., Liu, M., Wang, Z., Li, S., Deng, Y., He, N., 2021. Point-of-care diagnostics for infectious diseases: From methods to devices. Nano Today 37, 101092.

Weingart, O.G., Eyer, K., Lüchtenborg, C., Sachsenheimer, T., Brügger, B., van Oostrum, M., Wollscheid, B., Dittrich, P.S., Loessner, M.J., 2019. In vitro quantification of botulinum neurotoxin type A1 using immobilized nerve cell-mimicking nanoreactors in a microfluidic platform. *Analyst* 144, 5755–5765.

Whitesides, G.M., 2006. The origins and the future of microfluidics. *Nature* 442, 368–373.

Xiao, W., Xiao, M., Fu, Q., Yu, S., Shen, H., Bian, H., Tang, Y., 2016. A portable smart-phone readout device for the detection of mercury contamination based on an aptamer-assay nanosensor. *Sensors* 16, 1871.

Yanik, A.A., Huang, M., Kamohara, O., Artar, A., Geisbert, T.W., Connor, J.H., Altug, H., 2010. An optofluidic nanoplasmonic biosensor for direct detection of live viruses from biological media. *Nano Lett.* 10, 4962–4969.

Ye, W., Yu, M., Wang, F., Li, Y., Wang, C., 2022. Multiplexed detection of heavy metal ions by single plasmonic nanosensors. *Biosens. Bioelectron.* 196, 113688.

Yildirim, D.U., Ghobadi, A., Ozbay, E., 2021. Nanosensors based on localized surface plasmon resonance, in: Denizli, A. (Ed.), *Plasmonic Sensors and Their Applications*, Wiley, New Jersey, US, pp. 23–54.

Yilmaz, E.G., Ece, E., Erdem, Ö., Eş, I., Inci, F., 2023. A sustainable solution to skin diseases: Ecofriendly transdermal patches. *Pharmaceutics* 15, 579.

Yılmaz, G.E., Saylan, Y., Göktürk, I., Yılmaz, F., Denizli, A., 2022. Selective amplification of plasmonic sensor signal for cortisol detection using gold nanoparticles. *Biosensors* 12, 482.

Zhai, J., Li, H., Wong, A.H.H., Dong, C., Yi, S., Jia, Y., Mak, P.I., Deng, C.X., Martins, R.P., 2020. A digital microfluidic system with 3D microstructures for single-cell culture. *Microsystems Nanoeng.* 6, 6.

Zhang, Z., Ma, P., Ahmed, R., Wang, J., Akin, D., Soto, F., Liu, B., Li, P., Demirci, U., 2022. Advanced point-of-care testing technologies for human acute respiratory virus detection. *Adv. Mater.* 34, 2103646.

Zhou, B., Xiao, X., Liu, T., Gao, Y., Huang, Y., Wen, W., 2016. Real-time concentration monitoring in microfluidic system via plasmonic nanocrescent arrays. *Biosens. Bioelectron.* 77, 385–392.

Zhou, J., Tao, F., Zhu, J., Lin, S., Wang, Z., Wang, X., Ou, J.Y., Li, Y., Liu, Q.H., 2019. Portable tumor biosensing of serum by plasmonic biochips in combination with nano-imprint and microfluidics. *Nanophotonics* 8, 307–316.

Zilberman, Y., Sonkusale, S.R., 2015. Microfluidic optoelectronic sensor for salivary diagnostics of stomach cancer. *Biosens. Bioelectron.* 67, 465–471.

Zordan, M.D., Grafton, M.M.G., Acharya, G., Reece, L.M., Cooper, C.L., Aronson, A.I., Park, K., Leary, J.F., 2009. Detection of pathogenic *E. coli* O157:H7 by a hybrid microfluidic SPR and molecular imaging cytometry device. *Cytom. Part A* 75A, 155–162.

6 Surface-Enhanced Raman Spectroscopy- Based Plasmonic Nanosensors for the Detection of Threats

Najma Memon

6.1 INTRODUCTION

When a substance is exposed to laser light, it can cause various phenomena, such as a change in the frequency of the scattered photons due to the vibration of the sample. This scattering process involves the interaction of the incoming photon with the substance (in any state of matter), and the photon can either loose or gain some energy, shifting its frequency to the red or blue end of the spectrum. A red shift happens when the photon gives some of its energy to the substance, increasing its internal energy. This is called Stokes Raman scattering. A blue shift happens when the substance gives some of its internal energy to the photon. This is called anti-Stokes Raman scattering. These two processes, Stokes and anti-Stokes scattering, are both part of Raman scattering.

The frequency and intensity of the Raman peaks for different vibration modes reveal information about the bond strength and the types of atoms involved in these modes, making them unique for a specific material or molecule. As a result, Raman spectroscopy is an effective method for classifying substances and molecules, researching how they interact with the environment, and keeping track of changes that occur during chemical processes. Moreover, Raman spectroscopy can be utilized for quantitative analysis because the strength of Raman peaks is connected to the quantity or concentration of molecules. Raman scattering, on the other hand, is incredibly weak, typically only generating one Raman photon for every 106 to 1010 incoming photons. Raman scattering can be improved using techniques like the resonance Raman effect, nonlinear Raman effects, or the plasmon-enhanced Raman effect to get over this restriction.

The resonance Raman effect increases Raman scattering when the wavelength of the laser light matches the electronic absorption energy of the sample, making

DOI: 10.1201/9781003459316-6

Raman scattering much more likely, often by several orders of magnitude. Nonlinear Raman effects, such as stimulated Raman scattering and coherent anti-Stokes Raman scattering, happen when high-power lasers are used, leading to signal intensity that grows faster than laser power, thus improving the Raman signal.

Surface plasmon resonance (SPR), which produces intensely amplified and localized electric fields on the surfaces of metal nanostructures, is what causes the plasmon-enhanced Raman effect. Raman signals produced by samples placed in these strengthened electric fields are up to 106 times stronger than those produced by samples in a free condition, possibly enabling single-molecule sensitivity. The surface-enhanced Raman spectroscopy (SERS), tip-enhanced Raman spectroscopy (TERS), and shell-isolated nanoparticle-enhanced Raman spectroscopy (SHINERS) take advantage of this phenomenon. Plasmon-enhanced Raman spectroscopy (PERS) is the name given to several Raman methods collectively (X. Wang, Huang, Hu, Yan, & Ren, 2020). For details, readers are referred to the book chapter "Principles of Surface-Enhanced Raman Spectroscopy" by Le Ru and Etchegoin (2009).

This chapter mainly focuses on the implication of SERS and novel substrates that are reported in the literature for sensing threat agents. SERS has great potential for chemical and biological sensing applications because it shows little interference in aqueous solutions yet is selective and sensitive (Stiles, Dieringer, Shah, & Van Duyne, 2008). A detailed review of applications of SERS in analytical chemistry is authored by Fan, Andrade, and Brolo (2020). The review discusses advances in substrate technology and microfluidic SERS systems for applications to food, biomolecules, etc. Furthermore, peptide modified SERS has been reported for detection of Ricin and Abrin, which are alarming threats because both can be weaponized and there is no antidote for exposure (Farrell et al., 2016). SERS has been designated as one of the techniques that can detect solid samples, along with FTIR technique. It has also been used to detect homemade explosives, which actually have a very complex matrix. The authors concluded that coherent anti-Stokes Raman spectroscopy presents potential for enhanced capabilities in this field (Forbes, Krauss, & Gillen, 2020). SERS and other Raman scattering techniques are reviewed by Lister, Sellors, Howle, and Mahajan (2020) and Mogilevsky, Borland, Brickhouse, and Fountain Iii (2012) for defense and security applications.

Photonics or plasmonic SERS and other technique-based sensors are reported for threat agents (Akgönüllü et al., 2021; Garibbo & Palucci, 2014) showing the tremendous potential of this field. Functionalized gold nanoparticles alone are also reported for threat detection (Upadhyayula, 2012). SERS has the capability to detect as low as single molecules of dengue virus 2 using SERS nanoboxes (Farokhinejad et al., 2022). Researchers used 2,7-mercapto-4-methylcoumarin (MMC) as a Raman reporter to enhance the selectivity and they obtained a characteristic Raman signal of 1603 cm^{-1} for identification of DENV2 NS1. Figure 6.1 shows the platform's framework for detecting DENV2 in a single infected mosquito. Figure 6.1A is the diagram of architecture for integrating nanoyeast scFvs affinity probes and nanobox-based SERS nanotags to detect DENV2 NS1 in a single infected mosquito sample.

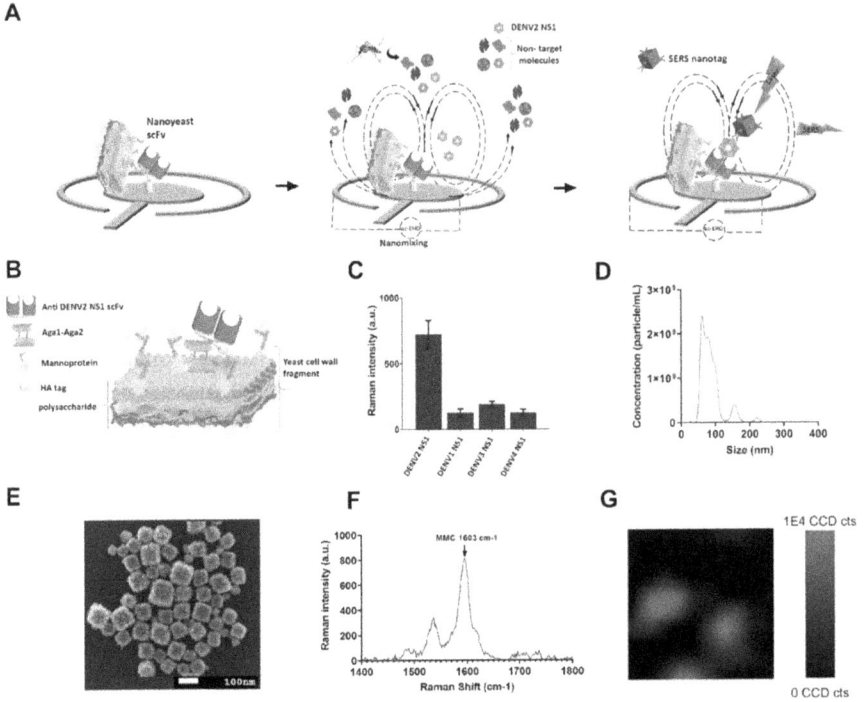

FIGURE 6.1   Framework of the platform for DENV2 detection in single infected mosquitoes. Reproduced with permission from Farokhinejad et al. (2022)

Figure 6.1B is the nanoyeast scFvs build schematic, Figure 6.1C is the nanoyeast scFvs HA8 serotype specificity for DENV2, and Figure 6.1D shows the use of NTA to characterize nanoyeast scFvs HA8. Figure 6.1E is a SEM image of the nanobox, and Figures 6.1F and G show SERS spectra and SERS mapping images, respectively. Figure 6.2 shows the evaluation of the platform's specificity for DENV2 NS1 detection. The photos in (A) and (B) show the matching SERS spectra and (C) shows the corresponding SERS mapping images for DENV2 NS1 and the corresponding negative controls (Zika, RBD, and PBS).

The field of SERS sensing is taken now to include artificial intelligence-based designs to enhance the detection capabilities of SERS for disease diagnostics, and biomolecule detection (Beeram, Vepa, & Soma, 2023) where diagnostic disease is coupled with statistical software and artificial intelligence.

SERS has shown tremendous progress toward the sensitive and selective detection of analytes in various fields, like biological samples, hazardous chemical and bio- analysis, and many others. This chapter is focused on applications of SERS in threat detection. Henceforth, both concepts—threat agents and SERS—will be briefly described, followed by a literature review specific to SERS for detecting threat agents.

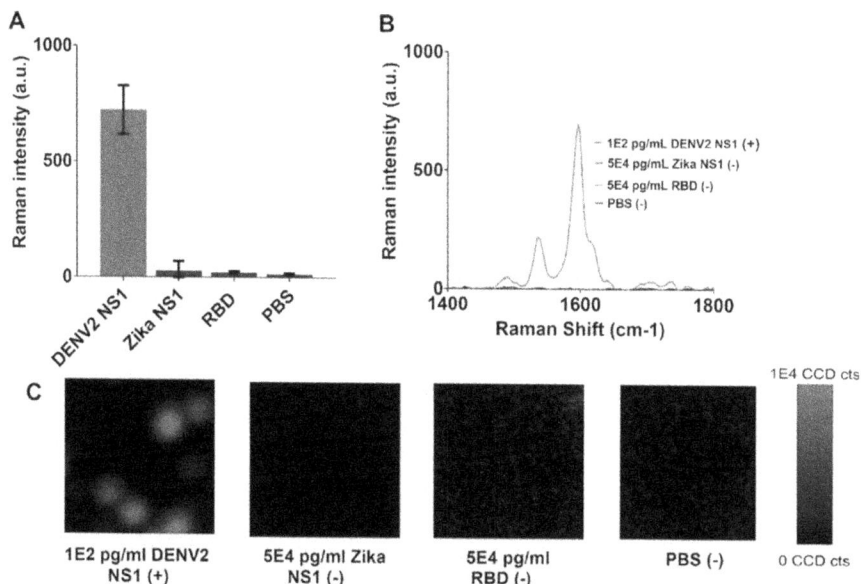

A

B

1E2 pg/mL DENV2 NS1 (+)
5E4 pg/mL Zika NS1 (-)
5E4 pg/mL RBD (-)
PBS (-)

C

1E2 pg/ml DENV2
NS1 (+)

5E4 pg/ml Zika
NS1 (-)

5E4 pg/ml
RBD (-)

PBS (-)

FIGURE 6.2   Scheme for the preparation of highly sensitive SERS substrate and selectivity of detection. Reproduced with permission from Farokhinejad et al. (2022)

6.2   THREAT AGENTS

The prospect of non-State actors, including terrorist groups and their supporters, gaining access to and using weapons and materials of mass destruction is a serious threat to international peace and security. Vladimir Voronkov (Under-Secretary-General, UN)

Humanity has faced various attacks of mass destruction which have involved the use of chemical, biological, nuclear, and radioactive elements as weapons. These kinds of attacks are serious threats to the security of countries, especially in the situation of cold wars. Threatening chemical and biological agents are defined under various treaties (Table 6.1). These treaties are signed by various countries to keep the globe safe, but keeping check on the malpractices or good practices of certain countries, to trace the detection of threats in suspected areas, has remained an integral part of inspection. Another threatening group of chemicals not listed in Table 6.1 is compounds that cause explosions (TNT and other similar compounds) or can cause mass destruction akin to chemical and biological threat agents (Hakonen, Andersson, Stenbæk Schmidt, Rindzevicius, & Käll, 2015). Nevertheless, reliable analytical methods that can see the target analytes (chemical, biological, radioactive, nuclear, or explosive) are real tools for inspectors to decide on malpractice or good practice.

There has been a range of analytical methodologies to detect and confirm suspect analytes (Cornish, 2007; Garibbo & Palucci, 2014; Hülseweh & Marschall,

TABLE 6.1
List of Treaties That Ban or Regulate the Use of Chemicals, and Bio- and Nuclear Agents (Cornish, 2007)

Treaty and its website	Year	Purpose	Definition
Nuclear Non-Proliferation Treaty (NPT) (https://treaties.unoda.org /t/npt)	1968	To prevent the spread of nuclear weapons and promote disarmament. It divides countries into two categories: Nuclear-armed states (the five recognized nuclear-weapon states), and non-nuclear-armed states. Non-nuclear-armed states commit not to develop nuclear weapons, while nuclear-armed states commit to disarmament.	Nuclear weapons: There are two primary types of nuclear weapon: Atomic Bomb (A-Bomb): An atomic bomb, also known as a fission bomb or nuclear bomb, relies on nuclear fission reactions. In an atomic bomb, uranium-235 or plutonium-239 isotopes are brought together to create a supercritical mass, initiating a chain reaction of nuclear fission events. This chain reaction releases a significant amount of energy in the form of an explosion. Thermonuclear Bomb (H-Bomb): A thermonuclear bomb, also known as a hydrogen bomb or fusion bomb, operates on the principle of nuclear fusion. It typically consists of a primary stage that uses a fission bomb to create the conditions necessary for nuclear fusion in a secondary stage. The fusion reactions in a hydrogen bomb release far greater energy than fission reactions, making these devices even more powerful than atomic bombs.
Comprehensive Nuclear-Test-Ban Treaty (CTBT) (https://www.ctbto.org/ our-mission/the-treaty)	1996	Bans all nuclear explosions for both civilian and military purposes. It has not yet entered into force as some key states have not ratified it.	
Iran Nuclear Deal (Joint Comprehensive Plan of Action or JCPOA) (https://www.eeas.europa .eu/eeas/nuclear -agreement-%E2%80 %93-jcpoa_en)	2015	Aimed to limit Iran's nuclear program in exchange for sanctions relief. However, it faced challenges and underwent changes in subsequent years.	

(Continued)

TABLE 6.1 (CONTINUED)
List of Treaties That Ban or Regulate the Use of Chemicals, and Bio- and Nuclear Agents (Cornish, 2007)

Treaty and its website	Year	Purpose	Definition
Biological Weapons Convention (BWC) (https://disarmament .unoda.org/biological -weapons/)	1975	It prohibits the development, production, and acquisition of biological weapons. It also aims to promote cooperation in the peaceful use of biology and biotechnology.	Biological weapons: Biological weapons can take various forms. Pathogens: Disease-causing microorganisms, such as bacteria (e.g., anthrax), viruses (e.g., smallpox), and fungi (e.g., mycotoxins), can be weaponized to infect and harm individuals or populations. Toxins: Poisons produced by certain organisms, like botulinum toxin or ricin, can be used as biological weapons when purified and delivered effectively. Biological vectors: In some cases, insects or animals that transmit diseases, such as mosquitoes (for diseases like malaria or dengue fever), could be used as vectors to spread illness.
Chemical Weapons Convention (CWC) (https://www.opcw.org/ chemical-weapons -convention)	1997	It prohibits the production, stockpiling, and use of chemical weapons and their precursors. It establishes a framework for the destruction of existing stockpiles.	Chemical weapons: Chemical weapons can take various forms, including: Nerve agents: These chemicals disrupt the normal functioning of the nervous system, leading to muscle paralysis and, in severe cases, respiratory failure. Examples include sarin and VX. Blister agents: These chemicals cause blistering and tissue damage to the skin, eyes, and respiratory tract. Mustard gas is a well-known blister agent. Choking agents: Choking agents, such as chlorine or phosgene, irritate the respiratory system, leading to severe breathing difficulties and potentially fatal lung damage. Blood agents: Blood agents interfere with the body's ability to transport oxygen, which can lead to suffocation. Examples include hydrogen cyanide and cyanogen chloride. Riot control agents: While not intended to cause death, riot control agents like tear gas can temporarily incapacitate individuals by causing eye and respiratory irritation.

2013). Recent trends for using plasmon-based sensors in connection with numerous techniques are rapidly evolving, whereby Raman spectroscopy has seen tremendous progress (López-López & García-Ruiz, 2014). This chapter, in the following sections, will talk about progress in Raman spectroscopy using plasmonic substrates.

6.3   WORKING PRINCIPLES AND INSTRUMENTATION OF SERS

Raman Spectroscopy: A photon's oscillating electromagnetic field interacts with a molecule to polarize the molecular electron cloud, which increases the molecule's energy by absorbing the photon's energy. The interactions of the photon with a molecule produces a virtual transient state between the two, which is unstable and gets separated by the emission of the photon immediately, which is termed as scattered light. The energy of the molecule typically remains constant during the photon's interaction, and the scattered photon has the same energy and wavelength as the incident photon. Elastic scattering, often known as Rayleigh scattering, is the most typical technique (Figure 6.3).

Raman scattering, an inelastic scattering mechanism where energy is exchanged between the molecule and the scattered photon, occurs in a relatively infrequent event (approximately 1 in 10 million photons). The scattered photon loses energy and its wavelength lengthens if the molecule gains energy from the photon during scattering (moving to a higher vibrational state). Stokes Raman scattering is what causes this (after G. G. Stokes). The scattered photon, on the other hand, receives energy and its wavelength shortens if the molecule loses energy by donating energy to the photon during the scattering (moving to a lower vibrational state). Stokes and Anti-Stokes are equally feasible processes, according to quantum mechanics. The Stokes scatter is more likely to occur when a group of molecules is present since the majority of them will be at the lowest vibrational state (Boltzmann distribution). As a result, the Stokes Raman scatter is always greater than the anti-Stokes scatter, which is why in Raman spectroscopy, the Stokes Raman scatter is typically measured.

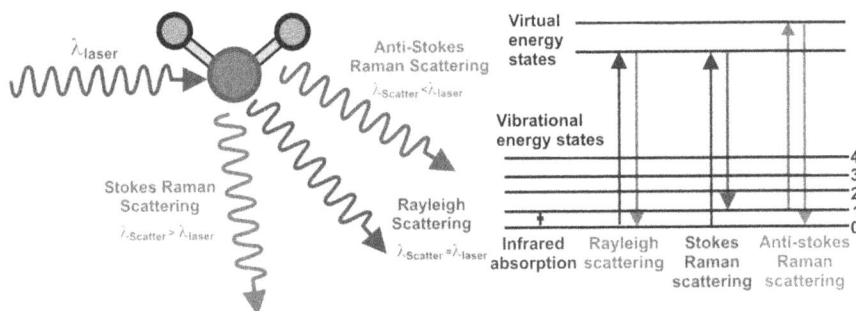

FIGURE 6.3   Scattering events when photon stikes a molecule (left) and energy states caused by interactions of photon with a mocleule. Open-source courtesy of *What is Raman Spectroscopy? | Raman Spectroscopy Principle* (edinst.com).

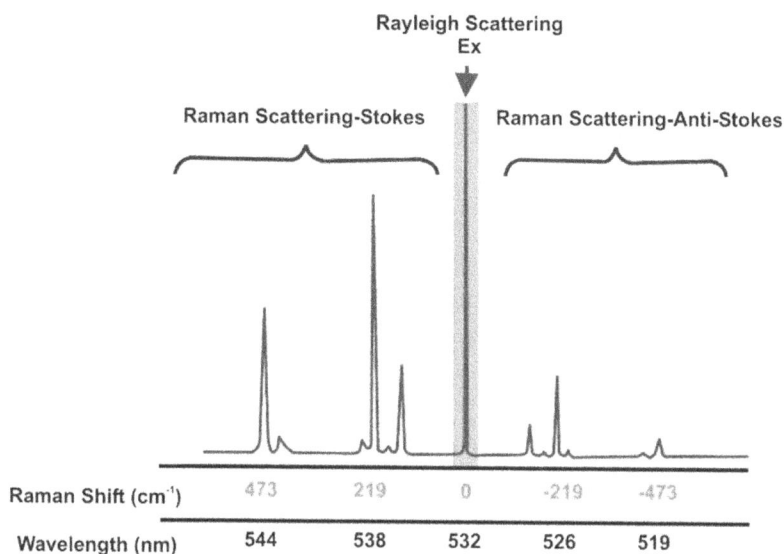

FIGURE 6.4 Graph elaboarting meaurement of Raman shift using 532 nm laser (open source, www.jasco-global.com).

Raman Shift: The Raman scattered light will have a different wavelength depending on the wavelength of the light that excites it. This means that the wavelength of the Raman scatter is not a useful number to compare spectra that are measured with different lasers. The Raman scatter position is then changed to a Raman shift that shows how far it is from the excitation wavelength (Figure 6.4).

Surface-Enhanced Raman Spectrometer: A Raman spectrometer comprises fundamental elements such as a light source, beam management optics, and a detector. In a common Raman spectroscopy detection technique, a spectrograph and image camera are employed to capture the spectrum in a single shot, eliminating the need for moving parts. In current portable Raman spectrometers, a laser acts as the excitation source for Raman scattering, alongside other essential components. Solid-state lasers, with popular wavelengths like 532 nm, 785 nm, 830 nm, and 1064 nm, are commonly used in modern Raman equipment. Shorter wavelength lasers provide good Raman cross section and produce stronger signals but are avoided due to interfering fluorescence possibilities. Therefore, many Raman systems opt for a 785 nm laser.

For laser energy signal processing (transmission and reception) to and from the sample, fiber optic cables are used. A notch filter is used to eliminate Rayleigh and anti-Stokes scattering. The dispersed Stokes light is then directed to a, typically, holographic grating which then reaches a CCD detector, which in turn captures the light, producing the Raman spectrum.

The schematic of a Raman spectrometer is illustrated in Figure 6.5. The depicted setup focuses the green illuminating laser onto a line on the sample in slit-scanning

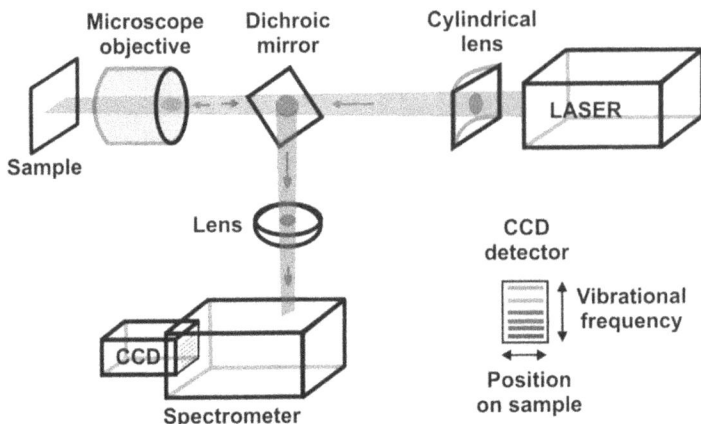

FIGURE 6.5 Schematic of Raman spectrometer. (Reproduced under Creative Commons Attribution 3.0 Unported (Downes & Elfick, 2010).)

mode. However, this configuration can be altered to a spot by removing the cylindrical lens. In spot mode, the Raman-shifted light (colored red) is separated from the laser light by a dichroic mirror and distributed along a vertical line on the two-dimensional CCD detector when a spot is illuminated on the sample. In slit-scanning mode, multiple spectra are acquired simultaneously, with each location along the sample's line resulting in a spectrum along the CCD detector.

6.4   SERS SUBSTRATES

As described earlier, in the introduction to Raman spectroscopy, Stokes and anti-Stokes shifts are very small fractions of the overall scattering phenomenon. Henceforth, poor sensitivity of assay methods has hampered the use of Raman spectroscopy in the detection of analytes at trace levels. SERS started its journey from roughened coinage (electrolithography od surface) (Albrecht & Creighton, 1977; Jeanmaire & Van Duyne, 1977). The EMR field was found to be significantly different at the surface of the metal, which, when it interacted with Raman active molecules, significantly enhanced the peak intensities of the molecules. After this observation, it took a long time to understand the underlying mechanism of the enhancement factor of coupling roughened coinage metals with Raman active molecules. The author (Stiles et al., 2008) described the evolution of the SERS process and claimed that it was the result of two factors: (a) An electromagnetic enhancement factor, and (b) a chemical enhancement component. The Raman scattering intensity is inversely proportional to the square of the induced dipole moment, μ_{ind}, which is the sum of the Raman polarizability and the intensity of the input electromagnetic field, E. This relationship explains how these two components are present. The local electromagnetic field is multiplied by ten when the localized surface plasmon resonance (LSPR) of a metal surface or nanoparticle with nanostructure is stimulated, for instance. The electromagnetic enhancement factor is of order 10^4 since Raman scattering approximately grows as E^4. Scientists hypothesized that the activation of adsorbate localized electronic resonances or metal-to-adsorbate charge-transfer resonances was responsible for the chemical enhancement factor of 10^2 (e.g., resonance Raman scattering). Also noteworthy is the fact that surface-enhanced resonance Raman scattering was feasible with combined SERS and resonance Raman scattering enhancement factors in the region of 10^9–10^{10}. However, it took until this past decade for the full potential of this improvement to be realized, which rendered SERS significantly more sensitive than standard Raman spectroscopy. Early systems had a lot of issues with reproducibility since the substrates created by electrochemical roughening were not well-defined. Modern nanotechnology developments and the 1997 discovery of single-molecule SERS (Nie & Emory, 1997), the final threshold of detection, have sparked a flood of new research and transformed SERS from a fascinating physical phenomenon into a reliable analytical method.

Since then, lots of substrates are developed and reported in the literature (Campion & Kambhampati, 1998) and many of them are commercially available. A wide range of strategies are adopted to develop labeled or label-free plasmonic surfaces to detect a variety of molecules. For details readers are referred to papers published by Stiles et al. (2008), Alvarez-Puebla and Liz-Marzan (2012), Fan et al. (2020), and Li, Huang, & Lu, (2020).

Figure 6.6 shows various parameters that affect the surface-enhanced Raman signals (Bharati & Soma, 2021). The most prominent factors are SERS substrates (SERS active materials and platforms), therefore, the substrates may be considered to be the heart of the SERS sensing systems and mainly responsible for the sensitivity and selectivity of assay methods. Despite of lot of work and enhanced understanding for developing SERS substrates, it is never simple to design highly efficient substrates. The distance of the plasmonic material and the analyte is crucial for the

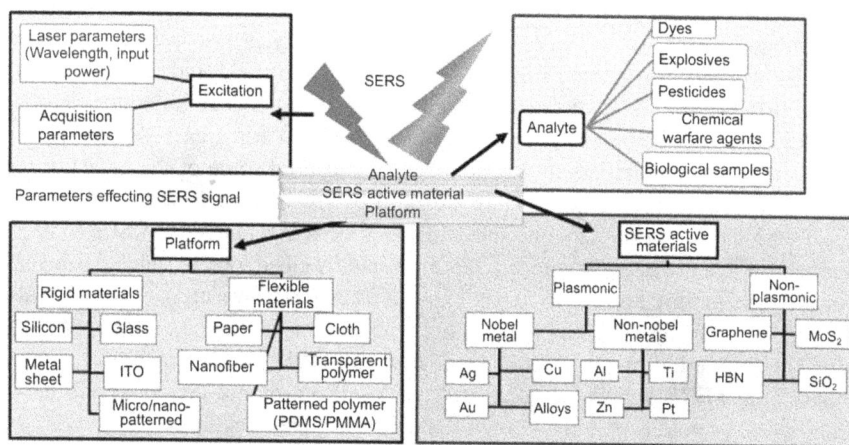

FIGURE 6.6 Factors that affect the signal in surface-enhanced Raman spectroscopic measurements. Reproduced from Bharati & Soma (2021) under Creative Commons Attribution 4.0 International License.

Raman enhancement factor. A tutorial review to control these factors is reported by Alvarez-Puebla and Liz-Marzan (2012) where delicacy of design is discussed. Another good read by X. Wang et al. (2020) shows the tremendous impact of the geometry of plasmonic particles on Raman signals. It has also been reported that plasmonic nanostructures—a 2D array of circular "nanopillars inside square nanoholes"—were able to polarize independent SERS substrates for the portable detection of chemical and biological molecules, in the sensing of 2,4-dintrotoluene (Sharma, Gupta, Das, & Dhawan, 2020).

The detection of chemical and explosive agents has seen some success as well. Liszewska et al. (2019) investigated various commercial substrates for the detection of explosives. Only Klarite 312 and GaN-pillars, two of the five analyzed SERS substrates, permitted trace analyses of all the evaluated explosive compounds. Depending on the explosive material and the Raman spectrometer utilized, the observed explosive concentrations in both cases ranged from a few to hundreds of g/ cm^2. According to our research, the best SERS substrates for explosives trace analysis are those which have hot spots that are equally and densely dispersed throughout the entire active area of the substrate.

6.5 SERS IN THE DETECTION OF CHEMICAL THREAT AGENTS

SERS has been widely used in the detection of chemical threat agents. Table 6.2 lists a few important reviews in the field. Chemical threat agents and treaties related to banning or regulated use for protection from threat chemicals are defined. These chemicals are small molecules with deadly properties and their sensitive detection in complex matrices and in all phases (vapor, liquids, and solids) is extremely important for the safety and security of the globe (Altmann, Oelze, & Niemeyer, 2013).

These reviews on the detection of chemical threat agents started appearing in the literature from 2014 onwards. Each review has seen remarkable sensitivity enhancement using SERS, whereas attomole sensitivity is also achieved in some cases. However, deployment of SERS sensors in the real field with high specificity with remote sensing (safe distance or unassisted by humans), and monitoring of air samples, are a few points of focus for further research in this area.

Nevertheless, researchers in this area have considered some of the issues of real field sensing to be overcome and have made progress in the gas sensing of chemical threats. Table 6.3 shows some literature specific to applications of SERS for the detection of chemical and explosive threat agents. A close look at Table 6.3 reveals that SERS is mainly applied to detect explosive agents. However, some reports also find that it can be applied to chemical threat agents like VX, tabun (Hakonen et al., 2016), and dimethyl methylphosphonate (Lafuente, Pellejero, et al., 2018), as well. Most of the SERS substrates used are structured plasmonic nanomaterials using either silver or gold as metals. Few of the substrates are functionalized as well, for example, monoethanolamine (MEA)—modified gold nanoparticles (Au NPs)—are reported for the sensing of TNT with a detection limit of 21.47 pM (Lin et al., 2020). In all cases, huge sensitivity enhancement was observed for SERS signals which is attributed to newly developed sensing platforms.

TABLE 6.2

Important Review Articles Related to the Application of SERS in the Sensing of Explosives and Chemical Threat Agents

Authors	Year	Review Title	Selected Concluding Comments	Ref
Bharati, M. S. S. and V. R. Soma	2021	Flexible SERS substrates for hazardous materials detection: recent advances	Flexible-based SERS substrates, including paper/cellulose, polymer nanofibers, 3D sponges, fabrics, etc., and their potential for on-site detection of explosives, pesticides, chemical warfare agents, drugs for homeland security, food safety, and medical fields. A summary of commercially available SERS detection systems using flexible substrates is also given.	(Bharati & Soma, 2021)
Lister A. P., Sellors W. J., Howle C. R. and Mahajan S.	2020	Raman Scattering Techniques for Defense and Security Applications	For Raman techniques to develop into common tools for this role, spectral libraries of materials of interests and close analogues must be developed and constantly maintained and updated as new hazards emerge. This library-building, in turn, requires that standardized methods be established to allow for the comparison of results between machines and laboratories across the globe.	(Lister et al., 2020)
Choi J., Kim J.-H., Oh J.-H., and Nam J.-M.	2019	Surface-enhanced Raman scattering-based detection of hazardous chemicals in various phases and matrices with plasmonic nanostructures	Chemical targets in the gaseous or vapor phase are arguably the hardest to identify with a SERS sensor. Due to the nature of both matrix and target analytes, SERS sensors that need to target airborne analytes must overcome their targets' high mobility and poor specificity. Because contemporary plasmonic substrates do not have highly specific binding affinity to air-phase analytes, additional care must be taken to properly equip the substrate to better capture or extract their targets from the air with chemical manipulation of the substrate surface or mechanical rearrangement to increase adsorption. Therefore, air-phase sensors require interacting surfaces that are structurally specialized in collecting the target material into the concentrated hot-spot areas.	(Choi, Kim, Oh, & Nam, 2019)

(Continued)

TABLE 6.2 (CONTINUED)

Important Review Articles Related to the Application of SERS in the Sensing of Explosives and Chemical Threat Agents

Authors	Year	Review Title	Selected Concluding Comments	Ref
Zapata, F., et al.	2016	Detection and identification of explosives by surface-enhanced Raman scattering	Functionalization makes SERS detection a bit more laborious because of the necessity to add an extra component to the SERS mixture (sample + NPs or metal foil). It is properly said that Raman spectroscopy is a non-destructive and non-invasive technique. However, that is not the case when using SERS. SERS is also non-destructive, but it is seriously invasive. This fact can be very unfavorable for forensic purposes in which the preservation of the evidence is mandatory.	(Zapata, López-López, & García-Ruiz, 2016)
Hakonen, A., et al.	2015	Explosive and chemical threat detection by surface-enhanced Raman scattering: A review	Several SERS studies have demonstrated attomolar sensitivities for TNT and DNT, which to the best of our knowledge is practically impossible using realistic competing techniques. New and progressing nanofabrication techniques are revolutionizing SERS and are likely to accelerate progress even further, as will the continued development of better and cheaper handheld Raman instruments.	(Hakonen et al., 2015)
López-López, M. and C. García-Ruiz	2014	Infrared and Raman spectroscopy techniques applied to identification of explosives	Stand-off explosives detection is still an emerging field, and the latest approaches point to the use of laser techniques as tools to overcome this challenge. Raman spectroscopy, aside from being a laser technique, is experiencing a reduction in the price of Raman spectrometers, which encourages their application to explosive detection.	(López-López & García-Ruiz, 2014)

TABLE 6.3

Recent Literature of SERS in the Sensing of Explosives and Chemical Threat Agents

Analyte	Active Surface/ Substrate	Technique	Comments	Ref
Nitro-aromatic compounds	Modified thin gold (Au) film on silicon (Si)	SERS	Factor of ~108 for the most prominent Raman mode	(Awasthi, Goel, Rai, & Dubey, 2022)
TNT, RDX, and PETN	Sputtered gold–silver alloy film on a roughened glass surface	Surface-enhanced Raman spectroscopy	Enhanced sensitivity with an acquisition time of only 10 s	(Guenther, Dowgiallo, & Branham, 2017)
VX and tabun	Flexible gold-covered Si nanopillars	Handheld Raman device (SERS)	Enrichment of target molecules in plasmonic hot spots with high Raman enhancement	(Hakonen et al., 2016)
2,4,6-trinitrophenol (TNP)	Silver nanopillar substrates	Handheld Raman device (SERS)	20 ppt, corresponding attomole amounts	(Hakonen et al., 2017)
Picric acid (PA) and 3-nitro-1,2,4-triazol-5-one (NTO)	3D biomimetic superhydrophobic Ag micro/nano-pillar array (AMA)	Surface-enhanced Raman spectroscopy	Sensitivity to rhodamine 6G (R6G) in a low concentration of 10–15 mol L–1, PA in a low concentration of 10–12 mol L–1, and NTO in concentration as low as 10–13 mol L–1	(He et al., 2017)
Trinitrotoluene	ZnO–Ag hybrids	Surface-enhanced Raman spectroscopy	Ultrasensitive detection	(He, Wang, Li, Chen, & Zhang, 2014)
Chemical warfare agent	Rough silver segments embedded in gold nanowires	Surface-enhanced Raman spectroscopy	Enhancement of 105–107	(Hoffmann et al., 2009)
Neurotoxic agents surrogate in gas phase	Au decorated mesoporous silica NPs	Microfluidic SERS (SERS)	Huge sorption capacity toward the gas phase	(Lafuente et al., 2019)

(Continued)

TABLE 6.3 (CONTINUED)
Recent Literature of SERS in the Sensing of Explosives and Chemical Threat Agents

Analyte	Active Surface/ Substrate	Technique	Comments	Ref
Methylphosphonate (DMMP) or 2-chloroethyl ethyl sulfide (CEES)	Core–shell Au@Ag nanorods individually encapsulated within a ZIF-8 framework	Surface-enhanced Raman scattering	Au@Ag nanoparticles amplify the Raman signal of molecules located near their surface, the ZIF-8 framework plays a key role in the trapping of analytes	(Lafuente et al., 2021)
Dimethyl methylphosphonate	Au@citrate nanoparticles (NPs) on etched silica	Surface-enhanced Raman scattering	3D-fractal microstructures and real-time gas phase detection	(Lafuente, Berenschot, et al., 2018)
Trinitrotoluene (TNT)	Monoethanolamine (MEA) - modified gold nanoparticles (AuNPs)	Surface-enhanced Raman scattering	Detection limit of 21.47 pM	(Lin et al., 2020)
4-Aminothiophenol, perchlorates (ClO4−), chlorates (ClO3−) and nitrates (NO3−)	Silver nanoparticles (AgNPs) immobilized on polyurethane (PU) sponge	Surface-enhanced Raman scattering	Detection limit for perchlorates, chlorates, and nitrates were down to 0.13, 0.13, and 0.11 ng, respectively	(J. Liu, Si, & Zhang, 2019)
P-aminothiophenol	Ag nanotriangles	Surface-enhanced Raman scattering	Detection limit 10–8 M	(Chong Wang, Liu, & Dou, 2016)
2,4-dinitrotoluene (DNT) vapor	Two-dimensional nanoparticle cluster arrays	Surface-enhanced Raman scattering	Superb sensitivity and selectivity, as well as a fast response time of <5 min.	(J. Wang, Yang, Boriskina, Yan, & Reinhard, 2011)

(Continued)

TABLE 6.3 (CONTINUED)
Recent Literature of SERS in the Sensing of Explosives and Chemical Threat Agents

Analyte	Active Surface/ Substrate	Technique	Comments	Ref
Gaseous 2-chloroethyl ethyl sulfide	CuO-coated SERS substrate	Surface-enhanced Raman scattering with Si wafer-based organic template-etching	Detection of sulfur mustards which are otherwise not detectable using simple SERS	(Xu et al., 2021)
2,4,6-trinitrotoluene, 2,4-dinitrotoluene, and 1,3,5-trinitrobenzene	Gold nanoparticles	SERS	Filter Paper Substrate	(Fierro-Mercado & Hernández-Rivera, 2012)
3-nitro-1,2,4-triazol-5-one (NTO)	Porous silica aerogels decorated with silver nanoparticles (SiO2–Ag hybrids)	Surface-enhanced Raman scattering	Limit of detection for NTO is 7.94 × 10–10 M	(W. Liu et al., 2020)
Explosives TNT, RDX, and PETN	Gold triangular nanoprisms (AuTNPs)	Surface-enhanced Raman scattering	Enhancement factor = ~6.0 × 106	(Liyanage et al., 2018)
2,4,6-trinitrotoluene (TNT), hexanitrostilbene (HNS), and 2,4,6-trinitrophenylmethylnitramine (tetryl)	Citrate-reduced silver nanoparticles	Surface-enhanced Raman scattering	Detection limits reached at ng/mL levels after complex formation of analytes with enolate ion of 3-mercapto-2-butanone	(Milligan, Shand, Graham, & Faulds, 2020)
Uranyl ion	Highly ordered ZnO-Ag arrays	SERS-microfluidic biosensor	In-the-field sensing platforms	(He, Zhou, Liu, & Wang, 2020)
Cyanide, 2-chloroethyl ethyl sulfide, phosphonates, Gram-positive, and Gram-negative bacteria	Gold and silver foil coupons	SERS	Detection of threat agents in water	(Spencer, Sylvia, Clauson, & Janni, 2002)

A few reports are also found in literature relating to the gas phase sensing of neurotoxic surrogate agents by Lafuente et al. (2019); Lafuente, Berenschot, et al. (2018); Lafuente et al. (2021); Lafuente, Pellejero, et al. (2018); Lafuente et al. (2020).

6.6   SERS IN THE DETECTION OF BIOLOGICAL THREATS

Biological pathogens and biotoxins are very effective in low doses, and they have no color or smell. They are not easily evaporated or absorbed through the skin. They can cause more damage than chemicals, in some cases, and they need only a few organisms to start a disease. Some of them take hours, days, or weeks to show symptoms, which can delay the diagnosis. It is also hard to tell them apart from the natural biological environment. It is important to know which bacteria are alive and which are dead to assess the biological threat, and high sensitivity is needed for biological detection and identification systems. There are not many commercial solutions for biological detection compared to chemical and nuclear detection, but many technologies are being studied (Hülseweh & Marschall, 2013).

SERS has attracted significant attention in the field of bacteria detection and identification, as well as drug susceptibility testing (Ravindranath, Wang, & Irudayaraj, 2011). There are multiple reviews published for the sensing of viruses (Savinon-Flores et al., 2021) and bacteria (Chaoguang Wang et al., 2019), pathogen diagnostics (Granger, Schlotter, Crawford, & Porter, 2016), biochemical sensing (Hankus & Cullum, 2006), and for SERS-based lateral flow immunoassay for point-of-care detection of SARS-CoV-2 (Yadav et al., 2021).

The anthrax biomarker poly-γ-d-glutamic acid (PGA) has been detected using PGA-conjugated gold nanoparticles as a SERS substrate with an LOD of a few picograms (Gao et al., 2015). Table 6.4 lists many applications of SERS in sensing of a range of possible biological threat agents, including *E. coli*, anthrax, and CoV-2. An interesting application of SERS in the sensing of CoV-2 is reported based on modified lateral flow immunoassay (LFIA) using gold nanoparticles with immobilized antibodies and 4-mercaptobenzoic acid as nanotag and registration of the nanotag binding by SERS spectrometry. A short assay time of 20 minutes with remarkable sensitivity was achieved (Serebrennikova et al., 2021).

6.7   PORTABLE RAMAN SPECTROMETERS

Raman spectrometry started its journey from a low signal intensity, due to the small fraction of Raman scattering and now it has reached maturity level where it has found a place in the on-site sensing of molecules of interest/threat agents.

Small handheld and portable Raman spectrometers have had a sharp rise in popularity over the past 15 years, as have the uses for them, leading to an expansion of the number of companies that make them. The use of these systems has increased, and sizeable new markets have appeared. Future developments in markets and a wider range of applications for Raman spectrometers are to be anticipated (Hargreaves, 2021).

TABLE 6.4
Recent Literature of SERS in the Sensing of Biological Threat Agents

Analyte	Active Surface/ Substrate	Technique	Comments	Ref
Staphylococcus aureus bacteria	AgNPs	Label-free SERS	Deep neural network application	(Ciloglu et al., 2021)
Escherichia coli	Silver (Ag) capped aluminum (Al) nanorods	Glancing angle deposition SERS	Raman signal of 107 was achieved considering R6G as a probe molecule	(Das et al., 2021)
Dengue virus (DENV2)	Plasmonic SERS nanoboxes	Surface-enhanced Raman scattering	Limit of detection of 500 fg of recombinant DENV2 NS1	(Farokhinejad et al., 2022)
Anthrax biomarker poly-γ-d-glutamic acid (PGA)	PGA-conjugated gold nanoparticles	(SERS)-based solenoid-embedded microfluidic device	LOD = 100 pg/mL	(Gao et al., 2015)
SARS-CoV-2 detection using reactive binding protein (RBD)	Au nanoparticles (NPs)	Surface-enhanced Raman scattering and Fish Model	SERS hot spots were constructed using aptamers and small molecules that can specifically bind to RBD at a concentration of 1.25 ng/mL.	(Huang et al., 2021)
Respiratory virus molecular signatures	Silver nanorod array	Surface-enhanced Raman scattering	Novel SERS assay can detect spectral differences between viruses, viral strains, and viruses with gene deletions in biological media.	(Shanmukh et al., 2006)
Yersinia pestis, *Francisella tularensis*, and *Bacillus anthracis*	Raman reporter-labeled gold nanoparticles	Colorimetric lateral flow assay (LFA) strips, Surface-enhanced Raman scattering	Tiny volume of pathogen sample (40 µL) is required for detection	(R. Wang et al., 2018)
SARS-CoV-2	Regular system	Miniaturized SERS with lateral flow immunoassay	Point-of-care (POC) testing arrangement	(Yadav et al., 2021)
Anthrax spores	Silver film over nanosphere	SERS handheld	Limit of detection (LOD) of ~2.6 × 103 spores	(Zhang, Young, Lyandres, & Van Duyne, 2005)
Staphylococcus aureus	Silver nanoflower silica core-shell structure (Ag NFs@SiO2)	Surface-enhanced Raman scattering	17 cfu/mL	(Zhu et al., 2021)

Yan and Vo-Dinh (2007) used a handheld Raman spectrometer for the screening of a wide range of chemical and biological substances for homeland defense applications. The field-deployable instrument is a fully integrated, adjustable, "point-and-shoot" Raman device that uses an avalanche photodiode (APD) for detection and an 830-nm diode laser for excitation. The work was excellent and paved the way for portable Raman devices, however, the system advised more optimization for field applications. In a different portable Raman spectrometer, the analyte was captured using antibodies and magnetic nanoparticles (Lesaicherre, Paxon, Mondello, Burrell, & Linsebigler, 2009).

Thermo Scientific's FirstDefender RM is a handheld Raman spectrometer with a 785-nm wavelength that can quickly identify unfamiliar solid and liquid samples in the field. A silver nanorod (AgNR) substrate with 1,2-di(4-pyridyl)ethylene provides sensitivity and effectiveness for SERS-based detection. For the first time, this portable handheld Raman spectrometer is used to detect and identify avian influenza A viruses (AIVs) using a multi-well AgNR SERS chip, because AIVs are considered a major emerging hazard to public health (Song, Driskell, Tripp, Cui, & Zhao, 2012).

A sensitive SERS substrate was created to enable handheld Raman spectrometers to detect gas-phase VX and HD. This is referred to as a nerve agent-specific handheld Raman spectrometer. The substrate was made up of quartz fibers enhanced with gold nanoparticles. Limits of detection (LOD) for VX and HD were 0.008 μg L^{-1} and 0.054 μg L^{-1}, respectively (Heleg-Shabtai, Sharabi, Zaltsman, Ron, & Pevzner, 2020). Microfluidic systems are currently exciting due to their benefits, which are effective detection, trace amounts of samples, low reagent consumption, convenience of use, and reduced danger of contamination which may result in false positives. The field of portable sensing has expanded. A recent review on microfluidics for sensing harmful substances is reported by Chen et al. (2022). The author concluded that microfluidic technology can combine biological and environmental sample separation and purification on the same chip. When examining the microscopic structural characteristics of substances, SERS technology has many benefits. The constructed microfluidic-SERS detection system can successfully supplement the current optical detection by integrating SERS technology into the microfluidic chip. The microfluidic-SERS detection system provides highly sensitive, quick, and high-throughput analysis while employing a smaller sample. The microfluidic-SERS detection system complies with the modern analytical technology development trend and satisfies current detection requirements.

6.8 CONCLUSION

SERS has evolved with very useful tools for the sensing of threat agents where much of the work is reported for chemical and biological threat agents, whereas, previously, much focus has remained on explosive agents and selected biological agents like anthrax, the Covid virus, and few others. The breadth of chemical and biological materials that can be used for mass destruction has always been a challenge which hampers the progress of predefined analyte detection systems. Therefore, most researchers focus on the detection of threat agents that are currently posing real threats. Consequently, most of the SERS methods reported in the literature are performed on a limited number of chemical threat agents or on their surrogate threat agents. Nevertheless, SERS has progressed remarkably in the field of sensitive and specific substrates which are seamlessly incorporated into field deployable portable systems. Homogenous plasmonic nanoparticles deposited onto various inactive substrates has long resolved the issue of reproducibility that was initially associated with roughened coin-based SERS. Nanolithography is an evolving area that can produce homogenously roughened surfaces with tenable surface properties. For example, one can create equally spaced nanostructures on metallic surfaces where geometry and depth of structure can also be regulated, depending upon the type of analyte. In turn, the surface can also be specifically responsive to the chosen analyte. Such 3D fractals as active platforms for SERS are reported for the gas sensing of G-nerve agents.

Other fields that have progressed in the detection of threat agents are the development of flexible substrates and microfluidic systems. Both paved the way for remotely controlled SERS sensing of threat agents in the real field. By introducing the SERS technology into the microfluidic chip, the formed microfluidic-SERS detection system can effectively expand the existing optical detection.

The future efforts of SERS in the detection of threat agents should be directed toward the development of gas-phase libraries of chemical threat agents listed by UN-CBRNE and integrated with microfluidic sensing platforms for the on-site remote sensing of threats. Also, similar approaches for biological threat agents using handheld SERS systems can be adopted. The robustness of real field sensing must be optimized to avoid false negative or positive results.

REFERENCES

Akgönüllü, S., Saylan, Y., Bereli, N., Türkmen, D., Yavuz, H., & Denizli, A. (2021). Plasmonic sensors for detection of chemical and biological warfare agents. *Plasmonic Sensors and their Applications*, 71–85.Plasmonic Sensors for Detection of Chemical and Biological Warfare Agents - Plasmonic Sensors and their Applications - Wiley Online Library

Albrecht, M. G., & Creighton, J. A. (1977). Anomalously intense Raman spectra of pyridine at a silver electrode. *Journal of the American Chemical Society*, 99(15), 5215–5217. https://doi.org/10.1021/ja00457a071

Altmann, H.-J., Oelze, S., & Niemeyer, B. (2013). Chemical agents – Small molecules with deadly properties. In *CBRN protection* (pp. 67–101). Ed. Andre Richardt, Birgit Hülseweh, Bernd Niemeyer, Frank Sabath, Wiley-VCH, Weinhem, Germany.

Alvarez-Puebla, R. A., & Liz-Marzan, L. M. (2012). Traps and cages for universal SERS detection. *Chemical Society Reviews*, 41(1), 43–51.

Awasthi, V., Goel, R., Rai, P., & Dubey, S. K. (2022). Detection of nitrogenous and nitro-aromatic compound with thin gold films roughened by cold argon plasma as SERS-active substrate. *Surfaces and Interfaces*, 28, 101556.

Beeram, R., Vepa, K. R., & Soma, V. R. (2023). Recent trends in SERS-based plasmonic sensors for disease diagnostics, biomolecules detection, and machine learning techniques. *Biosensors*, 13(3), 328.

Bharati, M. S. S., & Soma, V. R. (2021). Flexible SERS substrates for hazardous materials detection: Recent advances. *Opto-Electronic Advances*, 4(11), 210048.

Campion, A., & Kambhampati, P. (1998). Surface-enhanced Raman scattering. *Chemical Society Reviews*, 27(4), 241–250. https://doi.org/10.1039/A827241Z

Chen, J., Li, S., Yao, F., Bao, F., Ge, Y., Zou, M., . . . Chen, Q. (2022). Progress of microfluidics combined with SERS technology in the trace detection of harmful substances. *Chemosensors, 10*(11), 449.

Choi, J., Kim, J.-H., Oh, J.-W., & Nam, J.-M. (2019). Surface-enhanced Raman scattering-based detection of hazardous chemicals in various phases and matrices with plasmonic nanostructures. *Nanoscale, 11*(43), 20379–20391.

Ciloglu, F. U., Caliskan, A., Saridag, A. M., Kilic, I. H., Tokmakci, M., Kahraman, M., & Aydin, O. (2021). Drug-resistant Staphylococcus aureus bacteria detection by combining surface-enhanced Raman spectroscopy (SERS) and deep learning techniques. *Scientific Reports, 11*(1), 18444.

Cornish, P. (2007). *The CBRN system: Assessing the threat of terrorist use of chemical, biological, radiological and nuclear weapons in the United Kingdom.* Royal Institute of International Affairs., Catham House, UK.

Das, S., Goswami, L. P., Gayathri, J., Tiwari, S., Saxena, K., & Mehta, D. S. (2021). Fabrication of low cost highly structured silver capped aluminium nanorods as SERS substrate for the detection of biological pathogens. *Nanotechnology, 32*(49), 495301.

Downes, A., & Elfick, A. (2010). Raman spectroscopy and related techniques in biomedicine. *Sensors, 10*(3), 1871–1889.

Fan, M., Andrade, G. F., & Brolo, A. G. (2020). A review on recent advances in the applications of surface-enhanced Raman scattering in analytical chemistry. *Analytica Chimica Acta, 1097*, 1–29.

Farokhinejad, F., Li, J., Hugo, L. E., Howard, C. B., Wuethrich, A., & Trau, M. (2022). Detection of dengue virus 2 with single infected mosquito resolution using yeast affinity bionanofragments and plasmonic sers nanoboxes. *Analytical Chemistry, 94*(41), 14177–14184.

Farrell, M. E., Strobbia, P., Sarkes, D. A., Stratis-Cullum, D. N., Cullum, B. M., & Pellegrino, P. M. (2016). *The development of Army relevant peptide-based surface enhanced Raman scattering (SERS) sensors for biological threat detection.* Paper presented at the Smart Biomedical and Physiological Sensor Technology XIII.

Fierro-Mercado, P. M., & Hernández-Rivera, S. P. (2012). Highly sensitive filter paper substrate for SERS trace explosives detection. *International Journal of Spectroscopy, 2012*.

Forbes, T. P., Krauss, S. T., & Gillen, G. (2020). Trace detection and chemical analysis of homemade fuel-oxidizer mixture explosives: Emerging challenges and perspectives. *TrAC Trends in Analytical Chemistry, 131*, 116023.

Gao, R., Ko, J., Cha, K., Jeon, J. H., Rhie, G.-E., Choi, J., . . . Choo, J. (2015). Fast and sensitive detection of an anthrax biomarker using SERS-based solenoid microfluidic sensor. *Biosensors and Bioelectronics, 72*, 230–236.

Garibbo, A., & Palucci, A. (2014). Photonics for detection of chemicals, drugs and explosives. In *Photonics for safety and security* (pp. 114–144). World Scientific, Singapore.

Granger, J. H., Schlotter, N. E., Crawford, A. C., & Porter, M. D. (2016). Prospects for point-of-care pathogen diagnostics using surface-enhanced Raman scattering (SERS). *Chemical Society Reviews, 45*(14), 3865–3882.

Guenther, D., Dowgiallo, A.-M., & Branham, A. (2017). Trace-level detection of explosives using sputtered SERS substrates., *Spectroscopy Supplements, 32*, 8-17.

Hakonen, A., Andersson, P. O., Stenbæk Schmidt, M., Rindzevicius, T., & Käll, M. (2015). Explosive and chemical threat detection by surface-enhanced Raman scattering: A review. *Analytica Chimica Acta, 893*, 1–13. https://doi.org/10.1016/j.aca.2015.04.010

Hakonen, A., Rindzevicius, T., Schmidt, M. S., Andersson, P. O., Juhlin, L., Svedendahl, M., . . . Käll, M. (2016). Detection of nerve gases using surface-enhanced Raman scattering substrates with high droplet adhesion. *Nanoscale, 8*(3), 1305–1308.

Hakonen, A., Wang, F., Andersson, P. O., Wingfors, H. K., Rindzevicius, T., Schmidt, M. S., ... Boisen, A. (2017). Hand-held femtogram detection of hazardous picric acid with hydrophobic Ag nanopillar SERS substrates and mechanism of elasto-capillarity. *ACS Sensors, 2*(2), 198–202.

Hankus, M. E., & Cullum, B. M. (2006). *SERS probes for the detection and imaging of biochemical species on the nanoscale.* Paper presented at the Smart Medical and Biomedical Sensor Technology IV.

Hargreaves, M. (2021). Handheld Raman, SERS, and SORS. In *Portable spectroscopy and spectrometry* (pp. 347–376). Ed. Brooke W. Kammrath, Pauline E. Leary, Richard A. Crocombe, Wiley, UK.

He, X., Liu, Y., Xue, X., Liu, J., Liu, Y., & Li, Z. (2017). Ultrasensitive detection of explosives via hydrophobic condensation effect on biomimetic SERS platforms. *Journal of Materials Chemistry C, 5*(47), 12384–12392.

He, X., Wang, H., Li, Z., Chen, D., & Zhang, Q. (2014). ZnO−Ag hybrids for ultrasensitive detection of trinitrotoluene by surface-enhanced Raman spectroscopy. *Physical Chemistry Chemical Physics, 16*(28), 14706–14712.

He, X., Zhou, X., Liu, Y., & Wang, X. (2020). Ultrasensitive, recyclable and portable microfluidic surface-enhanced Raman scattering (SERS) biosensor for uranyl ions detection. *Sensors and Actuators B: Chemical, 311*, 127676.

Heleg-Shabtai, V., Sharabi, H., Zaltsman, A., Ron, I., & Pevzner, A. (2020). Surface-enhanced Raman spectroscopy (SERS) for detection of VX and HD in the gas phase using a hand-held Raman spectrometer. *Analyst, 145*(19), 6334–6341.

Hoffmann, J., Miragliotta, J., Wang, J., Tyagi, P., Maddanimath, T., Gracias, D., & Papadakis, S. (2009). *Scanning surface-enhanced Raman spectroscopy (SERS) of chemical agent simulants on templated Au–Ag nanowire substrates.* Paper presented at the Micro-and Nanotechnology Sensors, Systems, and Applications.

Huang, G., Zhao, H., Li, P., Liu, J., Chen, S., Ge, M., ... Li, S. (2021). Construction of optimal SERS hotspots based on capturing the spike receptor-binding domain (RBD) of SARS-CoV-2 for highly sensitive and specific detection by a fish model. *Analytical Chemistry, 93*(48), 16086–16095.

Hülseweh, B., & Marschall, H.-J. (2013). Detection and analysis of biological agents. In *CBRN protection* (pp. 211–241). Ed. Andre Richardt, Bernd Niemeyer, Birgit Hülseweh, Frank Sabath, Wiley-VCH, Weinhem, Germany.

Jeanmaire, D. L., & Van Duyne, R. P. (1977). Surface Raman spectroelectrochemistry: Part I. Heterocyclic, aromatic, and aliphatic amines adsorbed on the anodized silver electrode. *Journal of Electroanalytical Chemistry and Interfacial Electrochemistry, 84*(1), 1–20.

Lafuente, M., Almazán, F., Bernad, E., Urbiztondo, M. A., Santamaría, J., Mallada, R., & Pina, M. P. (2019). *SERS detection of neurotoxic agents in gas phase using microfluidic chips containing gold-mesoporous silica as plasmonic-sorbent.* Paper presented at the 2019 20th International Conference on Solid-State Sensors, Actuators and Microsystems & Eurosensors XXXIII (TRANSDUCERS & EUROSENSORS XXXIII).

Lafuente, M., Berenschot, E. J., Tiggelaar, R. M., Mallada, R., Tas, N. R., & Pina, M. P. (2018). 3D fractals as SERS active platforms: Preparation and evaluation for gas phase detection of G-nerve agents. *Micromachines, 9*(2), 60.

Lafuente, M., De Marchi, S., Urbiztondo, M., Pastoriza-Santos, I., Pérez-Juste, I., Santamaría, J., ... Pina, M. (2021). Plasmonic MOF thin films with Raman internal standard for fast and ultrasensitive SERS detection of chemical warfare agents in ambient air. *ACS Sensors, 6*(6), 2241–2251. https://doi.org/10.1021/acssensors.1c00178

Lafuente, M., Pellejero, I., Sebastián, V., Urbiztondo, M. A., Mallada, R., Pina, M. P., & Santamaría, J. (2018). Highly sensitive SERS quantification of organophosphorous chemical warfare agents: A major step towards the real time sensing in the gas phase. *Sensors and Actuators B: Chemical, 267*, 457–466.

Lafuente, M., Sanz, D., Urbiztondo, M., Santamaría, J., Pina, M. P., & Mallada, R. (2020). Gas phase detection of chemical warfare agents CWAs with portable Raman. *Journal of Hazardous Materials, 384*, 121279.

Le Ru, E. C., & Etchegoin, P. G. (2009). Chapter 1 – A quick overview of surface-enhanced Raman spectroscopy. In E. C. Le Ru & P. G. Etchegoin (Eds.), *Principles of surface-enhanced Raman spectroscopy* (pp. 1–27). Amsterdam: Elsevier.

Lesaicherre, M. L., Paxon, T. L., Mondello, F. J., Burrell, M. C., & Linsebigler, A. (2009). *Portable Raman instrument for rapid biological agent detection and identification.* Paper presented at the Next-Generation Spectroscopic Technologies II.

Li, Z., Huang, X., & Lu, G. (2020). Recent developments of flexible and transparent SERS substrates. *Journal of Materials Chemistry C, 8*(12), 3956–3969. https://doi.org/10.1039/D0TC00002G

Lin, D., Dong, R., Li, P., Li, S., Ge, M., Zhang, Y., . . . Xu, W. (2020). A novel SERS selective detection sensor for trace trinitrotoluene based on meisenheimer complex of monoethanolamine molecule. *Talanta, 218*, 121157.

Lister, A. P., Sellors, W. J., Howle, C. R., & Mahajan, S. (2020). Raman scattering techniques for defense and security applications. *Analytical Chemistry, 93*(1), 417–429.

Liszewska, M., Bartosewicz, B., Budner, B., Nasiłowska, B., Szala, M., Weyher, J. L., . . . Jankiewicz, B. J. (2019). Evaluation of selected SERS substrates for trace detection of explosive materials using portable Raman systems. *Vibrational Spectroscopy, 100*, 79–85.

Liu, J., Si, T., & Zhang, Z. (2019). Mussel-inspired immobilization of silver nanoparticles toward sponge for rapid swabbing extraction and SERS detection of trace inorganic explosives. *Talanta, 204*, 189–197.

Liu, W., Song, Z., Zhao, Y., Liu, Y., He, X., & Cui, S. (2020). Flexible porous aerogels decorated with Ag nanoparticles as an effective SERS substrate for label-free trace explosives detection. *Analytical Methods, 12*(33), 4123–4129.

Liyanage, T., Rael, A., Shaffer, S., Zaidi, S., Goodpaster, J. V., & Sardar, R. (2018). Fabrication of a self-assembled and flexible SERS nanosensor for explosive detection at parts-per-quadrillion levels from fingerprints. *Analyst, 143*(9), 2012–2022.

López-López, M., & García-Ruiz, C. (2014). Infrared and Raman spectroscopy techniques applied to identification of explosives. *TrAC Trends in Analytical Chemistry, 54*, 36–44.

Milligan, K., Shand, N. C., Graham, D., & Faulds, K. (2020). Detection of multiple nitroaromatic explosives via formation of a janowsky complex and SERS. *Analytical Chemistry, 92*(4), 3253–3261.

Mogilevsky, G., Borland, L., Brickhouse, M., & Fountain III, A. W. (2012). Raman spectroscopy for homeland security applications. *International Journal of Spectroscopy, 2012*, 808079. https://doi.org/10.1155/2012/808079

Nie, S., & Emory, S. R. (1997). Probing single molecules and single nanoparticles by surface-enhanced Raman scattering. *Science, 275*(5303), 1102–1106. https://doi.org/10.1126/science.275.5303.1102

Ravindranath, S. P., Wang, Y., & Irudayaraj, J. (2011). SERS driven cross-platform based multiplex pathogen detection. *Sensors and Actuators B: Chemical, 152*(2), 183–190.

Savinon-Flores, F., Mendez, E., Lopez-Castanos, M., Carabarin-Lima, A., Lopez-Castanos, K. A., Gonzalez-Fuentes, M. A., & Mendez-Albores, A. (2021). A review on SERS-based detection of human virus infections: Influenza and coronavirus. *Biosensors, 11*(3), 66.

Serebrennikova, K. V., Byzova, N. A., Zherdev, A. V., Khlebtsov, N. G., Khlebtsov, B. N., Biketov, S. F., & Dzantiev, B. B. (2021). Lateral flow immunoassay of SARS-CoV-2 antigen with SERS-based registration: Development and comparison with traditional immunoassays. *Biosensors, 11*(12), 510. https://www.mdpi.com/2079-6374/11/12/510

Shanmukh, S., Jones, L., Driskell, J., Zhao, Y., Dluhy, R., & Tripp, R. A. (2006). Rapid and sensitive detection of respiratory virus molecular signatures using a silver nanorod array SERS substrate. *Nano Letters, 6*(11), 2630–2636.

Sharma, Y., Gupta, S., Das, A., & Dhawan, A. (2020). *Polarization-independent SERS substrates for trace detection of chemical and biological molecules.* Paper presented at the Nanophotonics VIII.

Song, C., Driskell, J. D., Tripp, R. A., Cui, Y., & Zhao, Y. (2012). *The use of a handheld Raman system for virus detection.* Paper presented at the Chemical, Biological, Radiological, Nuclear, and Explosives (CBRNE) Sensing XIII.

Spencer, K. M., Sylvia, J. M., Clauson, S. L., & Janni, J. A. (2002). *Surface-enhanced Raman as a water monitor for warfare agents.* Paper presented at the Vibrational spectroscopy-based sensor systems.

Stiles, P. L., Dieringer, J. A., Shah, N. C., & Van Duyne, R. P. (2008). Surface-enhanced Raman spectroscopy. *Annual Review of Analytical Chemistry, 1*(1), 601–626. https://doi.org/10.1146/annurev.anchem.1.031207.112814

Upadhyayula, V. K. (2012). Functionalized gold nanoparticle supported sensory mechanisms applied in detection of chemical and biological threat agents: A review. *Analytica Chimica Acta, 715*, 1–18.

Wang, C., Liu, B., & Dou, X. (2016). Silver nanotriangles-loaded filter paper for ultrasensitive SERS detection application benefited by interspacing of sharp edges. *Sensors and Actuators B: Chemical, 231*, 357–364.

Wang, C., Meloni, M. M., Wu, X., Zhuo, M., He, T., Wang, J., . . . Dong, P. (2019). Magnetic plasmonic particles for SERS-based bacteria sensing: A review. *AIP Advances, 9*(1).

Wang, J., Yang, L., Boriskina, S., Yan, B., & Reinhard, B. M. (2011). Spectroscopic ultra-trace detection of nitroaromatic gas vapor on rationally designed two-dimensional nanoparticle cluster arrays. *Analytical Chemistry, 83*(6), 2243–2249.

Wang, R., Kim, K., Choi, N., Wang, X., Lee, J., Jeon, J. H., . . . Choo, J. (2018). Highly sensitive detection of high-risk bacterial pathogens using SERS-based lateral flow assay strips. *Sensors and Actuators B: Chemical, 270*, 72–79.

Wang, X., Huang, S.-C., Hu, S., Yan, S., & Ren, B. (2020). Fundamental understanding and applications of plasmon-enhanced Raman spectroscopy. *Nature Reviews Physics, 2*(5), 253–271. https://doi.org/10.1038/s42254-020-0171-y

Xu, W., Bao, H., Zhang, H., Fu, H., Zhao, Q., Li, Y., & Cai, W. (2021). Ultrasensitive surface-enhanced Raman spectroscopy detection of gaseous sulfur-mustard simulant based on thin oxide-coated gold nanocone arrays. *Journal of Hazardous Materials, 420*, 126668.

Yadav, S., Sadique, M. A., Ranjan, P., Kumar, N., Singhal, A., Srivastava, A. K., & Khan, R. (2021). SERS based lateral flow immunoassay for point-of-care detection of SARS-CoV-2 in clinical samples. *ACS Applied Bio Materials, 4*(4), 2974–2995.

Yan, F., & Vo-Dinh, T. (2007). Surface-enhanced Raman scattering detection of chemical and biological agents using a portable Raman integrated tunable sensor. *Sensors and Actuators B: Chemical, 121*(1), 61–66. https://doi.org/10.1016/j.snb.2006.09.032

Zapata, F., López-López, M., & García-Ruiz, C. (2016). Detection and identification of explosives by surface enhanced Raman scattering. *Applied Spectroscopy Reviews, 51*(3), 227–262.

Zhang, X., Young, M. A., Lyandres, O., & Van Duyne, R. P. (2005). Rapid detection of an anthrax biomarker by surface-enhanced Raman spectroscopy. *Journal of the American Chemical Society, 127*(12), 4484–4489.

Zhu, A., Jiao, T., Ali, S., Xu, Y., Ouyang, Q., & Chen, Q. (2021). SERS sensors based on aptamer-gated mesoporous silica nanoparticles for quantitative detection of Staphylococcus aureus with signal molecular release. *Analytical Chemistry, 93*(28), 9788–9796.

7 Surface Plasmon Resonance Nanosensors for the Sensing of Bacterial Threats

Neslihan Idil, Sevgi Aslıyüce, and Bo Mattiasson

7.1 INTRODUCTION

Biological threats are organisms or their toxins causing diseases, mortality, and morbidity in living organisms. These threats include those posed by biological agents and circumstances that include dangerous laboratory states (Thavaselvam and Vijayaraghavan, 2010). The explanation includes contagious and non-contagious diseases, biological agents (BA), which are present in the environment or diagnosed in animals and are causative agents for humans or weapons containing or emitting BA, which can be called weapons, and terrorist attempts with these, which are referred to as bioterrorism. A biological attack is the intentional release of BA into the environment which results in disease or mortality in living organisms. These agents exist in nature and can be modified to be more effective on emerging diseases, making them resistant to existing drugs, or enabling them to spread in the environment. They are preferred by terrorists due to their difficult detection and time required to cause disease (Clark and Pazdernik, 2016).

Bacterial threats are of global concern due to their importance in bioterrorism. Their usage as an agent in bioterrorism relies on their dangerousness, some of them are less hazardous and less likely to be preferred. Rapid and accurate detection of these bacterial threats has a great role in taking control of attacks. In this context, researchers working in this area try to determine the scope and size of the attack and apply an effective measurement (Das and Kataria, 2010). Clinical microbiology laboratories and the Centers for Disease Control and Prevention (CDC; Atlanta, GA, USA) have an organized network system for the reliable detection and reporting of BA.

CDC has classified BTA that can be applied in biological attacks into three categories (A, B, and C), which are determined by their spreading properties, severity of resultant diseases, susceptibility to BTA of the exposed population, obtainability and possibility of microbial growth, and method of their usage (aerosol, water, food,

DOI: 10.1201/9781003459316-7

etc.). Additionally, their application in the past, clinical and laboratory diagnostic criteria and possibilities, and availability of treatment strategies and vaccines all carry significance for this classification. Agents in Category A pose the highest risk, while agents in Category C are considered to be new threats that could be used as biological weapons in the future (CDC, 2018).

BTA of primary importance (Category A) are easily spread in the environment, are transmissible from person to person, and have a high risk of mortality. They cause serious social problems and specific safety plans focusing on elimination goals are required. The following BTA in this category are *Bacillus anthracis, Clostridium botulinum, Yersinia pestis,* and *Francisella tularensis* (Pohanka, 2019). In terms of biological warfare history, in 1979, a fatal accident in the Soviet Military Union occurred (Nikolakakis et al., 2023). Bacteria was accidentally transmitted via aerosol from the laboratory and anthrax spores were dispersed on the wind. Conclusively, doctors reported civilians dying of pulmonary anthrax. *Bacillus anthracis* spores were intentionally released and spread through contaminated US Postal Service mail in 2001 (Hughes and Gerberding, 2002).

BTA of secondary importance (Category B) cause moderate morbidity and relatively low mortality. They can spread relatively easily, so surveillance system development is needed and the capacity of specific diagnostic criteria should be enhanced. The following BTA in this category are: *Brucella melitensis, Brucella abortus, Clostridium perfringens, Salmonella* sp., *Shigella* sp., *Escherichia coli* O157:H7, *Burkholderia mallei, Burkholderia psedomallei, Chlamydia psittaci, Coxiella burnetti, Rickettsia prowazekii, Vibrio cholera,* and *Staphylococcus aureus* (Pohanka, 2019). BTA of lowest importance (Category C) are easy to grow and spread and they can endanger a wide population. Because of these characteristics, they have the potential to have a major impact on health and a high probability of becoming biological weapons in the future. The following bacterial threat in this category is Multidrug-resistant *Mycobacterium* sp. (Das and Kataria, 2010).

Most of the above-mentioned BTA have been commonly identified by conventional methods including microbiological assays and biochemical tests, immunoassays, and molecular analysis (Zeng et al., 2018). Immunoassays and molecular analysis are superior for their accurate and rapid detection of microorganisms. Immunoassays show low sensitivity and poor specificity because cross-reactivity occurs, which leads to false positive results. Although molecular methods are more sensitive than immunologic techniques, their complex procedures and high costs are disadvantages for their translation into POC platforms.

Many microorganisms found in food, water, and the environment can be detected with mentioned conventional methods. However, these strategies are time-consuming and typically take about two to three days, so they are not usually applicable. Besides, quantitative evaluation of microorganisms has a tremendous impact and most of the conventional methodologies have shortcomings in enumeration. On the other hand, traditional methods have low sensitivity, weak specificity, and require mostly cultivation. They are also laborious and need trained personnel. In the case of some bacterial threats, including *Vibrio cholerae*, they may be present in a food, water, or the environment, in a Viable but non-Culturable state (VBNC). Microorganisms in VNBC have an active metabolism, however, they are not culturable on routine bacteriological media (Nnachi et al., 2022).

Biological agents can be detected by detectors, which may detect the responsible agent later in forensics. In addition, some of them can carry information about who the responsible agents are and which precautions can be applied. Others estimate the dimensions of the threat, while others still can predict the risk to personnel against the agent (Grotte, 2001). The first detector system, known as detect-to-treat, does not give timely warnings to prevent exposure to the agent but allows early diagnosis. The second one, identified as detect-to-protect, carries attack information and provides timely alerts for the bacterial threat to prevent infection. Detect-to-treat platforms can detect bacterial threats within a few hours, while detect-to-treat biosensors perform analytical detection in a laboratory-dependent manner. Detect-to-protect platforms can detect samples within minutes, independent of the laboratory and personnel. Biosensors are analytical detection systems that have all the features of both systems (Gooding, 2006).

The RAMP® Biowarfare Detection System is applicable for rapid, on-site, diagnostic analysis for BTA, including anthrax and botulinum toxin. This system consists of a cartridge containing a target-specific immunochromatographic assay and a portable fluorescence-based reader. After the sample to be tested is loaded into the well, fluorescently dyed latex particles coated with antibodies combine with the specific antigen. The limitation of this method is that no quantitative data can be obtained when the test is positive. In this context, it is recommended to use the results in combination with other methods of confirmation. Reported application fields for this platform are water security and building structures (RAMP-200). Alexeter Biological Defense Systems TM Technologies (Alexeter Technologies, Wheeling, IL, USA) have lateral flow tests and hold promise for detecting anthrax, botulinum toxin, plaque, tularemia, brucellosis, and Staphylococcal enterotoxin B.

Biosensors are tools that combine a biological sensor and a physical transducer to get a measurable signal proportional to the concentration of the analyte. Nucleic acids, antibodies, peptides, aptamers, enzymes, whole cells and microorganisms, and molecularly imprinted polymers (MIPs) have been introduced as recognition molecules interacting with the target analyte (Farooq et al., 2018). When the sensing system recognizes the target molecule, the transducer senses this interaction. Therefore, the transducer converts one signal to the other in a physicochemical way (optical, piezoelectric, electrochemical, etc.) and an electronic output is obtained (Perumal and Hashim, 2014), (Herrera-Domínguez et al., 2023).

Affinity-based sensing platforms can be organized by labeled and label-free strategies based on the different transducers mentioned above. Non-labeled sensors have the advantage that they are not complex, do not require preliminary preparation steps, and are inexpensive due to the absence of labeling. However, they also have disadvantages, such as low sensitivity and non-specific coupling. Labeled ones are preferred over non-labeled ones with the possibility of pre-preparation by trained personnel. Labeling can be combined on optical, electrochemical, and magnetic platforms for selective and sensitive detection of the target molecule. However, in some cases, the non-specific coupling also affects the design of the interfacing surface (Gooding, 2006), (Koyappayil and Lee, 2021).

In recent years, public health awareness and safety precautions have increased the demand of researchers for the detection of microorganisms. There has been growing attention in the development of different approaches for the rapid, inexpensive, reliable, portable, user-friendly, sensitive, and specific detection of microorganisms. These platforms have been widely applied in areas such as medicine and biomedicine (Janith et al., 2023), food safety control (Balbinot et al., 2021), water quality (Zhou et al., 2018), and environmental monitoring (Sharma and Sharma, 2023). It is noteworthy to indicate that modified or previously uncharacterized agents have to be successfully detected in complex media without false results. On the other hand, many sensing platforms are able to detect the responsible agents directly from complex media at or below human risk levels. In this context, cost-effective analytical platforms have enabled detection of low concentrations with fast turnaround times. Bacterial threats can be found in complex media such as blood and stool samples, powder, food, water, and air. Their detection in these media is difficult and there are several challenges in the detection methodologies.

In this chapter, existing and emerging technologies in the detection of BTA are discussed and the various advances in SPR sensing are summarized. The achievements in integrating the proposed SPR platforms and the representative applications for the detection of BTA are examined and the challenges for the rapid and accurate detection of BTA were indicated.

7.2   SURFACE PLASMON RESONANCE (SPR)

SPR sensing platforms were applied to investigate their potential to detect bacterial threats (Idil et al., 2021). SPR is an optical sensing platform where the interaction between the target molecule and the designed sensor surface results in a change in the refractive index (RI) near the recognition layer. Polarized light is reflected through a prism onto a sensor chip coated with a thin metal (gold, silver, etc.) film on the surface. The light is reflected by the metal film acting as a mirror. When the incidence angle is changed and the intensity of the reflected light is observed, the reflected light is reflected over the minimum. At this angle of incidence, light reflects surface plasmons which produces a decrease in the intensity of the reflected light. At this angle of incidence, the light triggers the surface plasmons, therefore, a decrease in the intensity of the reflected light occurs. P-polarized light photons induce a wave-like oscillation of free electrons. Thus, the intensity of the reflected light is decreased, and these photons can interact with the free electrons of the metal layer (Figure 7.1) (Idil et al., 2023; Kumar and Rani, 2013).

SPR sensors have gained great attention for the detection of microorganisms in complex matrices. The usage of sensitive transducers providing a recognition signal has resulted in high analytical performance.

SPR systems can suffer from peak drift due to the effect of the strong magnetic field on sensitive biological samples. Overcoming these limitations, SPR systems also have advantages. The structure of the samples remains intact and unchanged.

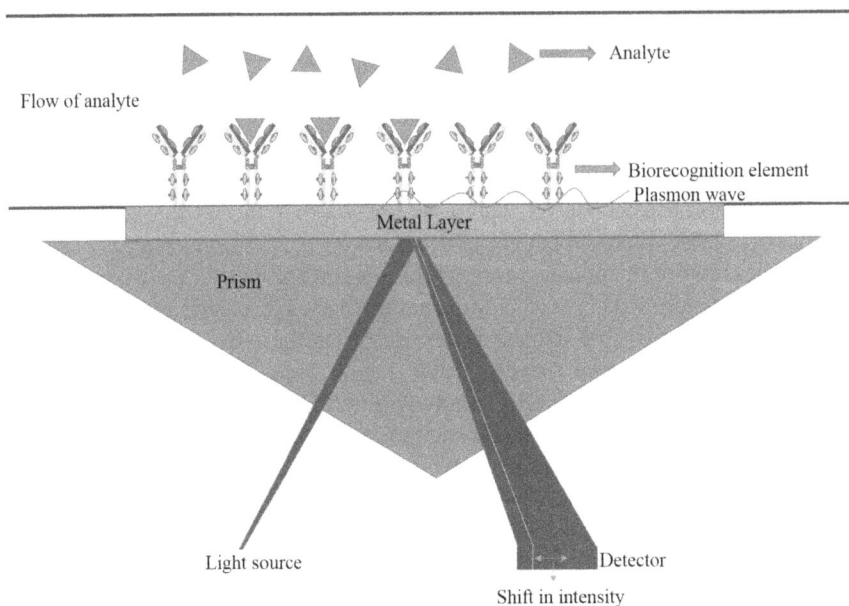

FIGURE 7.1 Schematic presentation of SPR sensing.

Real-time and sensitive detection can occur without the use of any radioactive or fluorescent agents. Once the target molecule is detected, the system can be regenerated and reused. Fast, accurate, and cost-effective recognition tools with hybrid sensor systems can be developed (J. H. Park et al., 2022).

7.3 BACTERIAL THREAT AGENTS (BTA)

Recent publications indicate that different sensing techniques and strategies have been applied for the detection of microorganisms. Colorimetric (Peng and Chen, 2019), fluorescence (Bahari et al., 2021), interferometric (Janik et al., 2021), SPR (Perçin et al., 2017a) (Kushwaha et al., 2018), LSPR (Li et al., 2022), Raman spectroscopy (Bräuer et al., 2022), and SERS (Chattopadhyay et al., 2019) sensing are basic strategies in optical sensing applied for the detection of microorganisms. Nanomaterials have been applied to enhance the performance of sensing systems providing much lower detection limits due to their large surface area (Idil et al., 2023). In this respect, carbon-, metallic-, and metal oxide nanoparticles (NPs), and nanocomposite materials are of great interest in the construction of optical sensing (Alafeef et al., 2020).

Apart from these, noble metal NPs have been easily incorporated with electrochemical, optical sensing. Noble metal NPs enable detection with the naked eye by visualizing the color change in the colorimetric detection process. These NPs provide signal amplification; therefore, higher sensitivity can be obtained in plasmonic systems. In this context, the surface of the chip is replaced with these NPs which are smaller than the wavelength of incident light and a localized SPR plasmon is released around them (Idil et al., 2023). On the other hand, antimicrobial peptides are promising tools to be integrated into the sensing systems because of their stability, multiple recognition regions for target microorganisms, and high specificity (Islam et al., 2022). MIP-based synthetic sensing approaches are also attractive to design systems with inexpensive preparation, high affinity, specificity, and mechanical and chemical stability. The synthesis of MIPs is based on the preparation of a mixture of target molecules and monomers, co-monomers, and cross-linkers. After the removal of the target molecule, recognition cavities complementary to the target molecule are formed, therefore, the target molecule can easily rebind onto the polymer surface using both geometrical accordance and chemical functionality (Aslıyüce et al., 2022). Several studies have been reported in the literature for SPR sensing platforms to detect BTA, based on the approaches described above. An overview of the representative examples of SPR sensing applications for the detection of BTA is given in Table 7.1.

7.3.1 CATEGORY A-BTA

7.3.1.1 *Bacillus anthracis*

Bacillus anthracis is a Gram-positive, spore-forming bacterium. It is the responsible agent for anthrax. It can survive for many years, withstanding adverse conditions such as high pressure, temperature, and pH in its spore state. Along with the transition to the vegetative growth stage, the anthrax toxin's protective antigen, lethal factor, and edema factor are produced (Mock and Mignot, 2003; Turk, 2007). *B. anthracis* carries two plasmids, pXO1 and pXO2, encoding the primary disease-causing factors, which are responsible for toxin production and capsule formation, respectively. Anthrax is usually transmitted to humans from herbivorous animals or their infected products. The disease course varies depending on whether the spores get into the body through the skin, or the respiratory or digestive tract. In cutaneous anthrax, spores entering through the skin form charcoal-like lesions on the skin. In gastrointestinal anthrax, spores are ingested orally and symptoms of nausea, vomiting, diarrhea, and fever are observed. In pulmonary anthrax, pores smaller than 5 µm in diameter pose a great danger when inhaled in aerosol form. If the disease is left untreated, the patient dies within 24 hours following an incubation period of about 4 days (Sweeney et al., 2012). *B. anthracis* is a Category A high-priority threat reported by CDC (Park et al., 2022). It is one of the most dangerous biological threats due to its transmission through the air and showing a fatal effect in a short time. Although *B. anthracis* and its virulence factors have been studied for a long time, the biological attacks in 2001 caused a great deal of fear and paved the way for researchers to develop rapid, label-free, and pre-purification-free detection methods for anthrax.

The monoclonal antibodies (mAbs) that create high sensitivity were preferred for use in the detection of *B. anthracis*. In this context, Wang et al. developed an SPR sensor able to detect *B. anthracis* spores without requiring labeling. The sensor

TABLE 7.1
Overview of the Representative Examples of SPR Sensing Applications For the Detection of BTA.

Category of BTA	Bacteria	Disease	Target	Principle of SPR/ Material	Ligand/ Substrate	LOD	Sample	References
Category A	B. anthracis	Anthrax	B. anthracis spore	Dextran	mAb 8G3	10^4 CFU/ mL	Buffer solution	(Wang et al., 2009)
	B. anthracis	Anthrax	Protective antigen	Dextran	Anti-PA mAb	12 fM	Spiked solid	(Ghosh et al., 2013a)
	B. anthracis	Anthrax	Protective antigen	Dextran	Anti-PA mAb	10 pg/mL	Human serum	(Ghosh et al., 2013b)
	B. anthracis	Anthrax	B. anthracis spore	Au/carboxylic groups	Anti-Bacillus anthracis spore antibody	1×10^3 spores/mL	PBS buffer solution	(Trzaskowski and Ciach, 2017)
	C. botulinum	Botulism	BoNT/A B F	OEG and BAT	Biotynilated pAb	1, 1, 0.5 ng/ mL	PBS buffer solution and honey	(Ladd et al., 2008)
	C. botulinum	Botulism	BoNT/B	CM3	VAMP2	$0.01\ LD_{50}/$ mL	Human serum	(Ferracci et al., 2011)
	C. botulinum	Botulism	BoNT/A	SNAP-25	mAb10F12	1.5 fM	Buffer solution	(Lévêque et al., 2013)
	C. botulinum	Botulism	BoNT/A	His6-SNAP-25	mAb10F12	0.1 LD50/ mL	Human serum	(Lévêque et al., 2014)
	C. botulinum	Botulism	BoNT/E	His6-SNAP-25	mAb11C3	0.01 LD50/ mL	Human serum	(Lévêque et al., 2015)
	C. botulinum	Botulism	BoNT/A	Dextran	Anti- rBoNT/ A-HCC	0.045 fM	Buffer solution	(Tomar et al., 2016)

(Continued)

TABLE 7.1 (CONTINUED)

Overview of the Representative Examples of SPR Sensing Applications For the Detection of BTA.

Category of BTA	Bacteria	Disease	Target	Principle of SPR/ Material	Ligand/ Substrate	LOD	Sample	References
	C. botulinum	Botulism	BoNT/A-LC	Four channel Au chip/ Nutravidine	Biotin-$(PEG)_5$-Peptit A	6.76 pg	Buffer solution	(Patel et al., 2017)
	Y. pestis	Plague	*Y. pestis*	Au/carboxylic groups	Anti-*Yersinia pestis* F1 antibody	1×10^3 CFU/mL	PBS buffer solution	(Trzaskowski and Ciach, n.d.)
	F. tularensis	Tularemia	*F. tularensis*	Glass/Au	IgG	1×10^2 cells/mL	PBS buffer solution	(Isakova et al., 2019)
Category B	*S. typhimurium*	Salmonellosis	*S. typhimurium*	Au/C18	pAbs	1.25×10^5 cells/mL	Milk and PBS buffer solution	(Mazumdar et al., 2007)
	S. typhimurium and *S. enteritidis*	Salmonellosis	*S. typhimurium* and *S. enteritidis*	Au/C18	Polyclonal anti-*Salmonella* antibody	2.5×10^5 cells/mL and 2.5×10^8 cells/mL	Milk	(Barlen et al., 2007)
	S. typhimurium	Salmonellosis	*S. typhimurium*	Carboxymethylated-dextran/Au	Aptamer	3×10^4 CFU/mL	Buffer solution	(Wang et al., 2017)
	S. paratyphi	Salmonellosis	*S. paratyphi*	MAH/Cu	MIP	1.4×10^6 CFU/mL	Apple juice	(Perçin et al., 2017b)
	S. typhimurium	Salmonellosis	*S. typhimurium*	Au/MNPs	mAb	4.7 and 5.2 log CFU/mL	Buffer solution and romaine lettuce samples	(Bhandari, 2022)

(Continued)

TABLE 7.1 (CONTINUED)
Overview of the Representative Examples of SPR Sensing Applications For the Detection of BTA.

Category of BTA	Bacteria	Disease	Target	Principle of SPR/Material	Ligand/Substrate	LOD	Sample	References
	B. abortus	Brucellosis	B. abortus CSP-31	CM5/Au	B. abortus CSP-31 antibody	0.05 pM	PBS buffer solution	(Gupta et al., 2011)
	B. melitensis	Brucellosis	Omp31	11-MUA/Au	Anti-Omp31	100 cell/mL	Buffer solution	(Saberi et al., 2016)
	B. melitensis	Brucellosis	IS71 gene	4-MBA/Au	complementary DNA	3.9 ng, 1.95 ng	Buffer solution/ real serum	(Sikarwar et al., 2017)
	B. melitensis B. abortus	Brucellosis	B. melitensis B. abortus	4-MBA/Au	Mouse IgG	1×10^2 cell/ mL	PBS buffer solution	(Hans et al., 2020)
	B. melitensis	Brucellosis	B. melitensis	Avidin and biotinylated B46 aptamers/Au	B46 aptamer	27 ± 11 cells/mL	PBS buffer solution/ spiked milk	(Dursun et al., 2022)
	S. aureus	Food poisoning	SEB	Au/peptide	Secondary antibodies	2.8 pg/mL	Milk	(Naimushin et al., 2002)
	S. aureus	Food poisoning	SEB	Carboxymethyldextran/ Au	SEB antibody	1.0 pM	Buffer solution	(Gupta et al., 2010)
	C. perfringens	Food poisoning	Clostridial epsilon prototoxin	CM5	mAb	–	Buffer solution	(Féraudet-Tarisse et al., 2017)

(Continued)

TABLE 7.1 (CONTINUED)

Overview of the Representative Examples of SPR Sensing Applications For the Detection of BTA.

Category of BTA	Bacteria	Disease	Target	Principle of SPR/ Material	Ligand/ Substrate	LOD	Sample	References
	E. coli O157:H7	Food safety threats	E. coli O157:H7	Au/Neutravidin	Anti E. coli O157:H7 antibody	1×10^2 – 10^3 CFU/mL	Spiked milk, apple juice, ground beef extract	(Waswa et al., 2007)
	E. coli O157:H7	Food safety threats	E. coli O157:H7	CM5	Anti-E. coli O157:H7 antibody	3×10^5 CFU/ mL	Buffer solution	(Si et al., 2011)
	E. coli O157:H7	Food safety threats	E. coli O157:H7	CM5	Rabbit anti-goat IgG polyclonal antibodies	3×10^4 CFU/ mL	Buffer solution	(Wang et al., 2011)
	E. coli O157:H7	Food safety threats	E. coli O157:H7	CM5	Lectin	3×10^3 CFU/ mL	Cucumber and ground beef sample	(Wang et al., 2013)
	E. coli O157:H7	Food safety threats	E. coli O157:H7	Au/carboxyl group	Anti O-157 antibody	6.3×10^4	Buffer solution	(Yamasaki et al., 2016)
	E. coli O157:H7	Food safety threats	E. coli O157:H7	AgNPs-rGO	AMP, Magainin I	5×10^2 CFU/mL	Water and juice	(Zhou et al., 2018)
	E. coli O157:H7	Food safety threats	E. coli O157:H7	AuNP/GO	-	142 CFU/ mL	-	(Jiang et al., 2023)
	B. pseudomallei	Melioidosis	Melioidosis antibody	11-MUA/Au	B. pseudomallei BipD protein	-	PBS buffer solution	(Dawan et al., 2011)

(Continued)

TABLE 7.1 (CONTINUED)

Overview of the Representative Examples of SPR Sensing Applications For the Detection of BTA.

Category of BTA	Bacteria	Disease	Target	Principle of SPR/ Material	Ligand/ Substrate	LOD	Sample	References
	B. pseudomallei	Melioidosis	B. pseudomallei	4-MBA/Au	Rabbit anti-rpGroEL rAb	-	PBS buffer solution	(Sikarwar et al., 2016)
	V. cholerae	Cholera	V. cholerae	Au/carboxylic groups	Anti-V. cholerae O1 Serovar Ogawa antibody	5×10^1 CFU/mL	PBS buffer solution	(Trzaskowski and Ciach, 2017)
	V. cholerae	Cholera	OmpW antigen	11-MUA/Au	Protein G/ anti-OmpW antibody	43 cells/mL	Buffer solution	(Taheri et al., 2016)
	V. cholerae	Cholera	V. cholerae	Amine-coupling/Au	Recombinant N1N2	-	Buffer solution	(Shin et al., 2021)

BoNT: Botulinum neurotoxin; SPR: Surface plasmon resonance; mAb: Monoclonal antibody; pAb: Polyclonal antibody; PA: Protective antigen; OEG: Oligo (ethylene glycol); BAT: Biotinylated alkanethiol; BoNT/A-LC: Botulinum neurotoxin type A light chain; 4-MBA: 4-mercaptobenzoic acid; CM: Carboxymethyldextran; 11-MUA: 11-mercaptoundecanoic acid; AuNp: Gold nanoparticle; GO: Graphene oxide; GNRs: Gold nanorods; AMP: Antimicrobial peptides; MNPs: Magnetic nanoparticles; MAH: N-methacryloyl-L-histidine methyl ester; SEB: Staphylococcal enterotoxin B; 11-MUA: 11-mercaptoundecanoic acid.

was composed of a monoclonal antibody (mAb) 8G3 targeting *B. anthracis* spores. The SPR sensor chip surface was modified by immobilizing goat anti-mouse IgG on the activated CM5 surface. The chip surface was regenerated after each analysis using 20 mM HCl. The lowest concentration at which *B. anthracis* spores could be detected was 10^4 CFU/mL (Wang et al., 2009). On the other hand, Ghosh et al. prepared an SPR sensor to recognize the *B. anthracis* protective antigen (PA) for indirect detection. The 3E5B8 mouse monoclonal antibody is generated against *B. anthracis* PA (anti-PA mAb). The specificity of the mAb is examined by integrating six different *Bacillus* sp. into the sensing system. The gold surface of the SPR sensor chip was modified with carboxymethyl dextran (CM) and the anti-PA mAb was immobilized on the modified chip surface. The LOD of the SPR sensor was reported to be 12 fM PA. *B. anthracis* Sterne (pXO1+, pXO2-), *B. cereus*, *B. licheniformis*, and *B. thuringiensis* strains at concentrations ranging from $10-10^5$ spores/g were spiked to the soil. Soil samples without spores of *Bacillus* sp. were used as negative control. It could be concluded that the anti-PA mAb immobilized SPR sensor was highly specific for PA produced by *B. anthracis* (Ghosh et al., 2013a). In another study, Ghosh et al. detected PA from human serum. They prepared monoclonal antibodies against *B. anthracis* PA and generated the SPR sensor chip surface as in the previous study. They prepared samples by adding PA toxin to human serum at concentrations between 1-10 ng/mL. The detection limit was found to be 10.0 and 1.0 pg/mL in the spiked serum and pure samples (Ghosh et al., 2013b).

7.3.1.2 *Clostridium botulinum*

C. botulinum was first isolated in 1897 by van Ermengem from salted raw ham that killed three people in Belgium. It is a Gram-positive, rod-shaped, strictly anaerobic, and spore-forming bacteria. *C. botulinum* spores can survive in the air for a long time and germinate in the presence of oxygen. However, vegetative cells susceptible to oxygen cannot survive due to the lethal effect of oxygen. Four different isolates of *C. botulinum*, Groups I–IV, can produce seven different types (A–G) of botulinum neurotoxins (BoNTs). Botulinum is one of the most powerful neurotoxins indicating a lethal effect even in very low amounts. In humans, it has a lethal effect of approximately 1–2 ng/kg of body weight. The toxin blocks the release of neurotransmitters in the nervous system, leading to paralysis and death following exposure to the toxin. BoNTs consist of two main components: A 50 kDa light chain (LC) and a 100 kDa heavy chain (HC). The LC functions as the proteolytic unit responsible for cleaving a specific soluble N-ethylmaleimide-sensitive factor activating protein receptor (SNARE) target protein at the nerve cell terminals which results in the inhibition of neurotransmitter release. The HC serves as a supporting element, aiding in the attachment to nerve cells and facilitating the transport of the LC into the cytosol of these nerve cells. *C. botulinum* can grow in food and water and release its toxins into the environment. Therefore, it can pose a serious threat to food and water safety. In addition to the fact that botulinum toxin shows its toxic effect through the ingestion of contaminated food and water, its easy production requires remarkable measures to be taken in a field such as biological warfare. The standard detection technique that is currently applied, the mouse-bioassay, frequently requires several days to verify

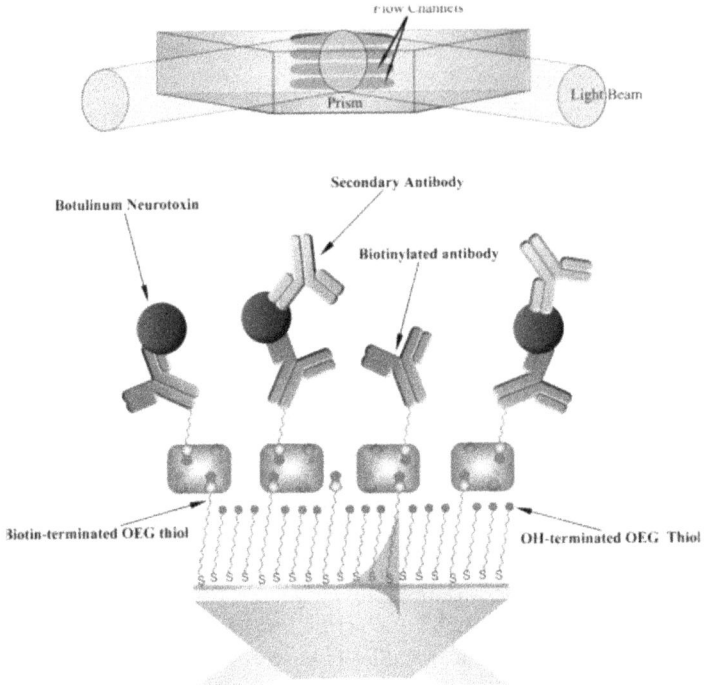

FIGURE 7.2   Schematic illustration of the custom-designed SPR. Reprinted with permission from Ladd et al. (2008).

botulism cases (Ravichandran et al., 2006). Besides, this method can only detect BoNT/A up to 20-30 pg (Lindström and Korkeala, 2006). The potential threat of bioterrorism involving BoNTs is addressed but rapid and accurate detection is crucial.

Ladd et al. have successfully identified the existence of BoNT/A, BoNT/B, and BoNT/F in both honey and buffer solutions. This was achieved by utilizing a custom-designed SPR system, featuring four distinct detection channels. Figure 7.2 illustrates the design of the biosensor, which features the four detection channels, based on the principles of wavelength modulation and the Kretschmann configuration within the attenuated total reflection method. In this configuration, a collimated light beam passes through an optical prism and contacts a thin metallic layer at a precise angle of incidence. Each of the flow channels emits reflected light that is meticulously collected by dedicated spectrophotometers. As the light interacts with the region close to the metal surface, it produces a distinctive trough in the observed light spectrum at a particular wavelength. Amendments to the sensor surface cause this characteristic dimple to shift. By monitoring these shifts, binding events occurring on the sensor surface can be accurately measured. The surface of the glass chip was initially coated with gold by the researchers. Subsequently, they functionalized it with oligo (ethylene glycol) (OEG) and biotinylated alkanethiol (BAT). Following that, they biotinylated polyclonal antibodies (pAbs) specific for BoNT/A/B/F and immobilized them on the surface. Toxin detection was performed separately for each serotype in the PBS buffer at concentrations ranging from 0.1 to 20 ng/mL, in the PBS buffer containing the toxin mixture, and in 20% honey samples. The LOD in the PBS buffer was 1 ng/mL for BoNT/A and BoNT/B, and 0.5 ng/mL for toxin BoNT/F. The reported analyses of mixtures containing three serotypes in the buffer and 20% honey solution were consistent with these values (Ladd et al., 2008).

Lévêque at al., have developed an SPR sensor to test BoNT/A endo protease activity. They first produced monoclonal antibodies (mAb10F12) that specifically recognize BoNT/A-induced snare protein SNAP-25 neo-epitopes. They then attached the recombinant SNAP-25 to the surface of an SPR sensor chip. After the sensor system was prepared, samples containing BoNT/A were injected into the substrate sensor for analysis. Barging of the substrate was monitored by binding of mAb10F12 to a SNAP-25 neo-epitope. The SNAP-25-chip assay could detect 55 fM BoNT/A within 5 minutes, and 0.4 fM within 5 hours. The LOD of 1.5 fM was reported (Lévêque et al., 2013). In their subsequent study, the same working group compared two different SPR sensor assemblies for the detection of BoNT/A in human serum. In one of these sensors, they immobilized SNAP-25 directly on the surface of the chip and mAb10F12 on the surface of the other chip. When BoNT/A was mixed with SNAP-25 in a solution, the resulting reaction products were collected on a chip that was immobilized with mAb10F12. This yielded a high sensitivity of 5 fM. However, it should be noted that this setup necessitated prior immunoprecipitation of BoNT/A. A significant sensitivity improvement was achieved by directly attaching SNAP-25 to the chip, reaching 0.5 fM in the presence of 10% serum. In all 11 patients with type A botulism, BoNT/A endoprotease activity was detected in the serum samples within five hours (Lévêque et al., 2014).

Lévêque et al. developed an SPR sensor to detect BoNT/E by coating the sensor surface with His-SNAP-25 and using mAb11C3 as the antibody for binding to the product. The LOD from the BoNT/E spiked serum was found to be 0.01 LD50/mL (Lévêque et al., 2015).

Tomar et al. described a label-free and real-time SPR detection approach for BoNT/A using SPR. They immobilized antibodies against the C terminal domain of recombinant botulinum neurotoxin A (rBoNT/A-HCC) fragment and synaptic vesicles (SV) on the sensor chip modified with carboxymethyldextran. In the optimization experiments of the sensing platform, temperatures between 10–37 °C and different pH ranging from pH 4–9 were examined. The optimum temperature and pH found for the binding of BoNT/A to the gold sensor surface were 25 °C and pH 7.5, respectively. The LOD of the method was found to be 0.045 fM (Tomar et al., 2016).

In a study, Patel et al. showed the feasibility of an SPR device for the rapid detection of BoNTs. The proposed method was based on the degradation of the substrate bound to the sensor chip by botulinum neurotoxin type A light chain (BoNT/A LC). The substrate was designed with biotin at one end for binding to neutravidin, a toxin specific sequence in the middle, and NH_2 at the other end for the binding of gold nanoparticles. The surface of the SPR gold chip is coated with neutravidin and bound to the substrate. When the toxin cleaves the substrate, the biotin-tipped segment separates it from the remainder of the molecule, so that a plasmon is formed based on the mass difference on the surface. Thus, toxin activity is detected by SPR. The LOD was reported as 6.76 pg/mL for BoNT/A LC. It could be concluded that the system's detection sensitivity aligned with the conventional mouse LD50 bioassay for BoNT/A (Patel et al., 2017).

7.3.1.3 Yersinia pestis

Y. pestis is a gram-negative bacterium and a member of Enterobacteriaceae family. It was first identified by Alexandre Yersin during the third pandemic plague in Hong Kong. *Y. pestis* evolved from *Yersinia pseudotuberculosis*, acquiring a life cycle transmitted by fleas and the ability to cause systemic infections. Critical genetic components, including plasmids pMT1 and pPCP1, along with the inherited plasmid pCD1, are fundamental in *Y. pestis* pathogenicity. Plasmid pCD1 encodes a type III secretion system that injects toxic Yersinia outer proteins (Yops) into host cells, resulting in the inhibition of immunity and damage to host cell structures (Yang, 2018). Plague is a zoonosis caused by *Y. pestis*. After an incubation period of about one week, the symptoms of the disease appear. These symptoms are usually characterized by a sudden onset of fever, chills, headache, body aches, weakness, vomiting, and nausea. The disease manifests in three forms, depending on the route of ingestion of the pathogen, (i) bubonic plague, (ii) septicemic plague, and (iii) pneumonic plague. The bubonic plague is the most prevalent manifestation of the disease, induced by the bite of a flea carrying an infection (Stenseth et al., 2008). The strongest form of the disease is pneumonic plague, which emerges by inhalation of *Y. pestis* and has a mortality rate of approximately 100% if left untreated. The rapid progression of the illness, its lethality, and its ability to be transmitted through aerosols have exacerbated concerns of *Y. pestis* as a bioweapon that can be intentionally released (Pechous et al., 2016).

Selecting appropriate antigens to produce antibodies utilized in therapy and diagnosis is a time-consuming and challenging process that necessitates numerous experiments. Broecker et al. employed a mixture of glycan microarray scanning, SPR, and saturation transfer difference (STD)-NMR to recognize epitopes of antibodies that are specific to the lipopolysaccharides of *Y. pestis*. They isolated monoclonal antibodies (mAbs) produced by clones 1B7, 1E12, and 3C11 and then analyzed them using denaturing SDS-PAGE. They then applied Glycan array microanalysis to elucidate the properties of these antibodies that determine their selectivity. They analyzed the binding properties and kinetics of the antibodies through SPR experiments. Here, they immobilized 10,000 RU of a-mouse IgG antibody on a commercial CM5 sensor chip. The samples were then passed over the chip and the changes were monitored (Broecker et al., 2014).

Hau et al. developed a lateral flow immunoassay (LFI) using two antigens to detect *Y. pestis*. LcrV, a protein produced by all pathogenic strains of *Y. pestis*, can be a challenge for assay specificity due to homologous proteins in other *Yersinia* species. F1, a unique marker for *Y. pestis*, is not consistently present in all pathogenic strains. The researchers then produced mAbs against this specific *Y. pestis* component. LFI with dual antigen detection and ELISA was developed. The mAbs were characterized by evaluating their binding kinetics using SPR. mAbs specific to LcrV demonstrated a narrow range of equilibrium dissociation constants, ranging from 0.3 to 4.5 nM. In contrast, mAbs targeting F1 showed a wider range, including values from 0.002 to 250 nM. The SPR analysis data reveals that a significant number of the generated mAbs in this study have a noteworthy affinity for their specific target antigens (Hau et al., 2022).

7.3.1.4 *Francisella tularensis*

Francisella tularensis is a Gram-negative, non-spore forming coccobacillus with aerobic properties. This pathogenic bacterium causes tularemia, a potentially fatal disease if not treated. Tularemia is also known as rabbit fever or deer fly fever. It was first reported in 1912 in ground squirrels in Tulare, California (McCoy, 1912). Tularemia is a zoonotic disease commonly found in animals, particularly rodents, which can transmit the disease to humans. The disease is most prevalent in North America and Northern Europe. Tularemia can be contracted through the bite or bodily fluids of infected animals, consumption of contaminated food or water, or inhalation of pathogen-containing dust particles. The typical incubation period is 2–10 days. The onset of the illness is typically marked by symptoms such as sore throat, weakness, loss of appetite, back pain, headache, or fever accompanied by chills and sweating. Subsequent symptoms vary according to the localization of the pathogen. Its strong pathogenicity and the diversity of its transmission route make *F. tularensis* a potential agent of bioterrorism (Farlow et al., 2005).

Isakova et al. designed a technique to detect *F. tularensis* by SPR spectrometry for the diagnosis of tularemia. They immobilized IgG antibodies against *F. tularensis* on the surface of a gold-coated glass sensor chip and identified the tularemia agent from aqueous solutions. The sensor system used had a flat two-chamber flow cell. Antibodies and bacterial cells were introduced into the SPR system at 25°C with pH 7.4 PBS buffer solution. The flow was maintained for about 2 minutes and after the required concentration was fully reached in the measuring cell, the liquid flow inside the cell was cyclized. The system was washed with PBS to examine the removability of the substances attached to the surface. They reported that the lowest concentration for the measurement of *F. tularensis* from the buffer solution was 1.0×10^2 cells/mL (Isakova et al., 2019).

Nualnoi et al. developed monoclonal antibodies (mAbs) specifically for *F. tularensis* to be used for POC diagnosis of tularemia. The binding properties of these mAbs were determined by SPR. They used two different assay methods for SPR experiments. As shown in Figure 7.3A, in the first method, biotinylated *F. tularensis* lipopolysaccharides (LPS) were attached to a streptavidin (SA)-coated sensor chip. Sensorgrams were performed by injecting monoclonal antibodies 1A4 IgG3, 1A4 IgG1, 1A4 IgG2b at varying concentrations at a flow rate of 30 µL/min for 180 seconds. Figure 7.3B displays sensorgrams and examines the binding of mAb 1A4 IgG3, demonstrating a stable affinity (KD) of 84 nM. Conversely, when investigating the 1A4 IgG1 and IgG2b subclasses (Figures 7.3C and 7.3D), a noteworthy shift in binding behavior was observed due to subclass conversion. Both 1A4 IgG1 and IgG2b indicated diminished overall binding to the immobilized LPS and faster

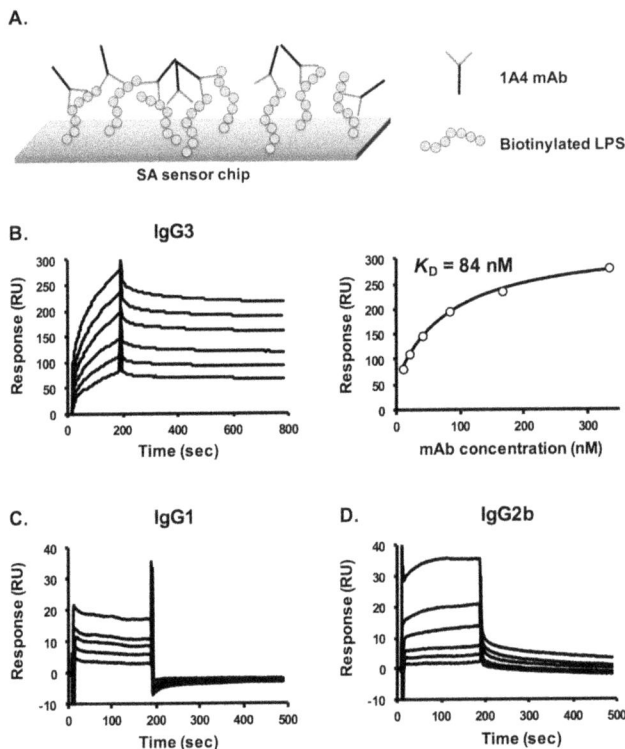

FIGURE 7.3 SPR analysis of LPS binding of mAbs 1A4 IgG3, IgG1, and IgG2b using a biotinylated LPS-streptavidin capture platform. Reprinted with permission from Nualnoi et al. (2018).

dissociation relative to 1A4 IgG3. Determining the KD values for IgG1 and IgG2b was unfeasible in this assay due to the high concentrations required. However, the outcomes demonstrated a significant decrease in LPS binding affinity for 1A4 IgG1 and IgG2b in comparison to the original IgG3 antibody.

In the second determination method, anti-mouse antibodies were covalently attached to the CM5 sensor chip using a mouse antibody capture kit through an amine-caplin reaction (Figure 7.4A). Each analysis cycle commenced with capturing antibodies, through the injection of 1 μg/mL of mAb 1A4 IgG3, IgG1, or IgG2b onto the sensor chip surface. This was carried out for 60 seconds at a flow rate of 30 μL/min, with one flow cell left untreated for reference purposes. Various concentrations of LPS were then injected onto the chip surface at a flow rate of 30 μL/min for 120 seconds. Sensorgrams were produced by adding different LPS concentrations to the surface of the chip. The top panel of Figure 7.4B illustrates this process. Figure 7.4A visually demonstrates the binding complex between mAb 1A4 and LPS that was captured in the experiment. Using a steady-state affinity model, the researchers established the KD values for LPS binding in each subclass of the 1A4 antibody, as

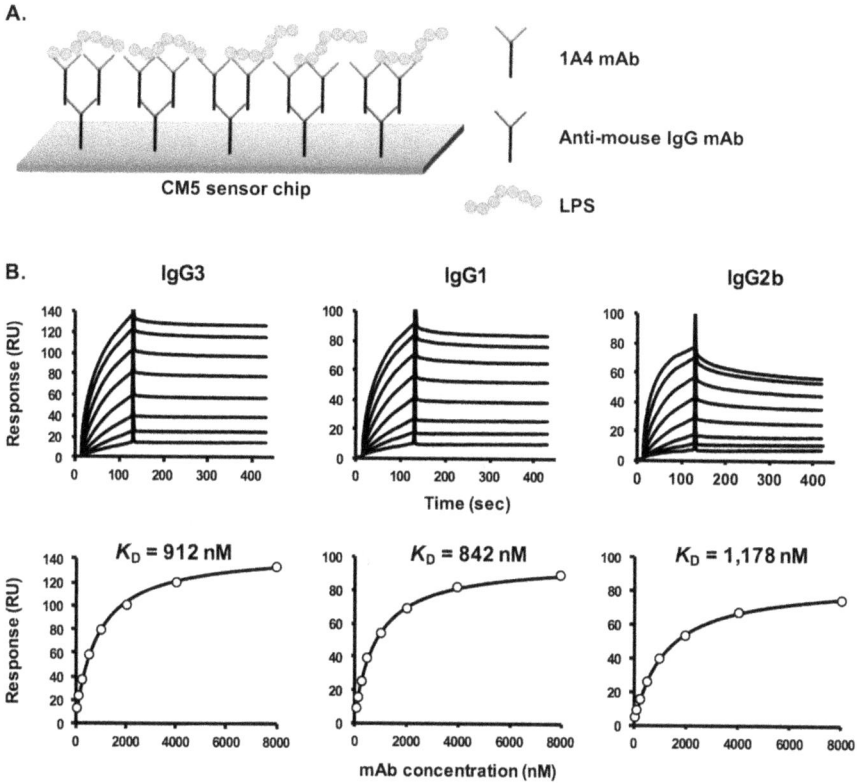

FIGURE 7.4 LPS binding affinity of each 1A4 subclass mAb analyzed by SPR using an antibody capture analysis approach. Reprinted with permission from Nualnoi et al. (2018).

depicted in the bottom panel of Figure 7.4B. Accordingly, they determined binding affinities for all 1A4 subclasses, with KD values of 912, 842 and 1.178 nM for 1A4 IgG3, IgG1, and IgG2b, respectively (Nualnoi et al., 2018).

7.3.2 Category B-BTA

7.3.2.1 *Salmonella* and *Shigella* sp.

Salmonella species are Gram-negative rod-shaped bacteria and one of the most globally distributed common pathogens. The strains included in this genus are associated with food-borne illness and eventual mortality. Transmission occurs through the consumption of contaminated food, and infected people may experience diarrhea, chills, fever, nausea, vomiting, abdominal pain, bloody stools, and headache (Perçin et al., 2017a). *Salmonella enterica* serovar *typhimurium* was identified as the responsible agent of the Oregon outbreak in what was reported to be the largest food-borne outbreak in the United States in 1984. It was concluded that the source of the outbreak was the failure of the personnel working in the food sector to comply with personal hygiene rules. Employees at the affected restaurant became ill before most of the customers (To et al., 1997). Therefore, it is significant to detect these strains in food rapidly and accurately for the prevention of food-borne diseases. In case of the outburst, detection has a key role in reducing the spreading.

Wang et al. reported that advances in nanotechnology have facilitated the development of label-free biosensors. Atomic force microscopy (AFM) was used to reveal the interactions between protein and aptamer, and SPR technology applied for bacterial detection was combined in their study. Proteins on the outer membrane of *S. typhimurium* and aptamer interactions were analyzed. *E. coli* was used as a control agent to verify the selectivity of two aptamers against target bacteria and no response could be obtained. The SPR chip surface was functionalized with carboxymethylated-dextran (CD) and aptamers were immobilized to create specificity. The LOD was reported as 3×10^4 CFU/mL (Wang et al., 2017).

In another study, magnetic nanoparticles (MNPs) were integrated into SPR sensing to detect *Salmonella typhimurium*. mAbs having an affinity towards flagellin were applied, in this respect; the first of mAbs was immobilized on chip surface and the second one was incorporated into the MNPs. Target bacteria was obtained from *S. typhimurium* containing romaine lettuce using a vacuum filter. The sensing mechanism is based on the interaction between flagellin and MNPs with the antibody-immobilized sensor surface. Direct, sequential two-step sandwich, and preincubation one-step sandwich methodologies were applied. Preincubation one-step sandwich assay is found as the most efficient sensing approach. The LOD was reported as 4.7 and 5.2 log CFU/mL in the buffer and romaine lettuce samples, respectively (Bhandari, 2022).

In our research group, we aimed to detect *Salmonella paratyphi*. For this purpose, an SPR sensing platform was prepared using microcontact-imprinting. Firstly, stamping of *S. paratyphi* was performed. Afterward, microcontact *S. paratyphi*-imprinted SPR chips were developed using the functional monomer N-methacryloyl-L-histidine methyl ester (MAH) preferred as a Cu^{+2} complexing agent. Functionalized SPR systems have complementary specific capturing regions to enhance selectivity towards amino acids present on the bacterial cell wall. On the other hand, the MAH-Cu^{+2} complex is able to bind the -OH groups found on the bacterial cell wall. The real-time bacterial recognition was carried out within the range of 2.5×10^6–15×10^6 CFU/mL. Selectivity was tested by competitive agents such as *Escherichia coli*, *Staphylococcus aureus*, and *Bacillus subtilis*. Imprinting efficiency was verified by the obtained responding results of imprinted and non-imprinted sensing systems. Real sample availability of the sensing platform was conducted in apple juice. Consequently, the LOD was found as 1.4×10^6 CFU/mL and developed imprinting based SPR sensing was a promising tool to detect *S. paratyphi* in food supplies or contaminated water (Perçin et al., 2017a).

An immunoassay-based SPR sensing was performed by the application of a sandwich strategy with a polyclonal antibody to detect *S. typhimurium* at the concentration of 1.25×10^5 cells/mL both in milk and PBS buffer. *Salmonella* serotypes are one of the most common food-borne pathogens causing salmonellosis related to public security (Mazumdar et al., 2007). Another publication of this group used polyclonal anti-*Salmonella* antibody to create a multiple SPR sensing detection of the involved serovars. The LOD was reported as 2.5×10^5 cells/mL and 2.5×10^8 cells/mL for *S. typhimurium* and *S. enteritidis*, respectively. The applicability of the sensing platform was verified using milk spiked samples (Barlen et al., 2007).

Shigella sp. is another pathogen member of the Enterobacteriaceae, gram-negative, rod-shaped, non-motile, facultatively anaerobic, non-spore forming bacteria. The somatic antigen in the lipoprotein structure, which has endotoxin properties is released by the lysis of *Shigella* sp. *S. dysenteriae* is the exception for the releasing exotoxin. Shigellosis is the disease affecting the gastrointestinal system causing edema, diarrhea, mucus formation, abscesses, and ulcerations in the mucosal epithelium of the large intestine (Ina et al., 2003).

In a previous study, an aptamer-immobilized SPR sensing was applied to detect *S. sonnei* virulent factor. The genes of *S. sonnei* encode invasion plasmid antigen H (IpaH) required for invasion. The specific aptamer IpaH17 has affinity against fM. The responses indicated that the proposed sensing system was employed in the IpaH concentration of 0-100 ng/mL. In conclusion, it was reported that selective and sensitive detection of IpaH was successfully performed in an hour (Song et al., 2017).

7.3.2.2 *Brucella* sp.

Brucella sp. are Gram-negative coccobacilli, non-spore forming, uncapsulated, non-toxin forming bacteria. *Brucella melitensis* and *Brucella abortus* are the causes of brucellosis. Brucellosis, one of the most common zoonotic diseases in the world, is particularly endemic in Mediterranean countries. Brucellosis is not only a cause of mortality in humans and animals, but also a problem affecting public health by causing socioeconomic losses.

Transmission occurs from animal to human (e.g., dairy industry workers, shepherds, farmers, slaughterhouse workers, and veterinarians), human to human (through blood and exudate, sexual contact, the placenta, the birth canal, and breast milk). Besides consuming contaminated food and drink (milk and dairy products, meat, and meat products), direct contact (cuts and abrasions on the skin, mucous membranes, or conjunctiva), and inhalation (contaminated aerosols) are the transmission factors. Fatigue, headache, myalgia, and fever are the major symptoms. Weight loss, abdominal pain, lymphadenopathy, hepatosplenomegaly, ulcerations, petechiae, and erythema nodosum can be listed as other symptoms.

Brucella sp. are facultative intracellular strains and can proliferate in monocytes/macrophages. They can easily be protected from host immune defenses due to this feature. *Brucella* species are reported as BTA due to their very low infectious dose (100 bacteria) and ability to spread via aerosols. Additionally, lyophilization of the bacteria is easy and therefore, it protects infective properties for several years.

Brucella sp. has been applied as a biological weapon by the USA, the UK, and the former USSR. The experiments were performed with *B. abortus* in aerosol form in the Caribbean. In the USA, in 1954, *B. suis* was loaded into rockets and artillery shells. The production and stockpiling of these agents continued until 1969, when the biological weapons program was officially terminated by the USA, and 1973 when the weapons were destroyed. In recent years, it has been reported that the vaccine strain, which is widely used in animals, may develop disease in humans and studies have been carried out on this strain.

Dursun et al. developed an SPR aptasensor to detect *B. melitensis* in spiked milk. They used biotin-gold sensor chips. One of the two channels of these sensor chips is coated with avidin and biotinylated B46 aptamers for specific detection, while the other channel is coated with avidin and a biotinylated scrambled sequence, which serves as a control for comparison. They then modified the Fe_3O_4 magnetic nanoparticles with amine groups using (3-aminopropyl) trimethoxysilane (APTES) and coupled 2-bromoacetyl bromide to these amine groups on the nanoparticle. They then prepared $Fe_3O_4@SiO_2@p$(PEGMA-GMA) particles by attaching poly(ethylene glycol) methacrylate (PEGMA) to bromine-terminated magnetic particles. Finally, amine-labeled aptamers were attached to epoxy groups on the magnetic nanoparticles. They mixed the aptamer-bound magnetic nanoparticles with milk samples spiked with bacterial cells. They then injected the spiked milk sample into the SPR sensor system. They first evaluated the efficiency of the SPR sensor using samples containing between 10^2 and 10^4 *B. melitensis* cells in PBS buffer and found a proportional increase in SPR signals with increasing concentration. The applicability of the sensor was then tested by adding *B. melitensis* cells directly to milk samples. A linear change in angle was observed for *B. melitensis* cells up to a concentration of 250 cells/mL, while the experiments with *E. coli* and *B. subtilis* showed insignificant signals. The LOD value they found for this sensor system is 27 $\pm$ 11 cells/mL (Dursun et al., 2022).

Sikarwar et al. developed a novel SPR immunosensor using a 4-mercaptobenzoic acid (4-MBA) modified gold (4-MBA/Au) SPR chip to detect *B. melitensis*. DNA probes derived from the IS711 gene were designed. The DNA probes consist of 34 nucleotides and were designed to detect the IS711 gene of *B. melitensis* with some modifications of the sequence complementary to the IS711 gene. Probe 1 and Probe 2 were used to detect different parts of the DNA target. The kinetics and thermodynamics of the interaction between the complementary DNA targets and probe 1 were found to be more efficient than that of probe 2 based on experimental data. For probe 1, The KD value was 15.3 pM and Bmax value was 81.02 m°, while for probe 2 these values were 54.9 pM and 55.29 m° for KD and Bmax, respectively. The detection limits of probe 1 and probe 2 were 3.9 ng and 1.95 ng, respectively. Real serum samples were also analyzed using DNA probes immobilized by SPR. *B. melitensis* was detected in less than ten minutes (Sikarwar et al., 2017).

Hans et al. created a sandwich ELISA method to detect whole cells of *B. abortus* and *B. melitensis,* using polyclonal rabbit IgG and mouse IgG. They assessed the validity of this method via SPR, using mouse IgG immobilized on an Au chip surface modified with 4-MBA.

To immobilize mouse IgG, the modified sensor chip was treated with EDC/NHS carboxyl groups. Subsequently, the surface was interacted with IgG in a PBS buffer. The KD values obtained from the kinetic analyses of *B. melitens*is and *B. abortus* were 16.48 pm and 0.42 pm, respectively. Correspondingly, the Bmax values were determined as 81.67 m° and 54.50 m°. It was also reported that both bacteria had a minimum detectable concentration of 10^2 cells/mL (Hans et al., 2020).

7.3.2.3 Staphylococcal Enterotoxin B (SEB)

SEB is a superantigenic toxin and known as powerful toxin threat and bioweapon. This offensive biological warfare agent is produced by one of the most important pathogenic bacteria Coagulase-positive *Staphylococcu aureus*. CDC reported SEB as an agent causing food poisoning which triggers diarrhea and vomiting and many attempts have been made to inactivate SEB (Bae et al., 2021).

Half of *S. aureus* strains collected from clinical isolates produce SEB. It causes nonspecific activation of the infected person's immune system by stimulating the release of cytokines such as interferon-gamma, interleukin-6, and tumor necrosis factor-alpha. SEB is of great importance due to its easy spread, late onset of symptoms, high mortality, and difficulty in distinguishing between natural and intentional intoxication. In this respect, rapid, accurate and selective detection of SEB has a crucial role in recognition, management and surveillance of SEB. (Ahanotu et al., 2006).

In bioterrorism, SEB is expected to disperse in aerosol form by inhalation and cause fatal pulmonary and systemic disease. In the 1960s, SEB was weaponized by the United States and the LD50* for SEB is approximately 0.02 µg/kg (Hu et al., 2021).

The commercial tools are large, however, easily applicable small devices were developed for the detection of SEB (Spreeta, Texas Instruments, Attleboro, MA). Gold-binding peptide was preferred to modify the sensor surface and the angle of reflectance is monitored at a fixed wavelength. The other SPR device is based on side-polished, single-mode optical fiber to track spectral changes at a fixed angle of incidence. It is able to directly detect SEB in milk samples and the sensitivity was enhanced with the usage of a secondary antibody. The generated sensing platform was able to detect 70 pM (2 ng/mL) SEB in 15 minutes. Secondary antibodies were chosen to obtain signal amplification and they can bind to the captured SEB. Third antibodies were applied to recognize the secondary antibodies and therefore, the LOD was reduced to 100 fM (2.8 pg/mL) in buffer (Naimushin et al., 2002).

Homola et al. developed wavelength modulation-based SPR sensing to detect SEB. Direct and sandwich assays were applied in buffer and milk samples. Sandwich approaches are based on the use of secondary antibodies bound to the captured toxin and applied in buffer and milk samples. The LOD was reported as 0.5 and 5.0 ng/mL for sandwich and direct detection assay, respectively. It was suggested that the polyclonal antibodies did not give significant cross-reactions against the biomolecules from milk that were attached to the chip surface and the designed system is applicable for SEB detection in milk (Homola et al., 2002).

Gupta et al. constructed a sensitive SPR sensing to quantify SEB. In this context, the SEB antibody was bound to a carboxymethyldextran modified sensor surface. After this immobilization step, BSA was used to block unbound NHS ester and prevent nonspecific binding on the sensor surface in the recognition step. In the assay of SEB detection, it was accomplished to get linearity in the concentration of 2.0–32.0 pM. The LOD was found as 1.0 pM and the detection time was reported less than 10 minutes. Selectivity of the sensing format was clarified by evaluating the responses against other toxins such as SEA, SEC, and SED with the interference of 20.01, 27.15, and 12.05%, respectively. (Gupta et al., 2010).

7.3.2.4 *Clostridium perfringens*

Clostridium perfringens is a Gram-positive, rod-shaped, anaerobic, capsulated, spore-forming and toxin-producing bacteria. *C. perfringens* spores are resistant to environmental stress and also survive normal cooking temperatures. Outbreaks are generally related to improperly heated or reheated gravy, poultry, or meats. It is responsible for gastroenteritis varying from diarrhea to necrotizing enterocolitis and myonecrosis (gas gangrene) (Wijnands et al., 2011). α-toxin is responsible for gas gangrene. The mechanism is based on the presence of α-toxin in the plasma membrane and forming gaps which result in the disruption of cellular activity. Wound infections occur by inoculation into an open wound. Food poisoning in humans has emerged from enterotoxin produced by *C. perfringens* tip A. Epsilon toxin can disperse through aerosol or respiratory and/or oral exposure (Uzal et al., 2014). The toxin has been announced as a bioterrorism agent by CDC and is the third most potent one after botulinum and tetanus toxin (Janik et al., 2019).

Tarisse and colleagues developed a sandwich immunoassay for the detection of Clostridial epsilon prototoxin (PεTX) from various media and investigated the affinity of the mAb using SPR. The kinetic properties of five antibodies were assessed via SPR biosensor technology, using epsilon toxins (PεTX2, PεTX5, PεTX6, PεTX7, and PεTX9) as the target antigens. Two independent kinetic measurements were conducted for each antibody at 25°C in a flowing buffer, utilizing a CM5 sensor chip. A sensor chip prepared with mouse IgG antibody was utilized. Individual mAbs were captured and exposed to different concentrations of epsilon toxin, followed by dissociation. A sensor chip attached with mouse IgG antibody was utilized. Individual monoclonal antibodies were captured and exposed to different concentrations of epsilon toxin, followed by dissociation. KD values were calculated based on the obtained data, indicating the efficacy of certain antibodies in neutralizing epsilon toxin. All antibodies displayed comparable values in the 10^{-9} M range, however, researchers emphasized that PεTX2 and PεTX5 exhibited marginally lower KD values of 9.6 × 10^{-9} and 9.8 × 10^{-9} M, respectively. Despite these consistent findings, there were discernible disparities observed in how these antibodies interacted with the antigen. PεTX2, PεTX6, and PεTX9 were found to exhibit slower dissociation rates, at around 1 × 10^{-4} s^{-1}. Conversely, PεTX5 and PεTX7 displayed faster association rates, with values of 1.2 × 10^5 and 2.0 × 10^5 $M^{-1}s^{-1}$, respectively. By utilizing these antibodies, it was possible to develop highly sensitive assays for various samples, including milk, tap water, serum, feces, and intestinal contents (Féraudet-Tarisse et al., 2017).

7.3.2.5 *Escherichia coli* O157:H7

Escherichia coli O157:H7 is Gram-negative, rod-shaped, facultative anaerobic, Shiga-like toxin-producing serotype of *E. coli*. It is also known as Enterohemorrhagic *E. coli* (EHEC), Verotoxin-producing *E. coli* (VTEC) and Shiga-like toxin-producing

E. coli (STEC). It causes food poisoning and hemorrhagic colitis. These toxins, which are active in the colon, bind to endothelial cells. The symptoms are intestinal cramps, and watery and bloody diarrhea. Additionally, kidney failure may emerge, and hemolytic uremic syndrome (HUS) disease occurs. Another disease is Thrombotic thrombocytopenic purpura (TTP) showing thrombocytic microangiopathy and is characterized by clotting form of blood in small blood vessels throughout the body (Piérard et al., 2012).

In 1975 and subsequently between 1978 and 1982, it was first isolated and identified by the Centers for Disease Control (CDC) from patients with diarrhea. In 1982, *E. coli* O157:H7 was identified in two outbreaks of gastroenteritis in Oregon and Michigan caused by the consumption of contaminated hamburgers with inadequate heat treatment. In 1996–97, 21 people died in Scotland because of this bacterium (Rangel et al., 2005).

Wang et al. have developed a biosensor utilizing lectins as bioreceptors for fast detection of *E. coli* O157:H7 through SPR technology. The recognition method is founded on the selective interaction of lectins with carbohydrate components present on the bacterial cell surface. The researchers employed five distinct lectin species to facilitate the process. The fabrication process of the SPR biosensor, as shown in Figure 7.5, began with activation of the sensor chip surface. This activation involved the injection of a mixture of 0.1 M NHS and 0.4 M EDC. Lectins at concentrations of 0.1, 0.5, and 1.0 mg/mL in 0.01 M sodium pH 4.5 acetate buffer were then applied to the chip surface. A solution of 1 M ethanolamine hydrochloride at pH 8.5 was injected to deactivate unreacted sites. The performance of the biosensor was evaluated by adding various concentrations of diluted *E. coli* O157:H7, ranging from 3.0×10^1 to 3.0×10^8 CFU/mL, for direct detection. The biosensor achieved a remarkable LOD of 3×10^3 CFU/mL for *E. coli* O157:H7 using the binding molecule derived from *T. vulgaris*. In the analyses of real samples, cucumber and ground beef samples

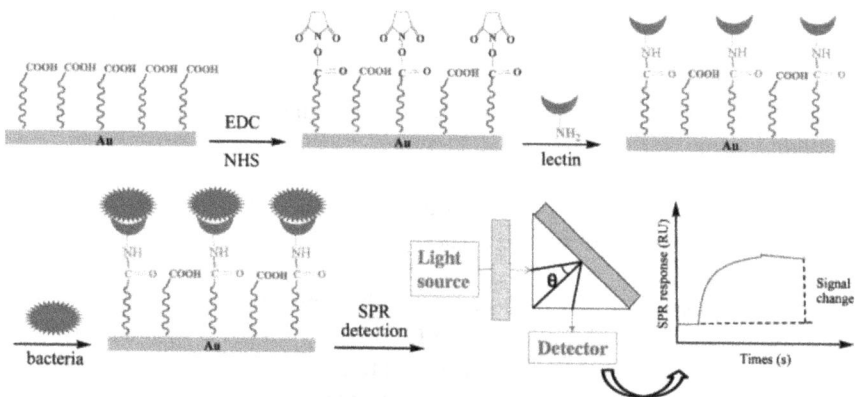

FIGURE 7.5 Schematic illustration of the lectin-based surface plasmon resonance biosensor for *E. coli* O157:H7 detection. Reprinted with permission from Wang et al. (2013).

spiked with different concentrations of *E. coli* O157:H7 were used. In analyses of food samples contaminated with *E. coli* O157:H7, the LOD of the SPR sensor was reported as 3.0 x 10^4 CFU/mL in cucumber samples and 3.0 x 10^5 CFU/mL in ground beef (Wang et al., 2013).

Wang et al. created a compact SPR sensor for the portable detection of *E. coli* O157:H7. The primary experimental arrangement comprised an all-in-one biosensor and a custom-made microfluidic chamber equipped with a three-way solenoid valve. The SPR gold chip surface was initially altered by the researchers through 3-mercaptopropionic acid (3-MPA). Subsequently, the *E. coli* O157:H7 antibody was anchored to the surface via NHS and EDC. SPR experiments were conducted involving *E. coli* solutions at concentrations of 0.25–4.0 x 10^4 CFU/mL. A correlation coefficient of 0.982 was observed. The LOD was calculated to be 1.87 x 10^3 CFU/mL (Wang et al., 2016).

Zordan et al. developed a hybrid microfluidic system that utilizes SPR and fluorescence imaging methods for identifying *E. coli* O157:H7 bacterial cells. This system comprises a range of gold nodes modified with various biomolecules, each targeting a specific agent. The array of biosensors is placed within a microfluidic channel to enable the targeted delivery of a sample via magnetic capture for evaluation. This sample is visualized by SPR from the bottom side of the biochip and by epi-fluorescence from the top side. The researchers first treated the bacteria with an antibody specific to an antigen on the surface of the bacterial cell membrane. They then pre-concentrated the samples by interacting them with magnetic particles modified with a carboxyl group. The samples were then fluorescently labeled, analyzed, and imaged. According to their results, they reported the number of *E. coli* O157:H7 cells bound to the sensor surface as 1,653, while the number of *E. coli* DH5-a cells in the control group was reported as 300. Finally, they reported the viability of *E. coli* O157:H7 bacterial cells captured by SPR as 97.7% (Zordan et al., 2009).

Yamasaki and co-workers prepared an SPR immunosensor for the identification of Shiga toxin-producing *Escherichia coli* (STEC). They also removed bacteria bound to the surface of the sensor chip using a gel displacement technique. The multi-analysis of *E. coli* O-antigens with this SPR immunosensor is schematically summarized in Figure 7.6. The researchers first purified rabbit-derived polyclonal antibodies against ten different STEC O-antigen types and immobilized them on the sensor chip. For immobilization, they modified the surface of the gold sensor chip with carboxyl groups. They completed the activation by esterifying the carboxyl groups with N-hydroxysuccinimide. They performed the detection experiments in the concentration range of STEC O157 cells, 10^5–10^8 CFU/mL, and found %ΔR values in the range of 0.012 ± 0.008 and 6.5 ± 0.5. The SPR immunosensor reported a detection limit of 6.3 × 10^4 cells for STEC O157 bacterial cells within a timeframe of 75 seconds. Notably, each of the ten O-antigens tested on STECs was specifically detected by the cognate pAb without cross-reactivity with others. The detected STEC O-157 bacterial cells were removed from the surface of the sensor chip using gelatin and agarose gel to maintain pAb functionality. This allowed the sensor chip to be reused (Yamasaki et al., 2016).

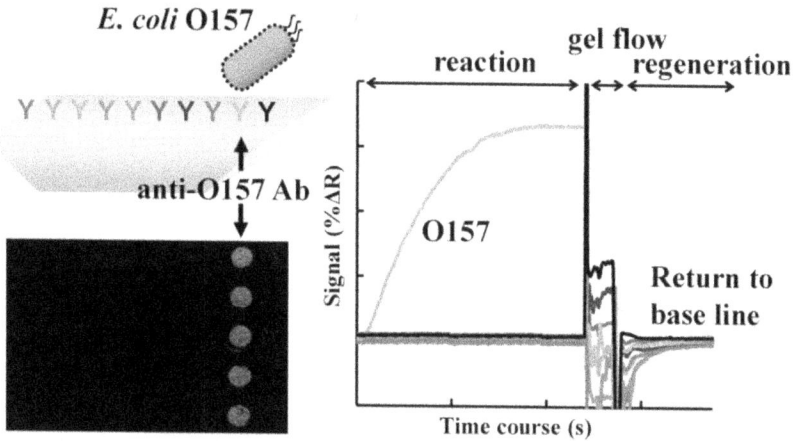

FIGURE 7.6 SPR immunosensor for multi-detection of *E. coli* O-antigens. Reprinted with permission from Yamasaki et al. (2016).

7.3.2.6 *Burkholderia* sp.

Burkholderia species are small, aerobic, encapsulated, non-spore-forming, gram-negative and have straight or slightly curved bacilli morphology. *Burkholderia mallei* and *Burkholderia pseudomallei* are bacteria that cause Ruam (Glanders) and melioidosis. Ruam is a disease of single-hoofed animals (such as horses, donkeys, mules, and camels) and rarely occurs in other domestic/wild animals and humans. *Burkholderia pseudomallei,* the causative agent of melioidosis, is a natural saprophyte commonly found in soil, stagnant water, and areas where rice is grown. Transmission occurs via contaminated soil and water by direct skin contact or by passage into the digestive tract and causes similar clinical symptoms (Godoy et al., 2003).

B. mallei was used in World Wars I and II to eliminate cavalry troops from combat. Today, there is no detailed information on the use of these agents as biological weapons. However, the fact that these agents have been studied for a period of time, are easy to produce, are not widespread, and have high mortality and morbidity when inhaled causes them to be taken into consideration (Singha et al., 2020).

Sikarwar et al. established an SPR sensing technique employing a gold chip conjugated with 4-MBA to identify *B. pseudomallei*. The devised methodology entailed observing the bond between the immobilized rabbit anti-rpGroEL rAb and the rpGroEL antigen (rpGroEL Ag). The detection mechanism consisted of immobilizing anti-rpGroEL rAbs on a gold SPR chip modified with 4-4-MBA through EDC-NHS activation. Throughout the investigation, pH 7.5 PBS was used as the working buffer, and they injected 75 μL of rpGroEL Ag and whole-cell sonicated crude Ag diluted to various concentrations (ranging from 1:51,200 to 1:100), with mixing occurring at a flow rate of 16.7 μL/h. The lowest concentration of rpGroEL Ag that resulted in a measurable response when interacting with anti-rpGroEL rAb immobilized on a 4-MBA/Au SPR chip was 0.07 nM. Kinetic evaluations indicated that the KD value was 14.77 pM, and the analyte's Bmax was 105.40 mol. The researchers also examined the changes in ΔG (Gibbs free energy), ΔH (enthalpy), and ΔS (entropy) and concluded that the interaction occurs spontaneously and releases energy through exothermic reactions, primarily due to changes in entropy (Sikarwar et al., 2016).

7.3.2.7 Vibrio cholerae

Vibrio sp. is a Gram-negative, curved rod-shaped, motile bacteria causing vibriosis. Vibriosis is a gastrointestinal tract disease leading to watery diarrhea, abdominal cramps, nausea, vomiting, fever, redness, and septicemia. It is known that the *Vibrio cholerae* reservoir may be nonagglutinable vibrions commonly found in surface waters and it is considered that these strains transform into virulent epidemic strains under certain conditions. Epidemic cholera outbreaks remain a public health concern due to high morbidity and mortality related to dehydration (Jin et al., 2013). The short incubation period, explosive epidemics, and the difficulty of treatment have led to the selection of *Vibrio cholerae* as a biological weapon. They cause a loss of workforce and place an excessive burden on healthcare organizations. International isolation, and disruption to trade and tourism can also increase these economic losses.

Contaminated water sources are a particularly important cause of cholera infection. In addition, poorly washed fruits, uncooked vegetables, and other foods can contain the bacteria that cause cholera. *V. cholerae* produces a powerful toxin which attaches to the small intestine. Then, it interferes with sodium and chloride when it binds to the intestinal walls. When the bacteria bind to the walls of the small intestine, the patient starts to secrete excessive water resulting in diarrhea and rapid loss of fluids and salts (Okello et al., 2019). Immobilization technology has been introduced to obtain lower detection limits and highly sensitive and selective sensing platforms using antibodies to eliminate non-specific binding Taheri et al. developed an immunosensor based SPR assay format to detect *V. cholerae* using an antibody with affinity to recombinant outer membrane protein (anti-OmpW). Recombinant OmpW antigen, present in the outer membrane of the bacteria and showing reactivity and specificity to anti-OmpW, was purified. Following that, Protein G was bound onto an 11-mercaptoundecanoic acid (11-MUA) self-assembled monolayer (SAM) via amine coupling. The hydroxyl groups of ethanolamine used in amine coupling to eliminate unbound NHS-ester groups are crucial to prevent non-specific binding. Protein G bound on SAM can create more freedom for immobilized antibodies to interact with the target antigen. Therefore, the anti-OmpW and OmpW antigens interacted with high bioaffinity using the advantage of immobilization on the protein G layer. It was revealed that sensitive detection of bacteria was reported as 43 cells/mL. (Taheri et al., 2016).

Bacteriophages are promising recognition elements used in combination with sensor systems for the detection of bacteria. In this study, an SPR sensor system was designed for the detection of *Vibrio cholera* by using the minor coat protein pIII domain (N1N2) of CTXφ phage. N1N2, which contains relevant parts for bacteriophage attachment and entry into the cell, was obtained from *Escherichia coli* by recombinant protein expression and isolation techniques. The protein bound to *V. cholerae*, but not to *E. coli* was K-12. Recombinant N1N2 was immobilized to the gold surface by amine-coupling. With *V. cholera* in the concentration range of 10^3 to 10^9 CFU/mL, protein and bacterial interactions were detected on the SPR system. The use of recognition molecules for the host bacterial cell in biosensor design helps to solve the problem of non-specific transient recognition of specific recombinant proteins to non-host bacterial cells (Shin et al., 2021).

7.4 CONCLUSION AND FUTURE PERSPECTIVES

Biological warfare has been a challenge since the dawn of life. Then, the emergence of infectious diseases and agents developing antimicrobial resistance have become a global concern. The increase in antibiotic resistance has led to the improvement of new strategies and technologies to combat these bacterial threats. In the past, bacterial threats have often been emphasized for use in biological warfare. Diseases caused by many highly virulent and infectious agents such as anthrax and plague, as well as diseases caused by the toxins of some bacterial agents, have emerged as threatening factors. Whether it is possible to create new biological weapons through advances in genetic engineering is still a matter of current research.

Many publications are indicating the availability of sensing platforms in the detection of BTA, however, a challenge still remains to construct accurate and efficient systems to detect them. It is important that sensor systems designed for the detection of BTA provide fast analysis time and a low detection limit. It is noteworthy to emphasize that low-cost, low-labor, portable, and miniaturized systems are attractive. On the other hand, the usage of receptors, ligands, and other advancements in SPR can improve the performance of designed sensors.

BTA can be detected in SPR sensing platforms using compatible recognition molecules. However, there are some limitations in BTA sensing and devices do not meet all of the requirements of applicable sensing systems. Scientific and financial attempts have to be made to create promising tools. The current state of BTA sensing will find applications across many different disciplines, especially in the field of security, in the near future.

REFERENCES

Ahanotu, E., Alvelo-ceron, D., Ravita, T., Gaunt, E., 2006. Staphylococcal enterotoxin B as a biological weapon: Recognition, management, and surveillance of staphylococcal enterotoxin. *Appl. Biosaf.* 11, 120–126.

Alafeef, M., Moitra, P., Pan, D., 2020. Nano-enabled sensing approaches for pathogenic bacterial detection. *Biosens. Bioelectron.* 165, 112276. https://doi.org/10.1016/j.bios.2020.112276.

Aslıyüce, S., Idil, N., Mattiasson, B., 2022. Upgrading of bio-separation and bioanalysis using synthetic polymers: Molecularly imprinted polymers (MIPs), cryogels, stimuli-responsive polymers. *Eng. Life Sci.* 22, 204–216. https://doi.org/10.1002/ELSC.202100106.

Bae, J.S., Da, F., Liu, R., He, L., Lv, H., Fisher, E.L., Rajagopalan, G., Li, M., Cheung, G.Y.C., 2021. Contribution of staphylococcal enterotoxin B to *Staphylococcus aureus* systemic infection. *J Inf. Dis.* 223. https://doi.org/10.1093/infdis/jiaa584.

Bahari, D., Babamiri, B., Salimi, A., Salimizand, H., 2021. Ratiometric fluorescence resonance energy transfer aptasensor for highly sensitive and selective detection of Acinetobacter baumannii bacteria in urine sample using carbon dots as optical nanoprobes. *Talanta* 221, 121619. https://doi.org/10.1016/J.TALANTA.2020.121619.

Balbinot, S., Srivastav, A.M., Vidic, J., Abdulhalim, I., Manzano, M., 2021. Plasmonic biosensors for food control. *Trends Food Sci. Technol.* 111, 128–140. https://doi.org/10.1016/j.tifs.2021.02.057.

Barlen, B., Mazumdar, S.D., Lezrich, O., Kämpfer, P., Keusgen, M., 2007. Detection of *Salmonella* by surface plasmon resonance. *Sensors* 7, 1427–1446. https://doi.org/10.3390/s7081427.

Bhandari, D., Chen, F.C., Bridgman, R.C., 2022. Magnetic nanoparticles enhanced surface plasmon resonance biosensor for rapid detection of *Salmonella typhimurium* in Romaine lettuce. *Sensors* 22, 475.

Bräuer, B., Thier, F., Bittermann, M., Baurecht, D., Lieberzeit, P.A., 2022. Raman studies on surface-imprinted polymers to distinguish the polymer surface, imprints, and different bacteria. *ACS Appl. Bio Mater.* 5, 160–171. https://doi.org/10.1021/ACSABM.1C01020.

Broecker, F., Aretz, J., Yang, Y., Hanske, J., Guo, X., Reinhardt, A., Wahlbrink, A., Rademacher, C., Anish, C., Seeberger, P.H., 2014. Epitope recognition of antibodies against a *Yersinia pestis* lipopolysaccharide trisaccharide component. *ACS Chem. Biol.* 9, 867–873. https://doi.org/10.1021/cb400925k.

Chattopadhyay, S., Sabharwal, P.K., Jain, S., Kaur, A., Singh, H., 2019. Functionalized polymeric magnetic nanoparticle assisted SERS immunosensor for the sensitive detection of *S. typhimurium*. *Anal. Chim. Acta* 1067, 98–106. https://doi.org/10.1016/J.ACA.2019.03.050.

Clark, D.P., Pazdernik, N.J., 2016. Biological warfare: Infectious disease and bioterrorism. *Biotechnology.* https://doi.org/10.1016/b978-0-12-385015-7.00022-3.

Das, S., Kataria, V.K., 2010. Bioterrorism: A public health perspective. *MJAFI* 66(3), 255–260. https://doi.org/10.1016.

Dawan, S., Kanatharana, P., Chotigeat, W., Jitsurong, S., Thavarungkul, P., 2011. Surface plasmon resonance immunosensor for rapid and specific diagnosis of melioidosis antibody. *Southeast Asian J. Trop. Med. Public Health* 42, 1168.

Dursun, A.D., Borsa, B.A., Bayramoglu, G., Arica, M.Y., Ozalp, V.C., 2022. Surface plasmon resonance aptasensor for *Brucella* detection in milk. *Talanta* 239, 123074. https://doi.org/10.1016/J.TALANTA.2021.123074.

Farlow, J., Wagner, D.M., Dukerich, M., Stanley, M., Chu, M., Kubota, K., Petersen, J., Keim, P., 2005. Francisella tularensis in the United States. *Emerg. Infect. Dis.* 11, 1835. https://doi.org/10.3201/EID1112.050728.

Farooq, U., Yang, Q., Ullah, M.W., Wang, S., 2018. Bacterial biosensing: Recent advances in phage-based bioassays and biosensors. *Biosens. Bioelectron.* 118, 204–216. https://doi.org/10.1016/j.bios.2018.07.058.

Féraudet-Tarisse, C., Mazuet, C., Pauillac, S., Krüger, M., Lacroux, C., Popoff, M.R., Dorner, B.G., Andréoletti, O., Plaisance, M., Volland, H., Simon, S., 2017. Highly sensitive sandwich immunoassay and immunochromatographic test for the detection of Clostridial epsilon toxin in complex matrices. *PLoS One* 12, e0181013. https://doi.org/10.1371/JOURNAL.PONE.0181013.

Ferracci, G., Marconi, S., Mazuet, C., Jover, E., Blanchard, M.P., Seagar, M., Popoff, M., Lévêque, C., 2011. A label-free biosensor assay for botulinum neurotoxin B in food and human serum. *Anal. Biochem.* 410, 281–288. https://doi.org/10.1016/J.AB.2010.11.045.

Ghosh, N., Gupta, G., Boopathi, M., Pal, V., Singh, A.K., Gopalan, N., Goel, A.K., 2013a. Surface Plasmon Resonance biosensor for detection of *Bacillus anthracis*, the causative agent of anthrax from soil samples targeting protective antigen. *Indian J. Microbiol.* 53, 48–55. https://doi.org/10.1007/S12088-012-0334-3.

Ghosh, N., Gupta, N., Gupta, G., Boopathi, M., Pal, V., Goel, A.K., 2013b. Detection of protective antigen, an anthrax specific toxin in human serum by using surface plasmon resonance. *Diagn. Microbiol. Infect. Dis.* 77, 14–19. https://doi.org/10.1016/J.DIAGMICROBIO.2013.05.006.

Godoy, D., Randle, G., Simpson, A.J., Aanensen, D.M., Pitt, T.L., Kinoshita, R., Spratt, B.G., 2003. Multilocus sequence typing and evolutionary relationships among the causative agents of melioidosis and glanders, *Burkholderia pseudomallei* and *Burkholderia mallei. J. Clin. Microbiol.* 41, 2068–2079. https://doi.org/10.1128/JCM.41.5.2068-2079.

Gooding, J.J., 2006. Biosensor technology for detecting biological warfare agents: Recent progress and future trends. *Anal. Chim. Acta* 559, 137–151. https://doi.org/10.1016/j.aca.2005.12.020.

Grotte, J.H., 2001. Frequently asked questions regarding biological detection. Institute for Defense Analyses, IDA Document D-2663.

Gupta, G., Kumar, A., Boopathi, M., Thavaselvam, D., Vijayaraghavan, R., 2011. Rapid and quantitative determination of biological warfare agent *Brucella abortus* CSP-31 using Surface plasmon resonance. *Anal. Bioanal. Electrochem* 3, 26–37.

Gupta, G., Singh, P.K., Boopathi, M., Kamboj, D.V., Singh, B., Vijayaraghavan, R., 2010. Surface plasmon resonance detection of biological warfare agent staphylococcal enterotoxin B using high affinity monoclonal antibody. *Thin Solid Films* 519, 1171–1177. https://doi.org/10.1016/j.tsf.2010.08.064.

Hans, R., Yadav, P.K., Sharma, P.K., Boopathi, M., Thavaselvam, D., 2020. Development and validation of immunoassay for whole cell detection of *Brucella abortus* and *Brucella melitensis. Sci. Reports* 101(10), 1–13. https://doi.org/10.1038/s41598-020-65347-9.

Hau, D., Wade, B., Lovejoy, C., Pandit, S.G., Reed, D.E., Demers, H.L., Green, H.R., Hannah, E.E., McLarty, M.E., Creek, C.J., Chokapirat, C., Arias-Umana, J., Cecchini, G.F., Nualnoi, T., Gates-Hollingsworth, M.A., Thorkildson, P.N., Pflughoeft, K.J., Aucoin, D.P., 2022. Development of a dual antigen lateral flow immunoassay for detecting *Yersinia pestis. PLoS Negl. Trop. Dis.* 16, e0010287. https://doi.org/10.1371/JOURNAL.PNTD.0010287.

Herrera-Domínguez, M., Morales-Luna, G., Mahlknecht, J., Cheng, Q., Aguilar-Hernández, I., Ornelas-Soto, N., 2023. Optical biosensors and their applications for the detection of water pollutants. *Biosensors* 13. https://doi.org/10.3390/bios13030370.

Homola, J., Dostálek, J., Chen, S., Rasooly, A., Jiang, S., Yee, S.S., 2002. Spectral surface plasmon resonance biosensor for detection of staphylococcal enterotoxin B in milk. *Int. J. Food Microbiol.* 75, 61–69. https://doi.org/10.1016/S0168-1605(02)00010-7.

Hu, N., Qiao, C., Wang, J., Wang, Z., Li, X., Zhou, L., Wu, J., Zhang, D., Feng, J., Shen, B., Zhang, J., Luo, L., 2021. Identification of a novel protective human monoclonal antibody, LXY8, that targets the key neutralizing epitopes of staphylococcal enterotoxin B. *Biochem. Biophys. Res. Commun.* 549, 120–127. https://doi.org/10.1016/J.BBRC.2021.02.057.

Hughes, J.M., Gerberding, J.L., 2002. Anthrax bioterrorism: Lessons learned and future directions. *Emerg. Infect. Dis.* 8, 1013. https://doi.org/10.3201/EID0810.020466.

https://emergency.cdc.gov/agent/agentlist-category.asp

https://www.cbrnetechindex.com/p/3447/Response-Biomedical-Corporation/RAMP-200-Biowarfare-Detection-System

https://www.rapidmicrobiology.com/supplier/alexeter-technologies-llc

Idil, N., Aslıyüce, S., Perçin, I., Mattiasson, B., 2023. Recent advances in optical sensing for the detection of microbial contaminants. *Micromachines* 14, 1668.

Idil, N., Bakhshpour, M., Perçin, I., Mattiasson, B., 2021. Whole cell recognition of staphylococcus aureus using biomimetic SPR sensors. *Biosensors* 11, 140. https://doi.org/10.3390/BIOS11050140.

Ina, K., Kusugami, K., Ohta, M., 2003. Bacterial hemorrhagic enterocolitis. *J. Gastroenterol.* 382(38), 111–120. https://doi.org/10.1007/S005350300019.

Isakova, A.A., Zharnikova, I.V., Zharnikova, T.V., Indenbom, A.V., 2019. Detection of the causative agent of tularemia by surface plasmon resonance. *Prot. Met. Phys. Chem. Surfaces* 55, 407–411. https://doi.org/10.1134/S2070205119020102.

Islam, M.A., Karim, A., Ethiraj, B., Raihan, T., Kadier, A., 2022. Antimicrobial peptides: Promising alternatives over conventional capture ligands for biosensor-based detection of pathogenic bacteria. *Biotechnol. Adv.* 55, 107901. https://doi.org/10.1016/j.biotechadv.2021.107901.

Janik, E., Ceremuga, M., Bijak, J.S., Bijak, M., 2019. Biological toxins as the potential tools for bioterrorism. *Int. J. Mol. Sci.* 20. https://doi.org/10.3390/ijms20051181.

Janik, M., Brzozowska, E., Czyszczoń, P., Celebańska, A., Koba, M., Gamian, A., Bock, W.J., Śmietana, M., 2021. Optical fiber aptasensor for label-free bacteria detection in small volumes. *Sensors Actuators B Chem.* 330, 129316. https://doi.org/10.1016/J.SNB.2020.129316.

Janith, G.I., Herath, H.S., Hendeniya, N., Attygalle, D., Amarasinghe, D.A.S., Logeeshan, V., Wickramasinghe, P.M.T.B., Wijayasinghe, Y.S., 2023. Advances in surface plasmon resonance biosensors for medical diagnostics: An overview of recent developments and techniques. *J. Pharm. Biomed. Anal. Open* 2, 100019. https://doi.org/10.1016/j.jpbao.2023.100019.

Jiang, S., Zhu, S., Qian, S., Xu, K., Liu, S., 2023. Detection of *E. coli* O157:H7 via GO-modified fiber optic SPR sensor with Au nanoparticle signal amplification. Proceedings Volume 12757, 3rd International Conference on Laser, Optics, and Optoelectronic Technology (LOPET 2023), 513–517. https://doi.org/10.1117/12.2690763.

Jin, D., Luo, Y., Zheng, M., Li, H., Zhang, J., Stampfl, M., Xu, X., Ding, G., Zhang, Y., Tang, Y.W. Quantitative detection of Vibrio cholera toxin by real-time and dynamic cytotoxicity monitoring. *J. Clin. Microbiol.* 51, 3968–3974. https://doi.org/10.1128/jcm.01959-13

Koyappayil, A., Lee, M.H., 2021. Ultrasensitive materials for electrochemical biosensor labels. *Sensors (Switzerland)* 21, 1–19. https://doi.org/10.3390/s21010089.

Kumar, H., Rani, R., 2013. Development of biosensors for the detection of biological warfare agents: Its issues and challenges. *Science Progress* 96, 294–308. https://doi.org/10.3184/003685013X13777066241280.

Kushwaha, A.S., Kumar, A., Kumar, R., Srivastava, M., Srivastava, S.K., 2018. Zinc oxide, gold and graphene-based surface plasmon resonance (SPR) biosensor for detection of *pseudomonas* like bacteria: A comparative study. *Optik (Stuttg).* 172, 697–707. https://doi.org/10.1016/j.ijleo.2018.07.066.

Ladd, J., Taylor, A.D., Homola, J., Jiang, S., 2008. Detection of botulinum neurotoxins in buffer and honey using a surface plasmon resonance (SPR) sensor. *Sensors Actuators B Chem.* 130, 129–134. https://doi.org/10.1016/J.SNB.2007.07.140.

Lévêque, C., Ferracci, G., Maulet, Y., Grand-Masson, C., Blanchard, M.P., Seagar, M., El Far, O., 2013. A substrate sensor chip to assay the enzymatic activity of Botulinum neurotoxin A. *Biosens. Bioelectron.* 49, 276–281. https://doi.org/10.1016/J.BIOS.2013.05.032.

Lévêque, C., Ferracci, G., Maulet, Y., Mazuet, C., Popoff, M., Seagar, M., El Far, O., 2014. Direct biosensor detection of botulinum neurotoxin endopeptidase activity in sera from patients with type A botulism. *Biosens. Bioelectron.* 57, 207–212. https://doi.org/10.1016/J.BIOS.2014.02.015.

Lévêque, C., Ferracci, G., Maulet, Y., Mazuet, C., Popoff, M.R., Blanchard, M.P., Seagar, M., El Far, O., 2015. An optical biosensor assay for rapid dual detection of Botulinum neurotoxins A and E. *Sci. Reports* 51(5), 1–9. https://doi.org/10.1038/srep17953.

Li, Y., Wang, X., Ning, W., Yang, E., Li, Y., Luo, Z., Duan, Y., 2022. Sandwich method-based sensitivity enhancement of Ω-shaped fiber optic LSPR for time-flexible bacterial detection. *Biosens. Bioelectron.* 201, 113911. https://doi.org/10.1016/j.bios.2021.113911.

Lindström, M., Korkeala, H., 2006. Laboratory diagnostics of botulism. *Clin. Microbiol. Rev.* 19, 298–314. https://doi.org/10.1128/CMR.19.2.298-314.2006.

Mazumdar, S.D., Hartmann, M., Kämpfer, P., Keusgen, M., 2007. Rapid method for detection of *Salmonella* in milk by surface plasmon resonance (SPR). *Biosens. Bioelectron.* https://doi.org/10.1016/j.bios.2006.09.004.

McCoy, G.W., 1912. Bacterium tularense, the cause of a plaguelike disease of rodents. *Public Health Bull.* 53, 17.

Mock, M., Mignot, T., 2003. Anthrax toxins and the host: A story of intimacy. *Cell. Microbiol.* 5, 15–23. https://doi.org/10.1046/J.1462-5822.2003.00253.X.

Naimushin, A.N., Soelberg, S.D., Nguyen, D.K., Dunlap, L., Bartholomew, D., Elkind, J., Melendez, J., Furlong, C.E., 2002. Detection of *Staphylococcus aureus* enterotoxin B at femtomolar levels with a miniature integrated two-channel surface plasmon resonance (SPR) sensor. *Biosens. Bioelectron.* 17, 573–584. https://doi.org/10.1016/S0956-5663(02)00014-3.

Nikolakakis, I., Michaleas, S.N., Panayiotakopoulos, G., Papaioannou, T.G., Karamanou, M., Michaleas, S., Papaioannou, T., 2023. The History of anthrax weaponization in the Soviet Union. *Cureus* 15(3). https://doi.org/10.7759/cureus.36800.

Nnachi, R.C., Sui, N., Ke, B., Luo, Z., Bhalla, N., He, D., Yang, Z., 2022. Biosensors for rapid detection of bacterial pathogens in water, food and environment. *Environ. Int.* 166, 107357. https://doi.org/10.1016/j.envint.2022.107357.

Nualnoi, T., Kirosingh, A., Basallo, K., Hau, D., Gates-Hollingsworth, M.A., Thorkildson, P., Crump, R.B., Reed, D.E., Pandit, S., AuCoin, D.P., 2018. Immunoglobulin G subclass switching impacts sensitivity of an immunoassay targeting *Francisella tularensis* lipopolysaccharide. *PLoS One* 13, e0195308. https://doi.org/10.1371/JOURNAL.PONE.0195308.

Okello, P.E., Bulage, L., Riolexus, A.A., Kadobera, D., Kwesiga, B., Kajumbula, H., Mulongo, M., Namboozo, E.J., Pimundu, G., Ssewanyana, I., Kiyaga, C., Aisu, S., Zhu, B., 2019. A cholera outbreak caused by drinking contaminated river water, Bulambuli. *BMC Infect. Dis.* 19, 1–8.

Park, C., Lee, J., Lee, D., Jang, J., 2022. Paper-based electrochemical peptide sensor for label-free and rapid detection of airborne *Bacillus anthracis* simulant spores. *Sensors Actuators B Chem.* 355, 131321. https://doi.org/10.1016/J.SNB.2021.131321.

Park, J.H., Cho, Y.W., Kim, T.H., 2022. Recent advances in surface plasmon resonance sensors for sensitive optical detection of pathogens. *Biosensors* 12. https://doi.org/10.3390/bios12030180.

Patel, K., Halevi, S., Melman, P., Schwartz, J., Cai, S., Singh, B.R., 2017. A novel surface plasmon resonance biosensor for the rapid detection of botulinum neurotoxins. *Biosensors* 7, 32. https://doi.org/10.3390/BIOS7030032.

Pechous, R.D., Sivaraman, V., Stasulli, N.M., Goldman, W.E., 2016. Pneumonic plague: The darker side of *Yersinia pestis*. *Trends Microbiol.* 24, 190–197. https://doi.org/10.1016/j.tim.2015.11.008.

Peng, H., Chen, I.A., 2019. Rapid colorimetric detection of bacterial species through the capture of gold nanoparticles by chimeric Phages. *ACS Nano* 13, 1244–1252. https://doi.org/10.1021/acsnano.8b06395.

Perçin, I., Idil, N., Bakhshpour, M., Yılmaz, E., Mattiasson, B., Denizli, A., 2017a. Microcontact imprinted plasmonic nanosensors: Powerful tools in the detection of salmonella paratyphi. *Sensors (Switzerland)*. https://doi.org/10.3390/s17061375.

Perçin, I., Idil, N., Bakhshpour, M., Yılmaz, E., Mattiasson, B., Denizli, A., 2017b. Microcontact imprinted plasmonic nanosensors: Powerful tools in the detection of *Salmonella paratyphi*. *Sensors* 17, 1375. https://doi.org/10.3390/S17061375.

Perumal, V., Hashim, U., 2014. Advances in biosensors: Principle, architecture and applications. *J. Appl. Biomed.* 12, 1–15. https://doi.org/10.1016/j.jab.2013.02.001.

Piérard, D., De Greve, H., Haesebrouck, F., Mainil, J., 2012. O157:H7 and O104:H4 Vero/Shiga toxin-producing *Escherichia coli* outbreaks: Respective role of cattle and humans. *Vet. Res.* 43, 13. https://doi.org/10.1186/1297-9716-43-13.

Pohanka, M., 2019. Current trends in the biosensors for biological warfare agents assay. *Materials* 12, 2303. https://doi.org/10.3390/ma12142303.

Rangel, J.M., Sparling, P.H., Crowe, C., Griffin, P.M., Swerdlow, D.L., 2005. Epidemiology of *Escherichia coli* O157:H7 outbreaks, United States, 1982–2002. *Emerg. Infect. Dis.* 11, 603–609. https://doi.org/10.3201/eid1104.040739.

Ravichandran, E., Gong, Y., Al Saleem, F.H., Ancharski, D.M., Joshi, S.G., Simpson, L.L., 2006. An initial assessment of the systemic pharmacokinetics of botulinum toxin. *J. Pharmacol. Exp. Ther.* 318, 1343–1351. https://doi.org/10.1124/JPET.106.104661.

Saberi, F., Kamali, M., Taheri, R.A., Ramandi, M.F., Bagdeli, S., Mirnejad, R., 2016. Development of surface plasmon resonance-based immunosensor for detection of *Brucella melitensis*. *J. Braz. Chem. Soc.* 27, 1960–1965. https://doi.org/10.5935/0103-5053.20160085.

Sharma, K., Sharma, M., 2023. Optical biosensors for environmental monitoring: Recent advances and future perspectives in bacterial detection. *Environ. Res.* 236, 116826. https://doi.org/10.1016/j.envres.2023.116826.

Shin, H.J., Hyeon, S.H., Cho, J.H., Lim, W.K., 2021. Minor coat Protein pIII domain (N1N2) of bacteriophage CTXϕ confers a novel surface plasmon resonance biosensor for rapid detection of *Vibrio cholerae*. *Microbiol. Biotechnol. Lett.* 49, 510–518. https://doi.org/10.48022/mbl.2109.09013.

Si, C.Y., Ye, Z.Z., Wang, Y.X., Gai, L., Wang, J.P., Ying, Y.B., 2011. Rapid detection of *Escherichia coli* O157: H7 using surface plasmon resonance (SPR) biosensor. *Spectros. Spect. Anal.* 31, 2598–2601. https://doi.org/10.3964/J.ISSN.1000-0593(2011)10-2598-04.

Sikarwar, B., Sharma, P.K., Kumar, A., Thavaselvam, D., Boopathi, M., Singh, B., Jaiswal, Y.K., 2016. Surface plasmon resonance immunosensor for the detection of *Burkholderia pseudomallei*. *Plasmonics* 11, 1035–1042. https://doi.org/10.1007/S11468-015-0139-4.

Sikarwar, B., Singh, V.V., Sharma, P.K., Kumar, A., Thavaselvam, D., Boopathi, M., Singh, B., Jaiswal, Y.K., 2017. DNA-probe-target interaction based detection of *Brucella melitensis* by using surface plasmon resonance. *Biosens. Bioelectron.* 87, 964–969. https://doi.org/10.1016/J.BIOS.2016.09.063.

Singha, H., Shanmugasundaram, K., Saini, S., Tripathi, B.N., 2020. Serological survey of humans exposed to *Burkholderia mallei*–infected equids: A public health approach. *Asia-Pacific J. Public Heal.* 32, 274–277. https://doi.org/10.1177/1010539520930500.

Song, M.-S., Sekhon, S.S., Shin, W.-R., Rhee, S.-K., Ko, J.H., Kim, S.Y., Min, J., Ahn, J.-Y., Kim, Y.-H., 2017. Aptamer-immobilized surface plasmon resonance biosensor for rapid and sensitive determination of virulence determinant. *J. Nanosci. Nanotechnol.* 18, 3095–3101. https://doi.org/10.1166/JNN.2018.14697.

Stenseth, N.C., Atshabar, B.B., Begon, M., Belmain, S.R., Bertherat, E., Carniel, E., Gage, K.L., Leirs, H., Rahalison, L., 2008. Plague: Past, present, and future. *PLOS Med.* 5, e3. https://doi.org/10.1371/JOURNAL.PMED.0050003.

Sweeney, D.A., Hicks, C,W., Cui, X., Li, Y., Eichacker, P.Q., 2012. Anthrax infection. *Am. J. Respir. Crit. Care. Med.* 184, 1333–1341. https://doi.org/10.1164/RCCM.201102-0209CI.

Taheri, R.A., Rezayan, A.H., Rahimi, F., Mohammadnejad, J., Kamali, M., 2016. Development of an immunosensor using oriented immobilized anti-OmpW for sensitive detection of *Vibrio cholerae* by surface plasmon resonance. *Biosens. Bioelectron.* 86, 484–488. https://doi.org/10.1016/j.bios.2016.07.006.

Thavaselvam, D., Vijayaraghavan, R., 2010. Biological warfare agents. *J. Pharm. Bioallied Sci.* 2, 179. https://doi.org/10.4103/0975-7406.68499.

To, T.J., Sokolow, R., Mauvais, S., Birkness, K.A., Skeels, M.R., Horan, J.M., Foster, L.R., 1997. A large community outbreak of salmonellosis caused by intentional Contamination of restaurant salad bars. *JAMA* 278, 389–395.

Tomar, A., Gupta, G., Singh, M., Boopathi, M., Biodef, J.B., 2016. Surface plasmon resonance sensing of biological warfare agent botulinum neurotoxin A. Artic. *J. Bioterrorism Biodefense* 7, 1. https://doi.org/10.4172/2157-2526.1000142.

Trzaskowski, M., Ciach, T., 2017. SPR system for on-site detection of biological warfare. *Curr. Anal. Chem.* 13, 144–149.

Turk, B.E., 2007. Manipulation of host signalling pathways by anthrax toxins. *Biochem. J.* 402, 405–417. https://doi.org/10.1042/BJ20061891.

Uzal, F.A., Freedman, J.C., Shrestha, A., Theoret, J.R., Garcia, J., Awad, M.M., Adams, V., Moore, R.J., Rood, J.I., Mcclane, B.A., 2014. Towards an understanding of the role of *Clostridium perfringens* toxins in human and animal disease. *Future Microbiol.* 9, 361–377. https://doi.org/10.2217/fmb.13.168.

Wang, B., Park, B., Xu, B., Kwon, Y., 2017. Label-free biosensing of *Salmonella enterica* serovars at single-cell level. *J. Nanobiotechnology* 15, 1–11. https://doi.org/10.1186/s12951-017-0273-6.

Wang, D.B., Bi, L.J., Zhang, Z.P., Chen, Y.Y., Yang, R.F., Wei, H.P., Zhou, Y.F., Zhang, X.E., 2009. Label-free detection of B. anthracis spores using a surface plasmon resonance biosensor. *Analyst* 134, 738–742. https://doi.org/10.1039/B813038H.

Wang, S., Xie, J., Jiang, M., Chang, K., Chen, R., Ma, L., Zhu, J., Guo, Q., Sun, H., Hu, J., 2016. The development of a portable SPR bioanalyzer for sensitive detection of *Escherichia coli* O157:H7. *Sensors* 16, 1856. https://doi.org/10.3390/S16111856.

Wang, Y., Ye, Z., Si, C., Ying, Y., 2013. Monitoring of *Escherichia coli* O157:H7 in food samples using lectin based surface plasmon resonance biosensor. *Food Chem.* 136, 1303–1308. https://doi.org/10.1016/J.FOODCHEM.2012.09.069.

Wang, Y., Ye, Z., Si, C., Ying, Y., 2011. Subtractive inhibition assay for the detection of *E. coli* O157:H7 using surface plasmon resonance. *Sensors* 11, 2728–2739. https://doi.org/10.3390/S110302728.

Waswa, J., Irudayaraj, J., DebRoy, C., 2007. Direct detection of *E. coli* O157:H7 in selected food systems by a surface plasmon resonance biosensor. *LWT – Food Sci. Technol.* 40, 187–192. https://doi.org/10.1016/J.LWT.2005.11.001.

Wijnands, L.M., van der Mey-Florijn, A., Delfgou-van Asch, E., 2011. *Clostridium perfringens* associated food borne disease 53.

Yamasaki, T., Miyake, S., Nakano, S., Morimura, H., Hirakawa, Y., Nagao, M., Iijima, Y., Narita, H., Ichiyama, S., 2016. Development of a surface plasmon resonance-based immunosensor for detection of 10 major O-antigens on shiga toxin-producing *Escherichia coli*, with a gel displacement technique to remove bound bacteria. *Anal. Chem.* 88, 6711–6717. https://doi.org/10.1021/ACS.ANALCHEM.6B00797.

Yang, R., 2018. Plague: Recognition, treatment, and prevention. *J. Clin. Microbiol.* 56. https://doi.org/10.1128/JCM.01519-17.

Zeng, L., Wang, L., Hu, J., 2018. Current and emerging technologies for rapid detection of pathogens. *Biosensing Technol. Detect. Pathog. – A Prospect. W. Rapid Anal.* 73178, 6–19. https://doi.org/10.5772/INTECHOPEN.73178.

Zhou, C., Zou, H., Li, M., Sun, C., Ren, D., Li, Y., 2018. Fiber optic surface plasmon resonance sensor for detection of *E. coli* O157:H7 based on antimicrobial peptides and
AgNPs-rGO. *Biosens. Bioelectron.* 117, 347–353. https://doi.org/10.1016/j.bios.2018.06.
005.
Zordan, M.D., Grafton, M.M.G., Acharya, G., Reece, L.M., Cooper, C.L., Aronson, A.I.,
Park, K., Leary, J.F., 2009. Detection of pathogenic *E. coli* O157:H7 by a hybrid microfluidic SPR and molecular imaging cytometry device. *Cytom. Part A* 75A, 155–162.
https://doi.org/10.1002/CYTO.A.20692.

8 Plasmonic Nanosensors for Real-time Detection of Nerve Agents

Monireh Bakhshpour-Yucel, Ali Araz, Sinem Diken-Gur, Nese Lortlar Unlu, and Bilgen Osman

8.1 INTRODUCTION

Sensors, sophisticated analytical systems, that are comprised of a recognition element, a transducer, and a signal operator, are designed for real-time monitoring and detection of chemical or biological interactions. They utilize diverse transduction methods rooted in functional biorecognition elements' distinct properties such as electrochemistry, acoustics, plasmonics, optics, chemistry, physics, or mechanics (Chen et al., 2020). They are attractive due to their cost-effectiveness, ease of use, scalability, and suitability for point-of-care applications, driving significant efforts and resources towards enhancing highly sensitive and selective sensor systems, especially for molecule detection (Adam et al., 2023). Researchers are pursuing advancements in the field of sensor technology, focusing on enhancing receptor affinity, signal enhancement using nanomaterials, and integrating microfluidic chips to push the boundaries of sensor capabilities (Cui et al., 2020). Recent advancements in plasmonic optics, particularly in the area of surface plasmon resonance (SPR), have driven progress in various fields, including negative refractive index materials, integrated circuits, and optical metal surfaces, with a focus on coupling electromagnetic waves to charge intensity oscillations at dielectric-metal interfaces (surface plasmon polaritons, SPPs), and this research not only expands scientific horizons but also brings us closer to highly effective and widely applicable sensor solutions. The widely used attenuated total reflection method, pioneered by Kretschmann and Otto, along with compact coupling structures like optical gratings and waveguides, plays a central role in SPR integrated circuits (Kretschmann et al., 1968; Otto, 1968). Metal-based nanoparticles, especially gold, and silver have gained attention for their compatibility with synthetic metal-based molecular probes and remarkable optical properties linked to localized plasmon resonance (Kurbanoglu and Ozkan, 2018; Yang and Duncan, 2018).

Nanoparticles have diverse applications in nanobiotechnology, biomedicine, and various analytical and therapeutic techniques, driving advancements in meta-materials with optical negative refractive index properties (Diken Gür et al., 2019; Calcagno et al., 2022).

Surface plasmons, harnessing their distinctive optical qualities, have become a highly adaptable tool with applications spanning various multidisciplinary fields, from biology to renewable energy, owing to the increasing popularity of plasmonic sensors, which excel in detecting subtle environmental changes by amplifying local electromagnetic fields (Shalaev et al., 2007; Moon et al., 2022).

These sensor technologies can be broadly classified into two primary categories such as surface-enhanced spectroscopic sensors and SPR sensors. Surface-enhanced spectroscopic sensors comprise a range of techniques, including surface-enhanced Raman scattering (SERS) (Lindquist et al., 2019), surface-enhanced infrared absorption (SEIRA) (Zhou et al., 2020), and surface-enhanced fluorescence (SEF) (Huang et al., 2021).

Certain sensor methods have not only proven highly effective but also spurred the establishment of numerous manufacturing companies to meet rising demand across industries, highlighting their practical significance and strong commercial viability in modern scientific research and industrial applications (Homola, 2008; Wenger et al., 2017). Plasmonic-based nanosensors, known for their label-free, real-time detection and remarkable sensitivity, have wide-ranging applications in fields like homeland security, environmental monitoring, diagnostics, and pharmaceuticals. The field of plasmonic nanosensors has emerged as a promising avenue for achieving real-time detection of hazardous substances. Among these substances, nerve agents, classified as weapons of mass destruction, pose an alarming threat to both civilian populations and military due to their extreme toxicity and potential for malicious use. Early and accurate detection of nerve agents is critical for effective countermeasures, emergency response, and the preservation of human lives. This chapter delves into the captivating world of plasmonic nanosensors, highlighting their exceptional potential for addressing the challenges associated with nerve agent detection.

8.2   NERVE AGENTS

Chemical warfare agents (CWAs) are synthetic chemicals designed to harm humans, animals, and plants, and they can cause lethal effects (Urabe et al., 2014). CWAs encompass various substances, including toxic gases, types of smoke, incendiary mixtures, irritants, and asphyxiants (Chan et al., 2002). These chemicals can lead to rapid incapacitation, immediate death, or long-term health issues. Initially developed as weapons of war, CWAs were eventually banned by international conventions (Pacsial-Ong and Aguilar, 2013). Nerve agents, a subset of CWAs, are highly potent organophosphorus compounds that irreversibly inhibit acetylcholinesterase. Although rarely used in warfare, their extreme lethality and potential for terrorism have led to significant production and stockpiling (Polat et al., 2018).

CWAs can be categorized into nerve agents, incapacitating/behavior-altering agents, and blood or asphyxiant agents. Among these, nerve agents are particularly deadly and are a potential choice for terrorists targeting civilians (Tomchenko et al., 2005). Nerve agents often contain organophosphorus compounds, known for their extreme toxicity. While these compounds are used in agricultural pesticides and insecticides, they can be misused as CWAs.

Nerve agents, which are highly toxic organophosphorus compounds, pose a formidable and persistent threat to both military and civilian populations (Reuhs and Rounds, 2010). These deadly CWAs are leading to rapid and severe health consequences, including paralysis, convulsions, respiratory failure, and, ultimately, death by disrupting the normal functioning of the nervous system (Makinen et al., 2010). Consequently, there is an urgent need for reliable and rapid detection methods to safeguard against the devastating effects of nerve agents.

8.2.1 TYPES OF ORGANOPHOSPHATE NERVE AGENTS

Organophosphate (OP) compounds are highly effective inhibitors that can lead to the phosphorylation of acetylcholinesterase (AChE), causing an internal dealkylation reaction referred to as an "aging" reaction (Millard et al., 1999). AChE is an enzyme that catalyzes the hydrolysis of acetylcholine (ACh), a neurotransmitter (Fukuto, 1990) This process of phosphorylation takes place at the hydroxyl group of serine located within the active site of AChE, leading to the deactivation of acetylcholine.

OP nerve agents can be divided into two classifications referred to as G-series and V-series agents (Chauhan et al., 2008). The G-series nerve agents, Tabun (GA; ethyl N-dimethyl-phosphoramidocyanidate) (Weinbroum, 2004), Sarin (GB; 2-(fluoro-methyl-phosphoryl) oxypropane) (Balali-Mood and Saber, 2012), Soman (GD; 3,3-dimethyl-2-butyl methylphosphonofluoridate) (Stengl et al., 2008), and Cyclosarin (GF; fluoromethylphophoryloxycyclohexane) (Soltaninejad and Shadnia, 2014), were synthesized by Dr. Gerhard Schrader and coworkers in the 1930s in Germany (Vásárhelyi and Földi, 2007). Typically, the G-series agents exhibit a lethal concentration (LC50) of approximately 1 ppm over a 10-minute exposure period, while there may be differences in concentration and duration of exposure among the various agents within the series. The V-series, the other category of nerve agents, were initially synthesized in the 1950s in the United Kingdom by researchers investigating the use of organophosphate esters as pesticides. The name "V-series" signifies "venomous," and these agents are characterized by their low volatility and prolonged persistence on surfaces after application (Salem and Katz, 2016). Five main types of V-series agents are VE (O-ethyl-S-(2-diethylaminoethyl-) ethyl-phosphonothioate), VX (O-ethyl-S-[2(diisopropylamino)ethyl] methylphosphonothioate), VM (O-ethyl-S-(2-(diethylamino)ethyl)methylphosphorotioate), VG (O,O-diethyl-S-(2-diethyl aminoethyl)-phosphorotiate), and VR (S-[2-(diethylamino)ethyl] O hydrogen methylphosphonothioate). Among these, VX is the most well-known and is considered as the most lethal and toxic, with odorless and tasteless properties that make it difficult to detect physically (Tu, 1996). VX is also more potent than the G-series nerve agents due to its high stability, enhanced resistance to detoxification, and increased skin penetration ability (Solano et al., 2008).

Because of its extreme toxicity, using this agent as a sample in experiments is dangerous, so researchers have developed simulant compounds that replicate the physical and chemical characteristics of these nerve agents (Patil et al.,

2012). Examples of simulants used in lieu of nerve agents in research include methylphosphonic dichloride (MPDC), diisopropylfluorophosphate (DFP), diethyl chlorothiophosphate (DCTP), and dimethyl methylphosphonate (DMMP). These simulants are preferred because they are less toxic, can mimic the effects of nerve agents, and are employed in training for emergency response personnel (Nikmaram et al., 2015). DMMP is frequently employed as a simulant because of its stability, which is maintained by the presence of methyl methylphosphonate, even at elevated temperatures, such as 400°C (Lee et al., 2009).

8.2.2   TRADITIONAL APPROACHES FOR DETECTING NERVE AGENTS

Over the past five decades, various methods have been employed to detect organophosphorus compounds, including ion mobility spectrometry, liquid chromatography, and gas chromatography. These techniques often use liquid solvents to extract the compounds onto sorbent media, generating toxic chemical waste. However, they typically cannot independently detect all types of the CWAs and require combining with other methods for lower and more reliable detection limits (Black, 2010). Gas chromatography-mass spectrometry is one such method that can rapidly analyze samples with minimal preparation, making it suitable for field deployment with affordable operational costs. However, it has limitations in detecting low concentrations and non-volatile analytes and lacks sensitivity and automation for ease of use (Reichenbach et al., 2003). Liquid chromatography (LC) or high-performance liquid chromatography (HPLC) can directly analyze compounds without derivatization and separate mixtures, including isomers. LC and HPLC instruments often incorporate various detectors, such as ultraviolet-visible (UV-VIS) spectrometry, mass spectrometry (MS), and fluorescence spectrometry, to complete the analysis (Reuhs and Rounds, 2010). Ion mobility spectrometry (IMS) is a technique based on the movement of ionized gases. It can quickly detect CWAs in field conditions, offering real-time monitoring data without the need for intricate sample preparation procedures. Furthermore, the integration of both liquid and gas chromatography with mass spectrometry can enhance analytical capabilities by enhancing selectivity and minimizing false-positive results. However, these instruments require skilled operators, involve complex procedures, and demand specific environmental conditions, including temperature control and humidity management, making them inconvenient for field deployment (Puton and Namienik, 2016). They are also relatively large, consume substantial power, and rely on a consistent supply of high-purity gases, making them challenging to use in field settings (Varghese et al., 2015).

8.2.3   PLASMONIC NANOSENSORS FOR THE DETECTION OF NERVE AGENTS

In the ever-evolving landscape of sensor technology, plasmonic nanosensors have emerged as a groundbreaking and versatile class of sensors with transformative potential across various scientific disciplines. These nanosensors harness the exceptional optical properties of plasmonic nanoparticles, typically made of noble metals such as gold, silver, and copper, to achieve unparalleled sensitivity and specificity in the detection of analytes, ranging from molecules to biological entities. The convergence of nanotechnology and plasmonics has paved the way for the development of plasmonic nanosensors, which offer numerous advantages over conventional sensing technologies. Surface plasmons (SPs), a concept first introduced by Stern and Ferrell in 1960, represent coherent electron oscillations occurring on the surface of a material (Stern and Ferrell, 1960).

These oscillations are initiated by the excitation of electrons and photons with distinct energy levels. Within the domain of SPR sensors, two fundamental categories are distinguished: Propagating surface plasmon resonance (PSPR) and localized surface plasmon resonance (LSPR). PSPR sensors, which find their excitation in continuous metal nanofilms through techniques like prism couplers or gratings, possess the unique capability to propagate across metal/dielectric interfaces, covering significant distances. On the other hand, LSPR sensors harness non-propagating SPs on the surfaces of nanomaterials composed of metals. These LSPR characteristics can be finely adjusted by manipulating parameters such as size, shape, and composition, offering a remarkable degree of versatility and precision in the design of sensor systems. This duality in SP-based sensing technologies opens up a rich landscape of possibilities for diverse applications in fields ranging from fundamental research to practical industry solutions. SPR occurs when incident photons excite the collective oscillation of free electrons at the interface between a metal and a dielectric medium, resulting in a characteristic dip in the reflectivity or transmittance spectrum. LSPR, on the other hand, is observed in nanoscale metallic structures, where the collective electron oscillations are confined to the nanoparticle. Both SPR and LSPR are highly sensitive to changes in the refractive index in the vicinity of the sensor surface, making them ideal for detecting molecular binding events and alterations in the local environment (Mayer and Hafner, 2011; Willets and Van Duyne, 2007).

The fundamental principle underlying plasmonic nanosensors is the spectral shift in the resonance wavelength or peak intensity of the SPR or LSPR in response to changes in the refractive index. This shift is directly proportional to the concentration of analytes or the degree of interaction occurring at the sensor's surface. By functionalizing the surface of plasmonic nanoparticles with specific receptors, such as antibodies, aptamers, or molecularly imprinted polymers, plasmonic nanosensors can be tailored to recognize and bind to target analytes with high specificity (Fan et al., 2008; and Liu, 2018). As a result, they enable real-time, label-free, and highly sensitive detection, making them invaluable tools in various fields, including chemistry, biology, medicine, environmental monitoring, and beyond.

The fabrication of plasmonic nanosensors is a critical aspect of their performance and versatility. Researchers have developed a plethora of techniques to engineer plasmonic nanoparticles with precise control over size, shape, composition, and surface properties. These parameters profoundly influence the plasmonic properties of the nanoparticles and, consequently, their sensing capabilities.

Among the widely employed methods for nanoparticle synthesis, chemical reduction processes, such as the Turkevich method, seed-mediated growth, and chemical etching, stand out (Saha et al., 2012). These methods enable the synthesis of spherical or anisotropic nanoparticles with tunable sizes and plasmon resonances. Additionally, lithographic techniques, including electron beam lithography and nanoimprint lithography, provide exquisite control over nanoparticle shapes and arrangements, facilitating the fabrication of complex nanostructures with tailored plasmonic responses (Chen et al., 2015).

Surface functionalization is a crucial step in the fabrication process, as it imparts selectivity to plasmonic nanosensors. Various strategies, such as self-assembly, covalent binding, and physical adsorption, are employed to attach recognition molecules onto the nanoparticle surfaces. This functionalization can be designed to specifically target analytes of interest, rendering the plasmonic nanosensors highly selective and capable of distinguishing between different molecules, pathogens, or environmental pollutants (Grzelczak et al., 2008). Plasmonic nanosensors have found applications in a diverse array of fields, each benefiting from their unique capabilities (Jain et al., 2008; Haes et al., 2005).

Environmental monitoring benefits greatly from the deployment of plasmonic nanosensors, offering rapid, on-site detection of pollutants, heavy metals, and contaminants in water, soil, and air (Cao et al., 2020). Their real-time monitoring capabilities facilitate the assessment of environmental quality and help in addressing critical ecological concerns.

The food industry leverages plasmonic nanosensors for the detection of food-borne pathogens, allergens, and contaminants. These sensors enable food safety testing with high sensitivity and specificity, reducing the risk of foodborne illnesses and ensuring the quality of food products (Rodríguez-Quijada et al., 2020).

In materials science and nanotechnology, plasmonic nanosensors play a pivotal role in characterizing nanomaterials and studying their properties. They offer insights into the size, shape, and aggregation state of nanoparticles and facilitate investigations into phenomena such as plasmon coupling and energy transfer (Elghanian et al., 1997).

8.2.3.1 Surface Plasmon Resonance

SPR, which involves the resonant coupling of electromagnetic waves to charge oscillations at the interface of a dielectric and a metal, offers distinctive optical features. Plasmonic sensors find applications in diverse fields, including biology, chemistry, materials science, and renewable energy (Moon et al., 2022; Bakhshpour and Denizli, 2020; Zhu et al., 2023; Mascaretti et al., 2019). Plasmonic-based nanosensors, known for their high sensitivity and label-free, real-time detection capabilities, are employed in various applications (Yesudasu et al., 2021). SPR sensors, known for their remarkable sensitivity, are valuable tools for detecting interactions involving biological and chemical molecules, particularly low-molecular-weight substances (Bakhshpour-Yucel, 2023).

These sensors operate by measuring changes in the refractive index of the medium on their detection surfaces through the collective oscillations of free elec-trons, known as surface plasmon polaritons, in the metal's conduction band when exposed to an electromagnetic field (Aznar-Gadea et al., 2021). This results in a highly responsive electric field extending into the surrounding medium, making SPR sensors sensitive to shifts in the medium's refractive index (Opoku et al., 2023).

Consequently, SPR sensors are widely used in both research and practical applications due to their sensitivity, real-time monitoring capabilities, label-free nature, and versatility across various fields for studying molecular interactions and detecting specific molecules in complex samples (Agharazy Dormeny et al., 2020).

8.2.3.2 Surface-Enhanced Raman Spectroscopy

SERS is a powerful analytical technique that amalgamates the molecular specificity of traditional Raman spectroscopy with an impressive boost in sensitivity. This heightened sensitivity arises from the substantial enhancement of Raman signal intensity when molecules of interest are brought into proximity with plasmonic nanostructures. Initially discovered in the 1970s during experiments involving trace-level pyridine detection on roughened silver electrodes, SERS predominantly capitalizes on two enhancement mechanisms: electromagnetic and chemical (Tong et al., 2011; Kern and Martin, 2011; Yoshida et al., 2010). The electromagnetic enhancement takes center stage as it begins with the excitation of surface plasmons achieved by matching the excitation wavelength to the plasmon absorption profile of specific nanoparticles. This excitation generates a robust, evanescent electromagnetic field on the metal surface, leading to a substantial augmentation in the Raman modes of nearby molecules. The Raman intensity is directly linked to the square of the incident electromagnetic field amplitude. The resulting Raman scattered light, emitted in all directions, is gathered through a microscope objective and then detected, and further enhanced when the relevant Raman mode aligns with the plasmon resonance (Le Ru et al., 2008). The electromagnetic SERS enhancement exhibits a fourth-power dependency on the incident field amplitude, underscoring its remarkable amplification capability. Meanwhile, the chemical enhancement facet encompasses diverse effects, including chemical interactions between molecules and nanoparticles in their ground state, resonant excitation of charge-transfer states, and resonance Raman enhancement driven by higher electronic states of the molecule. A unified description of SERS encapsulates all these resonances, encompassing SPR, charge-transfer resonance, and molecular resonances, employing the Herzberg–Teller coupling principle (Lombardi and Birke, 2009).

8.2.4 APPLICATIONS OF PLASMONIC NANOSENSORS FOR THE DETECTION OF NERVE AGENTS

The use of nanomaterials in combination with sensing systems has created a revolutionary impact by overcoming the limitations of conventional methods and allowing the detection of compounds even at the lowest concentrations. Upon the plasmonic features of noble metals that lead them to adequately couple with electromagnetic radiation, electron plasma which potentially yields electromagnetic hot spots occurs, thus by integrating these particles into the sensor systems, the capability for the sensing of low molecular weight substances like nerve agents has been enhanced (Ayivi et al., 2023).

In a study of Zhao et al., to reach the sensitive sarin detection with SERS, the weak adsorption capacity of sarin on metal surfaces was overcome by functionalyzing the surface of Au coated Si nanoarray with 2-aminoethanethiol. Methanephosponic acid (MPA), a harmless simulated agent that can be used instead of sarin, was utilized to demonstrate the amidation induced interaction between the metal surfaces and the molecules with weak metal binding capacity. The limit of detection was determined down to ~1ppb with enhanced target-surface interaction of the SERS substrate (Zhao et al., 2016).

Juhlin et al. demonstrated a rapid and highly selective method with SERS technology by using Au nanopillars functionalized with 4-pyridine amide oxime (4-PAO) to detect nerve agents tabun, VX, and cyclosarin (Juhlin et al., 2020). In this study, 4-pyridine amide oxime, a chemical nerve agent antidote, was preferred as a probe molecule due to its high affinity to Au and capturing efficiency towards the target nerve agents via the oxime group. Depending on the covalent binding between the probe and the target, structural changes in the SERS probe emerged, which are seen as a difference in the SERS reflectance response. As such, selective determination of nerve agents in aqueous solutions was achieved. Briefly, SERS substrate was synthesized with a two-step method: (i) Au-coated-Si NP obtaining an etching of silicon by reactive gas mixture of SF6 and O_2, followed by Au deposition with a thickness of 200 nm by electron beam evaporation, and (ii) additional coating with a 4-PAO reporter molecule. Owing to the binding features of 4-PAO, while specific binding with NA was observed in the bind between the central oxygen atom of 4-PAO and phosphorous group of NA, on the other hand, a stable and strong SERS response was still displayed because intramolecular hydrogen bond present in 4-PAO stabilizes the planar Z configuration of carboxamide oxime. Thus, the obtained SERS signals could be shown to distinguish specific and non-specific binding of Tabun, Cyclosarin, and VX down to levels below ppm.

In another group, different SERS substrates, prepared as an AuNPs layer coated with citrate, were used to develop a cost-effective SERS method in order to achieve label-free and real-time detection of G-agents in the gas phase, even at low concentrations (Lafuente et al., 2018). However, due to the features of G-agents, such as persistence at moderate temperatures and extreme toxicity, experiments were realized with dimethyl methyl phosphonate (DMMP) as a substitute molecule, owing to its structural similarities with sarin. Citrate used in this study provided a trap for the target near the AuNPs, based upon its ability to form reversible H bonds with the target. With the surface modifications performed, DMMP detection was achieved for the first time at sub-ppm concentrations. To perform DMMP measurements in the gas phase, a device was designed using a homemade gas chamber (2.7×10^{-2} cm^3) in which the SERS substrate could be placed at a 90° angle to the 785 nm laser excitation beam. As a sample, DMMP included nitrogen stream was given to the SERS platform by the gas chamber (Figure 8.1).

After the gas passed from the Raman spectroscopy, the signature band of DMMP at 708 cm^{-1} was rapidly recorded upon the gas connection with the SERS substrate. Also citrate and DMMP shared some bands at 1028 cm^{-1} and 1081 cm^{-1}, specific for O-CH$_3$ stretching. However, they also had some bands at the same wavelength,

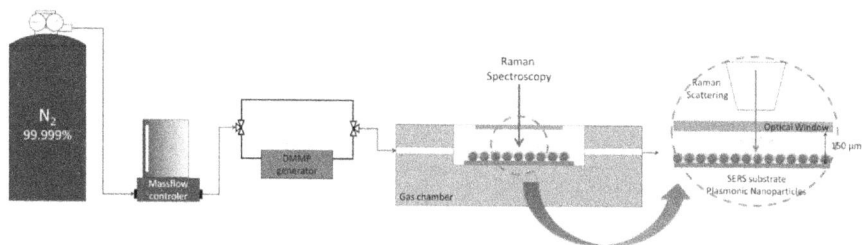

FIGURE 8.1 Experimental setup for DMMP detection in the gas phase with a SERS platform.

877 cm^{-1}, caused by different band vibrations, such as CH_3 bending for DMMP and C-COO stretching for citrate. According to the FTIR analysis of the saturated surface, at 1230 cm^{-1} band causing broad P-O stretching indicated that DMMP binding to the OH group of citrate-coated AuNPs can offer significant observations (Figure 8.2). According to the obtained data, it has been shown that the strength of citrate-DMMP interaction makes this method a proper candidate for the further detection of NAs.

Finally, the SERS spectrum of DMMP accumulation on the SERS substrate has been compared, depending on concentration ranges between 0 and 2.5 ppmV. The peak observed for the PC stretching of DMMP could be distinguished even at low concentrations (625 ppbV). They reached a limit of detection of 130 ppbV after calculating the standard deviation value for PC stretching peak intensity over five repeated experiments (Figure 8.3).

Hakonen et al. have ascertained the femtomole detection limits for two nerve gases, VX and tabun, by utilizing a handheld Raman device enhanced with an Au coated Si nanopillar SERS substrate (Hakonen et al., 2016). Owing to the high adhesion capability of the SERS substrate due to its nanopillar structure, enriched SERS spectra could be obtained with the increased accumulation of lower concentrations of target molecules in this method. They reached a LOD value of ~13 and ~670 fmol for VX and Tabun, respectively.

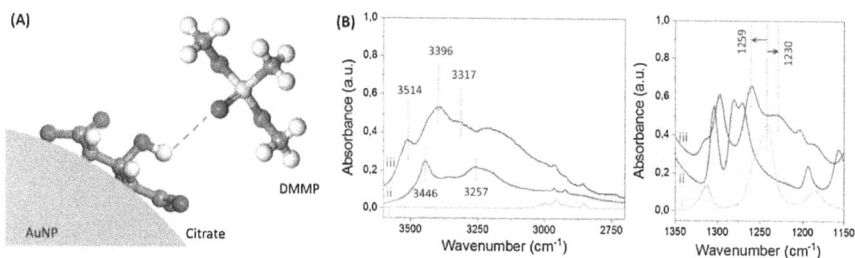

FIGURE 8.2 (A) Schematic illustration of DMMP and citrate binding. (B) FTIR analysis (i) DMMP in liquid phase, (ii) dihydrate salt of sodium citrate, (iii) DMMP saturated AuNPs-citrate monolayer coated on SiO$_2$/Si chip surface.

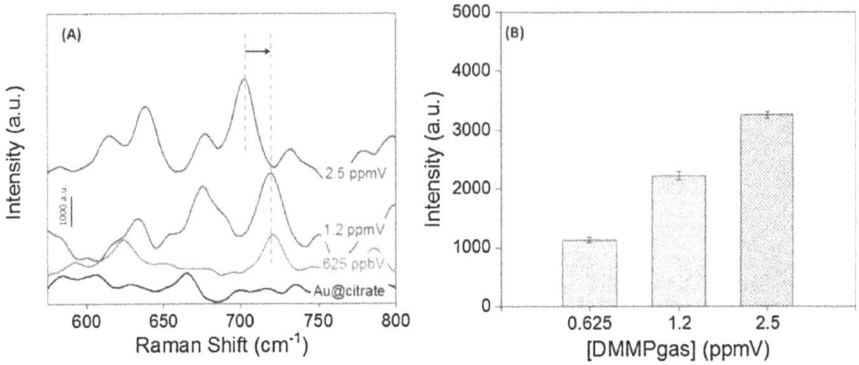

FIGURE 8.3 (A) SERS spectra of Au@citrate coating of SiO_2/Si chips during N_2 passage spoiled with different concentrations of DMMP. (B) Difference in the SERS intensity of PC stretching peaks, depending on the accumulation amount of DMMP on the SERS substrate, observed for different DMMP concentrations.

The main opportunity offered by this study is the ability to sensitively detect highly toxic chemical warfare agents in the required field with a small and portable device. Afterward, Hakonen et al. conducted a comprehensive study to detect cyclosarin and some narcotic compounds with a handheld spectrophotometer, based on SERS phenomena (Hakonen et al., 2018). In this study, they used Density Functional Theory (DFT) and 3D Finite Element Method (FEM) for observing the improved SERS intensity, due to the Ag nanopillar substrate. They utilized a simulation of dimer and trimer of nanopillars causing an electromagnetic field by FEM to observe the enhancement capacity of the SERS substrate (Figure 8.4).

When they compared the peaks obtained from simulation and from real-time measurements, they demonstrated the correlations between them (Figure 8.5). According to these findings, 3D FEM and DFT were suggested as efficient methods to ascertain

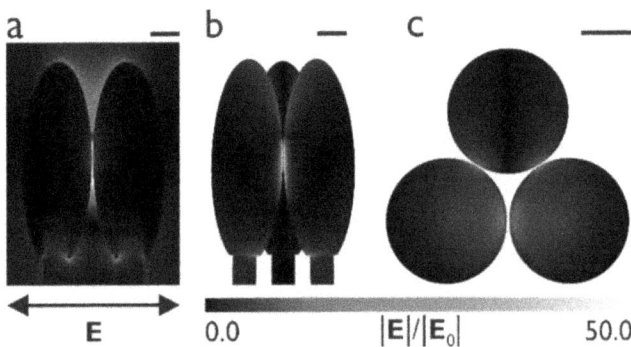

FIGURE 8.4 3D FEM simulation belongs to Ag covered Si nanopillars; (a) cross section field of a nanopillar dimer, (b) side view of trimer structure, (c) top view of nanopillar trimer, the gap distance between nanopillars ~3 nm.

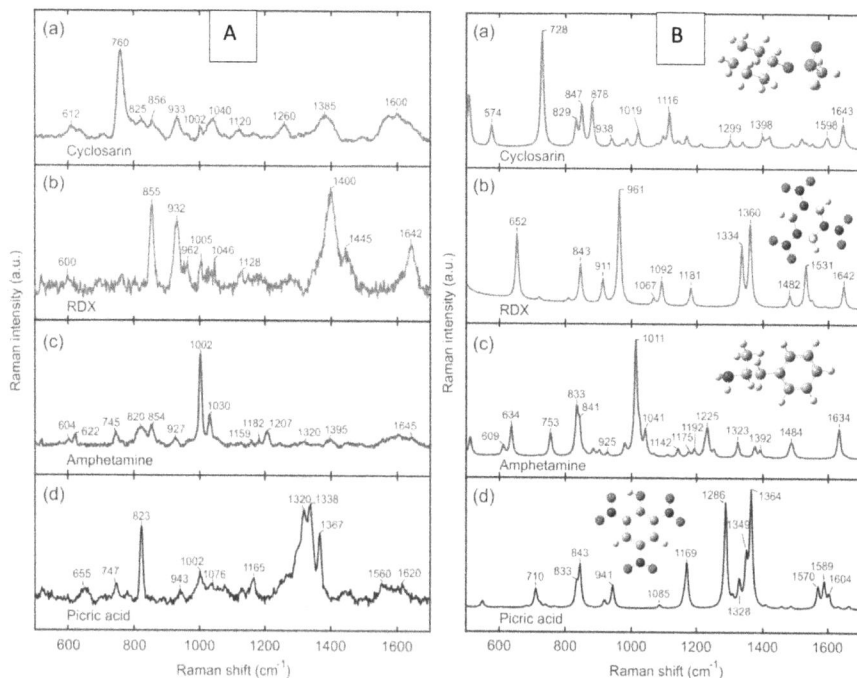

FIGURE 8.5 (A) Simulated spectral peaks for cyclosarin, RDX, amphetamine, and pitric acid acquired by DFT. (B) SERS peaks recorded by portable Raman spectrometer for cyclosarin (20 μg/mL), RDX (2 ng/mL), amphetamine (2 ng/mL), and picric acid (0.4 ng/mL).

the nature of the excitation of nanopillar dimer and trimer resonance, which leads to the immense enhancement of SERS intensity. Also, the SERS detection limit of cyclosarin could be measured as 40 ng for the first time, due to its complex structure.

Additionally, some pesticides which are classified as the esters of phosphoric acids, organophosphate pesticides (OPPs), such as chlorpyrifos, parathion, malathion, and dimethoate, show the same effects with nerve agents in living systems in response to AChE inhibitor. While considering their high toxicity, the extensive utilization of OPPs in agricultural processes can not be overemphasized. The spread of these hazardous pesticides into the environment, due to the inappropriate and extensive use of these OPPs, causes unprevented environmental damage and public health threat. The detection of these OPPs, which are widely used and have a similar effect to nerve agents, is also critical in taking the necessary precautions. Thus, nanomaterial integrated plasmonic sensors are also a powerful candidate for the rapid and sensitive detection of OPPs in low concentrations (Ayivi et al., 2023).

In a study of Yao and coworkers, a SPR-based sensor system that used surface molecular imprinting and molecular imprinted magnetic polymer nanoparticles (NPs) to enhance the sensitivity of the SPR, was reported on for its detection of organophosphate pesticide, chlorpyrifos (CPF) (Yao et al., 2013). CPF imprinted F_3O_4-dopamine NPs were synthesized by the self-polymerization technique to obtain CPF binding sides with a high recognition capacity. These NPs not only increase the sensitivity of the sensor by amplifying the SPR signal but also improve selectivity by presenting attractive recognition elements. For the detection of CPF-F_3O_4-dopamine NPs, the chip surface was fabricated with immobilized AChe (Figure 8.6).

Finally, the designed sensor demonstrated an excellent linear relationship between the SPR signal shift and the increased concentrations of CPF, at a range of 0.001–10 μM. Owing to the high surface to volume ratio of nanostructured molecular imprinted polymer and the increased number of recognition sides, a high sensitivity with a detection limit of 0.76 nM was achieved. When direct CPF detection efficiency was compared with the signals obtained by the method that used CPF- F_3O_4-dopamine NPs, an enhanced SPR response was observed (Figure 8.7).

Jia et al. used synthesized cube-shaped, polyarylene ether nitril coated AgNPs, with a size of ~300 nm, to cover a silicon substrate of SERS. Owing to the presence of Raman hot spots, depending on the enrichment of the target molecule on the surface after modification with super plasmonic nanoparticles, the chlorpyrifos pesticide was detected in low concentrations down to 10^{-10} M (Jia et al., 2020).

In a study of Dissanayaka et al., the researchers developed a sensing method for the detection of five different OPPs—paraoxon ethyl, parathion, fenthion, malathion, and ethion—by exploiting the benefits of the LPSR behavior of metal NPs

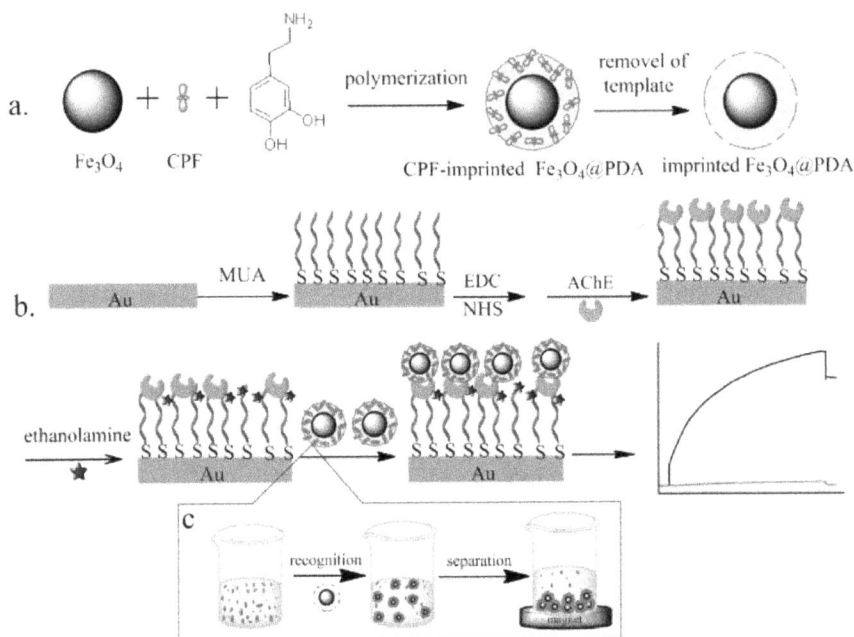

FIGURE 8.6 (a) Schematic representation of CPF imprinted F_3O_4-dopamine NPs synthesis, (b) preparation of SPR chip surface for CPF sensing, (c) recognition and separation steps of CPF from imprinted F_3O_4-dopamine NPs.

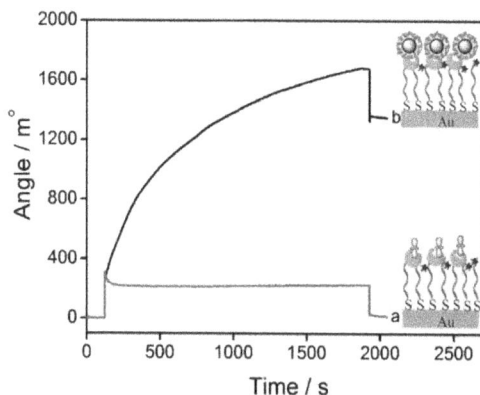

FIGURE 8.7   SPR response curve obtained for 10 μM CPF recognition (a) direct recognition of CPF, (b) recognition mediated with CPF imprinted F_3O_4-dopamine NPs.

(Dissanayaka et al., 2019). Briefly, in this study, changes in the LSPR band of colloidal shaped AuNPs, AgNPs, and Au-Ag bimetallic NPs, after the addition of five different OPPs separately in a solution, were characterized by UV visible spectrophotometer. Each noble metal has its unique LPSR band. Thanks to the wavelength shift observed in the LPSR band of the NP, depending on its interactions with each OPP, OPPs were detected and were distinguished from each other.

8.3   FUTURE PROSPECTS AND CHALLENGES

The future of plasmonic nanosensors is indeed promising, driven by continuous research endeavors aimed at pushing the boundaries of their capabilities. One key area of focus is enhancing sensitivity, with scientists striving to detect even the most minute concentrations of analytes. Simultaneously, researchers are exploring ways to enable multiplexing, allowing these sensors to simultaneously identify and quantify multiple substances within a single sample. These advancements are poised to revolutionize fields such as medical diagnostics and environmental monitoring.

Materials engineering and surface chemistry play pivotal roles in shaping the trajectory of plasmonic nanosensor development. Innovations in these domains are expected to expand the range of detectable analytes, making these sensors increasingly versatile and applicable across various scientific and industrial contexts. Notably, the integration of plasmonic nanosensors into lab-on-a-chip platforms and wearable devices stands out as a transformative prospect. Such integration could democratize access to cutting-edge sensing technologies, ushering in a new era of point-of-care diagnostics and personalized medicine.

Despite these exciting prospects, plasmonic nanosensors confront certain challenges on their path to widespread adoption. Reproducibility remains a concern, as ensuring consistent sensor performance is essential for reliable results. Moreover, issues related to stability and cost-effectiveness necessitate innovative solutions. Researchers are actively exploring ways to enhance the long-term stability of these sensors while optimizing their manufacturing processes to make them more accessible.

In the quest for greater selectivity, efforts are underway to mitigate interference from complex sample matrices and non-specific binding, which can compromise the sensor's accuracy and specificity. Achieving robust selectivity will be pivotal in ensuring that plasmonic nanosensors fulfill their potential as powerful tools in diverse applications, ranging from clinical diagnostics to environmental monitoring. So, the future of plasmonic nanosensors is poised to bring about transformative changes in various fields. With ongoing advancements and a concerted effort to address existing challenges, these sensors hold the promise of revolutionizing how we detect and analyze substances, ultimately leading to improved healthcare, environmental stewardship, and scientific discovery.

8.4 CONCLUSION

Plasmonic nanosensors stand at the forefront of scientific innovation, seamlessly blending the realms of nanotechnology, plasmonics, and sensing. Their remarkable capability to achieve ultrahigh sensitivity and precise detection spans a wide array of applications, ranging from critical healthcare diagnostics to environmentally conscious monitoring initiatives. Their profound impact on the analytical and diagnostic sciences is undeniable, as they empower researchers and professionals to unlock new frontiers in understanding the molecular intricacies of our world. The ongoing exploration of plasmonic nanosensors continues to unveil their boundless potential and confront existing challenges head-on. These tiny but mighty devices are set to redefine the landscape of sensing technology, offering ingenious solutions to intricate problems that were once insurmountable. As research evolves and these sensors mature, we can anticipate a future where plasmonic nanosensors not only drive innovation but also deepen our comprehension of the complex molecular universe, ultimately contributing to a brighter, more informed, and safer world.

REFERENCES

Adam, H., Gopinath, S.C.B., Md Arshad, M.K., Adam, T., Hashim, U., Sauli, Z., Fakhri, M.A., Subramaniam, S., Chen, Y., Sasidharan, S., Wu, Y.S., 2023. Integration of microfluidic channel on electrochemical-based nanobiosensors for monoplex and multiplex analyses: An overview. *J. Taiwan Inst. Chem. Eng.* 146, 104814.

Agharazy Dormeny, A., Abedini Sohi, P., Kahrizi, M., 2020. Design and simulation of a refractive index sensor based on SPR and LSPR using gold nanostructures. *Results. Phys.* 16, 102869.

Ayivi, R.D., Adesanmi, B.O., McLamore, E.S., Wei, J., Obare, S.O., 2023. Molecularly imprinted plasmonic sensors as nano-transducers: An effective approach for environmental monitoring applications. *Chemosensors.* 11, 203.

Aznar-Gadea, E., Rodríguez-Canto, P.J., Martínez-Pastor, J.P., Lopatynskyi, A.V., Chegel, R.A., 2021. Molecularly imprinted silver nanocomposites for explosive taggant sensing. *ACS Appl. Polym. Mater.* 3, 2960–2970.

Bakhshpour, M., Denizli, A., 2020. Highly sensitive detection of Cd(II) ions using ion-imprinted surface plasmon resonance sensors. *Microchem. J.* 159, 105572.

Bakhshpour-Yucel, M., 2023. SPR-based sensing of Lysozyme using Lyz-MIP-modified graphene oxide surfaces. *Chem. Pap.* 77, 2671–2678.

Balali-Mood, M., Saber, H., 2012. Recent advances in the treatment of organophosphorous poisonings. *Iran. J. Med. Sci.* 37(2), 74.

Black, R.M., 2010. History and perspectives of bioanalytical methods for chemical warfare agent detection. *J. Chromatogr. B* 878(17–18), 1207–1215.

Calcagno, M., D'Agata, R., Breveglieri, G., Borgatti, M., Bellassai, N., Gambari, R., Spoto, G., 2022. Nanoparticle-enhanced surface plasmon resonance imaging enables the ultrasensitive detection of non-amplified cell-free fetal DNA for non-invasive prenatal testing. *Anal. Chem.* 94, 1118–1125.

Cao, X., Ye, Y., Liu, S., Goldys, E.M., 2020. Nanoplasmonic sensors for biointerfacial science. *Chem. Soci. Rev.* 49(16), 5850–5884.

Chan, J., Yeung, R., Tang, S., 2002. An overview of chemical warfare agents. *Hong Kong. J. Emerg. Med.* 9(4), 201–205.

Chen, Y.T., Lee, Y.C., Lai, Y.H., Lim, J.C., Huang, N.T., Lin, C.T., Huang, J.J., 2020. Review of integrated optical biosensors for point-of-care applications. *Biosensors.* 10, 1–22.

Chen, W., et al., 2015. Nanoparticle assemblies: Dimensional control of multidimensional architectures. *Chem. Soci. Rev.* 44(8), 2681–2701.

Cui, F., Zhou, Z., Zhou, H.S., 2020. Review-measurement and analysis of cancer biomarkers based on electrochemical biosensors. *J. Electrochem. Soc.* 167, 037525.

Dissanayake, N.M., Arachchilage, J.S., Samuels, T.A., Obare, S.O., 2019. Highly sensitive plasmonic metal nanoparticle-based sensors for the detection of organophosphorus pesticides. *Talanta.* 200, 218–227.

Elghanian, R., Storhoff, J.J., Mucic, R.C., Letsinger, R.L., Mirkin, C.A., 1997. Selective colorimetric detection of polynucleotides based on the distance-dependent optical properties of gold nanoparticles. *Science.* 277(5329), 1078–1081.

Fan, J.A., et al., 2008. Self-assembled plasmonic nanoparticle clusters. *Science.* 328(5982), 1135–1138.

Fukuto, T.R., 1990. Mechanism of action of organophosphorus and carbamate insecticides. *Environ. Health Perspect.* 87, 245–254.

Grzelczak, M., Pérez-Juste, J., Mulvaney, P., Liz-Marzán, L.M., 2008. Shape control in gold nanoparticle synthesis. *Chem. Soci. Rev.* 37(9), 1783–1791.

Gur, S.D., Bakhshpour, M., Denizli, A., 2019. Selective detection of Escherichia coli caused UTIs with surface imprinted plasmonic nanoscale sensor. *Mater. Sci. Eng. C.* 104, 109869.

Haes, A.J., et al., 2005. A nanoscale optical biosensor: Real-time immunoassay in physiological buffer enabled by improved nanoparticle adhesion. *J. Physic. Chem. B.* 109(7), 304–310.

Hakonen, A., Rindzevicius, T., Schmidt, M.S., Andersson, P.O., Juhlin, L., et al., 2016. Detection of nerve gases using surface-enhanced Raman scattering substrates with high droplet adhesion. *Nanoscale.* 8, 1305–1308.

Hakonen, A., Wu, K., Schmidt, M.S., Andersson, P.O., Boisen, A., Rindzevicius, T., 2018. Detecting forensic substances using commercially available SERS substrates and hand-held Raman spectrometers. *Talanta.* 189, 649–652.

Homola, J., 2008. Surface plasmon resonance sensors for detection of chemical and biological species. *Chem. Rev.* 108, 462–493.

Huang, C.-T., Jan, F.-J., Chang, C.-C., 2021. A 3D plasmonic crossed-wire nanostructure for surface-enhanced Raman scattering and plasmon-enhanced fluorescence detection. *Molecules.* 26, 281.

Jain, P.K., Huang, X., El-Sayed, I.H., El-Sayed, M.A., 2008. Noble metals on the nanoscale: Optical and photothermal properties and some applications in imaging, sensing, biology, and medicine. *Acc. Chem. Res.* 41(12), 1578–1586.

Jia, K., Xie, J., He, X., Zhang, D., Hou, B., Li, X., Zhou, X., Hong, Y., Liu, X., 2020. Polymeric micro-reactors mediated synthesis and assembly of Ag nanoparticles into cube-like superparticles for SERS application. *Chem. Engi. J.* 395, 125123.

Jokanovi, M., 2009. Medical treatment of acute poisoning with organophosphorus and carbamate pesticides. *Toxicol. Lett.* 190(2), 107–115.

Juhlin, L., Mikaelsson, T., Hakonen, A., Schmidt, M.S., Rindzevicius, T., Boisen, A., Käll, M., Andersson, P.O., 2020. Selective surface-enhanced Raman scattering detection of Tabun, VX and Cyclosarin nerve agents using 4-pyridine amide oxime functionalized gold nanopillars. *Talanta.* 211, 120721.

Kanchan, A.J., Jason, T., Robert, H., Joseph, W., Wilfred, C., Ashok, W., 2005. A disposable biosensor for organophosphorus nerve agents based on carbon nanotubes modified thick film strip electrode. *Electroanalysis: An Int. J. Devoted Fundam. Asp. Pract. Electroanalysis.* 17(1), 54–58.

Kern, A.M., Martin, O.J.F., 2011. Excitation and reemission of molecules near realistic plasmonic nanostructures. *Nano. Lett.* 11(2), 482–487.

Kretschmann, E., Raether, H., 1968. Notizen: Radiative decay of non radiative surface plasmons excited by light. *Z. Für Naturforsch. A.* 23, 2135–2136.

Kurbanoglu, S., Ozkan, S.A., 2018. Electrochemical carbon based nanosensors: A promising tool in pharmaceutical and biomedical analysis. *J. Pharm. Biomed. Anal.* 147, 439–457.

Lafuente, M., Pellejero, I., Sebastián, V., Urbiztondo, M.A., Mallada, R., Pina, M.P., Santamaría, J., 2018. Highly sensitive SERS quantification of organophosphorous chemical warfare agents: A major step towards the real time sensing in the gas phase. *Sens. Actuators B Chem.* 267, 457–466.

Le Ru, E.C., Meyer, M., Blackie, E., Etchegoin, P.G., 2008. Advanced aspects of electromagnetic SERS enhancement factors at a hot spot. *J Raman Spectrosc.* 39(9), 1127–1134.

Lee, S.C., et al., 2009. The development of SnO2-based recoverable gas sensors for the detection of DMMP. *Sens. Actuators B Chem.* 137(1), 239–245.

Lindquist, N.C., de Albuquerque, C.D.L., Sobral-Filho, R.G., Paci, I., Brolo, A.G., 2019. High-speed imaging of surface-enhanced Raman scattering fluctuations from individual nanoparticles. *Nat. Nanotechnol.* 14, 981–987.

Lombardi, J.R., Birke, R.L., 2009. A unified view of surfaceenhanced Raman scattering. *Acc Chem Res.* 42(6), 734–742.

Mascaretti, L., Dutta, A., Kment, S., Shalaev, V.M., Boltasseva, A., Zbori, R., Naldoni, A., 2019. Plasmon-enhanced photoelectrochemical water splitting for efficient renewable energy storage. *Adv. Mater.* 31, 1805513.

Mayer, K.M., Hafner, J.H., 2011. Localized surface Plasmon resonance sensors. *Chem. Rev.* 111(6), 3828–3857.

Moon, G., Lee, J., Lee, H., Yoo, H., Ko, K., Im, S., Kim, D., 2022. Machine learning and its applications for plasmonics in biology. *Cell Rep. Phys. Sci.* 3, 101042.

Nikmaram, F., Najafpour, J., Ashrafi Shahri, M., 2015. Decontamination of DMMP by adsorption on ZnO, a computational study. *J. Phys. Theor. Chem.* 9(1), 11–15.

Opoku, G., Danlard, I., Dede, A., Kofi, E., 2023. Akowuah, design and numerical analysis of a circular SPR based PCF biosensor for aqueous environments. *Results Opt.* 12, 100432.

Otto, A., 1968. Excitation of nonradiative surface plasma waves in silver by the method of frustrated total reflection. *Z. Für Phys.* 216, 398–410.

Pacsial-Ong, E.J., Aguilar, Z.P., 2013. Chemical warfare agent detection: A review of current trends and future perspective. *Front. Biosci.* 5, 516–543.

Patil, L., et al., 2012. Detection of dimethyl methyl phosphonate–a simulant of sarin: The highly toxic chemical warfare–using platinum activated nanocrystalline ZnO thick films. *Sens. Actuators B Chem.* 161(1), 372–380.

Polat, S., Gunata, M., Parlakpinar, H., 2018. Chemical warfare agents and treatment strategies. *J. Turgut Ozal Med. Cent.* 25(4), 776-82.

Puton, J., Namienik, J., 2016. Ion mobility spectrometry: Current status and application for chemical warfare agents detection. *Trac. Trends Anal. Chem.* 85, 10–20.

Reichenbach, S.E., et al., 2003. Chemical warfare agent detection in complex environments with comprehensive two-dimensional gas chromatography. Chemical and Biological Sensing IV, International Society for Optics and Photonics.

Reuhs, B.L., Rounds, M.A., 2010. High-performance liquid chromatography, in *Food Analysis*, S. Suzanne Nielsen (ed), Springer, Boston, MA, 499–512.

Rodríguez-Quijada, C.A., et al., 2020. Plasmonic biosensors: Essential Insights into Their design, fabrication and applications. *Mater. Chem. Front.* 4(5), 1387–1403.

Saha, K., Agasti, S.S., Kim, C., Li, X., Rotello, V.M., 2012. Gold nanoparticles in chemical and biological sensing. *Chem. Rev.* 112(5), 2739–2779.

Salem, H., Katz, S.A., 2016. *Aerobiology: The Toxicology of Airborne Pathogens and Toxins.* Royal Society of Chemistry.

Shalaev, V.M., 2007. Optical negative-index metamaterials. *Nat. Photonics* 1, 41–48.

Solano, M.I., et al., 2008. Quantification of nerve agent VX-butyrylcholinesterase adduct biomarker from an accidental exposure. *J. Anal. Toxicol.* 32(1), 68–72.

Soltaninejad, K., Shadnia, S., 2014. History of the use and epidemiology of organophosphorus poisoning, in *Basic and Clinical Toxicology of Organophosphorus Compounds*, Mahdi Balali-Mood, Mohammad Abdollahi (eds); Springer, London, 25–43.

Stern, E.A., Ferrell, R.A., 1960. Surface plasma oscillations of a degenerate electron gas. *Phys. Rev.* 120, 130–136.

Stengl, V., et al., 2008. Zirconium doped titania: Destruction of warfare agents and photocatalytic degradation of orange 2 dye. *Open Process Chem. J.* 1(7), 1–7.

Tomchenko, A.A., Harmer, G.P., Marquis, B.T., 2005. Detection of chemical warfare agents using nanostructured metal oxide sensors. *Sens. Actuators B Chem.* 108(1–2), 41–55.

Tong, L.M., Zhu, T., Liu, Z.F., 2011. Approaching the electromagnetic mechanism of surface-enhanced Raman scattering: from selfassembled arrays to individual gold nanoparticles. *Chem Soc Rev.* 40(3), 1296–1304.

Tu, A.T., 1996. Basic information on nerve gas and the use of sarin by Aum Shinrikyo. *J. Mass Spectrom. Soc. Jpn.* 44(3), 293–320.

Urabe, T., Takahashi, K., Kitagawa, M., Sato, T., Kondo, T., Enomoto, S., Kidera, M., Seto, Y., 2014. Development of portable mass spectrometer with electron cyclotron resonance ion source for detection of chemical warfare agents in air. *Spectrochim. ActaA Mol. Biomol. Spectrosc.* 120, 437–444.

Vásárhelyi, G., Földi, L., 2007. History of Russia's chemical weapons. *Acad. Appl. Res. Mil. Sci.* 6(1), 135–146.

Varghese, S., et al., 2015. Two-dimensional materials for sensing: Graphene and beyond. *Electronics.* 4(3), 651–687.

Weinbroum, A.A., 2004. Pathophysiological and clinical aspects of combat anticholinesterase poisoning. *Br. Med. Bull.* 72(1), 119–133.

Wenger, T., Viola, G., Kinaret, J., Fogelstrom, M., Tassin, P., 2017. High-sensitivity plasmonic refractive index sensing using grapheme. *2D Mater.* 4, 025103.

Willets, K.A., Van Duyne, R.P., 2007. Localized surface plasmon resonance spectroscopy and sensing. *Annu. Rev. Phys. Chem.* 58, 267–297.

Yang, T., Duncan, T.V., 2021. Challenges and potential solutions for nanosensors intended for use with foods. *Nat. Nanotechnol.* 16, 251–265.

Yao, G.-H., Liang, R.-P., Huang, C.-F., Wang, Y., Qiu, J.-D., 2013. Surface Plasmon resonance sensor based on magnetic molecularly imprinted polymers amplification for pesticide recognition. *Anal Chem.* 85(24), 11944–11951.

Yesudasu, V., Pradhan, H.S., Pandya, R.J., 2021. Recent progress in surface plasmon resonance based sensors: A comprehensive review. *Heliyon.* 7, e06321.

Yoshida, K.-I., Itoh, T., Tamaru, H., Biju, V., Ishikawa, M., Ozaki, Y., 2010. Quantitative evaluation of electromagnetic enhancement in surface-enhanced resonance Raman scattering from plasmonic properties and morphologies of individual Ag nanostructures. *Phys Rev B.* 81, (11), 115406.

Zhang, J., Liu, J., 2018. Functionalized gold nanoparticles for biomedical applications. *Chem. Sci.* 9(35), 6904–6910.

Zhao, Q., Liu, G., Zhang, H., Zhou, F., Li, Y., Cai, W., 2017. SERS-based ultrasensitive detection of organophosphorus nerve agents via substrate's surface modification. *J Hazard Mater.* 15, 324(Pt B), 194–202.

Zhou, H., Hui, X., Li, D., Hu, D., Chen, X., He, X., Gao, L., Huang, H., Lee, C., Mu, X., 2020. Metal–organic framework-surface-enhanced infrared absorption platform enables simultaneous on-chip sensing of greenhouse gases. *Adv. Sci.* 7, 2001173.

Zhu, W., Yi, Y., Yi, Z., Bian, L., Yang, H., Zhang, J., Yu, Y., Liu, C., Li, G., Wu, X., 2023. High confidence plasmonic sensor based on photonic crystal fibers with a U-shaped detection channel. *Phys. Chem. Chem. Phys.* 25, 8583–8591.

9 Fluorometric Sensing of Chemical Threat Agents via Plasmonic Nanoparticles

Huma Shaikh, Abbas Aziz, and Tufail Hussain Sherazi

9.1 INTRODUCTION

Chemical threat agents, also known as chemical weapons, encompass poisonous substances intentionally employed to inflict pain or fatality upon individuals, contaminate the surroundings, or impede military activities [1]. The agents in question exhibit diverse manifestations, encompassing gaseous, liquid, or solid states, and are conventionally categorized according to their distinct features and resultant impacts. Detecting chemical threat agents holds significant importance in national security, public safety, and emergency response. The following is a comprehensive exposition of chemical threat agents and their respective detection modalities.

9.1.1 INDUSTRIAL TOXIC CHEMICALS

Certain frequently encountered industrial chemicals; including chlorine, ammonia, and hydrogen sulfide, have the potential to function as chemical threat agents when discharged in substantial volumes [2].

9.1.2 BLOOD AGENTS

These chemical substances disrupt the physiological process of oxygen transportation throughout the human body, resulting in asphyxiation. Hydrogen cyanide (HCN) and cyanogen chloride (CNCl) are blood agents [3].

9.1.3 CHOKING AGENTS

These chemicals harm the respiratory system, resulting in respiratory distress and, in some cases, fatality. Chlorine and phosgene are frequently encountered as choking agents.

DOI: 10.1201/9781003459316-9

9.1.4   BLISTER AGENTS

These compounds are known to induce significant irritation of the skin and eyes, potentially leading to blisters on the skin. Sulfur mustard, commonly called mustard gas, is a widely recognized example [3].

9.1.5   NERVE AGENTS

These are mostly organophosphorus compounds with a high toxicity level, which disrupt the normal functioning of the nervous system. Some examples of chemical warfare agents are sarin, soman, and VX. Detecting nerve agents is paramount due to their highly lethal nature [4].

Identifying chemical threat agents is crucial for protecting public health, national security, and the environment. It also plays a significant role in responding to acts of terrorism and ensuring adherence to international agreements prohibiting chemical weapons. The detection of chemical threat agents encompasses a diverse range of approaches, necessitating the integration of various technologies and tactics. The prompt and precise identification of potential chemical attacks or unintentional discharges is critical, as it enables timely alerts and efficient countermeasures.

9.2   DETECTION OF CHEMICAL WARFARE AGENTS

Specialized detectors have been specifically built to detect chemical warfare chemicals. The detectors exhibit a remarkable sensitivity towards particular chemicals, rendering them suitable for military and first-responder deployment. These devices can detect and identify hazardous substances swiftly [4].

Laboratory-based techniques, such as gas chromatography-mass spectrometry (GC-MS) and liquid chromatography-mass spectrometry (LC-MS), are employed to examine samples to detect the existence of chemical threat agents. These techniques exhibit a high level of accuracy and possess the capability to discern and identify precise chemical components [5].

A range of chemical sensors and detectors are employed to detect the existence of chemical threat agents. The sensors can be designed based on different principles, such as electrochemical, optical, and mass spectrometry [6]. These devices are specifically engineered to offer instantaneous data regarding the levels of chemicals present in the surrounding environment.

Certain biological detection systems employ living organisms or biomolecules to detect the existence of dangerous chemical substances. One illustrative instance involves the utilization of enzymes or antibodies in biosensors to elicit a response to particular substances, generating a quantifiable signal [7].

Remote sensing technologies, such as infrared spectroscopy and hyperspectral imaging, can remotely detect and ascertain the presence of chemical threat agents. These methodologies are particularly useful for monitoring expansive regions or evaluating the magnitude of a chemical discharge.

Among all methods, plasmonic nanoparticles using fluorometric sensing are in trend. Identifying chemical threat agents using plasmonic nanoparticles and a fluorometric sensing technique entails capitalizing on the distinct optical characteristics of plasmonic nanoparticles to ascertain and measure the existence of particular chemical compounds. In this technique, fluorescent molecules bind onto plasmonic nanoparticles. When the target agent forms a binding interaction with the functionalized nanoparticle, it can either suppress or amplify the fluorescence emitted by the fluorophores in proximity. The magnitude of this alteration in fluorescence can be measured and associated with the quantity of the specific chemical of interest.

The methodology mentioned above has significant value in various applications, including the identification of chemical danger agents, the monitoring of environmental conditions, and the advancement of biomedical diagnostics. This is primarily attributed to its notable sensitivity, selectivity, and capacity for expeditious detection. Scientists are currently engaged in ongoing investigations and efforts to refine and expand the effectiveness and scope of plasmonic nanoparticle-based sensing techniques.

The metallic nanoparticles of gold, silver and platinum are distinct from the nanoparticles of other metals, due to their extraordinary optical properties, and are therefore known as plasmonic nanoparticles. Plasmonic nanoparticles are different because of their exceptional light-absorbing and scattering properties with respect to their size and shape. By altering the composition, shape, and size of plasmonic nanoparticles; they can be tuned to respond optically in the ultraviolet, visible, and near-infrared regions of the electromagnetic spectrum. The color of nanoparticles can also be tuned by shifting the absorption and scattering regions of light. For instance, the color of spherical gold nanoparticles is ruby red because they reveal plasmon resonance in the green region of the spectrum; however, silver nanoparticles appear yellow because they show strong plasmon resonance in the blue region of the spectrum. Their excellent plasmonic properties allow them to exhibit localized surface plasmon resonance (LSPR), surface plasmon resonance (SPR), surface enhanced Raman (SER), and fluorescent phenomenon by slight variations in their fabrication [8]. Among all these, the fluorescent phenomenon is outstanding because it leads to extremely sensitive detection and imaging up to single-molecule level, with the help of strong emitter plasmonic nanoparticles. In spectroscopy, molecular phosphorescence and fluorescence are two distinct forms of luminescence. When fluorophore is eradiated using white light or monochromatic electromagnetic radiation; molecular fluorescence occurs. The resonant coupling of localized surface plasmons to fluorophore leads to the new phenomenon known as plasmon enhanced fluorescence (PEF). PEF causes greater polarization of the emitted radiation, angular distribution, and emission intensity [9]. It expands the field of fluorescence sensing by including relatively weak quantum emitters and overcoming photo-bleaching. Thus, it is leading towards the novel class of photo-stable probes by hyphenating quantum emitters with metal nanostructures using core-shell phenomenon in which the core is comprised of metal nanostructures. These photo-stable probes are developed by coupling localized surface plasmons in plasmonic nanoparticles with fluorophores at their excited states. The enhancement in the local field provides better brightness of molecular emission and enhanced sensitivity for trace level detection [10]. It is important to note here that a continuous transition from enhancement of fluorescence to quenching of fluorescence can be observed by changing the molecule-nanoparticle distance [11, 12]. This phenomenon was first explained in the pioneering work of Drexhage and his co-workers [13]. After their theory, many studies focused on the phenomenon of energy transfer to metal dielectric surfaces from excited molecules [14, 15]. It is important for the molecule attached at the surface of the plasmonic nanoparticle that it should allow the transfer of energy and show effective quenching, or form a complex at the surface of the plasmonic nanoparticle and behave as a new molecule with totally unique electronic states. Therefore, the dependence of the electromagnetic properties of plasmonic nanoparticles on metal-molecule distance is marked significant..

The most widely used fluorophores are molecules of organic dyes, due to their specialized chemical structure such as aromatic rings or conjugated carbon chains. Figure 9.1 shows the Jablonski diagram displaying energy states of molecules [16]. The molecules that are not absorbing energy remain confined to the ground state S_0 (Figure 9.1a). Figure 9.1b shows the spectral characteristics of a molecule related to absorption and emission of energy. The absorbed or emitted photons are shown by vertical arrows. The direction of the head shows absorption (upward) or emission (downward) of light. The color of the arrows shows the wavelength of photons. Figure 9.1c shows the time taken by the molecule in different fluorescence steps of excitation and emission. The thermally equilibrated molecules in their excited state generally produce fluorescence emission of S1, the vibrational state with the lowest energy [17]. The optical absorption process, resulting from the cross section, is usually the highest and fastest, occurring within femtoseconds. The isolated fluorophore emission is observed and defined by its "Q_0" (quantum yield) and "t" (lifetime). The brightest emissions are exhibited by the molecules that have largest quantum yields, like rhodamines. The optical properties of these molecules are characterized by estimation of k_r (radiative decay rate) and k_{nr} (non-radiative decay rate). Therefore, Q_0 can be calculated as;

$$Q_0 = \frac{k_r}{k_{r+} k_{nr}} \tag{9.1}$$

The non-radiative rate $k_{nr} = \Sigma\ k_d$ includes all courses which produce non-radiative decay to the electronic ground state [17]. The lifetime is broadly measured using equation 9.2.

$$\tau = \frac{1}{k_{r+} k_{nr}} \tag{9.2}$$

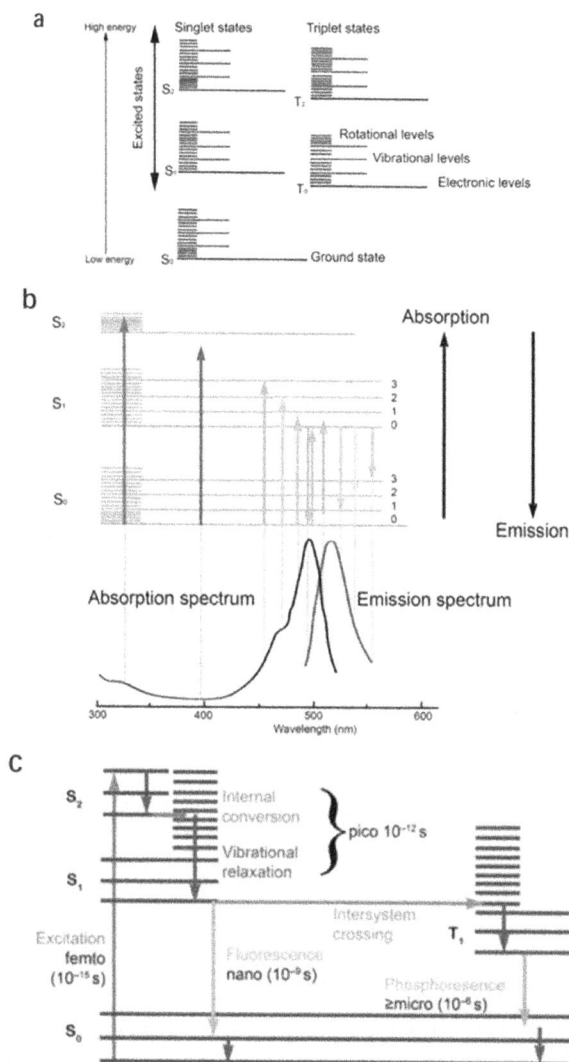

FIGURE 9.1 Jablonski diagram displaying energy states of molecules. Printed with the permission of Springer Nature [16].

9.3 DESIGN AND FABRICATION OF PLASMONIC NANOPARTICLE-BASED SEF SENSORS

The process of fabricating fluorometric sensors for chemical threat agents using plasmonic nanoparticles encompasses a sequence of meticulously regulated procedures, starting from the synthesis and functionalization of the nanoparticles, followed by their incorporation into a sensing platform, and concluding with the verification of their effectiveness. The sensors possess a notable sensitivity and specificity, rendering them highly important instruments for swiftly and precisely identifying chemical threat chemicals in various security and safety contexts. Ongoing research and development endeavors in this particular domain are focused on enhancing the performance of sensors, expanding the scope of detectable substances, and improving the practicality of their application in real-world scenarios. Three major steps for fabricating a Plasmonic nanoparticle-based fluorometric sensor are mentioned below.

9.3.1 Synthesis of Plasmonic Nanoparticles

The initial and essential stage in fabricating fluorometric sensors utilizing plasmonic nanoparticles involves the synthesis of the nanoparticles. Plasmonic nanoparticles are commonly comprised of noble metals, such as gold, silver, or other similar materials, primarily chosen for their distinctive optical characteristics, notably surface plasmon resonance (SPR).

The synthesis process commences by carefully choosing suitable chemical precursors. For example, in the context of gold nanoparticles, chloroauric acid is frequently employed ($HAuCl_4$). The precursors undergo a reduction in the presence of a reducing agent. An illustration of the reduction of $HAuCl_4$ to produce gold nanoparticles may be observed through the utilization of sodium borohydride ($NaBH_4$). The manipulation of reaction parameters, such as temperature, precursor concentration, and choice of stabilizing agents, enables control over the dimensions and morphology of the nanoparticles. The parameters mentioned above possess the potential to be adjusted to generate nanoparticles that exhibit precise characteristics. To mitigate the aggregation of nanoparticles, it is common practice to provide stabilizing reagents such as citrate or polyvinylpyrrolidone (PVP). These compounds also have a significant impact on the regulation of nanoparticle size. Following the synthesis process, it is customary to subject the nanoparticles to a purification step to eliminate any surplus reagents, byproducts, and stabilizing agents. This objective can be accomplished by employing either centrifugation or dialysis techniques.

9.3.2 Functionalization of Plasmonic Nanoparticles

After the successful synthesis and purification of plasmonic nanoparticles, they must be functionalized to enhance their specificity for detecting chemical threat agents. The functionalization process entails the attachment of molecules or ligands to the surface of a nanoparticle, which exhibit a strong affinity towards the target agents.

The selection of ligands is contingent upon the particular chemical threat agent one aims to detect. Ligands encompass a variety of molecules, including antibodies, aptamers (single-stranded DNA or RNA molecules exhibiting strong affinity for certain targets), and other receptor molecules recognized for their ability to bind to the target agent with notable selectivity. The plasmonic nanoparticles' surface is altered to incorporate specific binding sites for the selected ligands. Modifying nanoparticles frequently entails the utilization of bifunctional linkers, such as thiol-based compounds, to establish a connection between the nanoparticle surface and a functional group for ligand attachment. The ligands that have been chosen are subsequently affixed to the surfaces of the functionalized nanoparticles via covalent bonding or other binding processes. Achieving accurate ligand orientation and preservation of binding affinity necessitates meticulous management throughout this stage.

9.3.3 INTEGRATION INTO THE SENSING PLATFORM

Three approaches are commonly employed for platform integration.

The functionalized nanoparticles are evenly distributed inside a solution. Subsequently, the solution is subjected to the sample that contains the chemical threat agent being targeted. The fluorescence properties of nanoparticles undergo alterations due to their interactions with the target, which are subsequently quantified using a fluorometer or spectrofluorometer.

Surface-based assays can be developed by immobilizing functionalized nanoparticles onto solid substrates, such as glass slides or microplates. Upon the binding of the target agent to the ligands present on the nanoparticles' surface, fluorescence alterations are detected and measured.

The design of functionalized nanoparticles can incorporate the ability to aggregate upon interacting with the target agent. The aggregation process can be observed visually or spectroscopically, leading to alterations in fluorescence characteristics.

9.4 APPLICATIONS OF SEF SENSORS IN CHEMICAL THREAT MONITORING

SEF sensors, due to their extraordinary plasmonic properties, easy fabrication, robustness, and excellent selectivity and sensitivity are greatly desired for the detection of chemical threats in different matrices, such as food, water, air, etc. Tao Wang et al. [18] reported fluorescent magnetic nanoparticles ($MNP_{300}@SiO_2(FITC)$) for the detection of Ricin (Figure 9.2). Ricin is one of the most toxic plant proteins when ingested through contaminated food or water [19]. Therefore, it can be easily used as a toxic chemical warfare agent by international intelligence agents [20] and terrorists to perform destructive or assassination activities [21]. It has two peptide chains, i.e., ricin A (RTA) and ricin B (RTB). The RTB chain promotes the toxicity of the RTA chain. Furthermore, the RTB chain possesses an effective galactose binding site that results in its in vivo toxicity [19]. The SEF sensor shows excellent anti-interference capability. It was applied for the detection of RTB in tap water samples and edible oil. Zearalenone (ZEN) is neurotoxic and affects the reproductive system. It is equally toxic for both human beings and livestock. The Fusarium and Gibberella fungi are the main producers of this toxin [22]. ZEN is able to resist almost all food processing conditions which is why strict regulations are followed in Australia and countries of the European Union. The allowed residues of ZEN in corn and other grains are in the range of 50–350 μg/kg [23, 24]. Recently, a fluorescence sensor was developed for the detection of ZEN using Polyvinylpyrollidone-modified Zr-MOF

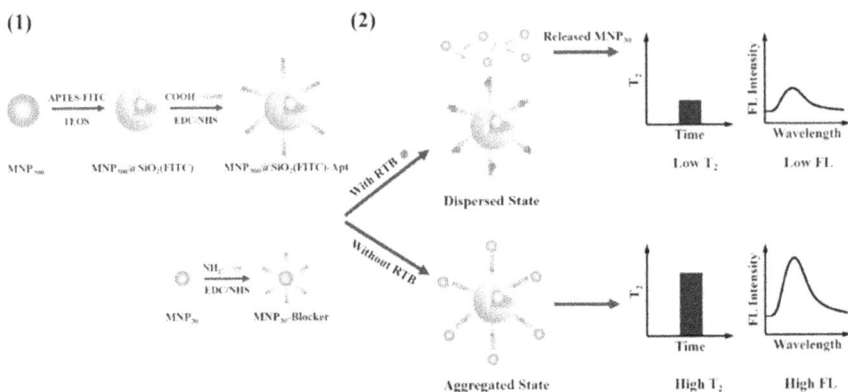

FIGURE 9.2　Schematic representation of fabrication of $MNP_{300}@SiO_2$(FITC) and fluorescence detection of ricin (RTB). Printed with the permission of Elsevier [18].

(PVP-UiO-67). The PVP-UiO-67 was able to detect ZEN as low as 7.44 nM with a linear range of 0.016 to 3.14 μM [25].

Aflatoxin B_1 (AFB_1) is produced by *Aspergillus parasiticus* and *Aspergillus flavus*. It is one of the most injurious toxins found in oil crops and grains [26]. The high exposure of AFB_1 can lead to liver cancer or liver cirrhosis. The maximum allowed level of AFB_1 in food stuffs is set as 12 ppb by the European Union [27]. Jinxiang Wei et. al. [28] recently reported on an aptasensor capable of detecting AFB_1 as a fluorescence sensor, as well as a surface-enhanced Raman scattering (SERS) sensor. The nanocomposite was composed of CdTe quantum dots responsible for fluorescence response and DTNB decorated gold nanoparticles for SERS response. The nanocomposite was capable of producing high SERS signals and low fluorescence. Upon interaction with AFB_1 its SERS signals decreased and fluorescence signals were enhanced. The limit of detection of developed methods was found as 0.094 pg/mL with a linear range of 10^{-4} to 10^3 mg/mL. Figure 9.3 shows the schematic representation of the dual-aptamer fluorescence sensor.

SEF sensing also plays an important role in the sensing of highly toxic metal ions, such as mercury and lead. Perish Chandra Ray et al. [29] reported for the first time on the use of an SEF sensor based on gold nanoparticles for the sensing of mercury in soil, fish, and water. The reported SEF sensor was ultra-sensitive and selective for mercury with a limit of detection of 2 ppt and a linear range of 0.8–170 ppb. After this pioneering report, many researchers were attracted to SEF sensing, due to its high sensitivity. The fluorescent core-shell nanoparticles comprised of silver as core and silica as shell were also reported for the sensing of mercury. The said SEF sensor was able to detect mercury in the range of 0 to 100 nM. The silver nanocube-based fluorescent nanocomposite was also reported for the detection of copper. The said SEF sensor was able to produce 3-fold enhanced fluorescence for Cu^{2+} ions. The limit of detection was obtained as 0.3 μM for Cu^{2+} with a linear range of 1–5 μM [30]. Another study reported the fabrication of a simple nanostructure comprised of

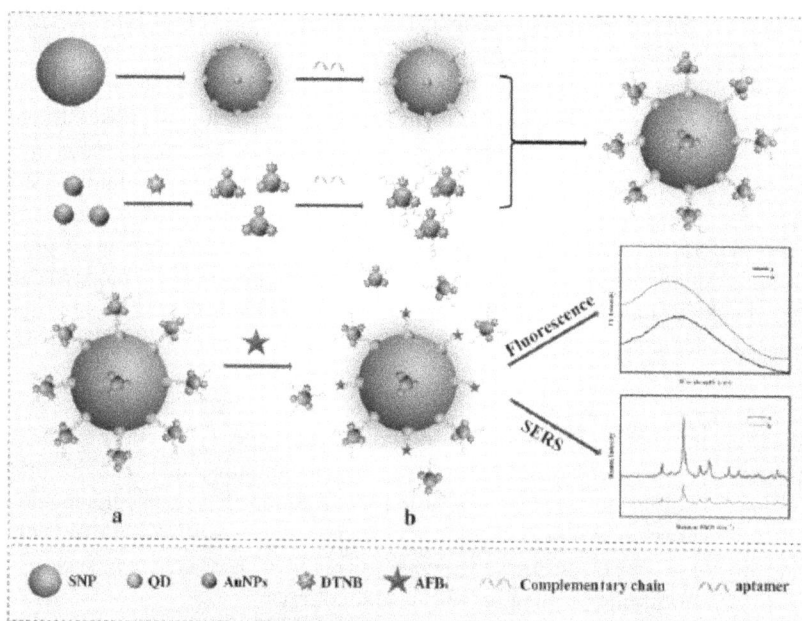

FIGURE 9.3 Schematic representation of sensing of AFB_1 using dual mode aptasensor. Printed with the permission of Elsevier[28].

silver nanoparticles modified with thiol-DNA [31]. The said SEF sensor was able to detect Hg^{2+} in the range of 1–80 nM. In another study, lead ions were detected using a $ZnFe_2O_4$@Au-Ag bifunctional nanocomposite. The detection limit of said SEF sensor was as low as 0.3 pM for lead with a linear range of 1–3 μM. The nanocomposite used in this study was easily recycled and repeatedly used. Among all the discussed studies, Darbha et al. [29] were able to produce the most sensitive SEF sensor for mercury. The sensing technology of SEF is flourishing exponentially due to its sensitivity, swiftness, and selectivity.

9.5 CONCLUSION

Plasmonic nanoparticle-based fluorometric sensors possess several advantageous characteristics; yet they also encounter certain obstacles and have the potential to be enhanced. The maintenance of stability and reproducibility in plasmonic nanoparticles, as well as the optimization of sensor performance, are crucial factors for considering the practical implementation of these technologies. The dependability of sensors can be influenced by the variability observed in the processes of nanoparticle manufacturing and functionalization. The investigation of multiplexed sensors with the ability to concurrently detect numerous analytes is a subject of scholarly inquiry. Another area of future development that is receiving attention is the integration of these sensors into portable and miniaturized platforms for on-site testing and diagnostics. The maintenance of biocompatibility in plasmonic nanoparticles is of utmost importance in biomedical contexts, as it serves to mitigate the risks associated with toxicity and immunological reactions. The design of surface changes and coatings must be meticulously crafted to ensure their safe application within living organisms. There is a pressing need for the development of sophisticated algorithms in real-time monitoring and data processing to facilitate prompt and precise measurements in time-critical domains, such as medical diagnostics and environmental monitoring. The establishment of uniform procedures for the synthesis and functionalization of plasmonic nanoparticles, as well as the development of standardized assay techniques, is of utmost significance for guaranteeing consistent sensor performance in diverse laboratory and industrial settings.

In conclusion, the use of plasmonic nanoparticle-based fluorometric sensors exhibits a remarkable degree of adaptability and efficacy, rendering them an invaluable instrument with extensive utility in diverse domains, such as environmental surveillance, biological investigation, assurance of food safety, and enhancement of security measures. As the field of research progresses, the exploration of issues and enhancements in sensor design is expected to result in further advancements and significant implications for the uses of this technology.

REFERENCES

1. Saylan, Y., S. Akgönüllü, and A. Denizli, Plasmonic sensors for monitoring biological and chemical threat agents. *Biosensors*, 2020. **10**(10): p. 142.
2. Ganesan, K., S.K. Raza, and R. Vijayaraghavan, Chemical warfare agents. *J Pharm Bioallied Sci*, 2010. **2**(3): p. 166–78.
3. Yue, G., et al., Gold nanoparticles as sensors in the colorimetric and fluorescence detection of chemical warfare agents. *Coordination Chemistry Reviews*, 2016. **311**: p. 75–84.
4. Burnworth, M., S.J. Rowan, and C. Weder, Fluorescent sensors for the detection of chemical warfare agents. *Chemicals*, 2007. **13**(28): p. 7828–7836.
5. Kumar, V., et al., Recent advances in fluorescent and colorimetric chemosensors for the detection of chemical warfare agents: A legacy of the 21st century. *Chemical Society Reviews*, 2023. **52**(2): p. 663–704.
6. Sempionatto, J.R., et al., Wearable ring-based sensing platform for detecting chemical threats. *ACS Sensors*, 2017. **2**(10): p. 1531–1538.
7. Wu, K., et al., Magnetic-nanosensor-based virus and pathogen detection strategies before and during COVID-19. *ACS Applied Nano Materials*, 2020. **3**(10): p. 9560–9580.
8. Anker, J.N., et al., Biosensing with plasmonic nanosensors. *Nature Materials*, 2008. **7**(6): p. 442–453.
9. Dong, J., et al., Recent progress on plasmon-enhanced fluorescence. *Nanophotonics*, 2015. **4**(4): p. 472–490.
10. Li, J.-F., C.-Y. Li, and R.F. Aroca, Plasmon-enhanced fluorescence spectroscopy. *Chemical Society Reviews*, 2017. **46**(13): p. 3962–3979.
11. Instrumentation for fluorescence spectroscopy, in *Principles of Fluorescence Spectroscopy*, J.R. Lakowicz, Editor. 2006, Springer US: Boston, MA. p. 27–61.
12. Anger, P., P. Bharadwaj, and L. Novotny, Enhancement and quenching of single-molecule fluorescence. *Physical Review Letters*, 2006. **96**(11): p. 113002.

13. Drexhage, K., H. Kuhn, and F.P. Schäfer, Variation of the fluorescence decay time of a molecule in front of a mirror. *Berichte der Bunsengesellschaft für physikalische Chemie*, 1968. **72**(2): p. 329–329.
14. Alivisatos, A.P., et al., Electronic energy transfer at semiconductor interfaces. I. Energy transfer from two-dimensional molecular films to Si(111). *The Journal of Chemical Physics*, 1987. **86**(11): p. 6540–6549.
15. Aslan, K., et al., Metal-enhanced fluorescence solution-based sensing platform. *Journal of Fluorescence*, 2004. **14**(6): p. 677–679.
16. Lichtman, J.W. and J.-A. Conchello, Fluorescence microscopy. *Nature Methods*, 2005. **2**(12): p. 910–919.
17. Introduction to fluorescence, in *Principles of Fluorescence Spectroscopy*, J.R. Lakowicz, Editor. 2006, Springer US: Boston, MA. p. 1–26.
18. Wang, T., et al., Magnetic relaxation switch and fluorescence dual-mode biosensor for rapid and sensitive detection of ricin B toxin in edible oil and tap water. *Analytica Chimica Acta*, 2022. **1232**: p. 340471.
19. Bozza, W.P., et al., Ricin detection: Tracking active toxin. *Biotechnology Advances*, 2015. **33**(1): p. 117–123.
20. Schep, L.J., et al., Ricin as a weapon of mass terror – Separating fact from fiction. *Environment International*, 2009. **35**(8): p. 1267–1271.
21. Fabbrini, M.S., et al., Plant ribosome-inactivating proteins: Progesses, challenges and biotechnological applications (and a few digressions). *Toxins*, 2017. **9**(10): p. 314.
22. Liao, Z., et al., Progress on nanomaterials based-signal amplification strategies for the detection of zearalenone. *Biosensors and Bioelectronics: X*, 2021. **9**: p. 100084.
23. Wu, Y., et al., Selection of a DNA aptamer for cadmium detection based on cationic polymer mediated aggregation of gold nanoparticles. *Analyst*, 2014. **139**(6): p. 1550–1561.
24. Li, R., et al., Dual quantum dot nanobeads-based fluorescence-linked immunosorbent assay for simultaneous detection of aflatoxin B1 and zearalenone in feedstuffs. *Food Chemistry*, 2022. **366**: p. 130527.
25. Zhang, X., et al., A novel fluorescence sensor for sensitive detection of zearalenone using a polyvinylpyrrolidone-modified Zr(IV)-based metal-organic framework. *Sensors and Actuators B: Chemical*, 2023. **395**: p. 134516.
26. Frangiamone, M., et al., In vitro and in vivo evaluation of AFB1 and OTA-toxicity through immunofluorescence and flow cytometry techniques: A systematic review. *Food and Chemical Toxicology*, 2022. **160**: p. 112798.
27. Wu, W., et al., Colorimetric liquid crystal-based assay for the ultrasensitive detection of AFB1 assisted with rolling circle amplification. *Analytica Chimica Acta*, 2022. **1220**: p. 340065.
28. Wei, J., et al., Satellite nanostructures composed of CdTe quantum dots and DTNB-labeled AuNPs used for SERS-fluorescence dual-signal detection of AFB1. *Food Control*, 2024. **156**: p. 110112.
29. Darbha, G.K., A. Ray, and P.C. Ray, Gold nanoparticle-based miniaturized nanomaterial surface energy transfer probe for rapid and ultrasensitive detection of mercury in soil, water, and fish. *ACS Nano*, 2007. **1**(3): p. 208–214.
30. Sun, B., et al., Metal-enhanced fluorescence-based multilayer core–shell Ag-nanocube@SiO2@PMOs nanocomposite sensor for Cu2+ detection. *RSC Advances*, 2016. **6**(66): p. 61109–61118.
31. Liang, L., et al., Metal-enhanced ratiometric fluorescence/Naked eye bimodal biosensor for lead ions analysis with bifunctional nanocomposite probes. *Analytical Chemistry*, 2017. **89**(6): p. 3597–3605.

10 The Application of Molecular Imprinting-Based Plasmonic Nanosensor Platforms for the Sensing of Chemical and Biological Threats

Rüstem Keçili and Adil Denizli

10.1 INTRODUCTION

Recognition elements play a crucial role in sensor platforms. These elements are responsible for the effective recognition and interaction of specific target compounds, providing the most appropriate designed and constructed sensor platform to sensitively detect and accurately determine the existence or concentration of these compounds (Wei et al. 2023; Justino et al. 2015; Van Dorst et al. 2010; Li et al. 2023; Xiao et al. 2023). Different types of recognition elements can be efficiently employed in sensor platforms, depending on the nature of the target compound. Various common recognition elements in many sensor systems include different biomolecules such as antibodies, enzymes, and nucleic acids (Santos et al. 2023; Yin et al. 2023; Kim et al. 2023; Bollella 2022; Shalileh et al. 2023). However, their physical and chemical stability is generally poor, which makes these biomolecules incompatible for application in harsh conditions (i.e., high temperatures, low and high pH values, etc.). To overcome these drawbacks, artificial receptors, which are biomimetic, can be successfully used. The biomimetic systems have a great ability to interact with the target compounds in a similar way to their natural counterparts.

Molecularly imprinted polymers (MIPs) are artificial and biomimetic materials which display outstanding recognition and selectivity behavior for the analytes of interest (Moreno-Bondi et al. 2023; Wang et al. 2023; Gu et al. 2016; Kupai et al. 2017; Keçili and Hussain 2018). MIPs can be successfully designed and prepared through the polymerization of an appropriate functional monomer and a cross-linker in the existence of a target molecule (template), enabling the creation of 3D cavities

DOI: 10.1201/9781003459316-10

that have selective binding regions for the target molecule during the polymerization stage. After removal of the template compound from the polymeric structure, this 3D cavity can effectively and specifically bind the target compound or its structural analogue (Keçili and Denïzli 2021; Wang and Zhang 2023; Say et al. 2016; Ayivi et al. 2023; Mabrouk et al. 2023). Non-imprinted polymer (NIP) as the control material is also prepared applying the same procedure without using the template. Figure 10.1 shows the preparation process of MIP and NIP.

MIPs have received high interest from researchers in recent years, due to their great superior qualities, which are briefly described in the following:

- *High Stability*: The high stability of MIPs makes them suitable for use in harsh conditions, such as industrial processes or biomedical applications.
- *High Selectivity*: MIPs are very specific in their recognition and binding to target molecules. This makes them useful in separating mixtures of similar molecules and detecting specific molecules in complex samples.
- *Cost-Effectiveness*: MIPs are cheaper to produce than other materials that can do similar things, such as antibodies or enzymes. They can be synthesized in large quantities using simple equipment and procedures.
- *Reusability*: MIPs can be used multiple times without losing their selectivity or binding capacity. This makes them an attractive alternative to other materials that are used only once and then discarded.
- *Tailored Features*: MIPs can be designed to have specific properties, such as size, shape, and functionality, which can be tailored to suit specific applications. This makes them versatile and adaptable to a wide range of uses.

Due to these features, MIPs are effectively applied in different fields including sensor platforms (Ayerdurai et al. 2023), catalysis (Keçili et al. 2010; Muratsugu et al. 2020; Keçili et al. 2012; Kirsch et al. 2009), drug delivery (Kakkar and Narula 2022; Nerantzaki et al. 2022; Zaidi 2020;

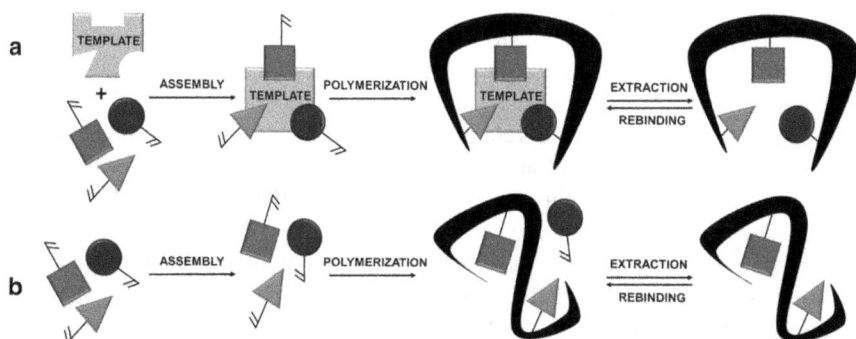

FIGURE 10.1   (a) The preparation process of MIP, and (b) NIP (Herrera-Chacón et al. 2021).

Keçili et al. 2023; Akgönüllü et al. 2023; Li et al. 2021), bioimaging (Vaněčková et al. 2020; Özçelikay et al. 2020; Inutsuka et al. 2019; Ren et al. 2021; Panagiotopoulou et al. 2017), and separation and purification (Gupta et al. 2023; Xu et al. 2023; Banan et al. 2022; Banan, Hatamabadi, et al. 2022; Zhou et al. 2023).

Nonetheless, although there is a rapidly growing number of reported research articles on this topic, to the best of our knowledge, there is no chapter that provides an up-to-date report. The aim of this chapter is to provide an overview of the latest research and developments of novel MIP-based plasmonic nanosensors for the sensitive detection of chemical (environmental pollutants, food contaminants, and explosives, etc.) and biological (viruses and bacteria, etc.) threats. The chapter starts with an introduction about molecularly imprinted polymers (MIPs) and the importance of these unique materials in the design and construction of novel and sensitive sensor platforms. In the next section, the main principles and combinations of plasmonic sensors with MIPs are presented. Then, the performance of MIP-based plasmonic nanosensors for the sensitive detection of chemical and biological threats is discussed. Finally, we draw our conclusion and future perspectives are also discussed.

10.2   PLASMONIC SENSOR PLATFORMS

Plasmonic sensor platforms are a type of optical sensor systems which rely on the interaction between light and surface plasmons for the sensing of analytes (Saylan et al. 2020; Akgönüllü and Denìzli 2023; Eskandari et al. 2022; Guo et al. 2015; Wang et al. 2023). These types of sensor platforms are highly sensitive and extensively applied for rapid and label-free detection (Ye et al. 2022; Jin et al. 2022; Rashid et al. 2021; Liu et al. 2023; Xianyu et al. 2021).

Molecularly imprinted plasmonic nanosensor platforms are mainly classified as surface plasmon resonance (SPR)-based molecularly imprinted plasmonic nanosensor platforms (Lazarević-Pašti et al. 2023; Özkan et al. 2019; Della Giustina et al. 2016) and surface-enhanced Raman spectroscopy (SERS)-based molecularly imprinted plasmonic nanosensor platforms (Guo et al. 2020; Tarannum et al. 2023).

The SPR technique is defined with three words which are "surface" where plasmonic phenomena occur, "plasmon", which refers to the collective oscillation of free electrons on the surface of a metal (commonly Au or Ag) when it interacts with incident light. This phenomenon occurs at the interface between a thin metal film and a dielectric material (usually glass) when specific conditions are met (Figure 10.2). And finally, "resonance", which refers to the transfer of energy from incident light to plasmons which exist on the surface. The SPR technique is widely and effectively used for the evaluation of the interactions that occur on the surface via surface plasmons provided by the resonance with incident light.

On the other hand, SERS is another powerful technique for the development of sensitive plasmonic sensor platforms (Liu et al. 2022; Restaino and White 2019). In conventional Raman spectroscopy technique, a laser is applied to the sample, and the detection of the scattered light is performed (Perevedentseva et al. 2024; Sha et al. 2023; Li et al. 2023). Most of the scattered photons have the same energy (frequency) as the incident laser light, resulting in Rayleigh scattering, which does

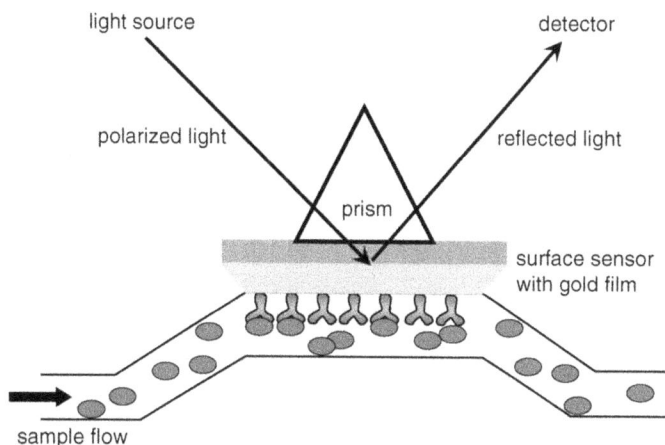

FIGURE 10.2 Schematic depiction of the SPR technique (Bakhtiar 2012).

not provide detailed chemical information. However, a small fraction of scattered photons undergo a Raman scattering process, where they gain or lose energy during interactions with molecular vibrations, resulting in a shift in their energy levels. These energy shifts create a unique fingerprint-like Raman spectrum that can be used effectively used for the identification of the molecules in the sample. SERS enhances the Raman scattering effect by leveraging the localized surface plasmon resonance (LSPR) phenomenon that occurs when incident light interacts with metallic nanoparticles (i.e., Au or Ag nanoparticles) on a surface (Liu et al. 2023; Sharifi et al. 2023). When the frequency of the incident light matches the natural frequency of the collective oscillations of free electrons on the metal's surface, surface plasmons are excited, creating intense electromagnetic fields near the metal surface. Molecules in close proximity to these intense electromagnetic fields significantly enhance Raman scattering signals. This is because the electromagnetic field amplifies the molecular vibrations, leading to much stronger Raman signals.

Integrating the molecular imprinting approach into SPR and SERS techniques is a very effective route for the fabrication of sensitive plasmonic sensor platforms toward the detection of chemicals such as environmental pollutants, food contaminants, and explosives, etc., and biological threats including virus and bacteria, etc.

10.3 MIP-BASED PLASMONIC SENSOR PLATFORMS TOWARD CHEMICAL THREATS

Chemical threats refer to the intentional or accidental release of hazardous compounds which significantly affect human health and the environment (Feraud et al. 2023; Manzo and Schiavo 2022). These threats can take various forms, such as environmental pollutants, food contaminants and chemical warfare agents, etc. Sensitive detection and mitigation of these chemical threats are crucial for public safety.

Aflatoxin food contamination is a serious and potentially life-threatening issue caused by the existence of aflatoxins that exhibit high toxicity and carcinogenicity (Wang et al. 2023). Aflatoxins are produced by certain molds which primarily belong to the *Aspergillus* species and can infect various crops, such as peanuts, corn, rice, wheat, and soybeans, etc. There are different types of aflatoxins, but there are four main types (Aflatoxin B1, B2, G1, and G2) that are of particular concern. Among these, Aflatoxin B1 (AFB1) has the highest toxicity and is classified as "Group 1 carcinogen" by the International Agency for Research on Cancer (IARC) (Marchese et al. 2018).

In a crucial piece of research published by the Denizli group (Akgönüllü et al. 2020), AFB1 was sensitively determined by using an MIP-based plasmonic sensor platform. In this research, precomplexation of AFB1 as the template and N-methacryloyl-L-phenylalanine (MAPA) as the functional monomer was carried out in the first step. Then, gold nanoparticles (AuNPs) were molecularly embedded on the imprinted layer, which was prepared using the UV-polymerization approach on the surface of the gold chip of the SPR system, as shown schematically in Figure 10.3.

The developed MIP-based plasmonic nanosensor platform displayed an outstanding sensing behavior for AFB1 in a linear concentration range varying from 0.0001 ng mL^{-1} to 10.0 ng mL^{-1}. The detection limit was found as 1.04 pg mL^{-1}. Furthermore, the sensing efficiency of the MIP-based plasmonic nanosensor platform was also evaluated using different food samples including corn and peanut. The MIP-based plasmonic nanosensor also showed high recovery values in the range from 96.63% to 105.94%.

Aflatoxin M1 (AFM1) is formed when aflatoxin B1 is metabolized by the liver in animals, particularly dairy cattle (Li et al. 2021). When cows consume feed contaminated with aflatoxin B1, it is partially converted into its metabolite AFM1, which can present in milk. Akgönüllü et al. developed a novel MIP-based plasmonic nanosensor platform for the sensitive detection of AFM1 in milk samples (Akgönüllü et al. 2021). In their study, the researchers used MAPA as the functional monomer for the

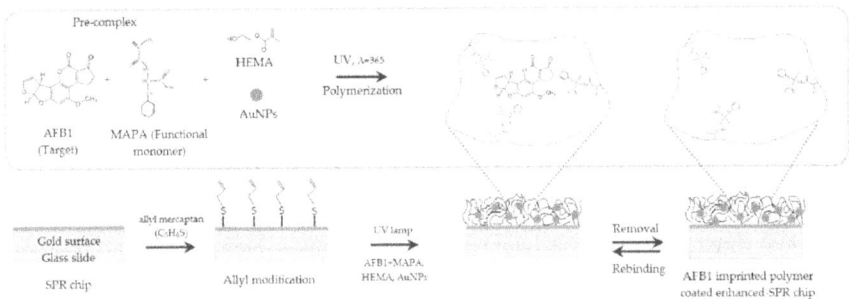

FIGURE 10.3 Representation of the construction of MIP-based plasmonic sensor platform toward AFB1 (Akgönüllü et al. 2020).

preparation of a gold nanoparticle-embedded MIP layer on the gold layer of the SPR chip. The fabricated MIP-based plasmonic nanosensor platform displayed excellent sensing efficiency for AFM1 in a linear concentration range between 0.0003 ng mL^{-1} and 20.0 ng mL^{-1}. The achieved detection limit was 0.4 pg mL^{-1}. In addition, the selectivity tests were performed using AFB1, ochratoxin A, and citrinin as competitive mycotoxins. The experimental results obtained from these tests demonstrated that the fabricated MIP-based plasmonic nanosensor platform also showed great selectivity toward the target mycotoxin AFM1.

In another work reported by the same research group (Akgönüllü et al. 2021), ochratoxin A (OTA) in dry fig samples was successfully recognized by applying an MIP-based plasmonic nanosensor. To this aim, the researchers firstly modified the gold surface of an SPR chip with allyl mercaptan. In the next step, OTA imprinted polymeric film was prepared using the functional monomer MAPA, cross-linker EGDMA, and initiator AIBN, respectively (Figure 10.4). The MIP-based plasmonic nanosensor displayed an outstanding sensing performance in a broad linear concentration range from 0.1 to 20 ng mL^{-1}. The obtained detection limit was 0.028 ng mL^{-1}.

In a work published by Atar and colleagues (Atar et al. 2015), a sensitive plasmonic nanosensor with an MIP layer was designed and constructed for the effective sensing of citrinin (CIT) in red yeast rice samples. In this work, N-methacryloyl-L-glutamic acid (MAGA) and 2-hydroxyethyl methacrylate (HEMA) were chosen as functional monomers for the preparation of a CIT selective MIP layer on the surface of the SPR chip. Prior to the preparation of the MIP layer, the modification of the surface of the SPR chip was done using allyl mercaptan. After characterization studies for the prepared imprinted SPR chip through atomic force microscopy (AFM), FT-IR, and contact angle measurements, the developed plasmonic nanosensor was successfully used for the detection of mycotoxin CIT in food samples. The sensing response of the plasmonic nanosensor toward CIT was linear in the concentration

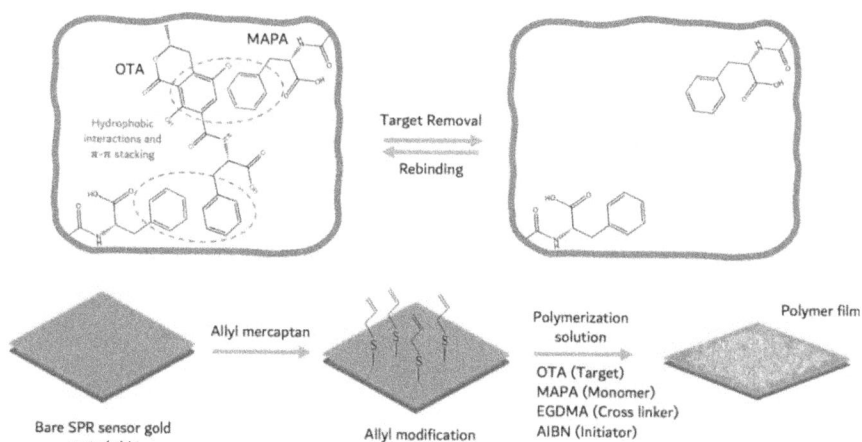

FIGURE 10.4 Schematic presentation of the preparation of an MIP-based plasmonic nanosensor platform for OTA (Akgönüllü et al. 2021).

range from 0.005 to 1.0 ng mL^{-1}. A very low detection limit (0.0017 ng mL^{-1}) was obtained.

Atrazine is a commonly used herbicide in agriculture (Hu et al. 2023). Because of its toxic feature, the detection of atrazine is crucial for protecting the environment, human health, and the sustainability of agriculture. Agrawal and co-workers developed a novel and sensitive plasmonic nanosensor toward atrazine, based on the molecular imprinting technique (Agrawal et al. 2016). In this research, the sensing probe of the plasmonic nanosensor was designed and constructed through a coating of Ag film (ca. 40 nm in thickness) over the unclad core of an optical fiber. In the next stage, phenol and HEMA were chosen as the functional monomers for the preparation of the selective MIP layer toward the target herbicide atrazine. The fabricated plasmonic nanosensor exhibited an outstanding recognition performance for atrazine in a linear concentration range which varied between 0 and 10^{-7} M. Detection and quantification limit values were found to be 1.92 × 10^{-14} and 7.61 × 10^{-14} M, respectively.

In another research study, published by Choi et al. (Choi et al. 2009), the sensitive detection of zearalenone, a mycotoxin produced by certain species of fungi, primarily from the genus *Fusarium*, was performed by applying the constructed MIP-based plasmonic nanosensor. For this purpose, the formation of a selective thin layer of imprinted polypyrrole toward zearalenone was conducted through electrochemical polymerization of pyrrole on the surface of the SPR chip, as schematically depicted in Figure 10.5. The MIP-based SPR sensor displayed a linear sensing response toward the target mycotoxin zearalenone in the concentration range from 0.3 to 3000 ng mL^{-1} and the detection limit was achieved as 0.3 ng mL^{-1}.

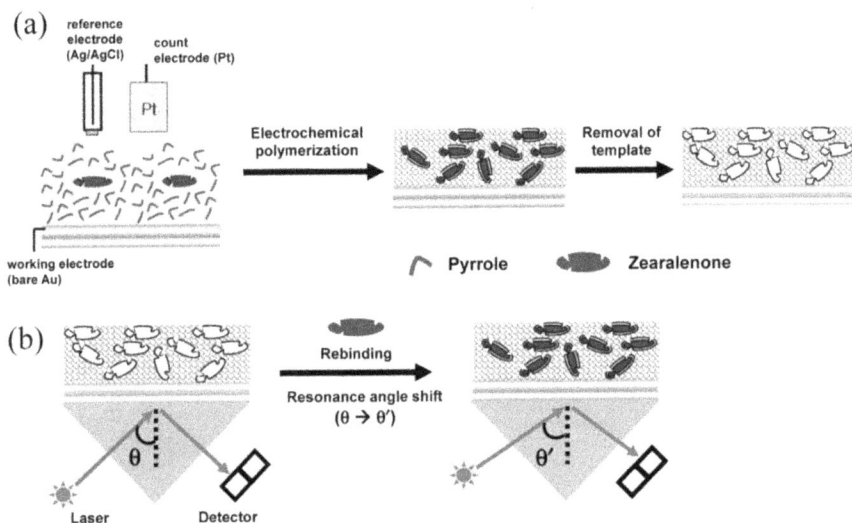

FIGURE 10.5 (a) Diagram of the construction of MIP-based SPR sensor platform for zearalenone, (b) The shift in resonance angle resulting from the binding of zearalenone to the layer of MIP on the gold surface (Choi et al. 2009).

Surface-enhanced Raman scattering (SERS) is another powerful approach, as previously mentioned, for the sensitive detection of chemical and biological threats. Castro-Grijalba and colleagues fabricated an SERS-based plasmonic nanosensor employing molecular imprinting technology for the sensitive and selective recognition of polycyclic aromatic hydrocarbons (PAHs) (Castro-Grijalba et al. 2020). In this report, the fabrication of an SERS-based plasmonic nanosensor comprises a two-step process. Firstly, a layer-by layer deposition process of gold nanoparticles on the surface of glass was performed. In the next step, the glass deposited with the gold nanoparticles was coated with a thin layer of selective MIP toward target PAHs including pyrene and fluoranthene. The constructed SERS-based plasmonic nanosensor with the MIP layer was utilized for the sensing of pyrene and fluoranthene in real samples, such as seawater and creek water. The successful combination of MIPs and SERS enabled very low detection limits (nM level) to be achieved, with an excellent improvement (ca. 100-fold) compared to the non-imprinted polymer (NIP)-based plasmonic sensor as the control.

10.4 MIP-BASED PLASMONIC SENSOR PLATFORMS TOWARD BIOLOGICAL THREATS

The sensitive detection of biological threats (i.e., viruses and bacteria) is of critical importance for public health, biosecurity, and biodefense (Brahmkhatri et al. 2021; (Upadhyayula 2012). MIP-based plasmonic sensors are promising platforms that can be successfully applied for the effective sensing of these biological threats. The research group of Denizli developed an MIP-based plasmonic sensor platform for the detection of *Escherichia coli (E. coli)* (Gür et al. 2019). In their research, Cu (II) ions were chosen to enable the interactions between the Au nanoparticles functionalized with amine groups on the plasmonic sensor surface and the cell wall of the *E. Coli* (Figure 10.6). The characterization of the fabricated SPR nanosensor

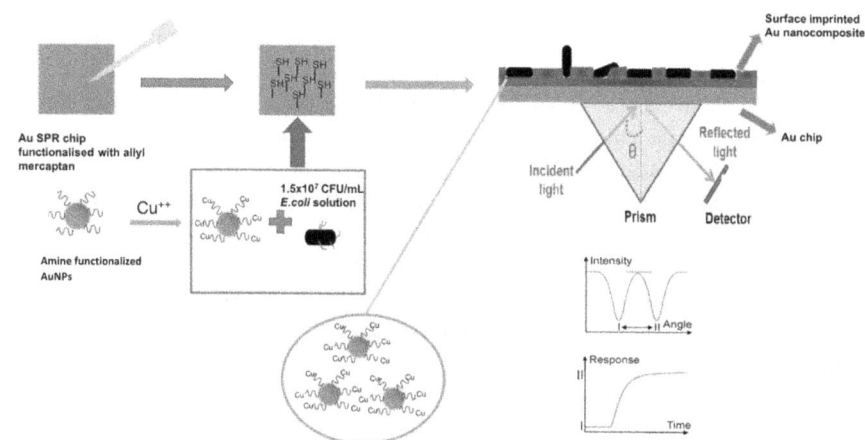

FIGURE 10.6   Representation of the fabrication of MIP-based plasmonic sensor platform and real-time detection of *E. Coli* (Gür et al. 2019).

was effectively carried out via scanning electron microscopy (SEM) contact angle measurements and AFM.

The real-time recognition of the target bacteria *E. Coli* was efficiently investigated through suspensions of *E. Coli* in the concentration range between 1×10^3 and 0.5×10^1 CFU mL^{-1}. In this reported research, the successful integration of the molecular imprinting approach with the great signal-enhancing features of the AU nanoparticles enabled the sensitive recognition of *E. Coli*. The achieved detection limit was 1 CFU mL^{-1}. The selectivity of the sensor platform toward *E. Coli* was also conducted. For this aim, *Pseudomonas aeruginosa, Klebsiella pneumoniae,* and *Staphylococcus aureus* were chosen. *Pseudomonas aeruginosa* and *Klebsiella pneumoniae* were preferred, considering their similar size, shape, and cell wall structure, while *Staphylococcus aureus* was preferred because of the different structure of its cell wall, size, and shape compared with the target bacteria *E. Coli*. The fabricated sensor platform displayed great selectivity for *E. Coli*.

In another interesting work, the same researchers constructed a novel MIP-based plasmonic nanosensor toward *E. Coli* (Özgür et al. 2020). In this work, a micro-contact imprinting approach was employed for the fabrication of a plasmonic nanosensor. For this purpose, N-methacryloyl-(L)-histidine methyl ester (MAH) was chosen as the functional monomer to design and develop the tailor-made recognition sites of the imprinted polymeric layer on the nanosensor for *E. coli*. In addition, Ag nanoparticles were embedded into the polymeric nanostructure to decrease the limit of detection value. The characterization of the developed plasmonic nanosensor was efficiently performed by SEM, AFM contact angle measurements, and ellipsometer. In this study, the detection limit was found as 0.57 CFU mL^{-1}.

In a crucial piece of research performed by the same group (Çimen et al. 2021), sensitive recognition of endotoxin was successfully carried out by the developed MIP-based plasmonic nanosensor. In the reported research, the endotoxin imprinted SPR sensor was designed and prepared. For this purpose, the modification of the sensor surface was carried out using allyl mercaptan. Then, MAH, 2-hydroxyethylmethacrylate (HEMA), and ethylene glycol dimethacrylate (EGDMA) were used as functional monomers and cross-linkers for the preparation of the imprinted layer on the sensor surface toward the target endotoxin. The developed MIP-based plasmonic nanosensor showed great sensing behavior for the target endotoxin. The achieved quantification and detection limit values were 0.078 and 0.023 ng mL^{-1}, respectively. In terms of the response time of the developed plasmonic nanosensor, binding and elution processes were completed within 14 minutes. Furthermore, the selectivity experiments were also conducted using hemoglobin and cholesterol as competitive biomolecules. The obtained selectivity results indicated that the endotoxin imprinted nanosensor exhibited high binding capacity toward endotoxin compared to the competitive hemoglobin and cholesterol biomolecules.

İdil et al. fabricated a new molecular imprinting-based plasmonic nanosensor for the sensitive recognition of *Staphylococcus aureus* which is another multi-drug-resistant pathogenic microorganism (İdil et al. 2021). In their research, firstly, glass slides were functionalized with amino groups. In the next step, *Staphylococcus aureus* was immobilized on the surface of the functionalized glass slides. Then, the modification of the gold surface of the SPR chips was conducted with allyl mercaptan. The microcontact imprinting approach was then employed for the preparation of the SPR sensor toward the target microorganism *Staphylococcus aureus.* The researchers used MAH, HEMA, and EGDMA as functional monomers and cross-linker, respectively. The sensing performance of the developed plasmonic nanosensor was evaluated in the concentration range between 1.0×10^2 and 2.0×10^5 CFU mL^{-1}. The developed plasmonic nanosensor showed excellent sensing behavior for *Staphylococcus aureus.* The detection limit was found as 1.5×10^3 CFU mL^{-1}. The selectivity of the fabricated MIP-based plasmonic nanosensor toward *Staphylococcus aureus* was also investigated using *Escherichia coli, Bacillus subtilis,* and *Salmonella paratyphi.* The data obtained from these investigations displayed that the developed plasmonic nanosensor could selectively detect *Staphylococcus aureus.* Furthermore, real sample experiments were also performed using milk samples. It has been confirmed that the developed plasmonic nanosensor can be efficiently applied for real samples.

Another molecular imprinting-based plasmonic nanosensor toward *Staphylococcus aureus* infection was constructed by Andersson and co-workers (Andersson et al. 2020). In this study, the aim was to develop a novel sensing approach which could efficiently, rapidly, selectively, and sensitively determine patients who suffered from *Staphylococcus aureus* infection, of which α-hemolysin is a main indicator. Thus, an α-hemolysin imprinted SPR nanosensor was designed and prepared using HEMA as the functional monomer. The results confirmed that the plasmonic nanosensor showed high affinity and sensitivity toward the target protein α-hemolysin with a low detection limit (0.022 μM).

Perçin and colleagues reported the fabrication of a plasmonic nanosensor based on the microcontact imprinting approach for the effective sensing of *Salmonella paratyphi*, which is one of the main globally distributed pathogenic microorganisms and causes food-borne diseases (Perçin et al. 2017). In their study, the researchers modified glass slides with 3-aminopropyl) triethoxysilane (APTES) in the first step. In the next step, the functionalization of SPR chips was carried out by allyl mercaptan. The target microorganism *Salmonella paratyphi* was then imprinted on the surface of the functionalized chip through the microcontact imprinting approach. MAH was chosen as the functional monomer for the formation of the imprinted layer on the chip surface. The microcontact imprinting-based plasmonic nanosensor was utilized for the efficient detection of *Salmonella paratyphi.* The obtained detection limit was 1.4×10^6 CFU mL^{-1}.

More recently, a novel coronavirus (COVID-19) was identified in late December 2019 in Wuhan, China (Wang et al. 2020; Huang et al. 2020). COVID-19 represents a new strain which had not been previously identified in humans. 89% and 82% of the nucleotide identity of this new virus is similar to severe acute respiratory syndrome (SARS)-like CoVZXC21 and human SARS-CoV, respectively (Chan et al. 2020). Thus, this new coronavirus is currently named as SARS-CoV-2. On March 11, 2020, the World Health Organization (WHO) declared SARS-CoV-2 as a pandemic (Cucinotta and Vanelli 2020). Scientists all over the world put great effort into the design and development of new sensing platforms which are quite rapid, accurate, reliable, and sensitive toward SARS-CoV-2 and also extensively available, especially since virus symptoms vary broadly and may not be observed for some days after infection. Recently, various sensing approaches were designed and investigated. In a research study carried out by Bajaj et al. (Bajaj et al. 2022), a sensitive plasmonic nanosensor based on the molecular imprinting approach was fabricated for the effective sensing of SARS-CoV-2. In the reported research, molecularly imprinted nanoparticles (MINPs) were prepared using derivatized glass beads. After derivatization of the glass beads with 3-aminopropyl) triethoxysilane (APTMS), the conjugation of SARS-CoV-2 on the surface of the glass beads with silane monolayers was performed. In the next stage, polymerization on the SARS-CoV-2 immobilized glass bead was carried out. After the elution process at 70°C, MINPs with high affinity toward the target virus SARS-CoV-2 were obtained. The characterization of the obtained MINPs was then conducted via atomic force microscopy (AFM) and dynamic light scattering. The prepared MINPs were then incorporated into the SPR chip for the sensitive detection of SARS-CoV-2. The detection process of SARS-CoV-2 was evaluated in the concentration range from 10^4 to 10^6 PFU mL^{-1}. The developed molecular imprinting-based plasmonic nanosensor exhibited great binding affinity for SARS-CoV-2. The achieved K_D value was 0.12 pM.

Cennamo et al. developed a proof of concept based on the integration of the molecular imprinting technique into a plasmonic optical fiber nanosensor for the rapid and highly sensitive on-site recognition of SARS-CoV-2 (Cennamo et al. 2021). In their study, the researchers firstly investigated the sensing performance of the fabricated plasmonic nanosensor toward the subunit 1 of the SARS-CoV-2 spike protein. Then, the obtained data from preliminary experiments, which were performed on SARS-CoV-2 virions were performed on the samples provided from nasopharyngeal swab samples and a physiological solution composed of 0.9% sodium chloride, and these were compared with the data achieved by employing RT-PCR. The results indicated that the constructed plasmonic optical fiber nanosensor exhibited higher sensitivity for SARS-CoV-2 as well as a relatively rapid response time (ca. 10 minutes).

In another interesting work reported by Bognar and colleagues (Bognár et al. 2022), the fabrication of an MIP-based plasmonic sensor platform was carried out for the effective recognition of the spike protein of SARS-CoV-2. In this work, microspotting of a characteristic nonapeptide GFNCYFPLQ of the spike protein of SARS-CoV-2 and other control peptides on the surface of gold SPR chips was effectively performed. In the next step, electropolymerization of scopoletin on the chip surface was carried out. The preparation of the MIP-based plasmonic sensor platform and recognition of the spike protein of SARS-CoV-2 is given in Figure 10.7. The developed MIP-based plasmonic sensor platform was successfully utilized for the effective and selective sensing of the target spike protein.

The recent published research on the construction of novel molecular imprinting-based plasmonic nanosensors toward chemical and biological threats are shown in Table 10.1.

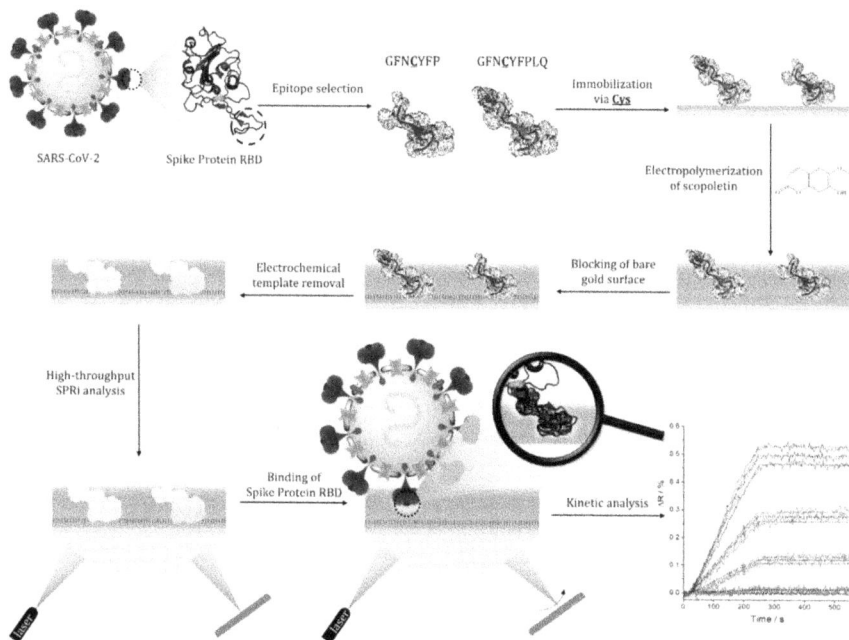

FIGURE 10.7 Representation of the preparation process of MIP-based plasmonic sensor platform and recognition of spike protein of SARS-CoV-2 (Bognár et al. 2022).

10.5 CONCLUSIONS AND FUTURE PERSPECTIVES

This chapter presents and highlights recent progress in the design and construction of MIP-based plasmonic nanosensor platforms toward chemical and biological threats. One of the major challenges of sensing processes is the limited selectivity and sensitivity of the sensor toward the analyte of interest. Therefore, the simultaneous improvement of these features is a crucial issue. The integration of MIPs into the plasmonic sensors is a promising approach for the detection of chemical threats (explosives, mycotoxins, PAHs, and pesticides, etc.) and biological threats, such as viruses and bacteria. Future perspectives can be summerized as follows:

Improved sensitivity of plasmonic nanosensor platforms: Great efforts and ongoing progress in nanofabrication techniques will enable the creation of plasmonic structures with even higher sensitivity. This will open new windows for the sensitive detection of a broad spectrum of target compounds at ultra-low concentrations.

Improved selectivity of plasmonic nanosensor platforms and detection of multiple compounds: The design and construction of novel MIP-based plasmonic sensors with the ability to simultaneously detect multiple target compounds with high selectivity are crucial. This can be achieved though

TABLE 10.1

Recently Published Research on the Construction of Novel Molecular Imprinting-Based Plasmonic Nanosensors toward Chemical and Biological Threats

Chemical Threats

Reference	MIP-Based Plasmonic Nanosensor Type	Target	Sample	Linear Concentration Range	Detection Limit
Çapar et al. 2023	SPR	Zearalenone	Rice grain	1–10 ng L^{-1}	0.33 ng L^{-1}
Çimen et al. 2021	SPR	Patulin	Apple juice	0.5–750 nmol	0.011 nmol
Bao et al. 2012	SPR	TNT	Acetonitrile	1×10^{-8} – 1×10^{-5} M	1×10^{-8} M
Holthoff et al. 2011	SERS	TNT	Acetonitrile	Not stated	3 μM
Zhao et al. 2019	SERS	Atrazine	Apple juice	0.005 –1.0 mg L^{-1}	0.0012 mg L^{-1}
Biological Threats					
Uzun et al. 2009	SPR	Hepatitis B surface antibody	Human serum	Not stated	208.22 mIU mL^{-1}
Bhalla et al. 2022	SPR	Receptor binding domain of the SARS CoV-2 spike protein of Alpha, Beta and Gamma variants of SARS CoV-2	Blood and nasal swabs	The shifts in wavelength for Alpha and Gamma variants were achieved for logistic response, while it is linear for Beta variant (100 fM–100 nM)	9.71 fM, 7.32 fM and 8.81 pM for Alpha, Beta and Gamma variants, respectively
Altuntaş et al. 2015	SPR	Adenovirus	Phosphate buffered saline-tween	0.01–20 pM	0.02 pM
Türkmen et al. 2022	SPR	Pseudomonas spp.	Chicken tenderloin	1×10^{2} - 1×10^{4} CFU mL^{-1}	0.5×10^{2} CFU mL^{-1}
Erdem et al. 2019	SPR	Enterococcus faecalis	Seawater	2×10^{4}–1×10^{8} CFU mL^{-1}	1.05×10^{2} CFU mL^{-1}

the integration of different plasmonic materials and preparation of more selective and sensitive MIPs.

Biomedical applications of plasmonic nanosensor platforms: Plasmonic nanosensors have potential in the field of healthcare. Future research may focus on the design and creation of more compact, portable, and point-of-care platforms for the rapid sensing of biological threats such as bacteria and viruses.

Environmental monitoring using plasmonic nanosensor platforms: With rapidly increasing environmental concerns, plasmonic sensors can play a pivotal role in the careful monitoring of environmental pollutants in soil, water, and air samples. The fabrication of robust, reusable, environmentally friendly, and field-deployable plasmonic nanosensor platforms will be a priority.

Integration of plasmonic nanosensor platforms with the Internet of Things (IoT): Efficient integration of plasmonic nanosensor platforms with the IoT will enable the monitoring and evaluation of the achieved real-time results and remote sensing applications.

In conclusion, MIP-based plasmonic nanosensor platforms represent a cutting-edge technology with immense potential. Their excellent ability to sensitively and selectively detect target compounds at the nanoscale has already led to significant progress in various fields.

REFERENCES

Agrawal, H., Shrivastav, A.M., and Gupta, B.D., 2016. Surface plasmon resonance based optical fiber sensor for atrazine detection using molecular imprinting technique. *Sensors and Actuators B: Chemical*, 227, 204–211.

Akgönüllü, S., Armutçu, C., and Deni̇Zli, A., 2021. Molecularly imprinted polymer film based plasmonic sensors for detection of ochratoxin A in dried fig. *Polymer Bulletin*, 79 (6), 4049–4067.

Akgönüllü, S., Bakhshpour, M., Topçu, A.A., Bereli, N., and Deni̇Zli, A., 2023. Molecularly imprinted polymeric carriers for controlled drug release. *In*: Sangita Das, Sabu Thomas and Partha Pratim Das, eds., *Elsevier eBooks*, Elsevier, Amsterdam, 85–103.

Akgönüllü, S. and Deni̇Zli, A., 2023. Plasmonic nanosensors for pharmaceutical and biomedical analysis. *Journal of Pharmaceutical and Biomedical Analysis*, 236, 115671.

Akgönüllü, S., Yavuz, H., and Deni̇Zli, A., 2020. SPR nanosensor based on molecularly imprinted polymer film with gold nanoparticles for sensitive detection of aflatoxin B1. *Talanta*, 219, 121219.

Akgönüllü, S., Yavuz, H., and Deni̇Zli, A., 2021. Development of gold nanoparticles decorated Molecularly Imprinted–Based plasmonic sensor for the detection of aflatoxin M1 in milk samples. *Chemosensors*, 9 (12), 363.

Altıntaş, Z., Pocock, J., Thompson, K.-A., and Tothill, I.E., 2015. Comparative investigations for adenovirus recognition and quantification: Plastic or natural antibodies? *Biosensors and Bioelectronics*, 74, 996–1004.

Andersson, T., Bläckberg, A., Lood, R., and Bergdahl, G.E., 2020. Development of a molecular imprinting-based surface plasmon resonance biosensor for rapid and sensitive detection of staphylococcus aureus alpha Hemolysin from human serum. *Frontiers in Cellular and Infection Microbiology*, 10. 1-8

Atar, N., Eren, T., and Yola, M.L., 2015. A molecular imprinted SPR biosensor for sensitive determination of citrinin in red yeast rice. *Food Chemistry*, 184, 7–11.

Ayerdurai, V., Cieplak, M., and Kutner, W., 2023. Molecularly imprinted polymer-based electrochemical sensors for food contaminants determination. *Trends in Analytical Chemistry*, 158, 116830.

Ayivi, R.D., Obare, S.O., and Wei, J., 2023. Molecularly imprinted polymers as chemosensors for organophosphate pesticide detection and environmental applications. *Trends in Analytical Chemistry*, 167, 117231.

Bajaj, A., Trimpert, J., Abdulhalim, I., and Altıntaş, Z., 2022. Synthesis of molecularly imprinted polymer nanoparticles for SARS-COV-2 virus detection using surface plasmon resonance. *Chemosensors*, 10 (11), 459.

Bakhtiar, R., 2012. Surface plasmon resonance spectroscopy: A versatile technique in a biochemist's toolbox. *Journal of Chemical Education*, 90 (2), 203–209.

Banan, K., Ghorbani-Bidkorbeh, F., Afsharara, H., Hatamabadi, D., Landi, B., Keçili, R., and Sellergren, B., 2022. Nano-sized magnetic core-shell and bulk molecularly imprinted polymers for selective extraction of amiodarone from human plasma. *Analytica Chimica Acta*, 1198, 339548.

Banan, K., Hatamabadi, D., Afsharara, H., Mostafiz, B., Sadeghi, H., Rashidi, S., Beirami, A.D., Shahbazi, M., Keçili, R., Hussain, C.M., and Ghorbani-Bidkorbeh, F., 2022. MIP-based extraction techniques for the determination of antibiotic residues in edible meat samples: Design, performance & recent developments. *Trends in Food Science and Technology*, 119, 164–178.

Bao, H., Wei, T., Li, X., Zhao, Z., He, C., and Zhang, P., 2012. Detection of TNT by a molecularly imprinted polymer film-based surface plasmon resonance sensor. *Chinese Science Bulletin*, 57 (17), 2102–2105.

Bhalla, N., Payam, A.F., Morelli, A., Sharma, P.K., Johnson, R., Thomson, A., Jolly, P., and Canfarotta, F., 2022. Nanoplasmonic biosensor for rapid detection of multiple viral variants in human serum. *Sensors and Actuators B: Chemical*, 365, 131906.

Bognár, Z., Supala, E., Yarman, A., Zhang, X., Bier, F.F., Scheller, F.W., and Gyurcsányi, R.E., 2022. Peptide epitope-imprinted polymer microarrays for selective protein recognition. Application for SARS-CoV-2 RBD protein. *Chemical Science*, 13 (5), 1263–1269.

Bollella, P., 2022. Enzyme-based amperometric biosensors: 60 years later … Quo Vadis? *Analytica Chimica Acta*, 1234, 340517.

Brahmkhatri, V., Pandit, P., Rananaware, P., D'Souza, A., and Kurkuri, M.D., 2021. Recent progress in detection of chemical and biological toxins in Water using plasmonic nanosensors. *Trends in Environmental Analytical Chemistry*, 30, e00117.

Çapar, N., Yola, B.B., Polat, İ., Bekerecioğlu, S., Atar, N., and Yola, M.L., 2023. A zearalenone detection based on molecularly imprinted surface plasmon resonance sensor including sulfur-doped g-C3N4/Bi2S3 nanocomposite. *Microchemical Journal*, 193, 109141.

Castro-Grijalba, A., Montes-García, V., Cordero-Ferradás, M.J., Coronado, E.A., Pérez-Juste, J., and Pastoriza-Santos, I., 2020. SERS-Based molecularly imprinted plasmonic sensor for highly sensitive PAH detection. *ACS Sensors*, 5 (3), 693–702.

Cennamo, N., D'Agostino, G., Perri, C., Arcadio, F., Chiaretti, G., Parisio, E.M., Camarlinghi, G., Vettori, C., Di Marzo, F., Cennamo, R., Porto, G., and Zeni, L., 2021. Proof of concept for a quick and highly sensitive on-site detection of SARS-COV-2 by plasmonic optical fibers and molecularly imprinted polymers. *Sensors*, 21 (5), 1681.

Chan, J.F., Kok, K.H., Zhu, Z., Chu, H., To, K.K.-W., Yuan, S., and Yuen, K.Y., 2020. Genomic characterization of the 2019 novel human-pathogenic coronavirus isolated from a patient with atypical pneumonia after visiting Wuhan. *Emerging Microbes & Infections*, 9 (1), 221–236.

Choi, S., Chang, H., Lee, N., Kim, J.-H., and Chun, H.S., 2009. Detection of mycoestrogen zearalenone by a molecularly imprinted Polypyrrole-Based Surface plasmon resonance (SPR) sensor. *Journal of Agricultural and Food Chemistry*, 57 (4), 1113–1118.

Çimen, D., Aslıyüce, S., Tanalp, T.D., and Deniẑli, A., 2021. Molecularly imprinted nanofilms for endotoxin detection using an surface plasmon resonance sensor. *Analytical Biochemistry*, 632, 114221.

Çimen, D., Bereli, N., and Deniẑli, A., 2021. Patulin imprinted nanoparticles decorated surface plasmon resonance chips for patulin detection. *Photonic Sensors*, 12 (2), 117–129.

Cucinotta, D. and Vanelli, M., 2020. WHO declares COVID-19 a pandemic. *Acta Biomedica: Atenei Parmensis*, 91 (1), 157–160.

Della Giustina, G., Sonato, A., Gazzola, E., Ruffato, G., Brusa, S., and Romanato, F., 2016. SPR Enhanced molecular imprinted sol–gel film: A promising tool for gas-phase TNT detection. *Materials Letters*, 162, 44–47.

Erdem, Ö., Saylan, Y., Cihangir, N., and Deniẑli, A., 2019. Molecularly imprinted nanoparticles based plasmonic sensors for real-time Enterococcus faecalis detection. *Biosensors and Bioelectronics*, 126, 608–614.

Eskandari, V., Kordzadeh, A., Zeinalizad, L., Sahbafar, H., Aghanouri, H., Hadi, A., and Ghaderi, S., 2022. Detection of molecular vibrations of atrazine by accumulation of silver nanoparticles on flexible glass fiber as a surface-enhanced Raman plasmonic nanosensor. *Optical Materials*, 128, 112310.

Feraud, M., O'Brien, J., Samanipour, S., Dewapriya, P., Van Herwerden, D., Kaserzon, S., Wood, I., Rauert, C., and Thomas, K.V., 2023. InSpectra – A platform for identifying emerging chemical threats. *Journal of Hazardous Materials*, 455, 131486.

Gu, Y., Yan, X., Liu, C., Zheng, B., Li, Y., Liu, W., Zhang, Z., and Yang, M., 2016. Biomimetic sensor based on molecularly imprinted polymer with nitroreductase-like activity for metronidazole detection. *Biosensors and Bioelectronics*, 77, 393–399.

Guo, L., Jackman, J.A., Yang, H., Chen, P., Cho, N., and Kim, D.H., 2015. Strategies for enhancing the sensitivity of plasmonic nanosensors. *Nano Today*, 10 (2), 213–239.

Guo, X., Li, J., Arabi, M., Wang, X., Wang, Y., and Chen, L., 2020. Molecular-imprinting-based surface-enhanced Raman scattering sensors. *ACS Sensors*, 5 (3), 601–619.

Gupta, Y., Beckett, L.E., Sadula, S., Vargheese, V., Korley, L.T.J., and Vlachos, D.G., 2023. Bio-based molecular imprinted polymers for separation and purification of chlorogenic acid extracted from food waste. *Separation and Purification Technology*, 327, 124857.

Gür, S.D., Bakhshpour, M., and Deniẑli, A., 2019. Selective detection of Escherichia coli caused UTIs with surface imprinted plasmonic nanoscale sensor. *Materials Science and Engineering: C*, 104, 109869.

Herrera-Chacón, A., Cetó, X., and Del Valle, M., 2021. Molecularly imprinted polymers – Towards electrochemical sensors and electronic tongues. *Analytical and Bioanalytical Chemistry*, 413 (24), 6117–6140.

Holthoff, E.L., Stratis-Cullum, D.N., and Hankus, M.E., 2011. A nanosensor for TNT detection based on molecularly imprinted polymers and surface enhanced Raman scattering. *Sensors*, 11 (3), 2700–2714.

Hu, Y., Jiang, Z., Hou, A., Wang, X., Zhou, Z., Qin, B., Cao, B., and Zhang, Y., 2023. Impact of atrazine on soil microbial properties: A meta-analysis. *Environmental Pollution*, 323, 121337.

Huang, C., Wang, Y., Li, X., Ren, L., Zhao, J., Hu, Y., Zhang, L., Fan, G., Xu, J., Gu, X., Cheng, Z., Takata, Y., Xia, J., Yuan, W., Wu, W., Xie, X., Yin, W., Li, H., Liu, M., Yan, X., Gao, H., Guo, L., Xie, J., Wang, G., Jiang, R., Gao, Z., Jin, Q., Wang, J., and Cao, B., 2020. Clinical features of patients infected with 2019 novel coronavirus in Wuhan, China. *The Lancet*, 395 (10223), 497–506.

İdil, N., Bakhshpour, M., Perçin, I., and Mattìasson, B., 2021. Whole cell recognition of Staphylococcus aureus using biomimetic SPR sensors. *Biosensors*, 11 (5), 140.

Inutsuka, T., Okamoto, M., and Yoshimi, Y., 2019. New approach for neuropharmacology profile: In-situ real-time neuropharmacology monitoring by imaging technique using the molecularly imprinted polymers (MIPs) probe. *Journal of Pharmacological and Toxicological Methods*, 99, 106595.

Jin, C., Wu, Z., Molinski, J., Zhou, J., Ren, Y., and Zhang, J.X.J., 2022. Plasmonic nanosensors for point-of-care biomarker detection. *Materials Today Bio*, 14, 100263.

Justino, C.I.L., Freitas, A.C., Pereira, R., Duarte, A.C., and Rocha-Santos, T., 2015. Recent developments in recognition elements for chemical sensors and biosensors. *Trends in Analytical Chemistry*, 68, 2–17.

Kakkar, V. and Narula, P., 2022. Role of molecularly imprinted hydrogels in drug delivery – A current perspective. *International Journal of Pharmaceutics*, 625, 121883.

Keçili, R. and DenìZli, A., 2021. Molecular imprinting-based smart nanosensors for pharmaceutical applications. *In: Elsevier eBooks*, Elsevier, Amsterdam 19–43.

Keçili, R. and Hussain, C.M., 2018. Recent progress of imprinted nanomaterials in analytical chemistry. *International Journal of Analytical Chemistry*, 2018, 1–18.

Keçili, R., Özcan, A., Ersöz, A., Hür, D., DenìZli, A., and Say, R., 2010. Superparamagnetic nanotraps containing MIP based mimic lipase for biotransformations uses. *Journal of Nanoparticle Research*, 13 (5), 2073–2079.

Keçili, R., Say, R., Ersöz, A., Hür, D., and DenìZli, A., 2012. Investigation of synthetic lipase and its use in transesterification reactions. *Polymer*, 53 (10), 1981–1984.

Keçili, R., Ünlüer, Ö.B., Ersöz, A., and Say, R., 2023. Molecularly imprinted polymers (MIPs) for biomedical applications. *In: Elsevier eBooks*, Kunal Pal, Sarika Verma, Pallab Datta, Ananya Barui, S.A.R. Hashmi, Avanish Kumar Srivastava, Elsevier, Amsterdam 745–768.

Kim, S., Lee, D., Ahn, K., Kim, J., and Kim, J., 2023. Damage-free remote SF6 plasma-treated CNTs for facile fabrication of electrochemical enzyme biosensors. *Applied Surface Science*, 628, 157386.

Kirsch, N., Hedin-Dahlström, J., Henschel, H., Whitcombe, M.J., Wikman, S., and Nicholls, I.A., 2009. Molecularly imprinted polymer catalysis of a Diels-Alder reaction. *Journal of Molecular Catalysis B-enzymatic*, 58 (1–4), 110–117.

Kupai, J., Razali, M., Büyüktıryakı, S., Keçili, R., and Székely, G., 2017. Long-term stability and reusability of molecularly imprinted polymers. *Polymer Chemistry*, 8 (4), 666–673.

Lazarević-Pašti, T., Tasić, T., Milanković, V., and Potkonjak, N.I., 2023. Molecularly imprinted plasmonic-based sensors for environmental contaminants—Current state and future perspectives. *Chemosensors*, 11 (1), 35.

Li, F., Lian, Z., Song, C., and Ge, C., 2021. Release of florfenicol in seawater using chitosan-based molecularly imprinted microspheres as drug carriers. *Marine Pollution Bulletin*, 173, 113068.

Li, H., Sheng, W., Haruna, S.A., Bei, Q., Wei, W., Hassan, Md.M., and Chen, Q., 2023. Recent progress in photoelectrochemical sensors to quantify pesticides in foods: Theory, photoactive substrate selection, recognition elements and applications. *Trends in Analytical Chemistry*, 164, 117108.

Li, M., Fink-Gremmels, J., Li, D., Tong, X., Tang, J., Nan, X., Yu, Z., Chen, W., and Wang, G., 2021. An overview of aflatoxin B1 biotransformation and aflatoxin M1 secretion in lactating dairy cows. *Animal Nutrition*, 7 (1), 42–48.

Li, Z., Deng, L., Kinloch, I.A., and Young, R.J., 2023. Raman spectroscopy of carbon materials and their composites: Graphene, nanotubes and fibres. *Progress in Materials Science*, 135, 101089.

Liu, C., Xü, D., Xuan, D., and Huang, Q., 2022. A review: Research progress of SERS-based sensors for agricultural applications. *Trends in Food Science and Technology*, 128, 90–101.

Liu, Y., Tan, G., Feng, S., Zhang, B., Liu, T., Wang, Z., Bi, Y., Yang, Q., Ren, H., Lv, L., Liu, W., Xia, A., and Zhao, Q., 2023. Localized surface plasmon resonance effect of V4c3-Mxene for enhancing photothermal reduction of CO_2 in full solar spectrum. *Separation and Purification Technology*. 326, 124726.

Liu, Y., Wang, F., Liu, Y., Cao, L., Hu, H.M., Yao, X., Zheng, J., and Liu, H., 2023. A label-free plasmonic nanosensor driven by horseradish peroxidase-assisted tetramethylbenzidine redox catalysis for colorimetric sensing H2O2 and cholesterol. *Sensors and Actuators B-Chemical*, 389, 133893.

Mabrouk, M.M., Hammad, S.F., Abdella, A.A., and Mansour, F.R., 2023. Tipps and tricks for successful preparation of molecularly imprinted polymers for analytical applications: A critical review. *Microchemical Journal*, 193, 109152.

Manzo, S. and Schiavo, S., 2022. Physical and chemical threats posed by micro(nano)plastic to sea urchins. *Science of the Total Environment*, 808, 152105.

Marchese, S., Polo, A., Ariano, A., Velotto, S., Costantini, S., and Severino, L., 2018. Aflatoxin B1 and M1: Biological properties and their involvement in cancer development. *Toxins*, 10 (6), 214.

Moreno-Bondi, M.C., Benito-Peña, E., Moya-Cavas, T., and Ruíz, J.M., 2023. Molecularly imprinted polymer-based biomimetic sensors for food analysis. *In: Elsevier eBooks*, 568–598.

Muratsugu, S., Shirai, S., and Tada, M., 2020. Recent progress in molecularly imprinted approach for catalysis. *Tetrahedron Letters*, 61 (11), 151603.

Nerantzaki, M., Michel, A., Petit, L., Garnier, M., Ménager, C., and Griffete, N., 2022. Biotinylated magnetic molecularly imprinted polymer nanoparticles for cancer cell targeting and controlled drug delivery. *Chemical Communications*, 58 (37), 5642–5645.

Özçelikay, G., Karadas-Bakirhan, N., Tok, T.T., and Özkan, S.A., 2020. A selective and molecular imaging approach for anticancer drug: Pemetrexed by nanoparticle accelerated molecularly imprinting polymer. *Electrochimica Acta*, 354, 136665.

Özgür, E., Topçu, A.A., Yılmaz, E., and Denìzli, A., 2020. Surface plasmon resonance based biomimetic sensor for urinary tract infections. *Talanta*, 212, 120778.

Özkan, A., Atar, N., and Yola, M.L., 2019. Enhanced surface plasmon resonance (SPR) signals based on immobilization of core-shell nanoparticles incorporated boron nitride nanosheets: Development of molecularly imprinted SPR nanosensor for anticancer drug, etoposide. *Biosensors and Bioelectronics*, 130, 293–298.

Panagiotopoulou, M., Kunath, S., Medina-Rangel, P.X., Haupt, K., and Bui, B.T.S., 2017. Fluorescent molecularly imprinted polymers as plastic antibodies for selective labeling and imaging of hyaluronan and sialic acid on fixed and living cells. *Biosensors and Bioelectronics*, 88, 85–93.

Perçin, I., İdil, N., Bakhshpour, M., Yılmaz, E., Mattìasson, B., and Denìzli, A., 2017. Microcontact imprinted plasmonic nanosensors: Powerful tools in the detection of salmonella paratyphi. *Sensors*, 17 (6), 1375.

Perevedentseva, E., Melnik, N.V., Muronets, E.M., Averyushkin, A.S., Karmenyan, A., and Elanskaya, I.V., 2024. Raman spectroscopy with near IR excitation for study of structural components of cyanobacterial phycobilisomes. *Journal of Luminescence*, 265, 120224.

Rashid, K.S., Hassan, Md.F., Yaseer, A.A., Tathfif, I., and Sagor, R.H., 2021. Gas-sensing and label-free detection of biomaterials employing multiple rings structured plasmonic nanosensor. *Sensing and Bio-Sensing Research*, 33, 100440.

Ren, X.-H., Wang, H., Li, S., He, X., Li, W., and Zhang, Y., 2021. Preparation of glycan-oriented imprinted polymer coating Gd-doped silicon nanoparticles for targeting cancer Tn antigens and dual-modal cell imaging via boronate-affinity surface imprinting. *Talanta*, 223, 121706.

Restaino, S.M. and White, I.M., 2019. A critical review of flexible and porous SERS sensors for analytical chemistry at the point-of-sample. *Analytica Chimica Acta*, 1060, 17–29.

Santos, A.C., Brandão, A.P., Hryniewicz, B.M., De Abreu, H.F.G., Bach-Toledo, L., Silva, S., Deller, A.E., Rogerio, V.Z., Rodrigues, D.S.B., Hiraiwa, P.M., Guimarães, B.G., Marchesi, L.F., De Oliveira, J.C., Gradia, D.F., Soares, F.L.F., Zanchin, N.I.T., De Oliveira, C.C., and Vidotti, M., 2023. COVID-19 impedimetric biosensor based on polypyrrole nanotubes, nickel hydroxide and VHH antibody fragment: Specific, sensitive, and rapid viral detection in saliva samples. *Materials Today Chemistry*, 30, 101597.

Say, R., Keçili, R., Denizli, A., and Ersöz, A., 2016. Biomimetic imprinted polymers: Theory, Design Methods, and Catalytic Applications . *In: Elsevier eBooks*, Songjun Li, Shunsheng Cao, A Sergey A. Piletsky and Anthony P.F. Turner, Elsevier, Amsterdam. 103–120.

Saylan, Y., Akgönüllü, S., and Denizli, A., 2020. Plasmonic sensors for monitoring biological and chemical threat agents. *Biosensors*, 10 (10), 142.

Sha, K.C., Shah, M.B., Solanki, S.J., Makwana, V.D., Sureja, D.K., Anuradha, G., Bodiwala, K.B., and Dhameliya, T.M., 2023. Recent advancements and applications of Raman spectroscopy in pharmaceutical analysis. *Journal of Molecular Structure*, 1278, 134914.

Shalileh, F., Sabahi, H., Golbashy, M., Dadmehr, M., and Hosseini, M., 2023. Recent developments in DNA nanostructure-based biosensors for the detection of melamine adulteration in milk. *Microchemical Journal*, 195, 109316.

Sharifi, M., Khalilzadeh, B., Bayat, F., Işıldak, İ., and Tajali, H., 2023. Application of thermal annealing-assisted gold nanoparticles for ultrasensitive diagnosis of pancreatic cancer using localized surface plasmon resonance. *Microchemical Journal*, 190, 108698.

Tarannum, N., Khatoon, S., Yadav, A., and Yadav, A.K., 2023. SERS-based molecularly imprinted polymer sensor for highly sensitive norfloxacin detection. *Journal of Food Composition and Analysis*, 119, 105281.

Türkmen, D., Yilmaz, T., Bakhshpour, M., and Denizli, A., 2022. An alternative approach for bacterial growth control: Pseudomonas spp. imprinted polymer-based surface plasmon resonance sensor. *IEEE Sensors Journal*, 22 (4), 3001–3008.

Upadhyayula, V.K.K., 2012. Functionalized gold nanoparticle supported sensory mechanisms applied in detection of chemical and biological threat agents: A review. *Analytica Chimica Acta*, 715, 1–18.

Uzun, L., Say, R., Ünal, S., and Denizli, A., 2009. Production of surface plasmon resonance based assay kit for hepatitis diagnosis. *Biosensors and Bioelectronics*, 24 (9), 2878–2884.

Van Dorst, B., Mehta, J., Bekaert, K., Rouah-Martin, E., De Coen, W., Dubruel, P., Blust, R., and Robbens, J., 2010. Recent advances in recognition elements of food and environmental biosensors: A review. *Biosensors and Bioelectronics*, 26 (4), 1178–1194.

Vaněčková, T., Bezděková, J., Han, G., Adam, V., and Vaculovičová, M., 2020. Application of molecularly imprinted polymers as artificial receptors for imaging. *Acta Biomaterialia*, 101, 444–458.

Wang, C., Horby, P., Hayden, F.G., and Gao, G.F., 2020. A novel coronavirus outbreak of global health concern. *The Lancet*, 395 (10223), 470–473.

Wang, L., Ahmad, W., Wu, J., Wang, X., Chen, Q., and Ouyang, Q., 2023. Selective detection of carbendazim using a upconversion fluorescence sensor modified by biomimetic molecularly imprinted polymers. *Spectrochimica Acta Part A: Molecular and Biomolecular Spectroscopy*, 284, 121457.

Wang, L. and Zhang, W., 2023. Molecularly imprinted polymer (MIP) based electrochemical sensors and their recent advances in health applications. *Sensors and Actuators Reports*, 5, 100153.

Wang, Y., Tan, Y., Zhao, C., and Hou, Z., 2023. On-chip plasmonic nanosensor based on multiple Fano resonances in rectangular coupled systems. *Optics Communications*, 549, 129915.

Wang, Y., Wang, X., and Li, Q., 2023. Aflatoxin B1 in poultry liver: Toxic mechanism. *Toxicon*, 233, 107262.

Wei, L.-N., Luo, L., Wang, B., Lei, H., Guan, T., Shen, Y.-D., Wang, H., and Xu, Z.-L., 2023. Biosensors for detection of paralytic shellfish toxins: Recognition elements and transduction technologies. *Trends in Food Science and Technology*, 133, 205–218.

Xianyu, Y., Su, S., Hu, J., and Yu, T., 2021. Plasmonic sensing of β-glucuronidase activity via silver mirror reaction on gold nanostars. *Biosensors and Bioelectronics*, 190, 113430.

Xiao, Y., Zhang, T., and Zhang, H., 2023. Recent advances in the peptide-based biosensor designs. *Colloids and Surfaces B: Biointerfaces*, 231, 113559.

Xu, Y., Tan, Y., Majeed, Z., Nie, F., Zheng, K., Li, Z., Yang, L., Zhao, C., and Li, C., 2023. Hybrid molecularly imprinted polymers for targeted separation and enrichment of 10-hydroxycamptothecin in *Camptotheca acuminata* Decne. *Natural Product Research*, 1–10.https://doi.org/10.1080/14786419.2023.2228981

Ye, W., Yu, M., Wang, F., Li, Y., and Wang, C., 2022. Multiplexed detection of heavy metal ions by single plasmonic nanosensors. *Biosensors and Bioelectronics*, 196, 113688.

Yin, M., Lin, H., Zhang, L., Wei, X., Sun, Y., Luo, Y., Yang, H., Deng, C., and Xu, D., 2023. Antibody-assisted MIL-53(Fe)/Pt-based electrochemical biosensor for the detection of the nicotine metabolite cotinine. *Bioelectrochemistry*, 153, 108470.

Zaidi, S.A., 2020. Molecular imprinting: A useful approach for drug delivery. *Materials Science for Energy Technologies*, 3, 72–77. https://doi.org/10.1016/j.mset.2019.10.012

Zhao, B., Feng, S., Hu, Y., Wang, S., and Lu, X., 2019. Rapid determination of atrazine in apple juice using molecularly imprinted polymers coupled with gold nanoparticles-colorimetric/SERS dual chemosensor. *Food Chemistry*, 276, 366–375.

Zhou, P., Li, X., Zhou, J., Li, W., and Shen, L., 2023. A molecular imprinting polymer based on computer-aided design: Selective enrichment and purification of synephrine from the extract waste liquid of Citrus aurantium L. *Reactive & Functional Polymers*, 190, 105647. https://doi.org/10.1016/j.reactfunctpolym.2023.105647

11 The Detection of Chemical and Biological Threat Agents Using Nanomaterial-Based Plasmonic Nanosensors

Süleyman Aşır, Deniz Türkmen, Mamajan Ovezova, Ilgım Göktürk, Gaye Ezgi Yılmaz, and Fatma Yılmaz

11.1 INTRODUCTION

Chemical and biological threat agents are harmful materials that destructively affect people and cause disasters. They are less expensive and easier to obtain than radiological or nuclear weapons. One of the most significant risks to the human population, world peace, and societal stability is chemical warfare agents (CWAs). To ensure the safety and preservation of humanity, their detection and measurement are therefore of the utmost importance. In recent years, significant advances have been made in supramolecular chemistry, analytical chemistry, and molecular sensors, which have restored our ability to detect CWAs (Kumar et al., 2022a).

Threats emerge from various directions as the nature of international conflicts changes. Exposure to biological or chemical warfare agents through toxic contamination can be a form of terrorism or vandalism. The Organization for the Prohibition of Chemical Weapons (OPCW) has confirmed multiple attacks in Syria that used sulfur mustard, sarin, and the toxic dual-use of chlorine. Analytical techniques have been extensively researched for detecting chemical threat agents in matrices such as organic solvents, air samples, soil, wipes, and other solid materials. However, because of these compounds' significant volatility and reactivity, it is frequently impossible to detect intact agents. Persistent biomarkers can be used as proof in certain situations (de Bruin-Hoegée et al., 2023). Many CWAs can bind covalently with proteins, forming long-lasting adducts. For example, it has been shown that histidine protein adducts are formed when humans are exposed to sulfur mustard (Hemme et al., 2021). Similarly, after exposure to nerve agents like sarin and Novichok A-234, phosphorylated serines or tyrosines are generated (John and Thiermann, 2021). The

 DOI: 10.1201/9781003459316-11

main protein markers that indicate the presence of chlorine are mono- and dichlorinated tyrosine adducts (Nishio et al., 2021).

Using animals and plants to spread diseases to gain political, religious, or economic control over individuals and society is bioterrorism. Biological agents such as bacteria, viruses, or toxins applied in bioterrorism also cause significant health problems. Biochemical and immunological-based diagnostics of bioagents, including bacteria, viruses, and toxins, often take a long time (hours to days) with traditional detection methods that require molecular techniques like polymerase chain reaction (PCR) or cellular methods like cell culture. In addition, a specialist must then diagnose the pathogen or toxin. Developing new sensors is inevitable, as biological threat agents are dangerous, and modern detection methods are insufficient for rapid diagnosis. For this purpose, the advantages of nanomaterials have been exploited in recent years.

Materials with individual building pieces smaller than 100 nm are called nanomaterials. The synthesis, assembly, and production of nanomaterials have advanced significantly, and equally important are the prospective uses for these materials in a wide range of technologies. The hazards caused by biological and chemical threat agents and the limited availability of modern methods for detecting these agents form the basis of the need to develop new sensors. In recent years, due to their advantages, nanomaterial-based plasmonic sensors have been widely used in applications to detect chemical and biological threat agents. Nanomaterial-based biosensors, which have benefits such as high surface/volume ratio, dispersibility, favorable physical and chemical characteristics, and distinctive nanoscale interactions, are increasingly providing consequences with higher sensitivity and accuracy with low sample volume, short analysis time, and low analysis cost (Rowland et al., 2016).

A wide range of standoff and point detection systems of fair marketable quality are employed by the armed services, such as ion mobility spectrometry (IMS), flame photometry, gas chromatography (GC), and surface acoustic waves (SAW) (Kumar et al., 2022b). Furthermore, devices depend on techniques such as electrochemical sensors (Apak et al., 2023; Jyoti et al., 2023), molecular imprinted polymeric (MIP) sensors (Karadurmus et al., 2022), biosensors (Diauudin et al., 2019), chemo-resistive sensors (Park et al., 2019), and surface plasmon resonance (SPR) (Saylan et al., 2020) and are in advanced stages of development. Plasmonic sensors are typically made up of sensing components that comprise metal or metal-dielectric nanostructures that create surface plasmons and recognition elements that are selective to the target analyte. When a sample containing the target analyte is presented to the sensor, the target analyte is collected by the recognition component attached to the sensing surface of the unit, which alters the refractive index in the near-surface area (Homola, 2008).

Herein, we provide an overview of the biological and chemical threat agents, and after a brief introduction to the working principles of the plasmonic sensors, the applications of plasmonic sensors used for detecting a broad range of biological and chemical threat agents will be extensively discussed. The final section summarizes concluding remarks about the identification of biological and chemical threat agents using nanomaterial-based plasmonic nanosensors and future perspectives.

11.2 OVERVIEW OF CHEMICAL AND BIOLOGICAL THREAT AGENTS

Nowadays, chemical terrorization severely threatens human safety globally, practically outweighing the effects of using the most advanced firearms. Compared to biological and nuclear threats, chemical warfare (CW) is possibly among humanity's most dreadful Weapons of Mass Destruction (WMD) to ever have evolved. It is also relatively easy to produce chemical weapons that, in small quantities, can cause mass casualties. In this chapter, the properties of various CW agents, available detection equipment, and disadvantages of current detection methods are mentioned, and nanomaterial-based plasmonic nanosensors, one of the latest detection methods today, are discussed. CW agents are synthetic compounds that are highly toxic and can be dispersed as gases, liquids, or aerosols. They can also be adsorbed into particles and used as a powder. The definition of a chemical agent can be described as "a chemical intended for use in military operations to kill, seriously injure, or immobilize humans due to their physiological effects", since such CW agents have mortal or disabling consequences on people (Smart, 1997). Chemical agents can be categorized as nerve agents, vesicants, choking agents, blood agents, riot control agents, psychotomimetic agents, and toxins (Aas, 2003). Also, according to their volatility, they are classified as persistent and non-persistent agents. Chlorine, phosgene, and hydrogen cyanide are non-persistent agents because they are more volatile. However, the less evaporative agents, like sulfur mustard and VX, are lasting agents. According to their chemical structures, they might be categorized as organophosphorus (OP), organosulfur and organofluorine compounds, and arsenic.

Toxic chemicals, for instance, phosgene, sulfur mustard, and lewisite, engendered 100,000 fatalities and 1.2 million injuries during the first World War I (WWI) (Lackner and Burghofer, 2019). The cost-effectiveness of chemical weapons, especially when utilized toward centralized forces or communities, is one of the reasons for their production and use. CW agents like chlorine, phosgene, and cyanide are easy to obtain because they have common usage in several chemical or pharmaceutical industry production procedures. When the targets of terrorism are industrial facilities, stockpiles, or shipping facilities, a terrorist act can appear via the delivery of toxic chemicals. The outcome of the deliberate release of CW agents differs significantly according to various variables, comprising toxicity, volatility, concentration, exposure route, duration, and environmental conditions (Patočka and Fusek, 2004).

Nerve agents that affect the functioning of the nervous system are classified as OP compounds. Gerhard Schrader developed the first noted nerve agent, tabun (GA), by investigating the evolution of OP insecticides in the 1930s. Later, a group of nerve agents called G-agents, including sarin (GB) and soman (GD), were produced (Dogaroiu, 2003). Ultimately, V-agents, a more stabilized choice over G-agents, were created. Sulfur including OP-based VX is more influential, stable, less volatile, and less water-soluble than sarin. It acts through direct skin contact and remains in the surroundings for some weeks after release. Nerve agents, which can prompt death in minutes to hours following exposure, are deadlier than other CW agents, and their mode of action is well documented in the literature (Bajgar et al., 2009).

Blister agents or vesicants causing skin injuries, similar to those from burns, are also toxic compounds. These substances, which affect the upper respiratory tract, cause pulmonary oedema when inhaled. There are two types of vesicants, mustards and arsenic, and the most essential substance in this sort of CW agent is sulfur mustard. Nitrogen mustards (HN-1, HN-2, and HN-3) and lewisite (L_1, L_2, and L_3) are present in the other members (Shakarjian et al., 2009).

Blood agents are cyanide group compounds that interfere with bodily processes by obstructing the body tissue's normal uptake of oxygen. While these agents can impede the generation of blood components, the blood itself is usually unaffected (Bhuvaneswari et al., 2019). This family's two primary CW agents are hydrogen cyanide (HCN) and cyanogen chloride (CNCl).

Choking agents are effective in a person's respiratory tract, namely in the nose, throat, and especially the lungs. In severe cases, the lungs fill with fluids, the membranes inflate, and the lack of oxygen causes death. As a result, these toxins "choke" unprotected people. The most well-known members of this family are chlorine and phosgene, diphosgene, nitric oxide, and perfluoroisobutylene (PFIB) (Zellner and Eyer, 2020).

Riot control agents irritate the eyes, causing them to close and irritate the upper respiratory tract. They are commonly known as irritants, lachrymators, and nuisances, with the general public typically referring to them as tear gases. Many tear-inducing drugs have undergone CW testing (Quiroga-Garza et al., 2023). Lachrymators, sternutators, and vomiting agents are three recognised categories of riot control agents. Only three chemical agents, 2-Chloroacetophenone (CN), 2-Chlorobenzylidene Malononitrile (CS), and Dibenz[b,f][1,4]oxazepine (CR), are of substantial significance among the numerous chemical compounds utilized as riot control agents. Psychotomimetic agents do not cause any significant disturbances of the autonomic nervous system or other severe disabilities but do cause alterations in thought, perception, and mood (Fusek et al., 2020). Thus, this class of agents typically consists of substances that, when taken in low doses (<10 mg), can result in circumstances resembling psychotic diseases or other symptoms of the central nervous system, such as sensory loss, paralysis, hallucinations, etc. Numerous different substances fall under the category of psychedelics. Lysergic acid diethylamide (LSD) is the one of those that is most well-known. Other psychedelics include psilocybin, ibogaine, and harmine, but none has the same strength as LSD.

Toxins are poisonous chemical compounds produced naturally by living things like bacteria, fungi, terrestrial animals, or aquatic animals. Their classification is based on their chemical structure, molecular weight, source, preferred targets in the body, and mechanism of action. As a result, they can be categorized into two groups: Protein toxins, which are made up of lengthy, folded chains of amino acids, and non-protein toxins, which are often tiny compounds with intricate chemical compositions (Anderson, 2012; Yu et al., 2020). Based on how they work, they are classified as cardiotoxins, dermatotoxins, hepatotoxins, neurotoxins, etc. Neurotoxins like botulinum, tetanus toxin, and other toxins, such as staphylococcal enterotoxin, are among the most potent.

German spies infected Romanian cavalry battalions with glanders during the First World War, which prompted many nations to research bacteriological weapons. The issue of using bacteria as weapons appeared in earnest in the early 1930s. In 1933, the article "Bacteriological and Gas War in the Next World War" was published in Shenbao, an influential newspaper, declaring that germ warfare threatened human life. The report, however, also emphasized the drawbacks of employing microbes as weapons, particularly in the delivery of lethal germs, which can be challenging and imprecise.

Bacteria, viruses, and toxins are examples of bioagents, and they don't necessarily need to be spread on purpose to pose a danger to people's health. Assays based on biochemical and immunological detection and molecular techniques, like PCR, or cellular methods, like cell culture, are traditional detection methods. These methods usually require ample time (hours or days) and skilled clinicians who can almost guess which infection or toxin is present (Rowland et al., 2016). A comprehensive explanation of pathogen characteristics is available in the NATO Handbook on the Medical Aspects of NBC Defensive Operations (NBC-nuclear, biological, and chemical) (States et al., 2001).

Bacteria are single-celled organisms that multiply by invading host tissue or causing disease by releasing toxins. The variety of diseases that bacteria cause and the infectious doses they produce indicate the diversity of bacteria. For instance, the case fatality rate of raw inhalation anthrax is greater than 90%, but one of the six species of Brucella that causes brucellosis has a 2–5% fatality rate. Contrarily, brucellosis can be spread by as little as 10 organisms, while the infectious dose of *B. anthracis* is supplied by 10.000 spores. Since cholera, typhoid, and dysentery are gastrointestinal illnesses caused by contaminated water, it was dangerous if enemy troops contaminated water supplies. Especially cholera bacteria are effective due to their short incubation period, causing an epidemic by spreading rapidly and resulting in a high death rate (Schillinger, 2023).

Viruses are infectious agents that need a host to propagate and are very good at finding new hosts (Rowland et al., 2016). Viruses have gained more media attention as pandemic breakouts have become more likely in recent years. Because there are currently restricted viral correspondents to antibiotics, care for the illnesses they prompt is often palliative. The West African Ebola epidemic started in 2014 with a high mortality rate and ominous symptoms, highlighting the potency of virus-borne diseases. Around the world, the Spanish Flu of 1918–19 is thought to have killed 20–50 million people. Every year, a new vaccine must be created as new influenza strains emerge, due to mutation and genetic recombination. The first Severe Acute Respiratory Syndrome (SARS) case appeared in 2003. Middle East Respiratory Syndrome (MERS) first appeared in 2012, as well. A pneumonia outbreak of unknown origin was mentioned in Wuhan, Hubei Province, China, in December 2019. The coronavirus 2 (SARS-CoV-2) is the name of the virus that induces severe acute respiratory syndrome. On March 12, 2020, the World Health Organization proclaimed a pandemic due to the hundreds of fatalities caused by COVID-19, a coronavirus disease (Ciotti et al., 2020). Due to their strong infectivity and illness severity, biological warfare agents have the potential to cause significant harm to or disable a huge number of people (Dechtman et al., 2023). There is currently no approved or effective therapy for some conditions, for example, those caused by viruses identified as bioterror threats (Bugert et al., 2020).

The Biological Toxin Weapon Convention (BTWC), which was created in 1972 and went into effect in 1975, aims to prevent the creation of such weapons while maintaining a system of checks and balances with regard to scientific research and weaponization. It was the first intergovernmental convention to forbid the use of biological weapons and toxins for anything other than peaceful or protective objectives. It also forbade their creation, stockpiling, and transfer. However, this treaty has many flaws that make it challenging to implement in the current situation (Jha, 2023).

Toxins are classified according to one of their common characteristics which is that they cannot reproduce and transfer to new hosts. Toxins can come from various sources, including animals, plants, and bacteria. The deadliest poison known to man is botulinum neurotoxin (BoNT), a protein toxin created by the bacterium *Clostridium botulinum*. Ricin, a less harmful protein found in the seeds of the castor bean plant, has been used in assassinations in the past. The protein toxin called Staphylococcal enterotoxin B (SEB), created by the *Staphylococcus aureus* bacteria is ineffectual at low concentrations and requires high quantities to be lethal. Trichothecenes (T-2 toxins), ochratoxin, and aflatoxin are mycotoxins that fungi can create. Some T-2 toxins are 400 times more toxic than mustard gas, and their effects can be seen in minutes rather than hours or days after exposure. Although antitoxins have been produced for several toxins, their application in non-prophylactic conditions is constrained by the need for quick administration (Bhaskar and Sant, 2019).

11.3 PLASMONIC SENSORS

The scientific investigation of surface plasmons, which are created when incoming light interacts with electrons to create surface-bound electromagnetic waves, is known as plasmonics. Plasmonic sensors have been developed for more than 40 years, and thousands of academic papers, patents, and dozens of marketable products have arisen. The reason for this is that these sensors show better performance compared to conventional ones in many respects, which includes real-time monitoring to reveal the interaction dynamics for observing different biological relationships within biomolecules, label-free identification, higher reuse ability, short response time, simple sampling, and the inclusion of a few electronic components. Plasmonic sensors do, nevertheless, have several disadvantages, like a nonspecific binding surface (which might be improved by placing the analyte-specific layer onto the plasmonic surface), mass transportation problems, steric handicap when binding, and the potential for incorrect interpretation at routine events.

Plasmons appeal to the progress of recent technologies with various applications since they can arrange light to nanoscale volumes. Transmission, raman scattering, and fluorescence may all be significantly improved by the powerful fields produced by plasmons. Thermoplasmonics, or considerable localized heating in metal nanostructures, is another effect that plasmons can produce. Progress in nanoplasmonics encompasses nanoscale sensor substrates and all facets of associated optical technologies, comprising SPR, localized surface plasmon resonance (LSPR), and surface-enhanced Raman scattering (SERS).

SPR, a commonly used optical method, tracks the refractive index change of a sensor layer following target molecule attachment. Unlabeled detection of specific small molecules using SPR presents substantial challenges, demanding significant signal enhancement for precise and reliable measurements. Integrating metal nanoparticles (NPs), such as gold (Au) NPs, into SPR sensor fabrication has demonstrated significant efficacy in both augmenting the signal and establishing plasmonic resonant coupling between NPs and the chip surface (Bhardwaj et al., 2020).

Electromagnetic waves parallel to the metal/nonconductor interface are called surface plasmons. These oscillations of electric charge are specific to the region where a metallic layer meets an insulating interface. The oscillations are extremely dependent on any modification in this border, such as the attachment of any species to the metal layer, because the wave is on the surface between the metal and the external medium (Kravets et al., 2018). Figure 11.1 illustrates the schematic representation of the detection tenet for the examined propagating surface plasmon resonance (pSPR) and LSPR systems.

Minor alterations in the refractive index of dielectric materials can lead to substantial variations in the resonance conditions of these surface plasmons. Conventional

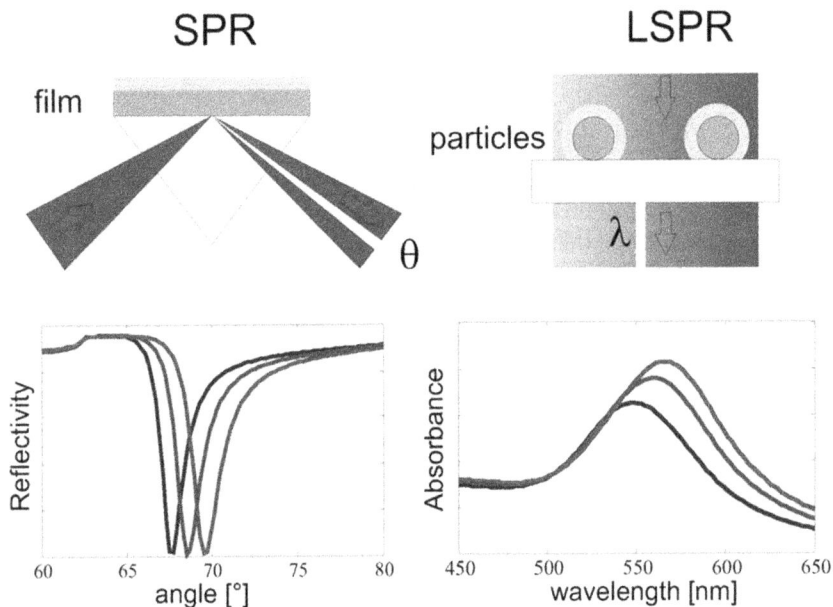

FIGURE 11.1 The diagram depicts the conceptual illustration of the sensing mechanism for the pSPR (left) and LSPR (right) systems (Jatschka et al., 2016).

TABLE 11.1

The Detection of Chemical and Biological Threat Agents Based on SPR

Warfare Agent Type	Threat Agent	Material	LOD	Ref.
Chemical	Explosives (2,4,6-trinitrotoluene)	AuNPs immunolabeled with TNT-specific antibodies	10 ppt	Kawaguchi et al., 2008
Chemical	Explosives (2,4,6-trinitrotoluene)	Peptide matrix	620 parts-per-trillion	Komikawa et al., 2020
Chemical	Diethylchloro-phosphate	Zirconium-phosphate functionalized AuNPs	5.10^{-5}	Newman et al., 2007
Chemical	Heavy metal Arsenic (III)	Alloyed AuFeZnSe quantum dots@gold nanorod nanocomposite	0,01 µg/L (69.12 pM)	Adegoke and Daeid, 2022
Chemical	Heavy metal Mercury (II)	DNA-AuNPs Probe	2.7 pM	Zhang et al., 2023
Biological	Plant toxin (ricin, abrin)	Well-oriented mAbs via Protein G	0.6 ng/mL	Luo et al., 2022
Biological	Bacterial pathogens (*E. coli*)	Silver NPs	0.576 CFU/mL	Özgür et al., 2020
Biological	Bacterial pathogens (*Brucella*)	Antibodies conjugated with functionalized AuNPs	10^3 CFU/mL	Hans et al., 2023

SPR sensors employing flat metal films generate surface plasmon polaritons capable of propagating over tens to hundreds of microns along the metal-dielectric interface (Homola 2003, 2008). Several chemical and biological threat agents can be detected based on surface plasmon resonance sensors (Table 11.1).

11.3.1 PLASMONIC SENSORS USED FOR BIOLOGICAL THREAT AGENTS

Biological threat agents in bioterrorism attacks refer to the deliberate release of bacteria, viruses, toxins, or other microbes intended to inflict disease or death upon humans, animals, or plants. These agents are particularly concerning because they are hard to detect, cost-effective to produce or acquire, can spread rapidly, and may remain undetectable in hosts for hours or days. These agents can be categorized into three main classes: Category A, Category B, and Category C, as outlined in Figure 11.2.

MAIN BIOLOGICAL THREAT AGENTS					
Group A		**Group B**		**Group C**	
Diseases	Agents	Diseases	Agents	Diseases	Agents
Anthrax	*Bacillus anthracis*	Brucellosis	*Brucella spp.*	Emerging infectious	Nipahvirus and
Botulism	*Clostridium botulinum* toxin	Epsilon toxin	*Clostridium perfringens*	diseases	Hantavirus
Plague	*Yersinia pestis*	Glanders	*Burkholderia mallei*		
Smallpox	*Variola major*	Melioidosis	*Burkholderia pseudomallei*		
Tularemia	*Francisella tularensis*	Psittacosis	*Chlamydia psittaci*		
		Q fever	*Coxiella burnetii*		
		Ricin toxin	*Ricinus communis*		

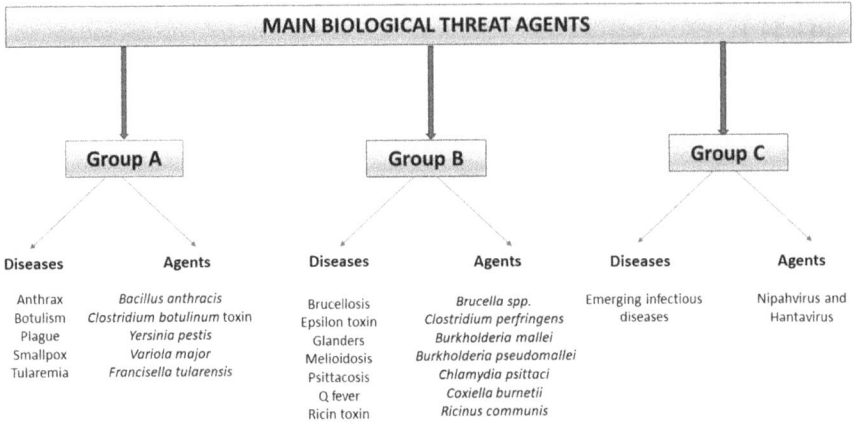

FIGURE 11.2 Main biological threat agents.

Category A agents pose a risk to national security due to their potential for rapid transmission, high mortality rates, significant public health impact, and their ability to induce widespread panic. Category B agents are moderately easy to disseminate but tend to result in lower mortality rates. Meanwhile, Category C agents consist of emerging pathogens that could be engineered for mass dissemination because they are readily available, simple to produce and spread, or they carry a high risk of causing severe health impacts or death (Eni, 1967; Saylan et al., 2020).

Biological threat agents represent a significant danger in attacks. Detecting these agents has become an area of heightened interest due to their potentially catastrophic impact. Easily dispersed through soil, air, and water, these harmful substances can be modified to enhance their dispersal capabilities and to resist current medications. Alarmingly, the accessibility of these biological warfare agents to terrorist groups and smaller nations has underscored the urgent need for easily deployable, real-time detection tools suitable for both military personnel and civilians (Saylan et al., 2020). Traditional methods for detecting microorganisms or their byproducts often fall short in speed and precision. Typically, these methods neither provide real-time results nor conclude within a single day. While numerous analytical techniques have been developed to address this challenge, many still face significant limitations, in terms of their reagent stability, detection speed, reliability, sensitivity, and user-friendliness (Gupta et al., 2010; Iqbal et al., 2000). The creation of quick-response sensor systems specifically tailored to detect biological warfare materials is of paramount importance. Fast and accurate detection is crucial, not just for preventing potential terrorism threats but also for bolstering internal security (Upadhyayula, 2012). Such detection systems should be swift, portable, and highly specific to the biological hazard. Plasmonic sensors have emerged as a viable solution due to their ability to detect target pathogens in real-time without requiring labels. Suitable for both manned and unmanned operations, these sensors can function effectively across diverse environments (Adducci et al., 2016). Incorporating nanomaterials into the crafting of nanosensors has further streamlined existing detection methodologies for on-site applications. Plasmonic nanoparticles (PNPs) are particularly noteworthy because of their unique attributes, such as adjustable size, heightened sensitivity, and efficient pathogen removal. These particles are increasingly being employed to detect trace elements, pathogens, and other applications (Das et al., 2017; Zeng et al., 2016). A typical plasmonic nanosensor design for detecting biological threats incorporates a recognition agent, NPs, and a signal transduction component. Their unique optical and electrical properties make PNPs especially attractive for sensor applications, offering advantages like long-term stability, biocompatibility, and cost-effectiveness (Brahmkhatri et al., 2021).

As an example, Sharma et al., used a real-time, label-free SPR technique to screen different monoclonal antibodies of the Ebola virus (mAb1, mAb2, and mAb3) for their affinity to the recombinant nucleoprotein of Ebola (EBOV-rNP) (Sharma et al., 2020). In another study, an LSPR nanobioprobe was designed, derived from specific immunological interactions between monoclonal antibody-conjugated AuNPs and diphtheria toxoids to detect diphtheria toxoids [56] rapidly. In another study, Sikarwar et al. established a method for the detection of *B. pseudomallei* using SPR by monitoring the interaction of the rpGroEL antigen (rpGroEL Ag) with immobilized rabbit antibody (anti-rpGroEL rAb) (Sikarwar et al., 2016). Yaghubi et al. evolved a new nanobiosensor for the first time with the LSPR bioprobe design to perform immuno-diagnostics of *E. coli* O157:H7 using AuNPs conjugated to chicken antibody (Yaghubi et al., 2020). In another study, Huynh et al. developed an immunosensor based on surface plasmon resonance imaging (SPRi) practising monoclonal antibody against *Y. pestis* F1 antigen, using *Y. pestis* as a model organism (Huynh et al., 2015). In another study, Mi et al. developed a sandwich-structured SERS biosensor for the simultaneous detection of multiple pathogenic bacteria, consisting of non-intrusive SERS probes for bacterial labeling and ConA-functionalized magnetic NPs for bacterial extraction (Mi et al., 2023). Ma et al. investigated the simultaneous detection of *S. typhimurium* and *S. aureus* using an SERS-based sandwich structure. Researchers used gold/silver nanodimer particles mediated by Raman signaling molecule-tagged aptamer arrays as SERS signal probes (Ma et al., 2021). Weng et al. fabricated integrated nanosensors comprising MNP@Ag core-shell NPs, customized with a Raman internal standard and an aptamer for bacterial capture. These all-in-one nanosensors successfully utilised the precise capture, magnetic separation, and ratiometric SERS detection of *E. coli* in diverse liquid food samples (Weng et al., 2021).

Although plasmonic nanosensors offer myriad advantages, the need for further improvement persists. Consistency, reproducibility, and scalability challenges must be addressed when integrating nanomaterials into sensor devices. Continuing research and innovation in this field are essential to developing practical tools for the timely detection and prevention of bioterrorism threats.

11.3.2 PLASMONIC SENSORS USED FOR CHEMICAL THREAT AGENTS

Recently, chemical threat agents (chemical warfare agents (CWAs), explosive compounds, radioactive compounds, heavy metal and halide ions, etc.) have become an important problem for the protection of the environment and society (Yue et al., 2016). Depending on the type and amount of agent used, chemical threat agents cause various symptoms in their victims. For example, some high chemical threat agents cause death by interfering with the nervous system, while others inhibit breathing and cause suffocation (Shea, 2003).

CWAs are synthetic or naturally derived chemical compounds that can penetrate the skin easily and quickly, causing incapacitation or even fatality. According to the literature, soman (GD), mustard (HD), and O-ethyl-S-diisopropyl amino methyl phophonothiolate (VX) are commonly used CWAs (Araújo et al., 2021). Depending on how they affect the human body, CWAs can be divided into several classes. Seven categories have been established: Nerve agents, suffocating/blood agents, vesicant agents, cytotoxic proteins, pulmonary agents, inactivating agents, and tear-forming agents. Vesicant, nerve, and asphyxiant/blood agents have high toxicity and are the most well-known CWAs (Akgönüllü et al., 2021). CWAs are one of the most prominent chemical threat agents that pose a threat to the human population, social stability, and peace. Therefore, the detection and quantification of CWAs are of great importance for the safety and protection of humanity (Kumar et al., 2022a).

A common alternative to G-series nerve agents is dimethyl methyl phosphonate (DMMP). Lafuente et al. created a SERS-based sensor that can detect DMMP in the gas phase at an incredibly low concentration (130 parts/billion) in real-time and without labeling. In the construction of the SERS-based plasmonic sensor, AuNPs are coated with a citrate layer (Figure 11.3). Thus, it is stated that the citrate layer acts effectively in the retention and adsorption of DMMP close to the surface of AuNPs through reversible hydrogen bond interactions around plasmonic AuNPs, which have an important role in the continuous operation of the SERS detection platform (Lafuente et al., 2018).

An explosive might be a pure chemical compound or a combination of numerous explosive substances. Depending on how quickly they decompose, they can be classified as low or high (Tanvanit et al., 2013). Explosives are grouped into three main groups according to their chemical composition. These groups are nitrate esters (PETN; Pentaerythritol tetranitrate, NG; nitroglycerin), nitramines (HMX; octah

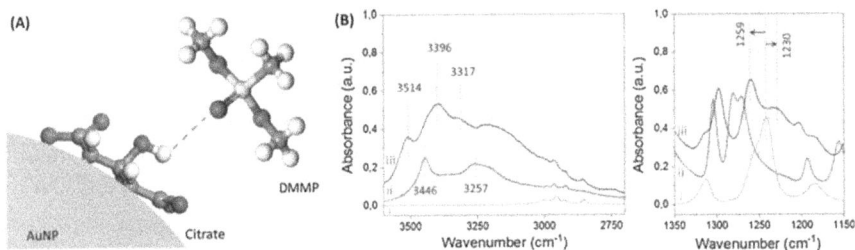

FIGURE 11.3 (A) Occurrence of PO⋯H hydrogen bond interactions between the PO functional group of DMMP and the citrate coating layer of AuNPs. (B) FTIR spectra of (i) DMMP in liquid form, (ii) sodium citrate dihydrate salt, and (iii) Au@citrate monolayers accumulated on SiO₂/Si chips when saturated with DMMP vapor phase at 3600–2700 cm⁻¹ and 1350–1150 cm⁻¹ respectively (Lafuente et al., 2018).

ydro-1,3,5,7-tetranitro-1,3,5,7-tetrazosin, RDX; hexahydro-1,3,5-trinitro -1,3,5-tri-azine), and nitroaromatics (DNT; dinitrotoluene, TNT; 2,4,6-trinitrotoluene) (Bhanot et al., 2020).

Wang created an SPR sensor chip functionalized with a low-cost organic ligand, 3-aminopropyltriethoxysilane (APTES), to rapidly detect explosive TNT. In the study, the detection limit (LOD) of the APTES functionalized SPR sensor was stated to be 134 ppb. It was also stated that the sensor exhibited excellent selectivity towards TNT over four types of TNT analogues (RDX, 4-nitrobenzoyl-glycyl-glycine, DNP-glycine, and 2,6-dinitrotoluene) (Wang, 2021). The study results reveal that the developed sensor has application potential and is promising for the rapid, selective, and sensitive detection of TNT explosives in the liquid phase.

Riskin et al. reported an SPR sensor for the sensitive determination of RDX with imprinted composites of bisaniline cross-linked AuNPs. They stated that the sensor's detection limit is 12 fM. The study used solid and non-planar Kemp acid as an imprinting substrate to create imprinted regions for RDX. They demonstrated that the electropolymerization of AuNPs in the presence of Kemp acid yielded an imprinted compound with strong binding affinities for RDX. They stated that the Kemp acid-imprinted Au-NP composite led to the sensitive detection of RDX by the SPR sensor (Riskin et al., 2010).

To detect explosives like TNT and RDX in their vapor phase, Bharadwaj and Mukherji developed a fibre optic sensor that leverages the localized surface plasmon resonance (LSPR) of AuNPs. To achieve volatile field-based excitation of localized surface plasmons of AuNPs, they fixed AuNPs on a fiber optic sensor probe bent in a U-shape. Following that, to create binding sites for nitro-based explosive chemicals, fixed AuNPs were functionalized with the three most commonly used receptor molecules (4-mercaptobenzoic acid (4-MBA), cysteamine, and L-cysteine). How these probes responded to the vapors of various explosive analytes was investigated. According to this study, AuNP-coated probes treated with L-cysteine and cysteamine showed high selectivity towards TNT. The developed sensor's detection limit was stated to be around 10 ppb. From the results, it appears that for 2,4-DNT, a widespread degradation product of TNT, 4-MBA-modified probes have the highest change of absorbance. It has been noted that the sensor's sensitivity to DNT may be because DNT has a more significant vapor pressure than TNT at an ambient temperature (Bharadwaj and Mukherji, 2014). In aqueous environments, metallic ions Hg^{2+} and Pb^{2+} are soluble and stable. Even at low concentrations, these extremely harmful metallic ions can induce a variety of ailments affecting the brain system, liver, and kidneys (Amirjani and Haghshenas, 2018).

Yuan et al. developed a sensitive, selective, simple, cheap, and portable fiber optic SPR sensor for Hg^{2+} determination using 4-mercapto pyridine (4-MPY) functionalized AuNPs (AuNPs/4-MPY) as a signal amplification tag (Yuan et al., 2019). They indicated that the lowest concentration detected for the heavy metal ion Hg^{2+} in the study was as low as 8 nM. This value is less than the maximum level the World Health Organization (WHO) allows in drinking water. The researchers' study showed that the sensor had good selectivity and recovery (108–112%) for Hg^{2+} over other widespread metal ions in water.

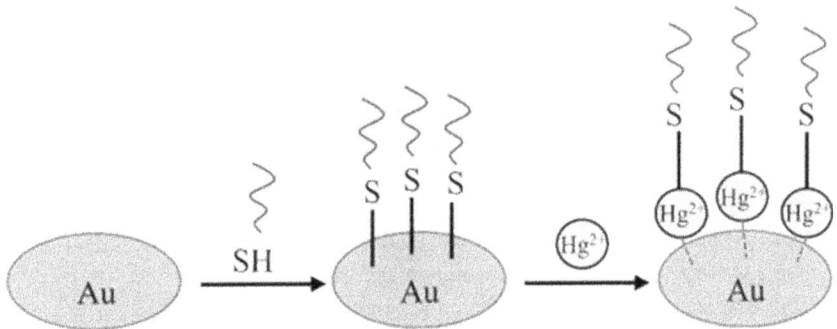

FIGURE 11.4　Mechanism of the modified chip's interaction with the mercury ions (Jiang et al., 2023).

Jiang et al. conducted another study on detecting Hg^{2+} ions with plasmonic sensors. The study presents a five-layer Kretschmann prism-junction structure for an angle-modulated SPR sensor. To absorb Hg^{2+} ions, 1,6-hexanedithiol was subjected to chemical modification on the sensor chip (Figure 11.4). It was also demonstrated that the sensor's sensitivity might be improved two-fold using a silver/gold composite sheet for high-sensitivity detection. The investigation demonstrated that the sensor displayed a linear connection between specific solution concentrations of Hg^{2+} (Figure 11.5). As a result, the sensor is reported to have a detection limit of 80 ppb and a sensitivity of 1.25×10^{-2} °/mg/L. The study also reported that even if the reuse of the substrate is possible to some extent, the elution/regeneration of 1,6-hexanedithiol remains a problem to be solved (Jiang et al., 2023).

Amirjani et al. developed a plasmonic-based assay kit that detects Pb^{2+} by immobilizing AgNPs functional with L-tyrosine on polyethylene terephthalate (PET) film. L-tyrosine has three functional sections. According to the study, even in the presence of other interfering metal ions, the carboxylic acid (-COOH) and hydroxyl

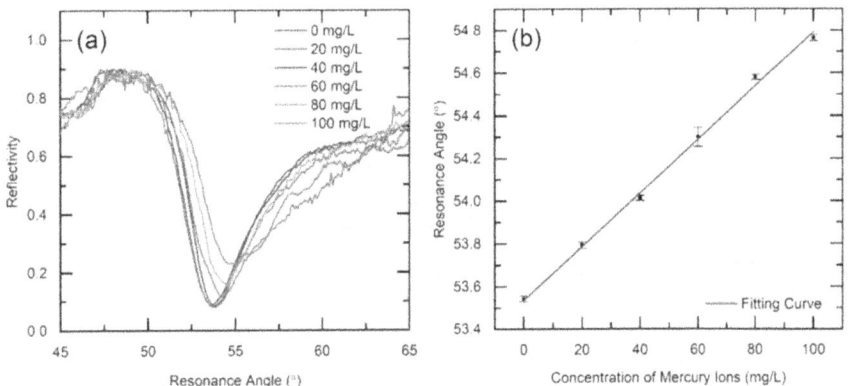

FIGURE 11.5　(a) SPR curves for various Hg^{2+} concentrations in modified chips, and (b) Resonance angle and Hg^{2+} concentration fitting curve in modified chips (Jiang et al., 2023).

(-OH) groups of L-tyrosine demonstrated a preferential affinity for Pb^{2+} ions. With an LOD as low as 1 nM and a dynamic range of 1 nM–1000 nM, this newly designed PET kit was able to detect Pb^{2+} ions in situ. Research has also been conducted to demonstrate the transformability of the test kit. It has been stated that the sensor's response remains stable even after three consecutive runs. Consequently, Amirjani et al. emphasized that their proposed methodology could create simple, sensitive sensors and devices (Amirjani et al., 2022).

A metalloid naturally occurring in certain groundwater, arsenic (As) is one of the most harmful environmental poisons. Additionally, it is a typical industrial contaminant (Bjørklund et al., 2022). Mustapha Kamil et al. developed an SPR sensor with a sensing layer consisting of aqueous ferric oxide ($Fe_2H_2O_4$) for the detection of As(V) and As(III) ions. Detection values of arsenic ions were reported as As(III) $1.083°·ppb^{-1}$ and As(V) $0.922°·ppb^{-1}$. The detection limit for both ions was reported to be 0.6 ppb (Mustapha Kamil et al., 2022).

11.4 CONCLUSIONS AND FUTURE PERSPECTIVES

Chemical and Biological threat agents cause disease and/or death in humans, animals, and vegetation. In the history of research in the domains of public health and medical sciences, managing disease outbreaks, developing efficient treatment procedures, and promoting general human health and well-being have always been given top priority. Generally speaking, classification in terms of the physiological effects that chemical threat agents have on people has been practised for many years. At lower concentration levels, they might be used to disturb the peace and serenity of civilians. It is very simple to achieve the deliberate release of such agents during national or international events by bringing them to the event site through water bottles, cold drink cans, ampules, or even pens, etc. An international threat is posed by bioterrorism, which is the intentional use of microbiological agents or their poisons as weapons. Biological weapons have an enormous and possibly uncontrollable destructive potential that can affect civilian populations even after war. One can refer to bacteria as a weapon to subtly target humankind if poisonous gas makes it possible to assault the citizens of an enemy state. Due to the numerous variables involved in its prospective deployment, it is impossible to quantify its risk. Early identification and clinical diagnosis of these potential threat compounds are challenging due to their numerous unique characteristics.

The threat posed by chemical and biological threat agents and also the limitations of a process to recognize them support the necessity for the sustained design of sensing systems. To detect chemical and biological threat agents, creating unique sensory concepts that support smart sensing capabilities in military and homeland security applications is imperative. As a detection device, a smart sensor should have important features such as speed, sensitivity, selectivity, and portability, and more importantly, it should be simple in identifying the target analyte. Emerging nanomaterial-based sensors, particularly those developed by utilizing plasmonic particles as a sensing component, offer many desirable features needed for threat agent detection. To develop new sensors, bioagent detection has recently benefited from the growing use of nanomaterials among scientists. Nanomaterials' small size and special features make them suitable for use in portable and multiplexed sensors. The use of detection schemes based on intrinsic nanomaterial properties and biological recognition mechanisms is increasing daily. This chapter began with a summary of chemical and biological threat agents before highlighting the variety of recently developed nanoscale plasmonic sensors for their detection. Also, key conceptual issues and descriptions of the highest category threat agents were reviewed.

REFERENCES

Aas, P., 2003. The threat of mid-spectrum chemical warfare agents. *Prehosp. Disaster Med.* 18, 306–312. https://doi.org/10.1017/S1049023X00001254

Adducci, B.A., Gruszewski, H.A., Khatibi, P.A., Schmale, D.G., 2016. Differential detection of a surrogate biological threat agent (Bacillus globigii) with a portable surface plasmon resonance biosensor. *Biosens. Bioelectron.* 78, 160–166. https://doi.org/10.1016/j.bios. 2015.11.032

Adegoke, O., Daeid, N.N., 2022. Alloyed AuFeZnSe quantum dots@gold nanorod nanocomposite as an ultrasensitive and selective plasmon-amplified fluorescence OFF-ON aptasensor for arsenic (III). *J. Photochem. Photobiol. A Chem.* 426, 113755. https://doi.org/10.1016/j.jphotochem.2021.113755

Akgönüllü, S., Saylan, Y., Bereli, N., Türkmen, D., Yavuz, H., Denizli, A., 2021. Plasmonic sensors for detection of chemical and biological warfare agents. *Plasmonic Sensors their Appl.* 71–85. https://doi.org/10.1002/9783527830343.ch4

Amirjani, A., Haghshenas, D.F., 2018. Ag nanostructures as the surface plasmon resonance (SPR)-based sensors: A mechanistic study with an emphasis on heavy metallic ions detection. *Sensors Actuators, B Chem.* 273, 1768–1779. https://doi.org/10.1016/j.snb. 2018.07.089

Amirjani, A., Kamani, P., Hosseini, H.R.M., Sadrnezhaad, S.K., 2022. SPR-based assay kit for rapid determination of Pb^{+2}. *Anal. Chim. Acta* 1220, 340030. https://doi.org/10.1016/ j.aca.2022.340030

Anderson, P.D., 2012. Bioterrorism: Toxins as weapons. *J. Pharm. Pract.* 25, 121–129. https:// doi.org/10.1177/0897190012442351

Apak, R., Üzer, A., Sağlam, Ş., Arman, A., 2023. Selective electrochemical detection of explosives with nanomaterial based electrodes. *Electroanalysis* 35, 1–12. https://doi. org/10.1002/elan.202200175

Araújo, J.C., Fangueiro, R., Ferreira, D.P., 2021. Protective multifunctional fibrous systems based on natural fibers and metal oxide nanoparticles. *Polymers (Basel).* 13, 1–25. https://doi.org/10.3390/polym13162654

Bajgar, J., Fusek, J., Kassa, J., Kuca, K., Jun, D., 2009. Chemical aspects of pharmacological prophylaxis against nerve agent poisoning. *Curr. Med. Chem.* 16, 2977–2986. https:// doi.org/10.2174/092986709788803088

Bhanot, P., Celin, S.M., Sreekrishnan, T.R., Kalsi, A., Sahai, S.K., Sharma, P., 2020. Application of integrated treatment strategies for explosive industry wastewater—A critical review. *J. Water Process Eng.* 35, 101232. https://doi.org/10.1016/j.jwpe.2020. 101232

Bharadwaj, R., Mukherji, S., 2014. Gold nanoparticle coated U-bend fibre optic probe for localized surface plasmon resonance based detection of explosive vapours. *Sensors Actuators, B Chem.* 192, 804–811. https://doi.org/10.1016/j.snb.2013.11.026

Bhardwaj, H., Sumana, G., Marquette, C.A., 2020. A label-free ultrasensitive microfluidic surface Plasmon resonance biosensor for Aflatoxin B1 detection using nanoparticles integrated gold chip. *Food Chem.* 307, 125530. https://doi.org/10.1016/j.foodchem.2019. 125530

Bhaskar, A.S.B., Sant, B., 2019. Chapter 3: Toxins as biological warfare agents. Handbook on Biological Warfare Preparedness. Elsevier Inc. 33–64. https://doi.org/10.1016/B978-0-12-812026-2.00003-7

Bhuvaneswari, R., Nagarajan, V., Chandiramouli, R., 2019. Exploring adsorption mechanism of hydrogen cyanide and cyanogen chloride molecules on arsenene nanoribbon from first-principles. *J. Mol. Graph. Model.* 89, 13–21. https://doi.org/10.1016/j.jmgm.2019.02.008

Bjørklund, G., Rahaman, M.S., Shanaida, M., Lysiuk, R., Oliynyk, P., Lenchyk, L., Chirumbolo, S., Chasapis, C.T., Peana, M., 2022. Natural dietary compounds in the treatment of arsenic toxicity. *Molecules* 27. https://doi.org/10.3390/molecules27154871

Brahmkhatri, V., Pandit, P., Rananaware, P., D'Souza, A., Kurkuri, M.D., 2021. Recent progress in detection of chemical and biological toxins in Water using plasmonic nanosensors. *Trends Environ. Anal. Chem.* 30. https://doi.org/10.1016/j.teac.2021.e00117

Bugert, J.J., Hucke, F., Zanetta, P., Bassetto, M., Brancale, A., 2020. Antivirals in medical biodefense. *Virus Genes* 56, 150–167. https://doi.org/10.1007/s11262-020-01737-5

Ciotti, M., Ciccozzi, M., Terrinoni, A., Jiang, W.C., Wang, C.B., Bernardini, S., 2020. The COVID-19 pandemic. *Crit. Rev. Clin. Lab. Sci.* 57, 365–388. https://doi.org/10.1080/10408363.2020.1783198

Das, R., Vecitis, C.D., Schulze, A., Cao, B., Ismail, A.F., Lu, X., Chen, J., Ramakrishna, S., 2017. Recent advances in nanomaterials for water protection and monitoring. *Chem. Soc. Rev.* 46, 6946–7020. https://doi.org/10.1039/c6cs00921b

de Bruin-Hoegée, M., Lamriti, L., Langenberg, J.P., Olivier, R.C.M., Chau, L.F., van der Schans, M.J., Noort, D., van Asten, A.C., 2023. Verification of exposure to chemical warfare agents through analysis of persistent biomarkers in plants. *Anal. Methods* 15, 142–153. https://doi.org/10.1039/D2AY01650H

Dechtman, I.-D., Ankory, R., Sokolinsky, K., Krasner, E., Weiss, L., Gal, Y., 2023. Clinically evaluated COVID-19 drugs with therapeutic potential for biological warfare agents. *Microorganisms* 11, 1577. https://doi.org/10.3390/microorganisms11061577

Diauudin, F.N., Rashid, J.I.A., Knight, V.F., Wan Yunus, W.M.Z., Ong, K.K., Kasim, N.A.M., Abdul Halim, N., Noor, S.A.M., 2019. A review of current advances in the detection of organophosphorus chemical warfare agents based biosensor approaches. *Sens. Bio-Sensing Res.* 26, 100305. https://doi.org/10.1016/j.sbsr.2019.100305

Dogaroiu, C., 2003. Chemical warfare agents. *Rom. J. Leg. Med.* 11, 132–140. https://doi.org/10.4103/0975-7406.68498

Eni., 1967. 済 無. *Angewandte Chemie International Edition* 6(11), 951–952.

Fusek, J., Dlabkova, A., Misik, J., 2020. Chapter 14: Psychotomimetic agent BZ (3-quinuclidinyl benzilate), handbook of toxicology of chemical warfare agents. INC. 203–213 https://doi.org/10.1016/B978-0-12-819090-6.00014-3

Gupta, G., Singh, P.K., Boopathi, M., Kamboj, D.V., Singh, B., Vijayaraghavan, R., 2010. Molecularly imprinted polymer for the recognition of biological warfare agent staphylococcal enterotoxin B based on Surface Plasmon Resonance. *Thin Solid Films* 519, 1115–1121. https://doi.org/10.1016/j.tsf.2010.08.054

Hans, R., Yadav, P.K., Zaman, M.B., Poolla, R., Thavaselvam, D., 2023. A rapid direct-differential agglutination assay for Brucella detection using antibodies conjugated with functionalized gold nanoparticles. *Front. Nanotechnol.* 5, 1–23. https://doi.org/10.3389/fnano.2023.1132783

Hemme, M., Fidder, A., van der Riet-van Oeveren, D., van der Schans, M.J., Noort, D., 2021. Mass spectrometric analysis of adducts of sulfur mustard analogues to human plasma proteins: Approach towards chemical provenancing in biomedical samples. *Anal. Bioanal. Chem.* 413, 4023–4036. https://doi.org/10.1007/s00216-021-03354-z

Homola, J., 2003. Present and future of surface plasmon resonance biosensors. *Anal. Bioanal. Chem.* 377, 528–539. https://doi.org/10.1007/s00216-003-2101-0

Homola, J., 2008. Surface plasmon resonance sensors for detection of chemical and biological species. *Chem. Rev.* 108, 462–493. https://doi.org/10.1021/cr068107d

Huynh, H.T.T., Gotthard, G., Terras, J., Aboudharam, G., Drancourt, M., Chabrière, E., 2015. Surface plasmon resonance imaging of pathogens: The Yersinia pestis paradigm microbiology. *BMC Res. Notes* 8, 1–8. https://doi.org/10.1186/s13104-015-1236-3

Iqbal, S.S., Mayo, M.W., Bruno, J.G., Bronk, B.V., Batt, C.A., Chambers, J.P., 2000. A review of molecular recognition technologies for detection of biological threat agents. *Biosens. Bioelectron.* 15, 549–578. https://doi.org/10.1016/S0956-5663(00)00108-1

Jatschka, J., Dathe, A., Csáki, A., Fritzsche, W., Stranik, O., 2016. Propagating and localized surface plasmon resonance sensing – A critical comparison based on measurements and theory. *Sens. Bio-Sensing Res.* 7, 62–70. https://doi.org/10.1016/j.sbsr.2016.01.003

Jha, A., 2023. Technological advances and evolution of biowarfare: A threat to public health and security. *KnE Social Sciences* 2023, 401–416. https://doi.org/10.18502/kss.v8i14.13853

Jiang, G., Miao, Y., Wang, J., Shao, H., Chen, H., Tao, P., Wang, W., Yu, Q., Peng, W., Zhou, X., 2023. Specific detection of mercury ions based on surface plasmon resonance sensor modified with 1, 6-hexanedithiol. *Sensors Actuators A Phys.* 356, 114343. https://doi.org/10.1016/j.sna.2023.114343

John, H., Thiermann, H., 2021. Poisoning by organophosphorus nerve agents and pesticides: An overview of the principle strategies and current progress of mass spectrometry-based procedures for verification. *J. Mass Spectrom. Adv. Clin. Lab* 19, 20–31. https://doi.org/10.1016/j.jmsacl.2021.01.002

Jyoti, J., Redondo, E., Alduhaish, O., Pumera, M., 2023. 3D-printed electrochemical sensor for organophosphate nerve agents. *Electroanalysis* 35, 139–144. https://doi.org/10.1002/elan.202200047

Karadurmus, L., Bilge, S., Sınağ, A., Ozkan, S.A., 2022. Molecularly imprinted polymer (MIP)-Based sensing for detection of explosives: Current perspectives and future applications. *TrAC - Trends Anal. Chem.* 155, 116694. https://doi.org/10.1016/j.trac.2022.116694

Kawaguchi, T., Shankaran, D.R., Kim, S.J., Matsumoto, K., Toko, K., Miura, N., 2008. Surface plasmon resonance immunosensor using Au nanoparticle for detection of TNT. *Sensors Actuators, B Chem.* 133, 467–472. https://doi.org/10.1016/j.snb.2008.03.005

Komikawa, T., Tanaka, M., Yanai, K., Johnson, B.R.G., Critchley, K., Onodera, T., Evans, S.D., Toko, K., Okochi, M., 2020. A bioinspired peptide matrix for the detection of 2,4,6-trinitrotoluene (TNT). *Biosens. Bioelectron.* 153, 112030. https://doi.org/10.1016/j.bios.2020.112030

Kravets, V.G., Kabashin, A.V., Barnes, W.L., Grigorenko, A.N., 2018. Plasmonic surface lattice resonances: A review of properties and applications. *Chem. Rev.* 118, 5912–5951. https://doi.org/10.1021/acs.chemrev.8b00243

Kumar, V., Kim, H., Pandey, B., James, T.D., Yoon, J., Anslyn, E.V., 2022a. Recent advances in fluorescent and colorimetric chemosensors for the detection of chemical warfare agents: A legacy of the 21st century. *Chem. Soc. Rev.* 52, 663–704. https://doi.org/10.1039/d2cs00651k

Kumar, V., Kim, H., Pandey, B., James, T.D., Yoon, J., Anslyn, E.V., 2022b. Recent advances in fluorescent and colorimetric chemosensors for the detection of chemical warfare agents: A legacy of the 21st century. *Chem. Soc. Rev.* 52, 663–704. https://doi.org/10.1039/d2cs00651k

Lackner, C.K., Burghofer, K., 2019. Medical aspects. In *The Networked Health-Relevant Factors for Office Buildings*, 147–156. https://doi.org/10.1007/978-3-030-22022-8_6

Lafuente, M., Pellejero, I., Sebastián, V., Urbiztondo, M.A., Mallada, R., Pina, M.P., Santamaría, J., 2018. Highly sensitive SERS quantification of organophosphorous chemical warfare agents: A major step towards the real time sensing in the gas phase. *Sensors Actuators, B Chem.* 267, 457–466. https://doi.org/10.1016/j.snb.2018.04.058

Luo, Li, Yang, J., Li, Z., Xu, H., Guo, L., Wang, L., Wang, Y., Luo, L., Wang, J., Zhang, P., Yang, R., Kang, W., Xie, J., 2022. Label-free differentiation and quantification of ricin, abrin from their agglutinin biotoxins by surface plasmon resonance. *Talanta* 238, 122860. https://doi.org/10.1016/j.talanta.2021.122860

Ma, X., Lin, X., Xu, X., Wang, Z., 2021. Fabrication of gold/silver nanodimer SERS probes for the simultaneous detection of Salmonella typhimurium and Staphylococcus aureus. *Microchim. Acta* 188:202, 1–9. https://doi.org/10.1007/s00604-021-04791-4

Mi, F., Guan, M., Wang, Y., Chen, G., Geng, P., Hu, C., 2023. Integration of three non-interfering SERS probes combined with ConA-functionalized magnetic nanoparticles for extraction and detection of multiple foodborne pathogens. *Microchim. Acta* 190:103, 1–11. https://doi.org/10.1007/s00604-023-05676-4

Mustapha Kamil, Y., Al-Rekabi, S.H., Abu Bakar, M.H., Fen, Y.W., Mohammed, H.A., Mohamed Halip, N.H., Alresheedi, M.T., Mahdi, M.A., 2022. Arsenic detection using surface plasmon resonance sensor with hydrous ferric oxide layer. *Photonic Sensors* 12, 1–8. https://doi.org/10.1007/s13320-021-0643-4

Newman, J.D.S., Roberts, J.M., Blanchard, G.J., 2007. Optical organophosphate/phosphonate sensor based upon gold nanoparticle functionalized quartz. *Anal. Chim. Acta* 602, 101–107. https://doi.org/10.1016/j.aca.2007.08.051

Nishio, T., Toukairin, Y., Hoshi, T., Arai, T., Nogami, M., 2021. Development of an LC–MS/MS method for quantification of 3-chloro-L-tyrosine as a candidate marker of chlorine poisoning. *Leg. Med.* 53. https://doi.org/10.1016/j.legalmed.2021.101939

Özgür, E., Topçu, A.A., Yılmaz, E., Denizli, A., 2020. Surface plasmon resonance based biomimetic sensor for urinary tract infections. *Talanta* 212, 120778. https://doi.org/10.1016/j.talanta.2020.120778

Park, S.Y., Kim, Y., Kim, T., Eom, T.H., Kim, S.Y., Jang, H.W., 2019. Chemoresistive materials for electronic nose: Progress, perspectives, and challenges. *InfoMat* 1, 289–316. https://doi.org/10.1002/inf2.12029

Patočka, J., Fusek, J., 2004. Chemical agents and chemical terrorism. *Cent. Eur. J. Public Health* 12, S75–S77. https://doi.org/10.21101/cejph.b0173

Quiroga-Garza, M.E., Ruiz-Lozano, R.E., Azar, N.S., Mousa, H.M., Komai, S., Sevilla-Llorca, J.L., Perez, V.L., 2023. Noxious effects of riot control agents on the ocular surface: Pathogenic mechanisms and management. *Front. Toxicol.* 5, 1–10. https://doi.org/10.3389/ftox.2023.1118731

Riskin, M., Tel-Vered, R., Willner, I., 2010. Imprinted au-nanoparticle composites for the ultrasensitive surface plasmon resonance detection of hexahydro-1,3,5-trinitro-1,3,5-triazine (RDX). *Adv. Mater.* 22, 1387–1391. https://doi.org/10.1002/adma.200903007

Rowland, C.E., Brown, C.W., Delehanty, J.B., Medintz, I.L., 2016. Nanomaterial-based sensors for the detection of biological threat agents. *Mater. Today* 19, 464–477. https://doi.org/10.1016/j.mattod.2016.02.018

Saylan, Y., Akgönüllü, S., Denizli, A., 2020. Plasmonic sensors for monitoring biological and chemical threat agents. *Biosensors* 10(10):142, 1–15. https://doi.org/10.3390/BIOS10100142

Schillinger, N., 2023. Microbic mass destruction – Biological warfare and epidemic prevention in Republican China. *East Asian Sci. Technol. Soc.* 17, 148–169. https://doi.org/10.1080/18752160.2021.1984009

Shakarjian, M.P., Heck, D.E., Gray, J.P., Sinko, P.J., Gordon, M.K., Casillas, R.P., Heindel, N.D., Gerecke, D.R., Laskin, D.L., Laskin, J.D., 2009. Mechanisms mediating the vesicant actions of sulfur mustard after cutaneous exposure. *Toxicol. Sci.* 114, 5–19. https://doi.org/10.1093/toxsci/kfp253

Sharma, P.K., Kumar, J.S., Singh, V.V., Biswas, U., Sarkar, S.S., Alam, S.I., Dash, P.K., Boopathi, M., Ganesan, K., Jain, R., 2020. Surface plasmon resonance sensing of Ebola virus: A biological threat. *Anal. Bioanal. Chem.* 412, 4101–4112. https://doi.org/10.1007/s00216-020-02641-5

Shea, D.A., 2003. CRS Report for Congress High-Threat Chemical Agents.

Sikarwar, B., Sharma, P.K., Kumar, A., Thavaselvam, D., Boopathi, M., Singh, B., Jaiswal, Y.K., 2016. Surface plasmon resonance immunosensor for the detection of Burkholderia pseudomallei. *Plasmonics* 11, 1035–1042. https://doi.org/10.1007/s11468-015-0139-4

Smart, J.K., 1997. History of chemical and biological warfare: An American perspective. In Medical Aspects of Chemical and Biological Warfare, Part I: Warfare, Weaponry, and the Casualty R. Zajtchuk (Ed), Borden Institute, Washington, DC, Department of the Army, Office of the Surgeon General, pp: 9–86.

States, U., John, A., Special, F.K., Bragg, F., Carolina, N., 2001. The medical aspects of NBC defensive operations AMedP-6 (B) Part I – Nuclear Part II – Biological Part III – Chemical. Department of the Navy Publication Department of the Air Force Joint Manual.

Tanvanit, P., Anotai, J., Su, C.C., Lu, M.C., 2013. Treatment of explosive-contaminated wastewater through the fenton process. *Desalin. Water Treat.* 51, 2820–2825. https://doi.org/10.1080/19443994.2012.750779

Upadhyayula, V.K.K., 2012. Functionalized gold nanoparticle supported sensory mechanisms applied in detection of chemical and biological threat agents: A review. *Anal. Chim. Acta* 715, 1–18. https://doi.org/10.1016/j.aca.2011.12.008

Wang, J., 2021. A simple, rapid and low-cost 3-Aminopropyltriethoxysilane (APTES)-based surface plasmon resonance sensor for TNT explosive detection. *Anal. Sci.* 37, 1029–1032. https://doi.org/10.2116/analsci.20N028

Weng, Y.W., Hu, X.-D., Jiang, L., Shi, Q.L., Wei, X.L., 2021. An all-in-one magnetic SERS nanosensor for ratiometric detection of Escherichia coli in foods. *Anal. Bioanal. Chem.* 413, 5419–5426. https://doi.org/10.1007/s00216-021-03521-2

Yaghubi, F., Zeinoddini, M., Saeedinia, A.R., Azizi, A., Samimi Nemati, A., 2020. Design of Localized Surface Plasmon Resonance (LSPR) biosensor for immunodiagnostic of E. coli O157:H7 using gold nanoparticles conjugated to the chicken antibody. *Plasmonics* 15, 1481–1487. https://doi.org/10.1007/s11468-020-01162-2

Sulikowska, M., Marek, A., Jarosz, Ł.S., Pyzik, E., Stępień-Pyśniak, D., Hauschild, T., 2024. Pathogenic Potential of Coagulase-Positive Staphylococcus Strains Isolated from Aviary Capercaillies and Free-Living Birds in Southeastern Poland. Animals 14, 295. https://doi.org/10.3390/ani14020295

Yuan, H., Ji, W., Chu, S., Liu, Q., Qian, S., Guang, J., Wang, J., Han, X., Masson, J.F., Peng, W., 2019. Mercaptopyridine-functionalized gold nanoparticles for fiber-optic surface plasmon resonance Hg_{2+} sensing. *ACS Sensors* 4, 704–710. https://doi.org/10.1021/acssensors.8b01558

Yue, G., Su, S., Li, N., Shuai, M., Lai, X., Astruc, D., Zhao, P., 2016. Gold nanoparticles as sensors in the colorimetric and fluorescence detection of chemical warfare agents. *Coord. Chem. Rev.* 311, 75–84. https://doi.org/10.1016/j.ccr.2015.11.009

Zellner, T., Eyer, F., 2020. Toxicology Letters 320, 73–79.https://doi.org/10.1016/j.toxlet.2019.12.005

Zeng, Y., Ren, J., Shen, A., Hu, J., 2016. Field and pretreatment-free detection of heavy-metal ions in organic polluted water through an alkyne-coded SERS test kit. *ACS Appl. Mater. Interfaces* 8, 27772–27778. https://doi.org/10.1021/acsami.6b09722

Zhang, W., Liu, G., Bi, J., Bao, K., Wang, P., 2023. In-situ and ultrasensitive detection of mercury (II) ions (Hg2+) using the localized surface plasmon resonance (LSPR) nanosensor and the microfluidic chip. *Sensors Actuators A Phys.* 349, 114074. https://doi.org/10.1016/j.sna.2022.114074

12 Plasmonic Sensors Based on Molecularly Imprinted Polymers for the Detection of Biological and Chemical Threats

Cem Esen and Francesco Canfarotta

12.1 INTRODUCTION

Chemical threats are hazardous substances which usually affect a person's health, and they can be naturally present in the environment or synthetically produced. On the other hand, biological threats are mainly viruses or bacteria that may take several days to affect the health of the infected host [1]. Understanding the type of chemical or biological threat an individual may have been exposed to is crucial to devise a suitable treatment strategy to minimize adverse effects [2].

Molecular imprinting is a technique for preparing polymeric materials with high selectivity towards a target molecule. The resultant molecularly imprinted polymers (MIPs) display features similar to natural antibodies, with the advantages of increased stability, shorter manufacturing time, and a much more reliable supply chain, as they do not rely on animal hosts or cells for their production. In addition, the synthetic nature of MIPs grants a better control over the chemistry of the resulting artificial binder, enabling the introduction of specific functionalities in the polymer that can be used to aid immobilization on sensor surfaces. Owing to these advantages, MIPs have become one of the most researched synthetic affinity reagents in the past three decades. Some MIP-based off-the-shelf products are now commercialized by established corporations such as Sigma-Aldrich and Biotage, whilst relatively new companies (MIP Discovery, Ligar, and Sixth Wave) now offer new alternatives in the cell and gene therapy and bioprocessing sectors. The recent applications of MIPs in plasmonic nanosensors for the detection of biological and chemical threats (BCTs) are discussed in this chapter, thus providing an overview of the most recent advancements in this field.

DOI: 10.1201/9781003459316-12

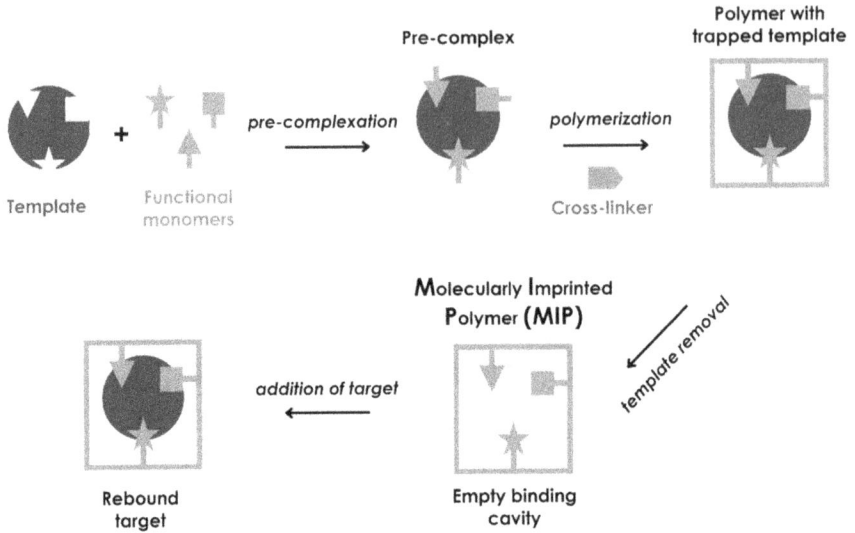

FIGURE 12.1   Schematic representation of the process of preparing MIPs [3].

12.2   OVERVIEW OF MOLECULARLY IMPRINTED POLYMERS

MIPs are obtained by a template-assisted synthesis method, which is composed of three steps; (1) pre-polymerization, (2) polymerization, and (3) template removal (Figure 12.1) [3]. In the first step, functional monomers and template molecules are brought into contact, and the former self-assemble around the template molecule via non-covalent interactions and according to their functionalities. The template molecule is either the exact molecule that needs to be detected or part thereof. Then, polymerization is triggered, and monomers are "frozen" in position around the template. In the final step, the template molecule is removed, leaving behind nano-scale cavities in the polymer, which are complementary to the template imprinted in the first place; these cavities will later act as artificial binding pockets and will enable the binding of the final target molecule [4, 5].

The molecular imprinting technique has received significant attention in the last two decades, as an alternative solution in the quest to develop sensitive and robust sensors [6]. In this context, MIPs have been employed as alternatives for bioreceptors to mitigate challenges related to antibody instability, relatively high cost, and incompatibility to harsh conditions [7-10].

12.3   PLASMONIC SENSORS

Typically, a sensor consists of three components: The recognition element, a transducer, and a detector. The analyte needs to be in close contact with the transducer as it translates the binding of the analyte into an observable and/or quantifiable signal [11, 12]. Owing to their advantages, such as high sensitivity and selectivity, rapid response, ease of operation, and in some cases portability, sensors have been employed in a broad range of applications [13, 14]. In recent years, plasmonic sensors have gained traction due to their outstanding electromagnetic control ability [15]. For target analyte detection, they have become the real-time, label-free detection platform of choice, especially for chemical and biological analysis, clinical diagnostics, and food safety [16, 17]. Plasmonic sensors are categorized into several subclasses, such as resonance, fluorescence, refraction, reflection, Raman scattering, chemiluminescence, dispersion, phosphorescence, and infrared absorption [18, 19]. These sensors are included in the group of optical sensors and are composed of sensing elements which contain metal or metal-dielectric nanostructures. A change in the refractive index is observed in the region close to the surface when a sample containing the target analyte is introduced, due to the binding of the target analyte by the receptor immobilized on the surface of the sensing element [13, 20]. Generally, surface plasmons are oscillations of electrons which are propagated at the interface of plasmonic materials such as noble metals and a dielectric medium (for instance, in an aqueous solution). The energy generated from these surface plasmons is a function of the wavevector of light and the dielectric constants of both the dielectric and the plasmonic material [16, 21, 22]. Plasmonic sensors rely on the evanescent field of these plasmons propagating through the surface of metals, which enables the detection of variations of the sample's dielectric constant at approximately 100 nm from the plasmonic surface. The interaction of the analyte with the evanescent field leads to a shift in the transmitted wavelength proportional to the changes in the sample's refractive index (Figure 12.2) [3, 23]. An important property of the surface plasmon resonance (SPR) effect is linked to sensitivity to changes in the refractive index surrounding the metallic structure, as well as the morphological characteristics of the nanostructure [24]. Furthermore, owing to their simple preparation, cost-effectiveness, and unique label-free property, plasmonic sensors with the characteristic SPR

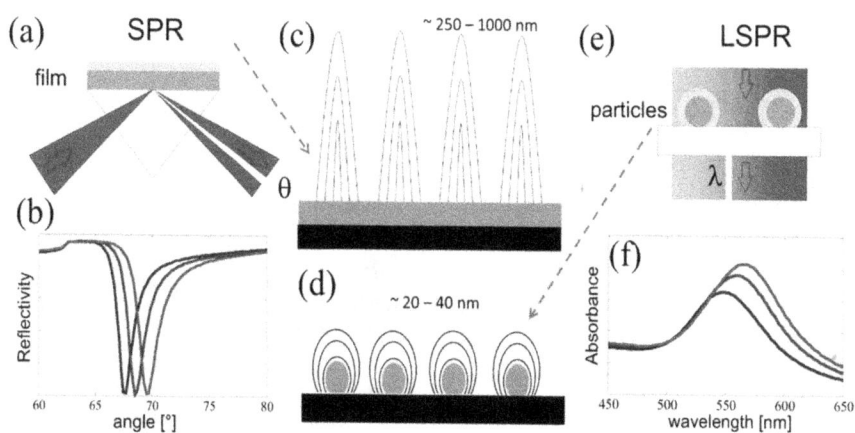

FIGURE 12.2 Schematic description of the mechanisms of SPR (a–c) and LSPR (d–f). λ: wavelength (nm) and θ: the angle of reflection (°) [3].

and Localized Surface Plasmon Resonance (LSPR) effect have become the ideal candidates for real-time monitoring applications [22, 25].

12.4 DETECTION OF BCTS USING MOLECULARLY IMPRINTED POLYMERS

Diverse analytical techniques have been utilized to detect different kinds of BCTs for a decade [13, 26–30]. The main challenges faced when detecting BCTs are sensitivity, selectivity, and robustness [31]. The synergistic coupling of plasmonic surfaces or SPR sensors with affinity reagents such as MIPs could improve the sensitivity of these sensors for a wide range of sensing applications [32]. MIPs have been employed as artificial receptors in sensing platforms on the basis of plasmon- and surface-enhanced spectroscopy [33], and the present work focuses on the detection of BCTs based on MIPs integrated with plasmonic sensors.

12.4.1 CHEMICAL THREATS

Chemical threats are listed in different categories depending on their effects and treatments. Organophosphate pesticides (OPPs) are the most extensively employed pesticides globally and may cause acute toxicity in humans, animals, and the environment [34]. OPPs are neurotoxins which potentially inhibit the enzyme acetylcholinesterase responsible for supporting the nervous system of all biological species, including humans. Thus, OPPs weaken neuromuscular transmission and impact the respiratory system [35]. The excessive use of OPPs may lead to the release of toxic compounds into the ecosystem which poses an environmental threat and a major public health risk [36]. Moreover, OPPs are also used as the surrogates of dimethyl 4-nitrophenylphosphate, commonly called paraoxon, as a toxic nerve agent simulant. MIP-based plasmonic sensors have been developed as rapid analytical tools for detection and monitoring of these threats [3, 6, 37]. A list of MIP-based plasmonic sensors for OPPs and their characteristics is presented in Table 12.1.

TABLE 12.1
MIP-Based Plasmonic Sensors for OPPs and Their Characteristics

Sensor Type	Analyte	Functional Monomer	Linear Range (mol L^{-1})	LOD (mol L^{-1})	Reference
SPR	Acephate	MAA	$(5\text{-}80) \times 10^{-13}$	1.14×10^{-13}	[38]
	Chlorpyrifos	Dopamine	$(1\text{-}1000) \times 10^{-9}$	7.6×10^{-10}	[39]
	Dimethoate	MATrp	$(4\text{-}436) \times 10^{-11}$	3.3×10^{-11}	[40]
SERS/ colorimetric dual sensor	Chlorpyrifos	MAA	$(2.85\text{-}285) \times 10^{-11}$	-	[41]

SPR: Surface plasmon resonance; SERS: Surface-enhanced Raman spectroscopy; MAA: Methacrylic acid; MATrp: N-methacryloyl-L-tryptophan methyl ester.

MIP-based plasmonic sensors possess a wider working range with a lower LOD and faster response than antibody-based plasmonic sensors in real samples [42].

Heavy metal pollution has been an extremely hot topic in scientific research for several decades. However, only a few examples of MIP-based plasmonic sensors for detection of heavy metals have been reported so far [43]. In one such example, Bakhshpour and Denizli developed an MIP-based plasmonic sensor for highly sensitive detection of Cd(II), which is one of the most toxic and carcinogenic heavy metals [44]. In this work, poly(hydroxyethyl methacrylate) thin-film, Cd(II) imprinted NPs and AuNPs were employed for developing the SPR sensor. Thanks to the signal enhancing properties of both AuNPs and imprinted NPs, a relatively low LOD value of $0.01 \ \mu g \ L^{-1}$ was obtained.

Another class of toxic chemicals includes polycyclic aromatic hydrocarbons (PAHs), which are pollutants containing two or more fused aromatic rings. Due to their abundant release from anthropogenic sources, PAHs are ubiquitous not only in aquatic and terrestrial ecosystems but also in the atmosphere. Mostly, PAHs are mutagenic, carcinogenic, teratogenic, and immunotoxic. Hence, because of their wide range of biological toxicity, detection and remediation of PAHs from the environment have become a global concern [45]. Recently, Castro-Grijalba et al. reported the first ever hybrid plasmonic MIP-based SERS sensor to detect PAHs [46]. This was developed by layer-by-layer deposition of MIP thin film onto AuNPs, using MAA and pyrene or fluoranthene as monomer and template molecules, respectively. The use of pyrene or fluoranthene as a template molecule to construct the Au@MIP thin films allowed its ultra-detection in the nM range with a 100-fold betterment compared with the non-imprinted polymer-based plasmonic sensors.

12.4.2 Biological Threats

Biological threats comprise microorganisms such as bacteria, viruses, and bacterial toxins; these may lead to microbial pollution, which in turn may cause diseases and potentially death for the host. Mycotoxins are highly toxic secondary metabolites of fungi and are reported to be teratogenic, mutagenic, or carcinogenic [47]. Therefore, it is of great importance to develop rapid, sensitive, and low-cost detection platforms for monitoring mycotoxins especially in foodstuffs [48]. Among mycotoxins, aflatoxin B1 (AFB1) has been proven to be a potent human carcinogen and is classified as a Group 1 biological toxin, and its accurate detection is of great importance [49]. In this context, Sergeyeva et al. developed a portable, highly selective, and sensitive plasmon-enhanced fluorescence sensor based on nanostructured MIP films with embedded silver nanoparticles (AgNPs) for the detection of AFB1 in cereal samples [50]. MIP films were synthesized using ethyl-2-oxocyclopentanecarboxylate and acrylamide as a dummy template and functional monomer, respectively. Using a dummy template instead of aflatoxin has the advantage of reducing health risks for the user, since aflatoxin is highly toxic. AgNPs were obtained in situ via the addition of $AgNO_3$ to the monomer solutions during the pre-heating step, followed by a UV-initiated polymerization procedure. Their results revealed that metal-enhanced fluorescence effectively increased the AFB1 quantum yield resulting in an improved

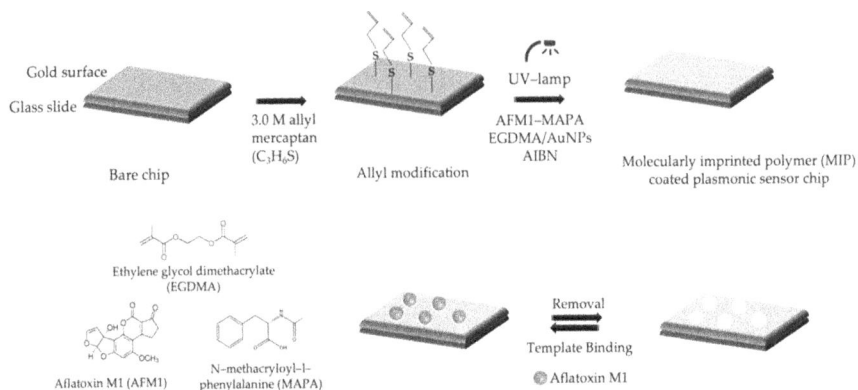

FIGURE 12.3 Scheme of the fabrication process of a plasmonic MIP sensor chip for AFM1 detection [52].

limit of detection (LOD) for AFB1 (0.3 ng mL^{-1}) and a dynamic linear range (LR) of 0.3–25 ng mL^{-1}.

In a similar fashion, Denizli and his team developed a rapid and sensitive SPR sensor for the selective detection of AFB1 in food samples using AFB1 MIP nanofilm with gold nanoparticles (AuNPs) [51]. The nanosensor exhibited a wide LR from 0.0001 to 10.0 ng mL^{-1} with an LOD of 1.04 pg mL^{-1}.

In another interesting study, a reliable MIP SPR nanosensor for the detection of aflatoxin M1 (AFM1), which is the major metabolite of AFB1, was designed by Akgönüllü et al. [52]. As depicted in Figure 12.3 [52], AuNPs were integrated into the MIP nanofilm to enhance the sensitivity of the plasmonic sensor. Subsequently, the proposed sensor was successfully applied for AFM1 detection in raw milk samples with a rather wide LR, from 0.0003 ng mL^{-1} to 20.0 ng mL^{-1} and an LOD of 0.4 pg mL^{-1}.

With regards to bacterial threats, *Escherichia coli* (*E. coli*) is the most studied microorganism which can cause bacterial infection, and it belongs to the class of gram-negative bacteria [53, 54]. This microorganism was one of several that Denizli's team explored and for which they built MIP SPR sensors [55–58]. For example, they developed a sensitive plasmonic MIP sensor for the selective detection of *E. coli* using the microcontact imprinting technique with the entrapment of AgNPs [58]. The sensor response was obtained in 20 minutes with a low LOD of 0.57 CFU mL^{-1} in aqueous solution, indicating high selectivity and sensitivity for *E. coli*.

Severe acute respiratory syndrome coronavirus 2 (SARS-CoV-2), which was the cause of the COVID-19 pandemic in 2020, is one of the current biological threats. Due to the urgent need for rapid, sensitive, selective, and cost-effective diagnostic tools for the detection of SARS-CoV-2, scientists have worked on this task since the onset of the pandemic. The first MIP-based SPR sensor for the detection of COVID-19 was reported by Cennamo et al. [59]. An innovative feature introduced by these authors was a plasmonic plastic optical fiber (POF) that enabled the detection of SARS-CoV-2 virions in aqueous media in approximately 10 minutes. As illustrated

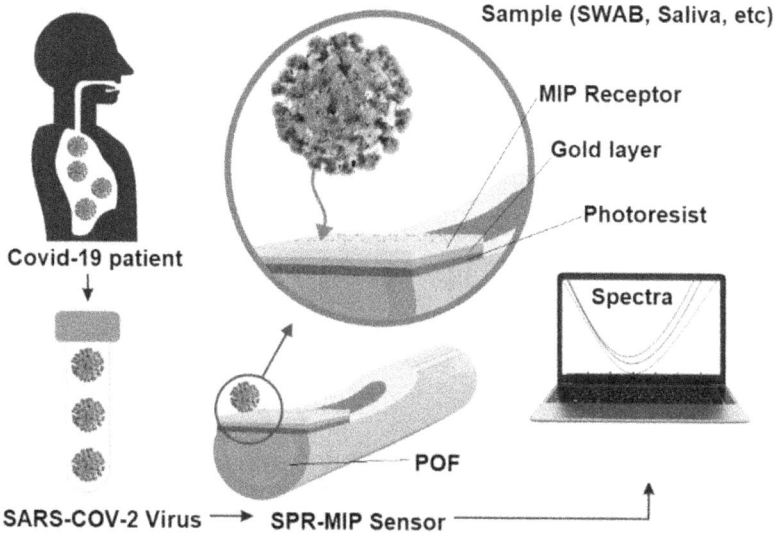

FIGURE 12.4   Schematic outline of the MIP-based SPR sensor for recognition of SARS-CoV-2 virus in different matrices [59].

in Figure 12.4 [59], an MIP nanofilm was integrated with an SPR-POF sensor for selective recognition of the S1 subunit of SARS-CoV-2 spike protein. Their tests performed on samples of nasopharyngeal (NP) swabs revealed that the proposed approach had the potential to be integrated at point-of-care (POC) for rapid and low-cost SARS-CoV-2 detection.

In another similar study, Bajaj et al. developed an MIP-based SPR sensor for selective SARS-CoV-2 detection [60], which was based on nanoMIPs manufactured via a solid-phase synthesis approach (Figure 12.5). SARS-CoV-2 detection was accomplished in the concentration range of $0.25–1.75 \times 10^6$ particles mL^{-1} with an LOD value of 10^4 particles. Furthermore, the obtained K_D of 0.12 pM confirmed the high binding affinity of the developed nanoMIPs towards SARS-CoV-2.

12.5   CONCLUSION

The severity of health issues associated with BCTs, as well as environmental pollutants, has led researchers to develop innovative sensors platforms to achieve early and accurate detection. Traditionally, antibodies are used in the development of such sensors, however, other affinity reagents may offer additional benefits. MIPs are one of the emerging artificial receptors which are known for their robustness, scalability, and excellent performance under extreme conditions. Electrochemical and optical sensors are arguably the most promising platforms to achieve accurate and (in some cases) label-free detection of BCTs. Electropolymerization presents the advantage of simultaneous synthesis and immobilization of the MIPs, however—from a commercial perspective—such an approach is not ideal in terms of scalability. On the

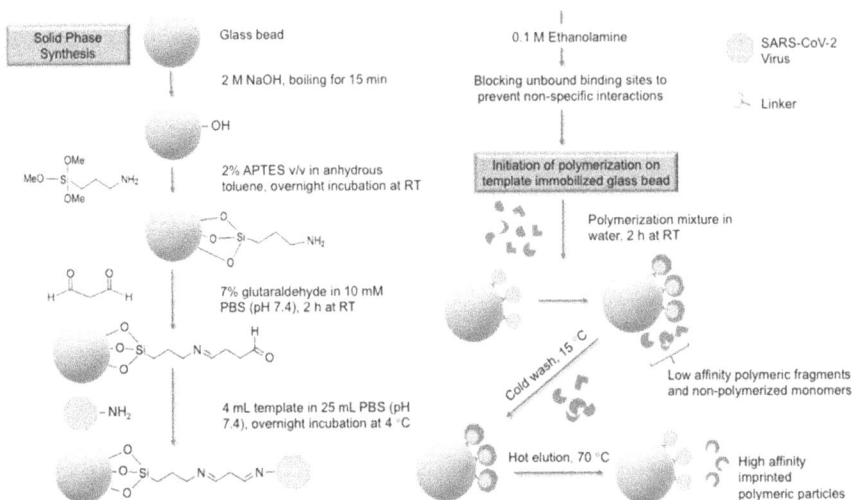

FIGURE 12.5   Scheme of the solid-phase synthesis of SARS-CoV-2 nanoMIPs [60].

other hand, the immobilization of pre-formed nanoMIPs onto sensor surfaces is completely scalable, thus enabling the mass-scale production of sensors. It is worth noting that MIP immobilization onto surfaces is a straightforward method and typically does not lead to loss of binding performance, as opposed to what often occurs when antibodies are immobilized on surfaces. Furthermore, the reduction in animal derived components is one of the obvious advantages MIPs can offer.

The most recent advances in the field of plasmonic sensors discussed in this work highlight the promise that MIP-based platforms hold, although probably further work on the reproducibility aspect of the MIP performance on sensors needs to be carried out. In light of the recent developments in electronics and MIP development processes, it is likely that we will witness a rise in multiplexed sensors and/or hybrid platforms capable of achieving lower LODs in biological matrices.

REFERENCES

1. Karakitsios, S., et al., Challenges on detection, identification and monitoring of indoor airborne chemical-biological agents. *Safety Science*, 2020. **129**.
2. Hakonen, A., et al., Explosive and chemical threat detection by surface-enhanced Raman scattering: A review. *Analytica Chimica Acta*, 2015. **893**: p. 1–13.
3. Alberti, G., et al., Trends in Molecularly Imprinted Polymers (MIPs)-based plasmonic sensors. *Chemosensors*, 2023. **11**(2).
4. Shen, Y., et al., Preparation and application progress of imprinted polymers. *Polymers (Basel)*, 2023. **15**(10).
5. Piletsky, S., et al., Molecularly imprinted polymers for cell recognition. *Trends in Biotechnology*, 2020. **38**(4): p. 368–387.
6. Ayivi, R.D., S.O. Obare, and J. Wei, Molecularly imprinted polymers as chemosensors for organophosphate pesticide detection and environmental applications. *TrAC Trends in Analytical Chemistry*, 2023. **167**.

7. Lowdon, J.W., et al., MIPs for commercial application in low-cost sensors and assays – An overview of the current status quo. *Sensors and Actuators B: Chemical*, 2020. **325**: p. 128973.
8. Barhoum, A., et al., Modern designs of electrochemical sensor platforms for environmental analyses: Principles, nanofabrication opportunities, and challenges. *Trends in Environmental Analytical Chemistry*, 2023. **38**.
9. Leibl, N., et al., Molecularly imprinted polymers for chemical sensing: A tutorial review. *Chemosensors*, 2021. **9**(6).
10. Ahmad, O.S., et al., Molecularly imprinted polymers in electrochemical and optical sensors. *Trends in Biotechnology*, 2019. **37**(3): p. 294–309.
11. Saylan, Y., et al., Advances in biomimetic systems for molecular recognition and biosensing. *Biomimetics (Basel)*, 2020. **5**(2).
12. Saylan, Y. and A. Denizli, Highly sensitive and selective plasmonic sensing platforms, in *Plasmonic Sensors and Their Applications*, A. Denizli, Editor. 2021, Wiley-VCH, Weinheim, p. 55–69.
13. Saylan, Y., S. Akgonullu, and A. Denizli, Plasmonic sensors for monitoring biological and chemical threat agents. *Biosensors (Basel)*, 2020. **10**(10).
14. Bhalla, N., et al., Opportunities and challenges for biosensors and nanoscale analytical tools for pandemics: COVID-19. *ACS Nano*, 2020. **14**(7): p. 7783–7807.
15. Cen, C., et al., High quality factor, high sensitivity metamaterial graphene-perfect absorber based on critical coupling theory and impedance matching. *Nanomaterials (Basel)*, 2020. **10**(1).
16. Masson, J.F., Portable and field-deployed surface plasmon resonance and plasmonic sensors. *Analyst*, 2020. **145**(11): p. 3776–3800.
17. Zhan, C., et al., Determining the interfacial refractive index via ultrasensitive plasmonic sensors. *Journal of the American Chemical Society*, 2020. **142**(25): p. 10905–10909.
18. Demirel, G., et al., Molecular engineering of organic semiconductors enables noble metal-comparable SERS enhancement and sensitivity. *Nature Communications*, 2019. **10**(1): p. 5502.
19. Wang, X., et al., Molecular imprinting based hybrid ratiometric fluorescence sensor for the visual determination of bovine hemoglobin. *ACS Sensors*, 2018. **3**(2): p. 378–385.
20. Homola, J., Surface plasmon resonance sensors for detection of chemical and biological species. *Chemical Reviews*, 2008. **108**(2): p. 462–493.
21. Wang, C. and C. Yu, Detection of chemical pollutants in water using gold nanoparticles as sensors: A review. *Reviews in Analytical Chemistry*, 2013. **32**(1): p. 1–14.
22. Ayivi, R.D., et al., Molecularly imprinted plasmonic sensors as nano-transducers: An effective approach for environmental monitoring applications. *Chemosensors*, 2023. **11**(3).
23. Li, D.-W., et al., Recent progress in surface enhanced Raman spectroscopy for the detection of environmental pollutants. *Microchimica Acta*, 2013. **181**(1–2): p. 23–43.
24. Condorelli, M., et al., Silver nanoplates paved PMMA cuvettes as a cheap and re-usable plasmonic sensing device. *Applied Surface Science*, 2021. **566**.
25. Esen, C. and S.A. Piletsky, Surface plasmon resonance sensors based on molecularly imprinted polymers, in *Plasmonic Sensors and Their Applications*, A. Denizli, Editor. 2021, Wiley-VCH, Weinheim, p. 221–236.
26. Öztürk, B.Ö. and S.K. Şehitoğlu, Pyrene substituted amphiphilic ROMP polymers as nano-sized fluorescence sensors for detection of TNT in water. *Polymer*, 2019. **183**.
27. Zhang, S., et al., Dual-state fluorescent probe for ultrafast and sensitive detection of nerve agent simulants in solution and vapor. *ACS Sensors*, 2023. **8**(3): p. 1220–1229.
28. Ma, K., et al., Protection against chemical warfare agents and biological threats using metal–organic frameworks as active layers. *Accounts of Materials Research*, 2023. **4**(2): p. 168–179.

29. Chen, L., D. Wu, and J. Yoon, Recent advances in the development of chromophore-based chemosensors for nerve agents and phosgene. *ACS Sensors*, 2018. **3**(1): p. 27–43.
30. Kim, K., et al., Destruction and detection of chemical warfare agents. *Chemical Reviews*, 2011. **111**(9): p. 5345–403.
31. Lieberzeit, P.A., et al., Artificial receptor layers for detecting chemical and biological threats. *Procedia Engineering*, 2010. **5**: p. 381–384.
32. Jahn, I.J., A. Muhlig, and D. Cialla-May, Application of molecular SERS nanosensors: Where we stand and where we are headed towards? *Analytical and Bioanalytical Chemistry*, 2020. **412**(24): p. 5999–6007.
33. He, H., et al., Controllably prepared molecularly imprinted core-shell plasmonic nanostructure for plasmon-enhanced fluorescence assay. *Biosensors and Bioelectronics*, 2019. **146**: p. 111733.
34. Costa, L.G., Organophosphorus compounds at 80: Some old and new issues. *Toxicological Sciences*, 2018. **162**(1): p. 24–35.
35. Kaushal, J., M. Khatri, and S.K. Arya, A treatise on Organophosphate pesticide pollution: Current strategies and advancements in their environmental degradation and elimination. *Ecotoxicology and Environmental Safety*, 2021. **207**: p. 111483.
36. Nicolopoulou-Stamati, P., et al., Chemical pesticides and human health: The urgent need for a new concept in agriculture. *Frontiers in Public Health*, 2016. **4**: p. 148.
37. Rebelo, P., et al., Molecularly imprinted plasmonic sensors for the determination of environmental water contaminants: A review. *Chemosensors*, 2023. **11**(6).
38. Wei, C., H. Zhou, and J. Zhou, Ultrasensitively sensing acephate using molecular imprinting techniques on a surface plasmon resonance sensor. *Talanta*, 2011. **83**(5): p. 1422–7.
39. Yao, G.H., et al., Surface plasmon resonance sensor based on magnetic molecularly imprinted polymers amplification for pesticide recognition. *Analytical Chemistry*, 2013. **85**(24): p. 11944–51.
40. Çakır, O. and Z. Baysal, Pesticide analysis with molecularly imprinted nanofilms using surface plasmon resonance sensor and LC-MS/MS: Comparative study for environmental water samples. *Sensors and Actuators B: Chemical*, 2019. **297**.
41. Feng, S., et al., Development of molecularly imprinted polymers-surface-enhanced Raman spectroscopy/colorimetric dual sensor for determination of chlorpyrifos in apple juice. *Sensors and Actuators B: Chemical*, 2017. **241**: p. 750–757.
42. Lazarević-Pašti, T., et al., Molecularly imprinted plasmonic-based sensors for environmental contaminants—Current state and future perspectives. *Chemosensors*, 2023. **11**(1).
43. Jalilzadeh, M., et al., Design and preparation of imprinted surface plasmon resonance (SPR) nanosensor for detection of Zn(II) ions. *Journal of Macromolecular Science, Part A*, 2019. **56**(9): p. 877–886.
44. Bakhshpour, M. and A. Denizli, Highly sensitive detection of Cd(II) ions using ion-imprinted surface plasmon resonance sensors. *Microchemical Journal*, 2020. **159**.
45. Patel, A.B., et al., Polycyclic aromatic hydrocarbons: Sources, toxicity, and remediation approaches. *Frontiers in Microbiology*, 2020. **11**: p. 562813.
46. Castro-Grijalba, A., et al., SERS-based molecularly imprinted plasmonic sensor for highly sensitive PAH detection. *ACS Sensors*, 2020. **5**(3): p. 693–702.
47. van Egmond, H.P., R.C. Schothorst, and M.A. Jonker, Regulations relating to mycotoxins in food: Perspectives in a global and European context. *Analytical and Bioanalytical Chemistry*, 2007. **389**(1): p. 147–57.
48. Zhou, Q. and D. Tang, Recent advances in photoelectrochemical biosensors for analysis of mycotoxins in food. *TrAC Trends in Analytical Chemistry*, 2020. **124**.

49. Upadhyayula, V.K., Functionalized gold nanoparticle supported sensory mechanisms applied in detection of chemical and biological threat agents: A review. *Analytica Chimica Acta*, 2012. **715**: p. 1–18.

50. Sergeyeva, T., et al., Highly-selective and sensitive plasmon-enhanced fluorescence sensor of aflatoxins. *Analyst*, 2022. **147**(6): p. 1135–1143.

51. Akgonullu, S., H. Yavuz, and A. Denizli, SPR nanosensor based on molecularly imprinted polymer film with gold nanoparticles for sensitive detection of aflatoxin B1. *Talanta*, 2020. **219**: p. 121219.

52. Akgönüllü, S., H. Yavuz, and A. Denizli, Development of gold nanoparticles decorated molecularly imprinted–Based plasmonic sensor for the detection of aflatoxin M1 in milk samples. *Chemosensors*, 2021. **9**(12).

53. Noormandi, A. and F. Dabaghzadeh, Effects of green tea on Escherichia coli as a uropathogen. *Journal of Traditional and Complementary Medicine*, 2015. **5**(1): p. 15–20.

54. Shoaie, N., M. Forouzandeh, and K. Omidfar, Voltammetric determination of the Escherichia coli DNA using a screen-printed carbon electrode modified with polyaniline and gold nanoparticles. *Mikrochim Acta*, 2018. **185**(4): p. 217.

55. Erdem, O., et al., Molecularly imprinted nanoparticles based plasmonic sensors for real-time Enterococcus faecalis detection. *Biosensors and Bioelectronics*, 2019. **126**: p. 608–614.

56. Percin, I., et al., Microcontact imprinted plasmonic nanosensors: Powerful tools in the detection of Salmonella paratyphi. *Sensors (Basel)*, 2017. **17**(6).

57. Erdem, Ö., et al., Comparison of molecularly imprinted plasmonic nanosensor performances for bacteriophage detection. *New Journal of Chemistry*, 2020. **44**(41): p. 17654–17663.

58. Ozgur, E., et al., Surface plasmon resonance based biomimetic sensor for urinary tract infections. *Talanta*, 2020. **212**: p. 120778.

59. Cennamo, N., et al., Proof of concept for a quick and highly sensitive on-site detection of SARS-CoV-2 by plasmonic optical fibers and molecularly imprinted polymers. *Sensors (Basel)*, 2021. **21**(5).

60. Bajaj, A., et al., Synthesis of molecularly imprinted polymer nanoparticles for SARS-CoV-2 virus detection using surface plasmon resonance. *Chemosensors*, 2022. **10**(11).

13 Explosive Detection Using Plasmonic Nanosensors

*Duygu Çimen, Sevgi Aslıyüce, Merve Çalışır,
Muhammed Erkek, and Nilay Bereli*

13.1 INTRODUCTION

Explosive detection stands as a cornerstone of public safety and security across various sectors, encompassing transportation, defense, and law enforcement (Singh, 2007). The prevailing techniques employed for explosive detection are often marred by time-intensive processes that may lack the requisite sensitivity and specificity (Wang, 2007). Consequently, the surge in pioneering sensing technologies capable of swiftly and accurately identifying explosives is noticeable (Liu et al., 2019). Among the array of emerging technologies, plasmonic nanosensors have surfaced as a prominent contender (Akgönüllü et al., 2023).

Plasmonic nanosensors harness the distinctive characteristics of surface plasmon resonance (SPR) to facilitate the detection and characterization of target analytes, notably including explosives, with precision and selectivity (Kawaguchi et al., 2008). SPR, an enthralling phenomenon, unfurls when electromagnetic waves, frequently manifesting as light, interact with unconfined electrons at the juncture between a metal and a dielectric medium. This interaction gives rise to electron oscillations with resonance, resulting in a distinct absorption trough that becomes evident within the range of reflectance or transmittance wavelengths (Homola and Piliarik, 2006).

Within the context of explosive detection, plasmonic nanosensors supervise the SPR to discern shifts in the refractive index proximate to the sensor's interface. A cooperative interaction takes place when an explosive molecule binds to a receptor molecule that is fixed on the surface of the sensor (Shankaran et al., 2004). This connection leads to a change in the nearby refractive index, consequently causing a modification in the wavelength of the SPR. This shift, meticulous in its quantification, establishes a tangible link with the presence and concentration of the explosive analyte.

The use of plasmonic nanosensors in explosive determination brings many advantages. One of them is that it is unrivaled in terms of sensitivity. The hallmark of plasmonic nanosensors lies in their exceptional sensitivity, engendered by the potent interaction between the plasmonic field and the analyte. This sensitivity threshold

DOI: 10.1201/9781003459316-13

facilitates the detection of even the faintest traces of explosive agents, thereby amplifying the overall effectiveness of security protocols (Divya et al., 2022). Another
advantage is real-time surveillance. A pivotal attribute characterizing plasmonic
nanosensors is their aptitude for real-time, label-free detection. This real-time
insight assumes paramount significance in scenarios necessitating expeditious identification of explosives. The pinnacle of the selectivity of plasmonic nanosensors
is another undeniable benefit. The versatility of plasmonic nanosensors emerges as
they are fashioned with bespoke receptors or ligands capable of selectively engaging
certain explosive molecules. This selectivity underscores the veracity of detection
outcomes, mitigating the prospects of false positives and increasing detection precision. Also, their ergonomics and portability take research one step further in terms
of ease of use. The form factor of plasmonic nanosensor apparatus can be tailored to
occupy minimal space, facilitating their seamless integration into portable detection
platforms (Das et al., 2022). This mobility assumes pronounced utility in field-based
applications and remote locales. Non-invasive screening is one of the advantages
that should be mentioned. Plasmonic nanosensors circumvent the need for exhaus-
tive sample preparation or chemical tagging, preserving the intrinsic integrity of the
sample under scrutiny.

Applications where plasmonic nanosensors are used in explosive determination
also vary. Aerodrome security is one example of the application of plasmonic nanosensors in the determination of explosives. The assimilation of plasmonic nanosensors into airport security infrastructure can revolutionize luggage and personal
belonging screening protocols, enabling the discernment of explosive residue. The
gamut of military applications reaps the benefits of plasmonic nanosensors in uncov-
ering concealed explosives present in vehicles, equipment, and structures (Fehlen et
al., 2022). Plasmonic nanosensors have also the potential to augment the safety apparatus governing public transportation networks, scrutinizing baggage, parcels, and
passenger possessions for explosive traces. Law enforcement agencies can harness
the capabilities of plasmonic nanosensors to dissect evidence pertinent to explosive
devices deployed in criminal activities. In this book chapter on the use of plasmonic
nanosensors in explosive detection, various explosive samples are first examined.
The chapter then provides an overview of different types of nanosensors, transition-
ing to plasmonic nanosensors, and presents examples of their applications.

13.2 EXPLOSIVES

Explosive substances decompose by generating a sudden and large amount of heat
and gas with a thermal and mechanical effect. These substances create a large
amount of pressure with the formation of gas in a closed system. Explosives release
excess heat during oxidation and at the same time a sudden reaction must occur. In
addition, they are substances that undergo a very rapid decomposition with a sudden
and self-sustained exothermic reaction, creating a very high pressure due to the heat
and gas released as a result of the reaction. The high pressure of explosive materials
increases the heat released at the time of explosion due to the high-pressure effect
resulting from the decomposition into gases with a much larger volume than the

solid or liquid state. An explosion occurs when a chemically unstable compound or mixture has a high reaction rate, changes in temperature and pressure occur, and large amounts of gaseous products are formed. Many explosives that are commonly used are made of organic compounds and are classified into six broad classes based on their chemical properties. There are many different types of explosive substances, ranging from those used in industry (like TNT and dynamite), to those used in the military (like C-4 and Semtex), to those used in mining and construction (like ANFO). Each type of explosive has different properties, including different rates of energy release, different sensitivity to ignition, and different levels of stability (Ali et al., 2015; Lefferts and Castell, 2015; Moore, 2007; Zapata and García-Ruiz, 2020).

13.2.1 NITROAROMATIC COMPOUNDS

Organic molecules with nitro groups connected to an aromatic ring, such as benzene, are known as nitroaromatic compounds. These substances are notorious for their detonative characteristics and are broadly applied for the manufacture of explosives and propellants. Popular nitroaromatics include trinitrotoluene (TNT), dinitrobenzene (DNB), hexanitrostilbene, and picric acid. Trinitrotoluene (TNT) is one of the more popular explosives and is characterized by a yellow, odorless solid that is safe and simple to use. It has been used in military situations, including in the making of artillery shells, bombs, and hand grenades, and in industrial activities such as demolition and development. Dinitrobenzene (DNB) is a nitroaromatic substance composed of three isomers, namely ortho-dinitrobenzene, meta-dinitrobenzene, and para-dinitrobenzene, which differ in the placement of the nitro groups on the benzene ring. This chemical has been utilized in the production of explosives and dyes. Hexanitrostilbene is a further potent nitroaromatic explosive. TNT has a lower detonation velocity and is more sensitive, making it suitable for specialized military and industrial applications. Picric Acid, also known as 2,4,6-trinitrophenol (TNP), is a yellow, crystalline solid that has a powerful explosive effect. It has been utilized as a military high explosive and in the manufacture of artillery shells and other munitions in the past. Despite its susceptibility to shock and friction, it has largely been supplanted by more secure explosives like TNT. It is crucial to bear in mind that the usage of these explosives necessitates great caution and expertise as they might be perilous.

Since these compounds are mostly used for military and industrial purposes, they are subject to stringent rules and are not usually accessible to the general public (Desai et al., 2022; Johnston et al., 2015; Ma et al., 2014).

13.2.2 ALIPHATIC NITRO COMPOUNDS

Organic compounds with a nitro-functional group connected to an aliphatic carbon chain are called nitro-aliphatic compounds. Organic chemistry classifies carbon chains either as aliphatic or aromatic based on the presence or absence of a benzene ring in their composition. Organic compounds with a nitro functional group connected to an aliphatic carbon chain are called aliphatic nitro compounds.

Aliphatic nitro compounds have a wide variety of properties and applications. Some of these compounds demonstrate high reactivity and are commonly used as intermediates in organic synthesis to create several functional groups. They can also be used as starting materials for the manufacturing of other beneficial compounds in the chemical industry. Nitromethane is a notable aliphatic nitro compound that has multiple industrial usages, for instance as a solvent, fuel enhancer, and for the manufacture of pharmaceuticals and agrochemicals. Nevertheless, a lot of aliphatic nitro compounds are prone to explosions, and care must be taken when dealing with or synthesizing them (Zhang et al., 2022).

13.2.3 NITRAMINES OR NITROSAMINES

Nitramines and nitrosamines are chemical compounds containing nitrogen. These compounds can be carcinogenic, leading to potential health and environmental risks. Nitramines are an organic compound with nitro and amine functional groups. They are often employed in the manufacture of explosive materials and fuels. Nitrates are also found in some pharmaceutical products and agricultural chemicals. Nitrosamines are a family of chemical compounds that contain the nitroso functional group. They can form through the reaction of secondary amines with nitrite compounds, often found in processed foods, as well as in certain industrial processes and environmental sources. Nitrosamines are known to be potentially carcinogenic, meaning they can cause cancer in humans. Various types of cancer, such as stomach, esophageal, and bladder cancer, have been linked to nitrosamines. Therefore, the U.S. Food and Drug Administration and the European Food Safety Authority have imposed limits on the amount of nitrosamines that are allowed in food, drugs, and consumer products, to minimize any potential health risks (Kayhomayun et al., 2020; Pavlov et al., 2019).

13.2.4 NITRATE ESTERS

Nitrate esters are a class of chemical compounds that are derived from nitric acid and alcohol through a process called esterification. These compounds, known for their explosive properties, are used in various applications, including repulsive gases and drugs. Pentaerythritol tetranitrate is a powerful explosive with high fragility and stability properties. It is used as a component in plastic explosives, especially in military and industrial applications. Nitroglycerin is a very sensitive explosive and was first used as dynamite. After, it was also used in the construction and mining industries. Nitroglycerin is also used medically to treat angina pectoris, a condition whose symptoms include chest pain caused by reduced blood flow. Ethylene glycol dinitrate is a less preferred explosive due to its sensitivity to changes and tendency to become unstable. It has been employed in military applications and has been known to be utilized as a component of homemade explosives. Nitroguanidine is a less sensitive nitrate ester explosive compared to others. It has been used in rockets and ammunition as a propellant component. These nitrate esters are characterized by their high energy content and the potential for rapid decomposition and explosive release of energy. Their instability and danger of accidental detonation necessitate that suitable handling and storage protocols be employed in their production and utilization. These compounds have been utilized for military, industrial, and therapeutic uses, and have been instrumental in numerous technological developments over time (Gao et al., 2015; Zhang et al., 2017).

13.2.5 Acid Salts

An acid salt is formed when a weak acid reacts with a strong base. One such acid salt is ammonium nitrate, which forms when ammonia and nitric acid react with one another. Acid salts have many different uses in the fields of chemistry and industry. Ammonium nitrate is a common example of a substance that can be used as a fertilizer and for explosives, thanks to its capacity to release nitrogen and oxygen gases when decomposed in certain conditions (Aguirre-Díaz et al., 2021; Diaz and Hahn, 2020).

13.2.6 Organic Peroxides

Organic peroxides are a class of chemicals with a peroxide group, which is an oxygen-oxygen single bond. These compounds are very reactive and could be dangerous owing to their susceptibility to heat, friction, and shock. While some organic peroxides are used in various industrial processes and as initiators in polymerization reactions, they can also be used for illicit purposes as explosives. Examples of organic peroxides include triacetone triperoxide and hexamethylene triperoxide diamine. Triacetone triperoxide (TATP), a potent explosive, has obtained notoriety due to its involvement in many terrorist occurrences. It is relatively easy to synthesize using common household chemicals, but it is extremely unstable and sensitive to heat, shock, and friction. TATP has been used in multiple terrorist assaults because of its relatively easy synthesis and its capacity to effect considerable destruction. Nevertheless, its high level of sensitivity to storage and handling conditions also poses a risk to those trying to create or use it. Hexamethylene triperoxide diamine (HMTD) is another organic peroxide explosive. Much like TATP, HMTD is reactive to heat, shock, and friction. It can be synthesized from basic ingredients, thus presenting a possible threat to safety and security (Krivitsky et al., 2019; Kuracina et al., 2021; Mahbub et al., 2023).

13.3 TYPES OF PLASMONIC NANOSENSORS

Plasmonic sensors exploit the remarkable phenomenon of plasmon resonance, where the interaction between light and electrons at the interface between a metal and a dielectric material generates an electromagnetic wave on the surface. This wave enables highly sensitive detection mechanisms for a variety of applications, including biosensing and chemical sensing (Lee et al., 2021).

Surface Plasmon Resonance (SPR) manifests in two well-defined manners: Localized Surface Plasmon Resonance (LSPR) and Propagating Surface Plasmon Polaritons (SPPs). LSPR arises when the size of a metallic nanostructure is smaller than the wavelength of incoming light, resulting in collective yet non-propagating oscillations of surface electrons within the metallic nanostructure. LSPR proves highly responsive to alterations in the refractive index of the encompassing environment, forming the basis for color-sensitive plasmonic sensors. Furthermore, LSPR concentrates the incident electromagnetic field around the nanostructure, influencing optical phenomena like fluorescence, Raman scattering, and infrared absorption. These effects give rise to phenomena such as Surface-Enhanced Raman Scattering (SERS). The localized electromagnetic field linked with LSPR extends into the adjacent environment and displays an exponential decay pattern when considering a dipole configuration. Conversely, SPPs represent propagating charge oscillations occurring on the surface of thin metal films. SPPs cannot be triggered by radiation from free space; instead, they necessitate momentum matching, often achieved through the periodic arrangement of a nanostructure, to resonate effectively. SPPs exhibit a keen sensitivity to variations in the refractive index of the nearby medium, thus serving as valuable signal transducers in sensor applications. Additionally, SPPs play a pivotal role in modulating radiation during SERS processes. The evanescent electromagnetic field generated by SPPs boasts a longer decay length in comparison to LSPR, enabling SPPs to be influenced by alterations occurring at a more considerable distance from the surface of the nanostructure (Li et al., 2014).

13.3.1 SURFACE PLASMON RESONANCE

One of the primary branches of plasmonic sensing is SPR, which entails exciting surface plasmons on a metal surface, usually gold or silver. These surface plasmons are incredibly sensitive to variations in the surrounding medium's refractive index. As biomolecules or chemical analytes adhere to the sensor surface, they produce modifications in the refractive index, leading to changes in the plasmon resonance wavelength. This displacement can be accurately measured, permitting in-situ tracking of molecular interactions. Sensors based on SPR provide numerous benefits, including detection that is devoid of labeling, heightened sensitivity, and the potential to analyze binding kinetics (Homola and Piliarik, 2006).

13.3.2 LOCALIZED SURFACE PLASMON RESONANCE

LSPR is another plasmonic sensing technique that has gained prominence. LSPR involves the excitation of plasmons in nanoscale metallic structures, such as nanoparticles or nanoholes. These localized plasmons generate strong electromagnetic fields in their immediate vicinity, which are highly sensitive to changes in the local environment. LSPR sensors can detect analyte binding events at the nanoscale and are particularly useful for studying single-molecule interactions and small analytes. The captivating optical properties of metal nanostructures are intimately tied to the phenomenon known as LSPR. It is an optical phenomenon that arises from the collective oscillations of the electron gas within metal nanostructures when surrounded by a dielectric medium. Silver and gold are the most favored noble metals for nanostructures in plasmonic applications. Although silver generates more intense LSPR bands than gold, gold is preferred due to the superior chemical stability of the nanomaterials (Malekzad et al., 2017).

When light interacts with metal nanostructures, certain incident photons are absorbed, while others are scattered in different directions. Both absorption and scattering phenomena are significantly amplified upon excitation of LSPR. Extinction measurements are frequently employed to describe systems with a high concentration of nanostructures. On the other hand, scattering measurements that utilize Dark Field or Total Internal Reflection spectroscopy offer a lower signal-to-background ratio compared to extinction. However, they are more valuable when dealing with samples featuring a low density of nanostructures. This sensitivity allows for the optical characterization of even individual nanomaterials. The very intense and tightly confined electromagnetic fields produced by LSPR serve as a highly sensitive tool for detecting very small changes in the surrounding dielectric environment, which makes them especially appealing for sensing applications (Sepúlveda et al., 2009).

13.3.3 SURFACE-ENHANCED RAMAN SPECTROSCOPY

SERS is a method that encompasses the dispersion of incoming laser energy with an outcome of spectral peaks attributed to the vibrational patterns of molecules. These peaks are shifted in frequency from the incident energy and serve as characteristic fingerprints for virtually all polyatomic species. Typical SERS substrates consist of roughened surfaces made from silver, copper, or gold. SERS necessitates the adsorption of analyte molecules onto the SERS substrate. Following adsorption, the Raman signal of the analyte is significantly enhanced, achieving signal intensities like those obtained with fluorescence. However, unlike fluorescence, which features broad absorption and emission bands, SERS yields narrow spectral peaks. This high spectral resolution enables simultaneous analysis of multiple components. Other benefits of SERS are sample manipulation, which is easy, swift analysis, in-situ identification of analytes, and the presence of sturdy, portable Raman spectrometers. However, SERS has limitations such as the need for close contact between the surface and the analyte, the difficulty of reusability, the degradation of the substrate over time, and the signal attenuation. Despite these limitations, SERS offers exceptional sensitivity and spectral selectivity, making it an attractive method for the detection of a wide range of chemical species, including heavy metals, toxic ions, nutrients, bacteria, pesticides, drugs, and explosives (Mosier-Boss, 2017).

13.4 PLASMONIC NANOSENSOR APPLICATIONS FOR THE DETECTION OF EXPLOSIVES

The sensors are primarily used in medical, environmental, and chemical diagnostics. However, they're also highly significant in detecting explosive vapors. Bai and colleagues conducted a study through which they developed, for the first time, an SPR-enhanced photothermal nanosensor. This nanosensor is highly sensitive and selective and can help detect TNT in situ. The aim of this study was twofold: Firstly,

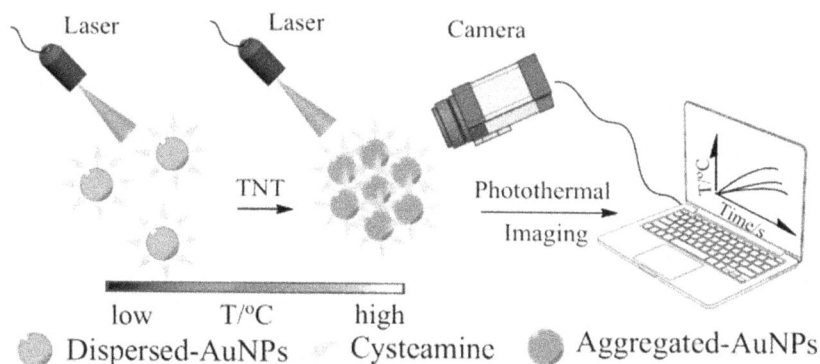

FIGURE 13.1   A schematic illustration of an SPR-enhanced photothermal nanosensor based on the TNT-induced aggregation of AuNPs (reproduced with permission from Bai et al., 2016).

to enrich amine with AuNPs and form a Meisenheimer complex of TNT molecules, and secondly to enhance the photothermal properties of these nanoparticles by using their SPR properties (Figure 13.1). Photothermal responses are affected by nitroaromatic compounds. The nanosensor has outperformed other methods (e.g., colorimetric methods) in analytically detecting TNT over a broad spectral range, thanks to the change in temperature difference and photothermal imaging (Bai et al., 2016).

In further research, Wang designed an SPR sensor chip using an organic ligand, 3-aminopropyltriethoxysilane (APTES). This was made possible by generating a Meisenheimer complex with TNT. The SPR chip, which was based on APTES, demonstrated a remarkable level of sensitivity and selectivity in identifying TNT explosives at the ppm level. It is expected that the sensor chip will be effective in the detection of TNT explosives due to its simplicity, speed, and low cost, and that it also has the potential to be used in portable applications (Wang, 2021).

Wang and colleagues synthesized three TNT-binding peptides in this study and derived anti-TNT monoclonal antibodies by amino sequencing. The authors developed a TNT-specific, highly sensitive, easy-to-detect sensor using an SPR sensor. This study has enabled the creation of a new sensor surface platform to develop and improve the performance of peptide-based TNT sensors (Figure 13.2). The study demonstrated that the TNT-binding peptide was specific and selective in its binding (Wang et al., 2018b).

In this research, Wang and colleagues developed an SPR sensor chip utilizing single-walled carbon nanotubes (SWCNTs) that had been modified with TNT recognition peptide (Figure 13.3). Carboxylic acid functionalized SWCNTs were set in position on the SPR Au chip surface, which was altered with APTES. The TNT recognition peptide TNTHCDR3 was immobilized on the SWCNT's surface via π-stacking interactions between the SWCNTs and aromatic amino acids. The surface of the sensor, which had been modified with peptide-SWCNTs, was assessed and confirmed via observation with atomic force microscopy (AFM). The SPR sensor

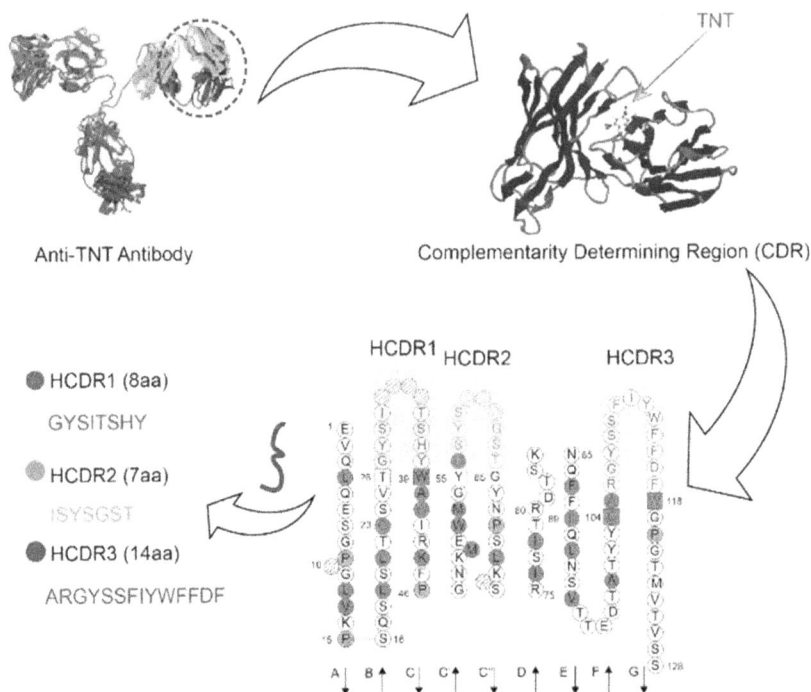

FIGURE 13.2 The amino acid sequence of the Complementarity Determining Region (CDR) in the anti-TNT monoclonal antibody was obtained (reproduced with permission from Wang et al., 2018b).

chip, which was a hybrid of peptide-SWCNTs, exhibited increased sensitivity, with a limit of detection (LOD) of 772 ppb (Wang et al., 2018a).

Riskin et al. developed an imaged compound comprising AuNPs crosslinked with bisaniline to detect RDX with high sensitivity. To create the imaged sites for RDX, a solid, non-planar imprinting base such as Kemp's acid was used. The authors demonstrated that changes in the dielectric properties of the gold-NP composite measurably influence the coupling between the local plasmon of the NPs and the surface plasmon wave by forming a π-donor-acceptor complex. This discovery enabled the detection of RDX at a bulk concentration of 12 fM (Riskin et al., 2010).

Zou et al. prepared LSPR-sensitive silver clusters embedded in porous silica by reverse microemulsion, which exhibited high LSPR sensitivity to both nitroaromatic and nitro-aliphatic explosives. The silver cluster structures were synthesized via microemulsion and embedded in porous silica shells, forming LSPR-sensitive hybrid nanostructures. These nanostructures show high sensitivity and selectivity to nitroaromatic and nitroaliphatic compounds and are particularly sensitive to commonly used high-explosive compounds. This sensitivity results from the combination of the adsorption properties of the porous silica matrix with the high LSPR sensitivity of the embedded Ag cluster structures. Such a hybrid structure provides a

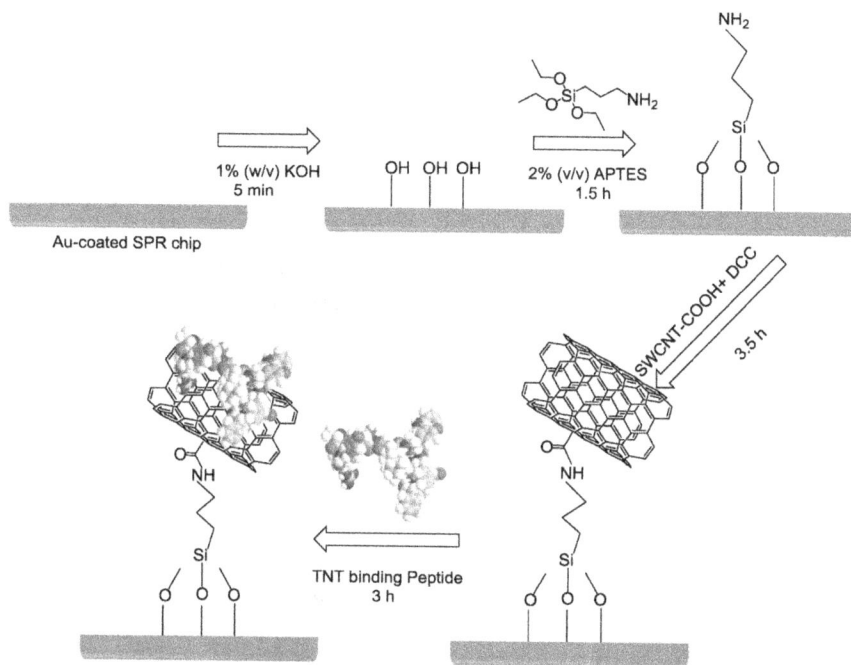

FIGURE 13.3   A schematic illustration of the immobilization procedure of the TNT-binding peptide TNTHCDR3 on the surface of a SWCNT-based SPR gold-coated chip (reproduced with permission from Wang et al., 2018a).

promising platform for engineering sensors by exploiting the concentration enrichment in porous silica (Zou et al., 2012).

Wang et al. devised a novel approach for detecting and distinguishing explosives with high sensitivity by employing a plasmonic substrate that is coated with metal-organic polyhedra (MOPs). The basis of this approach rests on the deliberate choice of assembly components, i.e., organic ligands and unsaturated metal sites contained within the MOP cage, which enables the precise control of weak molecular interactions. The MOP cage acts as a proficient receptor for the selective binding of explosives. The combination of the MOP cage and a plasmonic substrate with a surface-enhanced Raman scattering enhancement factor creates a sensor that exhibits good sensing proficiency for various explosive groups, including notably difficult aliphatic nitro-organic compounds (Wang et al., 2015).

Chegel et al. proposed sensor elements for detecting explosive analogues in liquid and vapor phases via molecularly imprinted acrylamide copolymer coating using localized surface plasmon resonance phenomena based on gold nanostructure arrays on glass substrates (nanochips). The nanochips demonstrated detection limits of 1 pM in aqueous solutions and 0.1 ppm in the gas phase for 4-nitrophenol. Vapor-phase sensing of 4-nitrotoluene, 1-nitronaphthalene, and 5-nitroisoquinoline using the plasmonic nanochips, which mimic 4-nitrophenol, revealed a partly selective response with signal saturation becoming evident after two minutes (Chegel et al., 2020).

Vendamani et al. have developed cost-effective and reproducible silver nanodendrite (AgND) substrates, with high-density stems and waves, via a straightforward electroless excavation process. The substrates are characterized by high-intensity impact sites, which facilitate efficient detection of molecules using SERS. Rhodamine 6G (R6G) molecules, at a concentration of 1 nM, were employed to evaluate the substrate's SERS activity. It was found to have an effective enhancement factor (EF) of around 108. AgNDs were subsequently utilized to detect RDX and ammonium nitrate explosives at concentrations up to 1 μM. Normally, an EF of approximately 104 would be obtained for RDX and AN. Deposition of Au nanoparticles on AgNDs doubled the sensitivity of 1 μM R6G. Reproducibility of measurements on this low-cost substrate was achieved, with an RSD of approximately 9% (Vendamani et al., 2020).

Recently, filter paper (FP)-based SERS substrates have gained significant attention due to their low cost, simple handling, and suitability for field applications, as compared to solid substrates. Byram et al. conducted a study in which they synthesized silver-gold alloy nanoparticles using liquid-based laser ablation. They further examined flexible base substrates such as sandpaper, printing paper, and NPs, and optimized the soaking time of NPs (5–60 minutes). Using optimized FP wetted for 30 minutes, SERS sensors were developed to detect small concentrations of pesticide (thiram: 50 nM), dye (Nile blue: 5 nM), and explosive (RDX-1,3,5-trinitroperhydro-1,3,5-triazine: 100 nM) molecules. The SERS effect resulting from gold nanoparticles was found to have an enhancement factor of approximately 105 and relative standard deviation values of less than 15% for all studied molecules. Furthermore, the optimized FP substrate successfully detected pesticide residues on the surface of a banana and RDX on a glass slide. This indicates a high potential the use of these substrates in future field applications (Byram et al., 2022).

Riskin et al. fabricated composites of molecularly processed gold nanoparticles on gold-coated glass substrates. The process involved electropolymerization of thioaniline-functionalized gold nanoparticles on a gold-coated surface using carboxylic acid as a sensor template for explosives. The template was then removed from the matrix to create the corresponding composites. The attachment of the analyte explosives to the gold nanoparticle matrix is monitored via SPR. This technique capitalizes on the electronic interaction between the localized plasmons of the gold nanoparticles and the surface plasmon wave to heighten the SPR responses generated by the dielectric changes of the matrices whenever distinct explosives bond with the matrices. Consequently, the treated matrices exhibit significant affinities with and selectivity for the treated explosives. For instance, gold nanoparticle matrices that are prepared using citric acid have a high sensitivity to pentaerythritol tetranitrate (PETN) or nitroglycerin. Meanwhile, matrices treated with maleic acid or fumaric acid yield high-affinity detection composites for ethylene glycol dinitrate (EGDN) and have achieved very low detection limits. This enables the detection of explosives sensitively and selectively (Riskin et al., 2011).

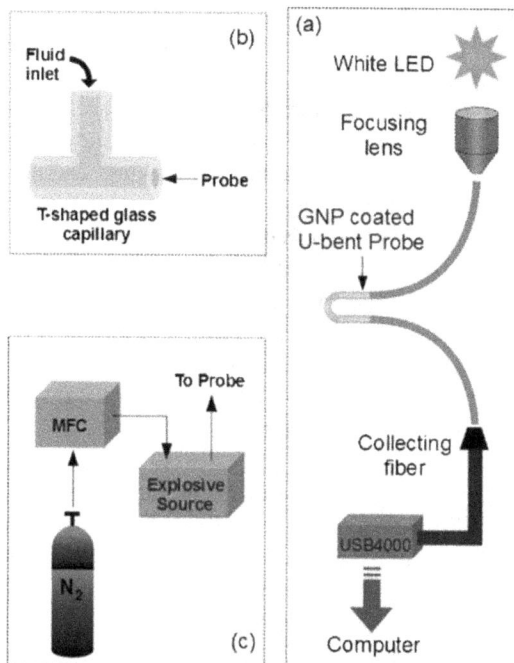

FIGURE 13.4 A schematic representation of fluid-based and vapor phase detection of explosives (reproduced with permission from Bharadwaj and Mukherji, 2014).

Bharadwaj and Mukherji, have developed a chemical sensor capable of detecting explosives, specifically TNT and RDX, in vapor form. The sensor uses the LSPR of gold nanoparticles that are attached to an optical fiber sensor probe with a core diameter of 200 µm and a bend diameter of 1.5 mm (Figure 13.4). The GNPs are immobilized on the probe to excite their localized surface plasmons based on the evanescent field. Functionalized receptor molecules, including 4-mercaptobenzoic acid (4-MBA), l-cysteine, and cysteamine, were used to immobilize gold nanoparticles and create binding sites for nitro-based explosive molecules. When explosive analytes bind to the gold nanoparticle surface groups, they induce refractive index changes in the surrounding environment. The gold nanoparticle-coated probe, which was modified with L-cysteine and cysteamine, exhibited a high level of selectivity towards TNT. The LSPR fiber-optic probe can detect TNT vapors in the low parts per billion range, with a determined detection limit (Bharadwaj and Mukherji, 2014).

Riskin and colleagues have developed a composite consisting of gold nanoparticles crosslinked with bisaniline on gold surfaces using electropolymerization. TNT is bound through the bisaniline bridging units, which enables the sensitive detection of TNT by tracking changes in the SPR reflection. Such changes occur due to the interaction between the local plasmons of the gold nanoparticles and the surface plasmon wave of the gold surface. The detection threshold of this approach for TNT analysis is approximately 10 pM. In addition, the bisaniline crosslinked gold nanoparticle compound's electropolymerization creates a molecularly suppressed framework for subsequent TNT binding in the presence of picric acid. The immobilized gold nanoparticle composite can detect TNT with a detection threshold equal to 10 fM, which enables TNT detection. Analysis of SPR reflectance changes in the presence of varying TNT concentrations indicates a two-stage calibration curve with highly responsive TNT detection by the imaged regions. The initial stage involves TNT identification at the imaged regions, followed by TNT detection at the unimaged π-donor bisaniline sites with reduced sensitivity. This labeled gold nanoparticle composite displays exceptional selectivity. Numerous techniques have been used to investigate the structure and properties of bisaniline crosslinked gold nanoparticle composites. It has been shown that the composite's ability to detect TNT with high sensitivity is due to changes in its dielectric features caused by the formation of π-donor-acceptor complexes between TNT and bisaniline units. These variations in the dielectric properties lead to changes in the conductivity of the gold nanoparticle framework (Riskin et al., 2009).

In another study, Singh et al. investigated the production of a SERS substrate with high efficiency. The substrate is composed of a rough microporous silicon layer and a thin layer of gold. The results indicate that the outstanding enhancement observed is due to the coupling between localized surface plasmons and long-lasting extended surface plasmons. Specially braided spongy porous Si was created using anodic electrochemical etching. Gold was deposited through electron beam evaporation. The SERS activity of the substrate was confirmed through the immobilization of 4-ATP. This was utilized to optimize the thickness of gold and achieve the greatest enhancement. The resultant substrate was then utilized in a photoacoustic (PA) sensor and illustrated a linear response to PA concentration within the range of 10^{-9} to 10^{-3} M. Usually, SERS-based sensors are calibrated through variation in peak height, but significant signal variability constrains their potential application. Therefore, we suggested performing SERS measurements utilizing peak ratio and demonstrated roughly two to three times superior variability in comparison to peak height variation (Singh et al., 2023).

13.5 CONCLUSION AND FUTURE PERSPECTIVES

In recent years, the threat of terrorism involving explosives has increased significantly. Explosive-based weapons are attractive to terrorists because of their simplicity, ease of use, and capacity to cause widespread damage. Detecting explosives remains a challenging task due to a variety of factors, including the low vapor pressure exhibited by most explosives, the constant emergence of new explosive compositions, and creative concealment and delivery methods. Trace detection of explosives typically involves the collection of vapor or particle samples, which are then analyzed using a highly sensitive sensor system. There is therefore a critical need for fast, sensitive, and portable platforms that will enable real-time detection of such threats. Plasmonic sensor-based platforms have emerged as valuable tools in this regard and have found their place in laboratories. These sensors offer sensitivity comparable to conventional platforms while being compact in size and cost-effective. As a result, they are not only suitable for routine laboratory use but also mobile and field portable systems. This makes them essential for rapid response to threat agents.

This chapter has focussed on their potential as a platform for explosives detection, with a special emphasis on plasmonic nanosensors capable of offering high sensitivity and selectivity. These platforms offer promising applications for enhancing safety, especially when compared to conventional methods involving labor-intensive sample preparation and sophisticated instrumentation. While plasmonic nanosensors are existing solutions for detecting explosives, there is still ample room for improvement.

REFERENCES

Aguirre-Díaz, L.M., Echeverri, M., Paredes-Gil, K., Snejko, N., Gómez-Lor, B., Gutiérrez-Puebla, E., Monge, M.Á., 2021. The effect of auxiliary nitrogenated linkers on the design of new cadmium-based coordination polymers as sensors for the detection of explosive materials. *Chem. – A Eur. J.* 27, 5298–5306. https://doi.org/10.1002/chem.202005166

Akgönüllü, S., Çalışır, M., Özbek, M.A., Erkek, M., Bereli, N., Denizli, A., 2023. Plasmonic nanosensors for chemical warfare agents, in: *Sensing of Deadly Toxic Chemical Warfare Agents, Nerve Agent Simulants, and Their Toxicological Aspects.* Elsevier, pp. 81–96. >https://doi.org/10.1016/B978-0-323-90553-4.00016-0

Ali, M.A., Geng, Y., Cavaye, H., Burn, P.L., Gentle, I.R., Meredith, P., Shaw, P.E., 2015. Molecular versus exciton diffusion in fluorescence-based explosive vapour sensors. *Chem. Commun.* 51, 17406–17409. https://doi.org/10.1039/C5CC06367A

Bai, X., Xu, S., Hu, G., Wang, L., 2016. Surface plasmon resonance-enhanced photothermal nanosensor for sensitive and selective visual detection of 2,4,6-trinitrotoluene. *Sensors Actuators B Chem.* 237, 224–229. https://doi.org/10.1016/j.snb.2016.06.093

Bharadwaj, R., Mukherji, S., 2014. Gold nanoparticle coated U-bend fibre optic probe for localized surface plasmon resonance based detection of explosive vapours. *Sensors Actuators, B Chem.* 192, 804–811. https://doi.org/10.1016/j.snb.2013.11.026

Byram, C., Rathod, J., Moram, S.S.B., Mangababu, A., Soma, V.R., 2022. Picosecond laser-ablated nanoparticles loaded filter paper for SERS-based trace detection of thiram, 1,3,5-Trinitroperhydro-1,3,5-triazine (RDX), and Nile blue. *Nanomaterials* 12. https://doi.org/10.3390/nano12132150

Chegel, V.I., Lopatynskyi, A.M., Lytvyn, V.K., Demydov, P.V., Martínez-Pastor, J.P., Abargues, R., Gadea, E.A., Piletsky, S.A., 2020. Localized surface plasmon resonance nanochips with molecularly imprinted polymer coating for explosives sensing. *Semicond. Physics, Quantum Electron. Optoelectron.* 23, 431–436. https://doi.org/10.15407/spqeo23.04.431

Das, C.M., Kong, K.V., Yong, K.-T., 2022. Diagnostic plasmonic sensors: Opportunities and challenges. *Chem. Commun.* 58, 9573–9585. https://doi.org/10.1039/D2CC03431J

Desai, V., Panchal, M., Dey, S., Panjwani, F., Jain, V.K., 2022. Recent advancements for the recognition of nitroaromatic explosives using Calixarene based fluorescent probes. *J. Fluoresc.* 32, 67–79. https://doi.org/10.1007/s10895-021-02832-y

Diaz, D., Hahn, D.W., 2020. Raman spectroscopy for detection of ammonium nitrate as an explosive precursor used in improvised explosive devices. *Spectrochim. Acta Part A Mol. Biomol. Spectrosc.* 233, 118204. https://doi.org/10.1016/j.saa.2020.118204

Divya, J., Selvendran, S., Raja, A.S., Sivasubramanian, A., 2022. Surface plasmon based plasmonic sensors: A review on their past, present and future. *Biosens. Bioelectron.* 11, 100175. https://doi.org/10.1016/j.biosx.2022.100175

Fehlen, P., Thomas, G., Posada, F.G., Guise, J., Rusconi, F., Cerutti, L., Taliercio, T., Spitzer, D., 2022. Gas sensing of organophosphorous compounds with Iii-V semiconductor plasmonics. *SSRN Electron. J.* https://doi.org/10.2139/ssrn.4178083

Gao, Y., Xu, W., Zhu, D., Chen, L., Fu, Y., He, Q., Cao, H., Cheng, J., 2015. Highly efficient nitrate ester explosive vapor probe based on multiple triphenylaminopyrenyl-substituted POSS. *J. Mater. Chem. A* 3, 4820–4826. https://doi.org/10.1039/C4TA05704J

Homola, J., Piliarik, M., 2006. Surface Plasmon Resonance (SPR) sensors, in: *Springer Ser Chem Sens Biosens.* Springer, Berlin, pp. 45–67. https://doi.org/10.1007/5346_014

Johnston, E.J., Rylott, E.L., Beynon, E., Lorenz, A., Chechik, V., Bruce, N.C., 2015. Monodehydroascorbate reductase mediates TNT toxicity in plants. *Science (80-.).* 349, 1072–1075. https://doi.org/10.1126/science.aab3472

Kawaguchi, T., Shankaran, D.R., Kim, S.J., Matsumoto, K., Toko, K., Miura, N., 2008. Surface plasmon resonance immunosensor using Au nanoparticle for detection of TNT. *Sensors Actuators B Chem.* 133, 467–472. https://doi.org/10.1016/j.snb.2008.03.005

Kayhomayun, Z., Ghani, K., Zargoosh, K., 2020. Surfactant-assisted synthesis of fluorescent SmCrO3 nanopowder and its application for fast detection of nitroaromatic and nitramine explosives in solution. *Mater. Chem. Phys.* 247, 122899. https://doi.org/10.1016/j.matchemphys.2020.122899

Krivitsky, V., Filanovsky, B., Naddaka, V., Patolsky, F., 2019. Direct and selective electrochemical vapor trace detection of organic peroxide explosives via surface decoration. *Anal. Chem.* 91, 5323–5330. https://doi.org/10.1021/acs.analchem.9b00257

Kuracina, R., Szabová, Z., Škvarka, M., 2021. Study into parameters of the dust explosion ignited by an improvised explosion device filled with organic peroxide. *Process Saf. Environ. Prot.* 155, 98–107. https://doi.org/10.1016/j.psep.2021.09.011

Lee, C., Lawrie, B., Pooser, R., Lee, K.G., Rockstuhl, C., Tame, M., 2021. Quantum plasmonic sensors. *Chem. Rev.* 121, 4743–4804. https://doi.org/10.1021/ACS.CHEMREV.0C01028/ASSET/IMAGES/MEDIUM/CR0C01028_M071.GIF

Lefferts, M.J., Castell, M.R., 2015. Vapour sensing of explosive materials. *Anal. Methods* 7, 9005–9017. https://doi.org/10.1039/C5AY02262B

Li, M., Cushing, S.K., Wu, N., 2014. Plasmon-enhanced optical sensors: A review. *Analyst* 140, 386–406. https://doi.org/10.1039/C4AN01079E

Liu, R., Li, Z., Huang, Z., Li, K., Lv, Y., 2019. Biosensors for explosives: State of art and future trends. *TrAC Trends Anal. Chem.* 118, 123–137. https://doi.org/10.1016/j.trac.2019.05.034

Ma, R.-M., Ota, S., Li, Y., Yang, S., Zhang, X., 2014. Explosives detection in a lasing plasmon nanocavity. *Nat. Nanotechnol.* 9, 600–604. https://doi.org/10.1038/nnano.2014.135

Mahbub, P., Hasan, C.K., Rudd, D., Voelcker, N.H., Orbell, J., Cole, I., Macka, M., 2023. Rapid and selective screening of organic peroxide explosives using acid-hydrolysis induced chemiluminescence. *Anal. Chim. Acta* 1255, 341156. https://doi.org/10.1016/j.aca.2023.341156

Malekzad, H., Sahandi Zangabad, P., Mirshekari, H., Karimi, M., Hamblin, M.R., 2017. Noble metal nanoparticles in biosensors: Recent studies and applications. *Nanotechnol. Rev.* 6, 301–329. https://doi.org/10.1515/NTREV-2016-0014/ASSET/GRAPHIC/J_NTREV-2016-0014_FIG_002.JPG

Moore, D.S., 2007. Recent advances in trace explosives detection instrumentation. *Sens. Imaging An Int. J.* 8, 9–38. https://doi.org/10.1007/s11220-007-0029-8

Mosier-Boss, P.A., 2017. Review of SERS substrates for chemical sensing. *Nanomater.* 7, 142. https://doi.org/10.3390/NANO7060142

Pavlov, J., Douce, D., Bajic, S., Attygalle, A.B., 2019. 1,4-Benzoquinone as a highly efficient dopant for enhanced ionization and detection of nitramine explosives on a single-quadrupole mass spectrometer fitted with a Helium-Plasma Ionization (HePI) source. *J. Am. Soc. Mass Spectrom.* 30, 2704–2710. https://doi.org/10.1007/s13361-019-02339-8

Riskin, M., Ben-Amram, Y., Tel-Vered, R., Chegel, V., Almog, J., Willner, I., 2011. Molecularly imprinted Au nanoparticles composites on Au surfaces for the surface plasmon resonance detection of pentaerythritol tetranitrate, nitroglycerin, and ethylene glycol dinitrate. *Anal. Chem.* 83, 3082–3088. https://doi.org/10.1021/ac1033424

Riskin, M., Tel-Vered, R., Lioubashevski, O., Willner, I., 2009. Ultrasensitive surface plasmon resonance detection of trinitrotoluene by a bis-aniline-cross-linked Au nanoparticles composite. *J. Am. Chem. Soc.* 131, 7368–7378. https://doi.org/10.1021/ja9001212

Riskin, M., Tel-Vered, R., Willner, I., 2010. Imprinted au-nanoparticle composites for the ultrasensitive surface plasmon resonance detection of hexahydro-1,3,5-trinitro-1,3,5-triazine (RDX). *Adv. Mater.* 22, 1387–1391. https://doi.org/10.1002/adma.200903007

Sepúlveda, B., Angelomé, P.C., Lechuga, L.M., Liz-Marzán, L.M., 2009. LSPR-based nanobiosensors. *Nano Today* 4, 244–251. https://doi.org/10.1016/J.NANTOD.2009.04.001

Shankaran, D.R., Gobi, K.V., Matsumoto, K., Imato, T., Toko, K., Miura, N., 2004. Highly sensitive surface plasmon resonance immunosensor for parts-per-trillion level detection of 2,4,6-trinitrophenol. *Sensors Actuators, B Chem.* 100, 450–454. https://doi.org/10.1016/j.snb.2004.02.010

Singh, N., Shrivastav, A.M., Vashistha, N., Abdulhalim, I., 2023. 3D plasmonic hot spots network via gold decorated deep micro-porous silicon exhibiting ultrahigh-SERS enhancement with application to explosives detection. *Sensors Actuators B Chem.* 374, 132813. https://doi.org/10.1016/j.snb.2022.132813

Singh, S., 2007. Sensors—An effective approach for the detection of explosives. *J. Hazard. Mater.* 144, 15–28. https://doi.org/10.1016/j.jhazmat.2007.02.018

Vendamani, V.S., Rao, S.V.S.N., Pathak, A.P., Soma, V.R., 2020. Robust and cost-effective silver dendritic nanostructures for SERS-based trace detection of RDX and ammonium nitrate. *RSC Adv.* 10, 44747–44755. https://doi.org/10.1039/d0ra08834j

Wang, C., Shang, J., Lan, Y., Tian, T., Wang, H., Chen, X., Gu, J.Y., Liu, J.Z., Wan, L.J., Zhu, W., Li, G., 2015. Metal-organic olyhedral cages immobilized on a plasmonic substrate for sensitive detection of trace explosives. *Adv. Funct. Mater.* 25, 6009–6017. https://doi.org/10.1002/adfm.201503071

Wang, J., 2021. A simple, rapid and low-cost 3-Aminopropyltriethoxysilane (APTES)-based Surface Plasmon Resonance Sensor for TNT explosive detection. *Anal. Sci.* 37, 1029–1032. https://doi.org/10.2116/analsci.20N028

Wang, J., 2007. Electrochemical sensing of explosives, in: *Counterterrorist Detection Techniques of Explosives.* Elsevier, pp. 91–107. https://doi.org/10.1016/B978-044452204-7/50023-7

Wang, J., Du, S., Onodera, T., Yatabe, R., Tanaka, M., Okochi, M., Toko, K., 2018a. An SPR sensor chip based on peptide-modified single-walled carbon nanotubes with enhanced sensitivity and selectivity in the detection of 2,4,6-Trinitrotoluene explosives. *Sensors* 18, 4461. https://doi.org/10.3390/s18124461

Wang, J., Muto, M., Yatabe, R., Tahara, Y., Onodera, T., Tanaka, M., Okochi, M., Toko, K., 2018b. Highly selective rational design of peptide-based surface plasmon resonance sensor for direct determination of 2,4,6-trinitrotoluene (TNT) explosive. *Sensors Actuators B Chem.* 264, 279–284. https://doi.org/10.1016/j.snb.2018.02.075

Zapata, F., García-Ruiz, C., 2020. Chemical classification of explosives. *Crit. Rev. Anal. Chem.* 51, 1–18. https://doi.org/10.1080/10408347.2020.1760783

Zhang, C., Bai, C., Ren, J., Chang, C., Yao, J., 2022. The promotion of nitromethane on solid–liquid fuel/air mixtures explosion characteristics under different ambient conditions. *Fuel* 322, 124190. https://doi.org/10.1016/j.fuel.2022.124190

Zhang, X., Zhu, D., Fu, Y., He, Q., Cao, H., Li, W., Cheng, J., 2017. Enhanced fluorescence of functionalized silica microsphere based on whispering gallery mode for nitrate ester explosives and hexogen vapour detection. *J. Mater. Chem. C* 5, 2114–2122. https://doi.org/10.1039/C6TC05642C

Zou, W., Liu, W., Luo, L., Zhang, S., Lu, R., Veser, G., 2012. Detection of nitro explosives via LSPR sensitive silver clusters embedded in porous silica. *J. Mater. Chem.* 22, 12474–12478. https://doi.org/10.1039/c2jm31770b

14 COVID-19 Detection with Plasmonic Nanosensors

Serhat Ünal, Yeşeren Saylan, Özge Altıntaş, Seçkin Kılıç, and Adil Denizli

14.1 INTRODUCTION

COVID-19 is a global epidemic of severe acute respiratory syndrome created by coronavirus 2. COVID-19 is the worst epidemic since World War II. This pandemic has had an important effect on health, environment, politics, and the economy, making the problems posed by the coronavirus more complex. It has become one of the toughest pandemics of the last century, with its high rate of infection and deadly consequences. There are thousands of diverse kinds of COVID moving around the world (Wu et al., 2020). Viruses always mutate. The important thing here is to provide rapid and early diagnosis, to prevent virus infection, and to design influential vaccines (Polack et al., 2020; Udugama et al., 2020). In this context, the usage of nanotechnology creates novel possibilities for treatment, diagnosis, and prevention of COVID-19. Nanosensors have a good potential to determine the lifespan, growth, and structure of a virus (Saylan and Denizli, 2020), because it is necessary to study at a similar dimension scale in order to better understand the interplay of virus (Hu et al., 2020; Laval et al., 2000). Nanosensors could bridge the gap between treatment and diagnosis in the fight against viruses (Saylan et al., 2019). They can also offer a range of solutions both inside and outside the host (Yakoubi and Dhafer, 2023). Plasmonic nanosensors have appeared as a strong and versatile platform for COVID-19 detection, offering the potential for highly sensitive, specific, and rapid diagnostics (Erdem et al., 2021). With further research and development, plasmonic nanosensors have the potential to contribute remarkably to the global efforts in combating the COVID-19 pandemic and future infectious disease outbreaks (Saylan and Denizli, 2021).

14.2 NANOSENSORS

The utilization of nanosensors across diverse sectors, including medical diagnostics, therapeutics, the food industry, and air and water quality monitoring, has witnessed a notable upsurge (Inci et al., 2020). The term nanosensors is a fusion of two

DOI: 10.1201/9781003459316-14

components: nano, which signifies a size range of 10^{-9} on the scale, and sensors, originated from the Greek word Sentire, which means "perceive" (Sze, 1994). In essence, nanosensors are devices designed to convert various physical stimuli, such as magnetic, electric, luminescent, or thermal effects, into electrical signals. When employed in any field of application, crucial parameters such as stability, resolution, selectivity, sensitivity, and calibration characteristics must be taken into account for their optimal functional performance.

14.2.1   The Principle of Nanosensors

The principle of nanosensors is based on the detection and transduction of biological, chemical, or physical signals at the nanoscale level (Saylan et al., 2020). Nanosensors typically consist of nanostructured devices or materials that can interact with the target analyte or stimulus and produce a measurable response. The principle behind nanosensors can vary depending on their specific design and application (Mousavi et al., 2018).

The basic principles are:

- *Readout and analysis:* Nanosensors generate a measurable signal or response that needs to be interpreted and analyzed. This can be achieved through various readout techniques, including optical spectroscopy, electrical measurements, mass spectrometry, or imaging methods (Zhao et al., 2016).
- *Amplification strategies:* Nanosensors often employ amplification strategies to enhance the signal and improve the detection limits. This can involve the use of nanomaterials with inherent signal amplification properties or the integration of signal amplification techniques, such as enzymatic reactions or nanoparticle labeling (Nocerino et al., 2022).
- *Recognition elements:* Nanosensors incorporate specific recognition elements that enable the selective detection of the target analyte. These elements can include antibodies, aptamers, enzymes, or molecularly imprinted polymers, which interact with the target analyte with high affinity and specificity (Choi and Choi, 2011).
- *Signal transduction:* Nanosensors employ various mechanisms to transduce the detected signal into a measurable output. This can involve the conversion of a physical stimulus, such as temperature or pressure, into an electrical signal or the interplay between the target analyte and the nanomaterial resulting in a change in optical, electrical, or mechanical properties (Volkert and Haes, 2014).
- *Surface effects:* Nanosensors often exploit the unique properties and high surface-to-volume ratio of nanomaterials. The increased surface area allows for improved interactions with the target, which leads to increase selectivity and sensitivity (Singh and Yadava, 2020).

14.2.2 Types of Nanosensors

Nanosensors can be broadly classified into two main categories. The first category consists of nanosensors with nanoscale dimensions, which can be further categorized into two types. The first type includes sensors whose three-dimensional characteristics, which are primarily in nanometric proportions. Examples of such sensors include quantum dots which are thin films of metal oxides and fluorescent nano-sized semiconductor crystals used for gas sensing, where one dimension is in the nanoscale range. The second type encompasses nanosensors with nanoscale measurements but without specific nanoscale dimensions. Another classification of nanosensors can be based on their energy requirements (Chakraborty et al., 2021). Active nanosensors, such as thermistors, rely on an external energy source to operate. In contrast, passive nanosensors, like piezoelectric nanosensors that measure strain, force, and pressure using the piezoelectric effect, do not require an additional energy source. Nanosensors can also be classified according to the type of signal they detect, which can be in the form of biological, chemical, and physical effects. Physical nanosensors encompass electromagnetic, optical, and mechanical nanosensors (Lim and Ramakrishna, 2006). These nanosensors are employed for measuring properties such as permeability, dielectric constants, luminescence, fluorescence, absorbance, viscosity, density, pressure, and force. On the other hand, chemical nanosensors are designed to identify foreign chemical particles, which measure the concentrations of specific substances and determine pH levels. Biological nanosensors, also known as biosensors, serve distinct purposes in detecting biological entities and analyzing biological interactions. In conclusion, nanosensors exhibit various classifications based on their dimensions, energy requirements, and the type of signal they detect. This diverse range of nanosensor classifications allows for their application in numerous fields, addressing different measurement and detection needs (Vo-Dinh et al., 2001).

14.2.3 Plasmonic Nanosensors

Plasmonic nanosensors are at the forefront of nanotechnology, offering exceptional sensitivity and remarkable capabilities in analyzing and detecting a wide variety of analytes (Özgür et al., 2020; Saylan, 2023; Saylan et al., 2022; Yılmaz et al., 2022). These nanosensors take advantage of the unique properties of plasmonics, a class of nanophotonics that is interested in the interaction between light and free electrons in metallic nanostructures (Atwater, 2007). The engineering and design of plasmonic nanosensors is tailored to target specific analytes. These nanosensors can selectively bind to target molecules and detect their presence by functionalizing their metal surfaces with specific recognition elements such as molecularly imprinted polymers, aptamers, and antibodies (Stewart et al., 2008; Whitney et al., 2005). With a combination of high sensitivity and remarkable selectivity, plasmonic nanosensors have a wide range of applications, such as the detection of dangerous substances, food safety, and environmental monitoring. One of the most important advantages of plasmonic nanosensors is their real-time and fast detection potential (Haes et al., 2004). In addition, they can provide on-site monitoring and maintenance point diagnosis by integrating into portable devices with their miniature dimensions. This feature is very important in emergencies and when resources are limited, as it facilitates timely response and decision making (Anker et al., 2008).

14.3 COVID-19

COVID-19, a novel epidemic disease originated by coronavirus 2, was first detected in East Asia in December 2019. Since then, it has become a global pandemic (Hsia, 2020). This acute respiratory tract infection is characterized by common symptoms such as cough, fever, headache, breathing difficulties, fatigue, and loss of smell and taste, which may appear within 1 to 14 days after exposure to the virus (Chams et al., 2020; Forchette et al., 2021; Yavuz and Ünal, 2020). Interestingly, about one third of those infected do not exhibit any symptoms, while most patients experience mild to moderate symptoms. However, a small percentage of cases can be severe, involving hypoxia, dyspnea, or more than 50 percent lung involvement (Harrison et al., 2020). Severe symptoms are more common in older individuals. Recovery from COVID-19 does not always mean the end of symptoms, as some people experience lingering effects known as long COVID, and organ damage has also been reported. Researchers are currently studying the long-term impact of the disease (Das et al., 2021; Hu et al., 2021). While the respiratory system is the primary target of COVID-19, other organs such as heart, kidneys, liver, muscles, skin, and the nervous system may also be affected, leading to various clinical manifestations (Gavriatopoulou et al., 2020). Understanding these multi-organ effects is crucial for managing and treating the disease effectively (Machhi et al., 2020).

14.3.1 THE IMPORTANCE OF COVID-19

COVID-19, an incredibly contagious illness, swiftly became a pandemic, spreading across the entire planet through human-to-human transmission. Researchers have identified 149 mutation sites in the virus, accounting for its rapid mutation and high contagiousness (Tang et al., 2021). Infections of the virus occur via indirect, direct, and close contact with an infected person, primarily by inhaling viral particles contained in airborne droplets (Prather et al., 2020). Additionally, the virus spreads through aerosols, with virus-laden droplets remaining viable for up to three hours, and through contaminated surfaces frequently touched by individuals in daily life (Lewis, 2020). Viruses generally exhibit a spherical shape, ranging from 60 to 140 nm diameter. This global pandemic has profoundly affected public health, safety, and well-being, as well as economies. The rapid worldwide spread, coupled with the lack of rapid diagnostic tests, has necessitated unprecedented emergency measures to combat this new viral threat (Huang et al., 2020).

14.3.2 UPDATES OF COVID-19

It has now been three years since the emergence of COVID-19. Recent research highlights the capability of the virus to spread both symptomatically and asymptomatically, with transmission occurring through contact with contaminated surfaces or respiratory droplets exchanged between individuals (Rahman et al., 2020). This exceptional level of contagiousness has led to rapid transmission, resulting in severe illness and loss of life.

In response, the World Health Organization declared COVID-19 a global pandemic in March 2020. Subsequently, several variants of the virus, such as Omicron (Thye and Law, 2022), Delta, Gamma, Beta, Alpha (Thye et al., 2021), and their subtypes, have emerged (Lambrou et al., 2022; Law et al., 2021; Loo and Letchumanan, 2022; Loo et al., 2022). This complex situation has had a significant impact on global healthcare systems, economies, and various aspects of society (Joseph and Ser, 2021; Kuai and Ser, 2021; Lee et al., 2021; Loo et al., 2021).

It is evident that COVID-19 has exposed unprecedented challenges and necessitated ongoing scientific, medical, and societal responses. It has underscored the importance of global cooperation and preparedness in the face of emerging infectious diseases. The ongoing efforts to combat this pandemic and its variants continue to shape the world's response to public health crises.

14.3.3 The Detection of COVID-19 Using Plasmonic Nanosensors

Recent studies have explored the use of plasmonic nanosensors for COVID-19 detection. Negahdari et al. introduced four distinct structures based on graphene-plasmonic nanosensors for this purpose (Negahdari et al., 2023). These structures included half-spheres constructed from aliminium (Al), gold (Au), silican dioxide (SiO_2), and graphene, as well as one-dimensional photonic crystal formats. In these photonic crystal formats, the manipulation of wavelengths and peak absorption played a critical role. A defect layer consisting of gallium-doped zinc oxide was strategically placed within the one-dimensional photonic crystal layers. This alteration effectively shifted the absorption peak wavelength range from 300 to 600 nm, enhancing detection capabilities. Additionally, the researchers suggested the use of this structural configuration as a refractive nanosensor. This plasmonic nanosensor demonstrates strong potential for COVID-19 detection within photonic integrated circuits, exhibiting a satisfactory sensitivity level of approximately 664.8 nm per refractive index unit.

In a study, a groundbreaking plasmonic nanosensor was introduced by Calvo-Lozano et al., offering rapid COVID-19 antibody detection in clinical samples, all without the need for signal amplification (Calvo-Lozano et al., 2021). This innovative nanosensor is built around a custom-designed multiantigen and achieves impressive detection limit value in the nanograms per milliliter (ng/mL) using polyclonal antibodies. What sets this nanosensor apart is its robust clinical validation process involving COVID-19-negative and -positive samples. These results conclusively demonstrate its efficacy as a top-tier diagnostic tool, showcasing both high sensitivity and specificity. They also emphasize that this nanosensor represents an accurate and user-friendly solution for dependable COVID-19 serology testing. It can be seamlessly integrated into laboratory settings or decentralized locations, making it invaluable for disease management and assessing immunological status during vaccination and treatment.

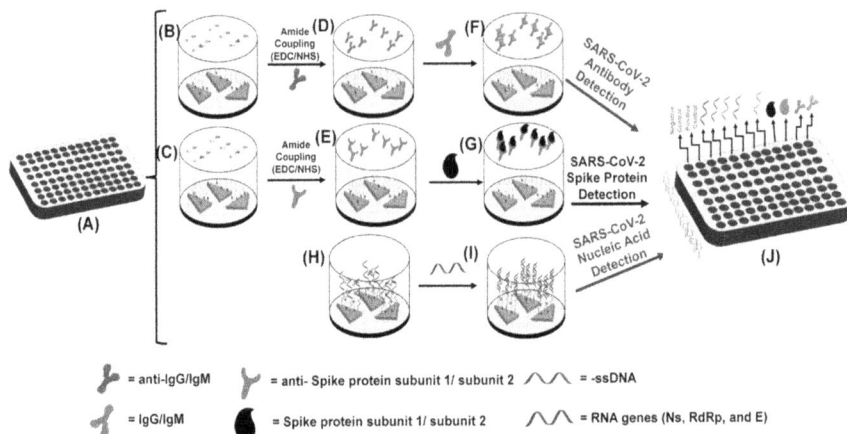

FIGURE 14.1 Scheme of well plate with glass coverslips and gold triangular nanoplates (A), surface of gold triangular nanoplates is modified with amide coupling (B, C), amide coupling for virus anti-IgG/IgM antigens (D), anti-spike protein subunit 1 and subunit 2 (E), detection of IgG/IgM antibodies (F), spike protein (G), nucleic acid (H) and multiplexed format of each well (J). Reproduced, adapted, and modified from Masterson et al. (2021).

In a notable study, Masterson et al. introduced a plasmonic nanosensor multiplex test for COVID-19 screening (Masterson et al., 2021). This advanced test has the remarkable capacity to quantitatively measure ten distinct biomarkers, achieving an exceptionally low detection limit value in attomolar (aM) (Figure 14.1). The plasmonic nanosensors exhibit outstanding specificity, a critical attribute to prevent false signals. This detection method demonstrates its high-throughput potential by directly quantifying IgM and IgG antibodies from plasma samples of COVID-19-positive patients in a single run. The robustness of this approach is underlined by the receiving operating characteristics, with areas under the curve reaching an impressive 0.999 and 0.997 for IgM and IgG antibodies, respectively. Statistical analysis, including the Mann–Whitney nonparametric test, yields a calculated p-value of <0.0001 when comparing COVID-19-positive patients to a control group of healthy people.

Funari et al. have developed a compelling proof-of-concept multiplex plasmonic nanosensor aimed at capturing humoral response in serums towards to various antigens (Funari et al., 2022). This plasmonic nanosensor functions by monitoring changes in the wavelength of the localized surface plasmon resonance (LSPR) peak in Au nanostructures when antibodies bind to immobilized antigens. To achieve this, antigens are initially anchored onto distinct detection parts using a mono-biotinylation system, taking advantage of the strong interaction between streptavidin and biotin. As depicted in Figure 14.2, the researchers confirm the versatility of this multiplex platform by successfully detecting the presence of three monoclonal antibodies specific to three different antigens. The circular spots with a reddish hue indicate areas of the substrate covered by Au nanospikes created through the process

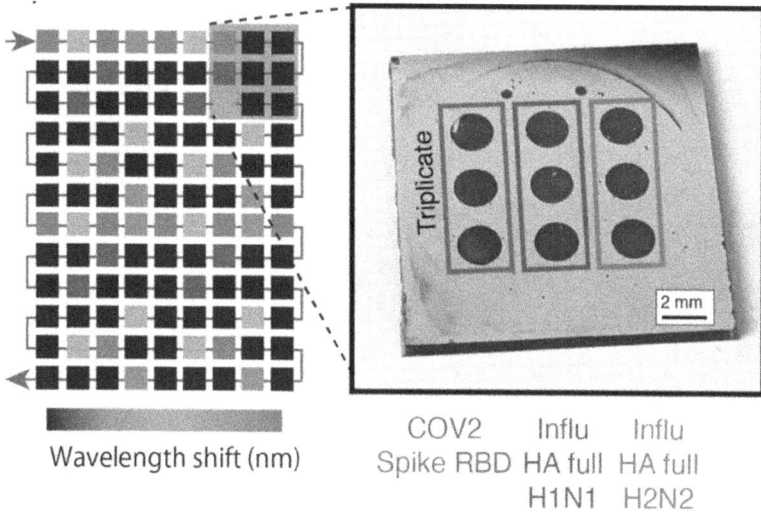

FIGURE 14.2 Illustration outlining the fundamental design of our multiplex bioassay platform and provides a photographic snapshot of a glass slide coated with gold. Reproduced, adapted, and modified from Funari et al. (2022).

of electrodeposition. Each vertical column in the array is treated with a distinct antigen, and this arrangement is replicated three times for each antigen, as denoted by the red, blue, and green boxes. These antigens include two hemagglutinins (HAs) derived from influenza viruses and the receptor binding domain (RBD) of the virus spike protein. Furthermore, they assess the humoral response in murine sera collected both before and after immunization with the virus spike protein. The results obtained align well with those from the Enzyme-Linked Immunosorbent Assay (ELISA) method. This plasmonic nanosensor showcases its proficiency in profiling multiple serum antibodies and exhibits the potential for future integration with microfluidic systems, enabling the creation of a high-throughput screening platform.

Li et al. have made significant strides in the development of a cutting-edge plasmonic nanosensor with several key attributes, as detailed in their study (Li et al., 2021). This innovative nanosensor, designed using a genetic algorithm intelligent program for automatic optimization, offers a range of advantages for on-site COVID-19 diagnosis (Figure 14.3). One of its standout features is its exceptional sensitivity, boasting an ultra-high sensitivity of 1.66% per nanometer. This sensitivity, combined with a broad detection range and adaptability to different measurement environments (both gas and liquid), enables precise and versatile COVID-19 detection. Moreover, the nanosensor's unique capability to recognize the infrared fingerprints of specific substances positions it as an optimum tool for screening mutant virus strains. The functionality enhances its utility in tracking and identifying new viral variants. In essence, this work not only provides a strong diagnostic instrument for label-free, rapid, and multi-functional COVID-19 detection but also contributes to advancing the broader field of high sensitivity sensing.

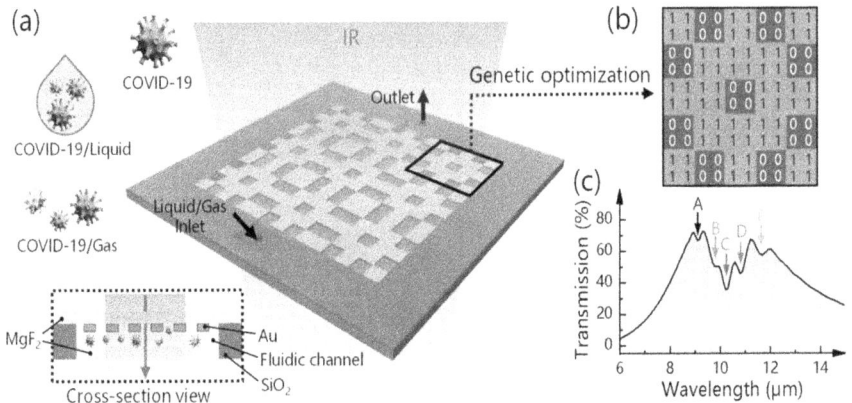

FIGURE 14.3 Three-dimensional plasmonic configuration of nanosensor for its operation (a), genetic algorithm program for optimizing metal structure (1 represents the presence of nano-metal, while 0 indicates its absence) (b), and metal structure's resonance position of characteristic vibration signals (c). Reproduced, adapted, and modified from Li et al. (2021).

A dual-functional plasmonic nanosensor has been developed, merging the advantages of both the plasmonic photothermal effect and loca LSPR sensing via Qiu et al. (Qiu et al., 2020). This innovative approach holds promise as an alternative solution for clinical COVID-19 diagnosis. The nanosensor leverages two-dimensional gold nanoislands (AuNIs) that have been functionalized with complementary DNA receptors. These receptors enable the sensitive detection of specific sequences from the SARS-CoV-2 through nucleic acid hybridization. To enhance sensing performance, the same AuNIs chip generates thermoplasmonic heat when illuminated at its plasmonic resonance frequency. The localized plasmonic photothermal heat serves a dual purpose. First, it raises the in-situ hybridization temperature, facilitating precise discrimination between two closely related gene sequences. Second, it significantly enhances the sensitivity of the nanosensor, enabling detection at extremely low concentrations, with a remarkably lower detection limit of just 0.22 pM. Furthermore, this nanosensor can accurately detect specific targets within a mixture of multiple genes. This study not only advances our understanding of the thermoplasmonic enhancement but also demonstrates its practical utility in nucleic acid tests and the diagnosis of viral diseases, particularly in the context of COVID-19 detection.

The introduced plasmonic nanosensor represents a critical advancement in the early-stage diagnosis of the COVID-19 virus by Akib et al. (Akib et al., 2021). In addition to its practical application, the nanosensor's performance was rigorously examined through numerical simulations involving various ligand-analyte interactions such as monoclonal antibodies and COVID-19 spike RBD. This configuration yielded impressive sensitivity results, with a detection capability of 183.33° per refractive index unit (RIU) in SPR angle (θSPR) and 833.33 terahertz per RIU in SPR frequency (SPRF) for COVID-19 virus spike RBD. Furthermore, another interaction between COVID-19 spike RBD and anti-spike proteins (IgM, IgG) was explored. In this scenario, the nanosensor achieved a sensitivity of 153.85°/RIU in SPR angle and 726.50 THz/RIU in SPRF for detecting the anti-spike protein. The final interaction was between a specific probe and COVID-19 virus single-standard ribonucleic acid (RNA). The nanosensor exhibited sensitivity of 140.35°/RIU in SPR angle and 500 THz/RIU in SPRF when detecting viral RNA. Notably, the detection sensitivity for the whole COVID-19 virus spike RBD exceeded that of the other two detection processes. The utilization of highly sensitive two-dimensional (2D) materials played a pivotal role in enhancing both the Goos-Hänchen shift detection sensitivity and plasmonic properties of the conventional SPR sensor. The addition of graphene layers provided an impressive (1 + 0.55) time sensitivity enhancement.

In another research paper, Uddin et al. introduced an SPR nanosensor, utilizing the Kretschmann configuration, which incorporates layers of silicon and $BaTiO_3$ on top of a Ag substrate (Uddin et al., 2021). This innovative configuration is designed for real-time detection of SARS-CoV-2 using thiol-tethered DNA as a ligand. Their study involves comprehensive numerical analysis, employing both transfer matrix theory and finite-difference time-domain (FDTD) techniques, to thoroughly characterize the sensor's response. They assess critical parameters such as sensitivity, full width at half maximum, and minimum reflection. The proposed nanosensor yields an impressive 7.6-fold increase in sensitivity for SARS-CoV-2 detection when compared to the standard Kretschmann configuration. Notably, this nanosensor consistently outperforms other competitive SPR configurations for both angular and wavelength interrogation, achieving a remarkable figure-of-merit of 692.28. Moreover, the researchers conducted simulations with various ligate-ligand pairs to evaluate the nanosensor's versatility and robust performance enhancement.

In another study, Lew et al. presented a colorimetric nanosensor designed for the detection of SARS-CoV-2 IgGs in patients' plasma (Lew et al., 2021). This nanosensor employs brief antigenic epitopes that have been linked to AuNPs. Four key linear B-cell epitopes, specifically situated on the spike (S) and nucleocapsid (N) proteins of the virus, were analyzed for their ability to bind to IgG antibodies. These epitopes serve as highly specific biological markers on the nanoparticle surface, enabling the identification of target antibodies (Figure 14.4). The nanosensor operates by relying on a particular bivalent binding interaction between SARS-CoV-2 antibodies and the epitope-functionalized AuNPs. This interaction prompts the aggregation of nanoparticles, causing a distinct change in the plasmon characteristics of the AuNPs within just 30 minutes of introducing the antibodies. This change can be easily observed by monitoring alterations in color. To enhance sensitivity, two epitopes were simultaneously attached to the AuNPs, resulting in improved detection capabilities. The assay showcases an impressive limit of detection, capable of measuring concentrations as low as 3.2 nanomolar (nM), which aligns with IgG levels observed in convalescent COVID-19-infected patients. To ensure the assay's reliability in real-world scenarios, a passivation strategy was developed to maintain its sensing accuracy when applied to human plasma. When tested against clinical plasma samples representing

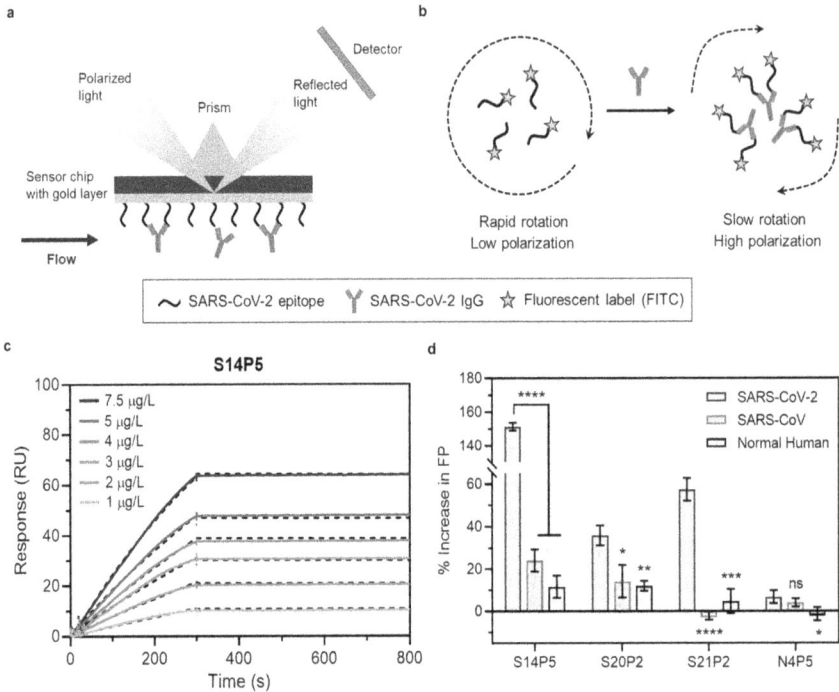

FIGURE 14.4 Scheme of SPR spectroscopy (a) and fluorescence polarization principles to assess interaction of virus IgG antibodies (b), SPR sensorgrams at different concentrations (c), relative increase in fluorescence polarization (d). Reproduced, adapted, and modified from Lew et al. (2021).

various degrees of illness severity, the optimized nanosensor assay effectively identifies virus infection with an exceptional 100% specificity and 83% sensitivity.

By combining magnetic separation techniques with three types of plasmonic NPs, Wen et al. have developed an innovative achromatic colorimetric nanosensor that offers greatly improved visual resolution (Wen et al., 2023). This nanosensor is designed for simultaneous detection of distinct pathogens such as COVID-19, *Staphylococcus aureus*, and *Salmonella typhimurium*. The nanosensor consists of three types of NPs, each with a different target pathogen affinity. Specifically, they use red gold NPs for the virus, yellow silver NPs for *Staphylococcus aureus*, and blue silver triangle NPs for *Salmonella typhimurium*. When these NPs are combined, they form a black color mixture. During the detection process, three corresponding magnetic probes are introduced into the NP mixture. In the presence of a target pathogen, it binds with the corresponding colored reporters and magnetic probes to create sandwich complexes. These complexes are then removed via magnetic separation, causing the nanosensor to shift from black to chromatic color. The presence of various target pathogens induces distinct colors. Therefore, by observing the color change or measuring absorption spectra, they can conveniently achieve the detection of multiple pathogens. Compared to most colorimetric nanosensors, the achromatic nanosensor offers a wide range of color changes, enhancing visual resolution and detection sensitivity.

A paper by Maghoul et al. introduces an optical modeling framework for COVID-19 detection within the bloodstream and extracts their optic characteristics through numerical computations (Maghoul et al., 2022). This groundbreaking work establishes a theoretical foundation for COVID-19 dependent on the latest experimental findings from the literature. It enables the simulation of the optical action of the virus under different physical conditions. To determine the optical properties of the COVID-19 system, the study simulates the reflectance of light by the structure at various geometrical sizes, including COVID-19 diameter and size of the spikes surrounding it. Remarkably, the reflectance spectra are found to be highly sensitive to changes in the coronavirus's geometry. Additionally, the research investigates the density of COVID-19 when incident light approaches from several sides of the sample. After that, the paper proposes that the nanosensor utilizing graphene, silicon, and gold nanodisks, showcases the effectiveness of these systems in determining COVID-19 within blood samples. This nanosensor design holds great potential for practical applications, including the development of wearable devices and nanoelectronic kits for the detection of the virus.

In a study by Santopolo et al., the researchers presented a unique AuNPs aggregation assessment, which exhibits the ability to differentiate between individuals in need of intensive care and those who have been released from an intensive care unit (ICU) (Santopolo et al., 2022). The process involves diluting a plasma sample devoid of platelets and introducing AuNPs. Notably, when these samples are obtained from patients in the ICU, the AuNPs tend to clump together more extensively. This clumping alters the color of the liquid suspension, and this change can be quantified by measuring the intensity of pixels. Although the precise factors causing these differing clumping patterns are not yet understood, control experiments confirm the essential role of proteins in samples for the test's functionality. Through principal component analysis, a strong connection emerges between the test results and established biomarkers commonly used to gauge the severity of patients' illnesses and predict their prognosis and inflammation levels. The findings in this study provide a foundation for the development of assays based on AuNPs clumping, which could categorize patients according to the severity of their illness. This advancement holds the potential to be invaluable in safely de-escalating care and optimizing the allocation of hospital resources.

In the last study, Moitra et al. have presented a rapid colorimetric nanosensor using AuNPs (Moitra et al., 2020). These AuNPs are enveloped with thiol-modified antisense oligonucleotides (ASOs) meticulously tailored to target the N-gene of the virus. This nanosensor can detect COVID-19-positive cases within a remarkably short time, starting from RNA samples isolation. The underlying concept of this test revolves around the specific clumping together of ASO-coated AuNPs when exposed to the RNA sequence unique to the virus. This interaction causes a shift in the way the AuNPs interact with light, known as SPR. Additionally, the introduction of RNaseH enzyme triggers the separation of the RNA strand from the RNA-DNA hybrid, leading to the formation of visible particles in the solution due to increased clumping of the AuNPs. The precision of this nanosensor has undergone rigorous examination, even when challenged with MERS-CoV viral RNA. It has demonstrated an impressive detection limit of 0.18 ng/μL of RNA, corresponding to viral load levels. This research signifies a significant leap forward in the realm of COVID-19 diagnostics. It offers a selective, easily discernible method for identifying the virus that can be observed with the naked eye, eliminating the need for complex laboratory apparatus.

14.4 CONCLUSION AND FUTURE PERSPECTIVES

Plasmonic nanosensors have emerged as powerful tools in the fight against COVID-19, offering rapid, sensitive, and versatile detection methods. These nanosensors have demonstrated their efficacy across various detection targets, including viral RNA, antibodies, and antigens. The diverse range of plasmonic nanosensor designs and principles including LSPR, SPR, and nanoplasmonic sensing, has enabled researchers to adapt and optimize detection strategies to meet specific diagnostic needs. These nanosensors have proven their worth in clinical settings, showing the potential for early-stage diagnosis, serological testing, and even monitoring the humoral response to vaccination. Their speed and sensitivity make them invaluable tools for efficient patient management and resource allocation. Moreover, the adaptability of plasmonic nanosensors extends beyond diagnostics, as they hold promise for applications in screening, epidemiological studies, and even mutant virus detection.

The future of plasmonic nanosensors for COVID-19 detection is promising, with several exciting avenues for exploration:

- Enhanced Sensitivity: Researchers will continue to refine and enhance the sensitivity of plasmonic nanosensors to enable even earlier and more accurate detection of the virus and its variants.
- Multiplex Detection: Advancements in multiplexing capabilities will enable the simultaneous detection of multiple viral targets, improving diagnostic efficiency.
- Point-of-Care Devices: The development of portable and user-friendly plasmonic nanosensor devices will bring COVID-19 testing closer to the point of care, enabling quicker decision making and reducing the burden on healthcare systems.
- Integration with Other Technologies: Integration with microfluidics, artificial intelligence, and smartphone apps will further streamline and automate the diagnostic process.
- Long-Term Surveillance: Plasmonic nanosensors have the potential for long-term viral surveillance, aiding in the monitoring and management of potential outbreaks.
- Antigenic Drift and Vaccine Development: These sensors will play a crucial role in monitoring antigenic drift in the virus, guiding vaccine development and adjustment.

- Low-Cost Solutions: Efforts to reduce the cost of these technologies will make them accessible to a wider range of healthcare settings and regions.

In summary, plasmonic nanosensors have demonstrated their value in the battle against COVID-19, and ongoing research and innovation in this field hold the potential to further revolutionize the approach to detection, monitoring, and management of viral diseases.

ACKNOWLEDGEMENT

Dr. Yeşeren Saylan and Özge Altıntaş greatly thank the Scientific and Technological Research Council of Turkey (TÜBİTAK) Directorate of Science Fellowships and Grant (BİDEB) 2247-D Program (Project No: 121C226).

REFERENCES

Akib, T.B.A., Mou, S.F., Rahman, M.M., Rana, M.M., Islam, M.R., Mehedi, I.M., Mahmud, M.P., Kouzani, A.Z., 2021. Design and numerical analysis of a graphene-coated SPR biosensor for rapid detection of the novel coronavirus. *Sensors* 21, 3491.

Anker, J.N., Hall, W.P., Lyandres, O., Shah, N.C., Zhao, J., Van Duyne, R.P., 2008. Biosensing with plasmonic nanosensors. *Nat. Mater.* 7, 442–453.

Atwater, H.A., 2007. The promise of plasmonics. *Sci. Am.* 296, 56–63.

Calvo-Lozano, O., Sierra, M., Soler, M., Estévez, M.C., Chiscano-Camón, L., Ruiz-Sanmartin, A., Ruiz-Rodriguez, J.C., Ferrer, R., González-López, J.J., Esperalba, J., 2021. Label-free plasmonic biosensor for rapid, quantitative, and highly sensitive COVID-19 serology: Implementation and clinical validation. *Anal. Chem.* 94, 975–984.

Chakraborty, U., Kaur, G., Chaudhary, G.R., 2021. Development of environmental nanosensors for detection monitoring and assessment. *New Front. Nano. Environ. Sci.*, 91–143.

Chams, N., Chams, S., Badran, R., Shams, A., Araji, A., Raad, M., Mukhopadhyay, S., Stroberg, E., Duval, E.J., Barton, L.M., 2020. COVID-19: A multidisciplinary review. *Front. Public Health.* 8, 383.

Choi, I., Choi, Y., 2011. Plasmonic nanosensors: Review and prospect. *IEEE J. Sel. Top. Quantum Electron.* 18, 1110–1121.

Das, K., Pingali, M.S., Paital, B., Panda, F., Pati, S.G., Singh, A., Varadwaj, P.K., Samanta, S.K., 2021. A detailed review of the outbreak of COVID-19. *Front. Biosci.* 26, 149–170.

Erdem, O., Es, I., Saylan, Y.E., Inci, F., 2021. Unifying the efforts of medicine, chemistry, and engineering in biosensing technologies to tackle the challenges of the COVID-19 pandemic. *Anal. Chem.* 94, 3–25.

Forchette, L., Sebastian, W., Liu, T., 2021. A comprehensive review of COVID-19 virology, vaccines, variants, and therapeutics. *Curr. Med. Sci.*, 411–15.

Funari, R., Fukuyama, H., Shen, A.Q., 2022. Nanoplasmonic multiplex biosensing for COVID-19 vaccines. *Biosens. Bioelectron.* 208, 114193.

Gavriatopoulou, M., Korompoki, E., Fotiou, D., Ntanasis-Stathopoulos, I., Psaltopoulou, T., Kastritis, E., Terpos, E., Dimopoulos, M.A., 2020. Organ-specific manifestations of COVID-19 infection. *Clin. Exp. Med.* 20, 493–506.

Haes, A.J., Zou, S., Schatz, G.C., Van Duyne, R.P., 2004. Nanoscale optical biosensor: Short range distance dependence of the localized surface plasmon resonance of noble metal nanoparticles. *J. Phys. Chem. B* 108, 6961–6968.

Harrison, A.G., Lin, T., Wang, P., 2020. Mechanisms of SARS-CoV-2 transmission and pathogenesis. *Trends Immunol.* 41, 1100–1115.

Hsia, W., 2020. Emerging new coronavirus infection in Wuhan, China: Situation in early 2020. *Case Study Case Rep.* 10, 8–9.

Hu, B., Guo, H., Zhou, P., Shi, Z.L., 2021. Characteristics of SARS-CoV-2 and COVID-19. *Nat. Rev. Microbiol.* 19, 141–154.

Hu, T.Y., Frieman, M., Wolfram, J., 2020. Insights from nanomedicine into chloroquine efficacy against COVID-19. *Nat. Nanotechnol.* 15, 247–249.

Huang, C., Wang, Y., Li, X., Ren, L., Zhao, J., Hu, Y., Zhang, L., Fan, G., Xu, J., Gu, X., 2020. Clinical features of patients infected with 2019 novel coronavirus in Wuhan, China. *Lancet* 395, 497–506.

Inci, F., Karaaslan, M.G., Mataji-Kojouri, A., Shah, P.A., Saylan, Y., Zeng, Y., Avadhani, A., Sinclair, R., Lau, D.T.Y., Demirci, U., 2020. Enhancing the nanoplasmonic signal by a nanoparticle sandwiching strategy to detect viruses. *Appl. Mater. Today* 20, 100709.

Joseph, R.J., Ser, H.L., 2021. Stories from the East: COVID-19 situation in India. *Prog. Microbes Mol. Biol.* 4, a0000213.

Kuai, Y.H., Ser, H.L., 2021. COVID-19 situation in Thailand. *Prog. Microbes Mol. Biol.* 4, a0000260.

Lambrou, A.S., Shirk, P., Steele, M.K., Paul, P., Paden, C.R., Cadwell, B., Reese, H.E., Aoki, Y., Hassell, N., Zheng, X.Y., 2022. Genomic surveillance for SARS-CoV-2 variants: Predominance of the Delta (B. 1.617. 2) and Omicron (B. 1.1. 529) variants—United States, June 2021–January 2022. *MMWR* 71, 206.

Laval, J.M., Mazeran, P.E., Thomas, D., 2000. Nanobiotechnology and its role in the development of new analytical devices. *Analyst* 125, 29–33.

Law, L.N.S., Loo, K.Y., Goh, J.X.H., Pusparajah, P., 2021. Omicron: The rising fear for another wave in Malaysia. *Prog. Microbes Mol. Biol.* 4, a0000261.

Lee, H.T., Loh, H.C., Ramlee, S.N.L., Looi, I., 2021. Oral dietary supplements use among healthcare workers during the COVID-19 pandemic in Malaysia. *Prog. Microbes Mol. Biol.* 4, a0000236.

Lew, T.T.S., Aung, K.M.M., Ow, S.Y., Amrun, S.N., Sutarlie, L., Ng, L.F., Su, X., 2021. Epitope-functionalized gold nanoparticles for rapid and selective detection of SARS-CoV-2 IgG antibodies. *ACS Nano* 15, 12286–12297.

Lewis, D., 2020. Is the coronavirus airborne? Experts can't agree. *Nature* 580, 175.

Li, D., Zhou, H., Hui, X., He, X., Mu, X., 2021. Plasmonic biosensor augmented by a genetic algorithm for ultra-rapid, label-free, and multi-functional detection of COVID-19. *Anal. Chem.* 93, 9437–9444.

Lim, T.C., Ramakrishna, S., 2006. A conceptual review of nanosensors. *Z. Naturforsch. A* 61, 402–412.

Loo, K.Y., Letchumanan, V., 2022. COVID-19: Understanding the new variants of concern. *Prog. Microbes Mol. Biol.* 5, a0000282.

Loo, K.Y., Letchumanan, V., Tan, L.T.-H., Law, J.W.F., 2021. Updated COVID-19 condition in Australia. *Prog. Microbes Mol. Biol.* 4, 1-9.

Loo, K.Y., Law, J.W.F., Tan, L.T.H., Letchumanan, V., 2022. South Africa's battle against COVID-19 pandemic. *Prog. Microbes Mol. Biol.* 5, a0000264.

Machhi, J., Herskovitz, J., Senan, A.M., Dutta, D., Nath, B., Oleynikov, M.D., Blomberg, W.R., Meigs, D.D., Hasan, M., Patel, M., 2020. The natural history, pathobiology, and clinical manifestations of SARS-CoV-2 infections. *J. Neuroimmune Pharmacol.* 15, 359–386.

Maghoul, A., Simonsen, I., Rostami, A., Mirtaheri, P., 2022. An optical modeling framework for coronavirus detection using graphene-based nanosensor. *Nanomater.* 12, 2868.

Masterson, A.N., Muhoberac, B.B., Gopinadhan, A., Wilde, D.J., Deiss, F.T., John, C.C., Sardar, R., 2021. Multiplexed and high-throughput label-free detection of RNA/spike protein/IgG/IgM biomarkers of SARS-CoV-2 infection utilizing nanoplasmonic biosensors. *Anal. Chem.* 93, 8754–8763.

Moitra, P., Alafeef, M., Dighe, K., Frieman, M.B., Pan, D., 2020. Selective naked-eye detection of SARS-CoV-2 mediated by N gene targeted antisense oligonucleotide capped plasmonic nanoparticles. *ACS Nano* 14, 7617–7627.

Mousavi, S.M., Hashemi, S.A., Zarei, M., Amani, A.M., Babapoor, A., 2018. Nanosensors for chemical and biological and medical applications. *Med Chem* 8, 2161–0444.1000515.

Negahdari, R., Rafiee, E., Kordrostami, Z., 2023. A sensitive biosensor based on plasmonic-graphene configuration for detection of COVID-19 virus. *Plasmonics*, 18 1325–1335.

Nocerino, V., Miranda, B., Tramontano, C., Chianese, G., Dardano, P., Rea, I., De Stefano, L., 2022. Plasmonic nanosensors: Design, fabrication, and applications in biomedicine. *Chemosensors* 10, 150.

Özgür, E., Saylan, Y., Bereli, N., Türkmen, D., Denizli, A., 2020. Molecularly imprinted polymer integrated plasmonic nanosensor for cocaine detection. *J. Biomater. Sci. Polym. Ed.* 31, 1211–1222.

Polack, F.P., Thomas, S.J., Kitchin, N., Absalon, J., Gurtman, A., Lockhart, S., Perez, J.L., Pérez Marc, G., Moreira, E.D., Zerbini, C., 2020. Safety and efficacy of the BNT162b2 mRNA Covid-19 vaccine. *N. Engl. J. Med.* 383, 2603–2615.

Prather, K.A., Wang, C.C., Schooley, R.T., 2020. Reducing transmission of SARS-CoV-2. *Science* 368, 1422–1424.

Qiu, G., Gai, Z., Tao, Y., Schmitt, J., Kullak-Ublick, G.A., Wang, J., 2020. Dual-functional plasmonic photothermal biosensors for highly accurate severe acute respiratory syndrome coronavirus 2 detection. *ACS Nano* 14, 5268–5277.

Rahman, H.S., Aziz, M.S., Hussein, R.H., Othman, H.H., Omer, S.H.S., Khalid, E.S., Abdulrahman, N.A., Amin, K., Abdullah, R., 2020. The transmission modes and sources of COVID-19: A systematic review. *Int. J. Surg. Open* 26, 125–136.

Santopolo, G., Clemente, A., Gonzalez-Freire, M., Russell, S.M., Vaquer, A., Barón, E., Aranda, M., Socias, A., Del Castillo, A., Borges, M., 2022. Plasma-induced nanoparticle aggregation for stratifying COVID-19 patients according to disease severity. *Sens. Actuators B Chem.* 373, 132638.

Saylan, Y., 2023. Unveiling the pollution of bacteria in water samples through optic sensor. *Microchem. J.* 193, 109057.

Saylan, Y., Akgönüllü, S., Denizli, A., 2020. Plasmonic sensors for monitoring biological and chemical threat agents. *Biosensors* 10, 142.

Saylan, Y., Akgönüllü, S., Denizli, A., 2022. Preparation of magnetic nanoparticles-assisted plasmonic biosensors with metal affinity for interferon-α detection. *Mater. Sci. Eng. B* 280, 115687.

Saylan, Y., Denizli, A., 2020. *Virus detection using nanosensors, nanosensors for smart cities.* Elsevier, pp. 501–511.

Saylan, Y., Denizli, A., 2021. Optical biosensors for virus detection, biosensors for virus detection. *IOP Publishing Bristol, UK*, 2-1–2-20.

Saylan, Y., Erdem, Ö., Ünal, S., Denizli, A., 2019. An alternative medical diagnosis method: Biosensors for virus detection. *Biosensors* 9, 65.

Singh, P., Yadava, R., 2020. *Nanosensors for health care, nanosensors for smart cities.* Elsevier, pp. 433–450.

Stewart, M.E., Anderton, C.R., Thompson, L.B., Maria, J., Gray, S.K., Rogers, J.A., Nuzzo, R.G., 2008. Nanostructured plasmonic sensors. *Chem. Rev.* 108, 494–521.

Sze, S.M., 1994. *Semiconductor sensors.* Wiley, p. 550.

Tang, P., Hasan, M.R., Chemaitelly, H., Yassine, H.M., Benslimane, F.M., Al Khatib, H.A., AlMukdad, S., Coyle, P., Ayoub, H.H., Al Kanaani, Z., 2021. BNT162b2 and mRNA-1273 COVID-19 vaccine effectiveness against the SARS-CoV-2 Delta variant in Qatar. *Nat. Med.*, 272136–2143.

Thye, A.Y.K., Law, J.W.F., 2022. A Variant of Concern (VOC) Omicron: Characteristics, transmissibility, and impact on vaccine effectiveness. *Prog. Microbes Mol. Biol.* 5, a0000280.

Thye, A.Y.K., Loo, K.Y., Tan, K.B.C., Lau, J.M.S., Letchumanan, V., 2021. v. *Prog. Microbes Mol. Biol.* 4, a0000243.

Uddin, S.M.A., Chowdhury, S.S., Kabir, E., 2021. Numerical analysis of a highly sensitive surface plasmon resonance sensor for SARS-CoV-2 detection. *Plasmonics* 16, 2025–2037.

Udugama, B., Kadhiresan, P., Kozlowski, H.N., Malekjahani, A., Osborne, M., Li, V.Y., Chen, H., Mubareka, S., Gubbay, J.B., Chan, W.C., 2020. Diagnosing COVID-19: The disease and tools for detection. *ACS Nano* 14, 3822–3835.

Vo-Dinh, T., Cullum, B.M., Stokes, D.L., 2001. Nanosensors and biochips: Frontiers in biomolecular diagnostics. *Sens. Actuators B Chem.* 74, 2–11.

Volkert, A.A., Haes, A.J., 2014. Advancements in nanosensors using plastic antibodies. *Analyst* 139, 21–31.

Wen, C.Y., Liang, X., Liu, J., Zhao, T.Y., Li, X., Zhang, Y., Guo, G., Zhang, Z., Zeng, J., 2023. An achromatic colorimetric nanosensor for sensitive multiple pathogen detection by coupling plasmonic nanoparticles with magnetic separation. *Talanta* 256, 124271.

Whitney, A.V., Elam, J.W., Zou, S., Zinovev, A.V., Stair, P.C., Schatz, G.C., Van Duyne, R.P., 2005. Localized surface plasmon resonance nanosensor: A high-resolution distance-dependence study using atomic layer deposition. *J. Phys. Chem. B* 109, 20522–20528.

Wu, F., Zhao, S., Yu, B., Chen, Y.M., Wang, W., Song, Z.G., Hu, Y., Tao, Z.W., Tian, J.H., Pei, Y.-Y., 2020. A new coronavirus associated with human respiratory disease in China. *Nature* 579, 265–269.

Yakoubi, A., Dhafer, C.E.B., 2023. Advanced plasmonic nanoparticle-based techniques for the prevention, detection, and treatment of current COVID-19. *Plasmonics* 18, 311–347.

Yavuz, S., Ünal, S., 2020. Antiviral treatment of COVID-19. *Turk. J. Med. Sci.* 50, 611–619.

Yılmaz, G.E., Saylan, Y., Göktürk, I., Yılmaz, F., Denizli, A., 2022. Selective amplification of plasmonic sensor signal for cortisol detection using gold nanoparticles. *Biosensors* 12, 482.

Zhao, Q., Huang, H., Zhang, L., Wang, L., Zeng, Y., Xia, X., Liu, F., Chen, Y., 2016. Strategy to fabricate naked-eye readout ultrasensitive plasmonic nanosensor based on enzyme mimetic gold nanoclusters. *Anal. Chem.* 88, 1412–1418.

15 Other Analytical Methods for the Detection of Biological and Chemical Threats

Simge Balaban Hanoglu, Nursima Ucar, and Suna Timur

15.1 INTRODUCTION

Today, biological and chemical threat agents are among the potential disasters that seriously threaten humanity. These threat agents range from the spread of disease to chemical attacks. Biological threat agents are hazardous substances created by living organisms or their products. These substances come from biological sources such as viruses, bacteria, fungi, poisonous plants, and animals (Pohanka et al., 2007). Chemical threats are posed by the malicious use of chemical substances. These can be gases, liquids, or solids and can have toxic, explosive, or destructive properties. Chemical threat agents include nerve agents, biological warfare agents, toxins, and explosive chemicals. If such substances fall into the wrong hands or are misused, they can lead to large-scale disasters (Yue et al., 2016). Over the years, various analytical techniques have been developed to detect the two main threat factors. There are various analytical methods for detecting these substances with high sensitivity and selectivity, such as mass spectroscopy and liquid chromatography (Jansson and Åstot, 2015; Strozier et al., 2016). These methods have disadvantages such as being laboratory dependent, requiring experts for analysis, and taking a long time. Therefore, the need for different analytical methods has increased and the development studies of different systems have accelerated.

Biosensor systems are excellent candidates to replace traditional analytical methods due to their many advantageous properties (Saylan et al., 2020; Upadhyayula, 2012). These systems not only meet the need for early and rapid detection of these agents but can also detect them at very low concentrations. In addition, biosensor systems have become a popular method of analysis because they can be designed as user-friendly, low-cost, and portable systems. Electrochemical and optical systems, which are types of biosensor systems, are often used for the detection of biological and chemical agents. In this chapter, electrochemical and optical biosensor systems

DOI: 10.1201/9781003459316-15

that have been developed in recent years for the detection of biological and chemical agents are discussed. Based on current developments in electrochemical and optical systems, their advantages and future directions are highlighted.

15.2 BIOLOGICAL AND CHEMICAL THREATS

Biological and chemical threat agents have harmful effects on the entire living population, causing disease and even death. These threatening agents, which spread rapidly in the soil, water, and air, have often been used intentionally and as agents of bioterrorism (Green et al., 2019). Biological warfare agents are cheaper than chemical warfare agents. They are not only more accessible, but also highly effective (Pohanka et al., 2007). According to Centers for Disease Control and Prevention (CDC, 2019), biological bioterrorism agents have three main classifications. Viral pathogens are defined as category A, including *Bacillus anthracis*, causing anthrax, *Yersinia pestis*, known as plague, *Clostridium botulinum*, a toxin known as botulism, and *Francisella tularensis*, which are filoviruses such as Ebola. It is known that the death rates as a result of the spread of these pathogens among the population are quite high and that this poses a risk to national security. Category B is less common than category A and has a lower mortality rate. *Brucella* species, the Epsilon toxin of *Clostridium perfringens*, *Burkholderia mallei*, *Burkholderia pseudomallei*, *Chlamydia psittaci*, *Coxiella burnetiid*, food safety threats such as *Salmonella* species, *Escherichia coli* O157:H7, and *Shigella*, and water safety threats such as *Vibrio cholerae* and *Cryptosporidium parvum* are category B. Emerging pathogens that have the potential to spread in the future, such as Nihap virus and hantavirus, are defined as category C. It was a biological agent, SARS-CoV-2, which affected the whole world within a short time (Carvalhais et al., 2021). Rapid diagnosis was very important because it was defined as a respiratory virus since the first day of its emergence and it is very rapidly transmitted through the air (Yu et al., 2020). Therefore, in recent years, studies on the identification and rapid detection of biological agents have focused on this virus.

Chemical agents are inactive molecules that have harmful effects on the health of living things. Chemical agents are divided into groups according to their effects on the human body. The main and most common agents can be summarized in three groups. These groups are nerve agents (G-type and V-type agents), vesicant/blister agents known as sulfur and nitrogen mustard, and finally, blood/suffocating agents including cyanogen chloride and hydrogen cyanide (Yue et al., 2016). The use of these agents is prohibited because they have lethal effects on living beings. Nevertheless, they continue to be used. Therefore, it is very important to detect these substances quickly and in low concentrations. Chemical and biological agents are summarized in Figure 15.1.

15.3 OTHER ANALYTICAL METHODS

One of the analytical methods used in the detection of chemical and biological agents are biosensor systems. Biosensors are basically analytical devices consisting of a

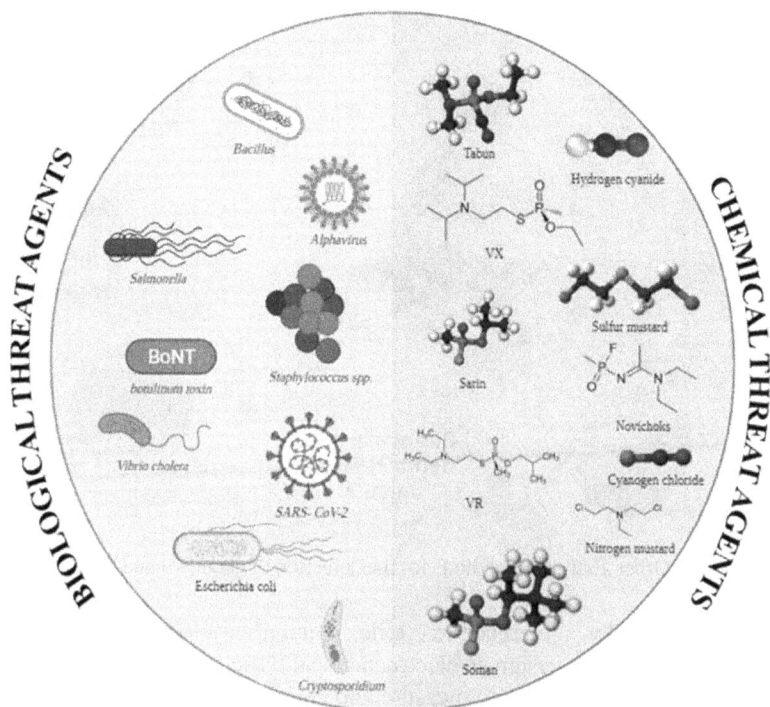

FIGURE 15.1 Some chemical and biological agents.

transducer, a bioreceptor, electronics, and a display. This device converts the reaction between the target analyte and the bioreceptor into measurable signals (Naresh and Lee, 2021). The World Health Organization (WHO) has reported that an ideal biosensor system should have some features, abbreviated as ASSURED (affordable, sensitive, specific, user-friendly, rapid, equipment-free, and deliverable). In addition to all these features, a biosensor system can be designed in small sizes and commercialized as point-of-care (POC) systems (Gubala et al., 2012). Many biosensor systems have been developed as POC systems for the detection of chemical and biological agents. The next section discusses electrochemical and optical biosensor systems that have been developed consistent with this goal (Figure 15.2).

15.3.1 ELECTROCHEMICAL BIOSENSORS

Electrochemical biosensors are analytical devices that convert biochemical processes into electrical signals (Li et al., 2017). In the electrochemical biosensor system, the interactions between the transducer, the target analyte, and the biological recognition element are converted into electric current or voltage. Electrodes are one of the most important components of electrochemical sensor systems. They play a particularly important role in sensor performance by controlling the flow of electrons and

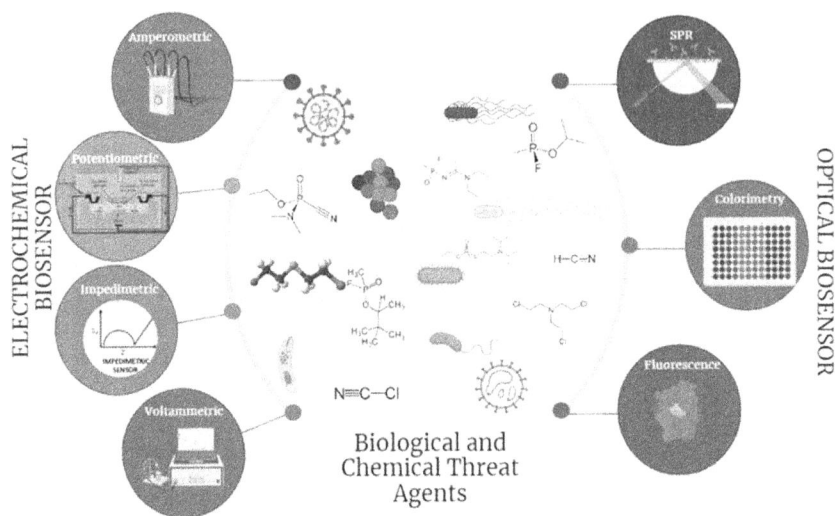

FIGURE 15.2   Other analytical methods for the detection of chemical and biological agents.

bioactivity. Essentially, a reference electrode, an auxiliary electrode, and a working electrode are required to design an electrochemical biosensor system. The reference electrode serves as a reference to measure and control the potential of the working electrode without current flowing. It must have a stable electrochemical potential (Dryden and Wheeler, 2015). The auxiliary electrode is provided a connection to the electrolytic solution so that a current can be applied to the working electrode (Grieshaber et al., 2008). Biological materials are fixed on the working electrode and the biochemical reaction takes place at this electrode. The reaction taking place at the surface is caused by a current difference with the transfer of electrons at the surface. The current difference varies depending on the concentration of the target analyte (Balkourani et al., 2021). Depending on the target analyte, different working electrodes such as gold, platinum, or carbon can be used. In addition to the analyte, many parameters such as the surface modification and the sensing capability of the electrode play an important role in electrode selection. Various electrochemical techniques can be used to measure the reaction signals in biosensors (Li et al., 2017). According to the principle of measurement, electrochemical biosensors can be classified as amperometric, potentiometric, impedimetric, and voltammetric (Singh et al., 2021a). In recent years, all these electrochemical measurement principles have been widely used to detect biological and chemical threats. Electrochemical biosensor platforms for the detection of biological and chemical agents are summarized in Table 15.1.

The amperometric biosensors are one of the most comprehensive analytical methods due to their simplicity and sensitivity. These biosensors measure current changes at the working electrode, which result from oxidation in a biochemical reaction (Chaubey and Malhotra, 2002). Usually, the operation of amperometric biosensors

TABLE 15.1
Some Electrochemical Biosensors Designed for the Detection of Chemical and Biological Agents

Target Agents	Sensor Type	Sensor Design	Results	Ref.
Salmonella	Amperometric	GE/CMCG/protein A/ anti-*Salmonella* Ab/BSA	• *Salmonella enterica* serovar Typhimurium can be detected with chronoamperometry. • LOD was determined as 10 CFU/mL with a detection time of 125 min. • Target analyte can be detected in milk samples.	Melo et al., 2021
E. coli	Amperometric	SPCE/Fe$_3$O$_4$@Au MNP-Ab-C/*E. coli* O157:H7/Ab-D+ HRP-Ab/ TMB	• Sandwich-type magnetic immunoassay was developed. • For *E. coli* O157:H7 strain, linear range was determined from 20 to 2 × 10^6 CFU/mL. • LOD was determined as 20 CFU/mL. • *E. coli* was detected in a milk sample.	Bazsefidpar et al., 2023
Brucella	Amperometric	SPGE/GO/Ppy/Ab$_1$/ BSA/*Brucella*/GO/ Fe$_3$O$_4$MB/Ab$_2$/BSA	• Sandwich-type sensors were designed. • LOD was determined as 2.2 × 10^{2}CFU/mL. • Serum samples were used as real samples and samples were prepared with a known concentration *Brucella* antigen.	Chen et al., 2020
Soman (GD), sarin (GB), VX, and tabun (GA)	Amperometric	GE/PoPD film/ AChE-ChOxCS-Glu	• A self-assembly bienzymatic electrochemical biosensor was designed. • LOD values for soman (GD), sarin (GB), VX, and tabun (GA) were determined as 2.9; 1.1; 3.1 and 41 nM, respectively.	Bu et al., 2020
Staphylococcus aureus and *E. coli* O157:H7	Potentiometric	SPCE/DenAu/PTAA/ Peptide	• SPCE electrode was prepared as four-channel. • Polypropylene paper was used for electrode printers. • Two target analytes can be detected on one electrode system. • Peptides were used for biorecognition molecules. • LOD values for *S. aureus* and *E. coli* were determined as 38 and 241 CFU/mL, respectively. • Milk samples were used for actual samples.	Xu et al., 2023

(Continued)

TABLE 15.1 (CONTINUED)
Some Electrochemical Biosensors Designed for the Detection of Chemical and Biological Agents

Target Agents	Sensor Type	Sensor Design	Results	Ref.
DFP	Potentiometric	Tattoo paper/Carbon ink/ PANi/OPH/Nafion/ PVA-based Hydrogel	• A wearable tattoo biosensor system was designed. • pH sensitive working electrode was performed with PANI electropolimerization. • LOQ was determined as 10 mM. • DFP can be detected on the body with wireless interface.	Mishra et al., 2018
SARS-CoV-2	Voltammetric, impedimetric	GE/DNA2 probe/ Peptide- DNA1 probe-AuNP/Methylene blue	• Papain-like enzyme SARS-CoV-2 protease (PLpro) was determined with the biosensor. • In the presence of PLpro, DNA1 fragments were removed from AuNP, and the fragment was combined with DNA2 in the electrode surface. • The feasibility of the biosensor was verified with EIS and SWV. • LOD was determined as 27.18 fM. • Blood, saliva, and swab samples can be used for detection.	Liang et al., 2023
SARS-CoV-2	Voltammetric, impedimetric	AN/AuNP/rGO/PMB/ P[Bvim]Br/SARS-CoV-2-S protein/P[Bvim]Br	• MIP-based biosensor was developed. • CV method was used for electropolymerization of [Bvim]Br. • CV and EIS methods were used for confirmation of modification of AN. • Analytical parameters were evaluated with SWV measurements. • S protein of SARS-CoV-2 can be detected from clinical serum samples. • LOD value was determined as 38 pg/mL.	Yang et al., 2023

(Continued)

TABLE 15.1 (CONTINUED)
Some Electrochemical Biosensors Designed for the Detection of Chemical and Biological Agents

Target Agents	Sensor Type	Sensor Design	Results	Ref.
SARS-CoV-2	Voltammetric, impedimetric	1. SPGE/MB-RNA Probe/ Cas13a-crRNA-S gene RNA 2. SPGE/Fc-RNA Probe/ Cas13a-crRNA-Orf1ab RNA	• This study proposed a CRISPR/Cas13a-powered electrochemical multiplexed biosensor for detecting SARS-CoV-2 RNA strands. • SWV technique was used for the detection of analytical parameters. • LOD values were determined: 2.5 ag/μL gor S gene and 4.5 ag/μL for Orf1ab gene. • SARS-CoV-2 S and Orf1ab genes were detected in both synthetic and clinical swab samples.	Kashefi-Kheyrabadi et al., 2023
SARS-CoV-2	Voltammetric	GCE//ZnS/grafen/pDNA/ tDNA	• CV and DPV techniques were used for characterization of the biosensor. • In this study, tests were performed for synthetic DNA samples and SARS-CoV-2 standard samples. • LOD values were determined as 4.453×10^{-20} M for DNA samples and 2.068×10^{-20} M for standard SARS-CoV-2 samples.	Sarwar et al., 2023
SARS-CoV-2	Voltammetric, impedimetric	SPCE/Cu$_7$S$_4$-Au/ MPBA/ SARS-CoV-2 spike protein/3-APBA	• MIP-based biosensor was developed. • CV technique was used for electropolymerization, EIS technique was used for characterization. • The biosensor was used for the detection of SARS-CoV-2 S protein in clinical serum samples. • S protein was bounded in pH 7-14, while the protein was removed in acidic pH (2-5). • LOD was determined as 1.76 pg/mL.	Yin et al., 2023

(Continued)

TABLE 15.1 (CONTINUED)

Some Electrochemical Biosensors Designed for the Detection of Chemical and Biological Agents

Target Agents	Sensor Type	Sensor Design	Results	Ref.
SARS-CoV-2	Voltammetric, impedimetric	SPCE/MNP-SARS-CoV-2 S1 or S2 protein/ SAR-CoV-2 S1 or S2 or cocktail Ab	• Cocktail antibody was purified from serum samples. • SARS-CoV-2 S1 and S2 proteins were conjugated with MNP. • SARS-COV-2 S1, S2 and cocktail Ab was used. • The biosensor can be used for the detection of original, alpha, delta, and beta variants. • LOD values were determined as 0.53–0.75 ng/mL for the antibody cocktail-based sensor compared with 0.93 ng/mL and 0.99 ng/mL for the platforms using anti-S1 and anti-S2, respectively.	Durmus et al., 2022
E. coli O157:H7 GXEC-N07	Voltammetric impedimetric	GCE/CFGO/CB/ EP01-BSA/GXEC-N07	• Phase EP01 was used as a recognition agent. • LOD was determined as 11.8 CFU/mL. • The biosensor has been applied to the quantitative detection of GXEC-N07 in fresh milk and raw pork. • Target bacteria can be detected in less than 30 min.	Zhou et al., 2023
E. coli O157:H7	Voltammetric, impedimetric	Au/Pt@BSA/GA/anti-E. coli Ab/BSAT/E.coli	• CV, EIS, and DPV techniques were applied to assess the sensing performance of the biosensor. • LOD was determined as 9 cfu/mL.	Hu et al., 2023
Staphylococcus aureus	Voltammetric, impedimetric	GCE/AuNPs/MCH modified MB-ssDNA/Cas12a/ crRNA	• Characterization and analysis of the electrode surface were evaluated with CV and EIS. • Under optimal conditions, LOD values were determined as 2.51 fg/µL for genomic DNA and 3 CFU/mL for S. aureus in pure cultures. • For S. aureus detection, milk samples were spiked with different concentrations of S. aureus DNA.	Huang et al., 2023

(Continued)

TABLE 15.1 (CONTINUED)
Some Electrochemical Biosensors Designed for the Detection of Chemical and Biological Agents

Target Agents	Sensor Type	Sensor Design	Results	Ref.
Shigella flexneri	Voltammetric, impedimetric	ITO electrode/P-Mel/PGA/DSS/CP-DNA/LT/RP	• DNA biosensor was designed. • Electrochemical characterization studies were performed with CV and EIS. • LOD values were determined as 7.4×10^{-22} mol/L in the complementary linear target of *S. flexneri* and 10 cells/mL in real *S. flexneri* sample. • In order to detect the bacteria, samples of food additives (meat, raw milk, bread, strip water, and salad) were examined.	Ali et al., 2023
Botulinum neurotoxin BoNT/A and BoNT/C	Voltammetric	Paper based SPE/AuNP/Sulphur/Peptide/MB	• The biosensor was integrated with a smartphone. • Measurements were performed with SWV. • LOD was determined as 10 pM. • The applicability of this biosensor was evaluated with spiked samples of orange juice.	Caratelli et al., 2021
BoNT-A	Impedimetric, voltammetric	GCE/H₂SO₄/Au-Gr-Cs/MPA/BoNT Ab/BSA/BoNT-A	• Carboxylic acid group was generated on the surface with H_2SO_4 using the CV technique. • LOD for immunosensor was determined as 0.11 pg/mL. • The performance of the immunosensor against BoNT/A in milk and serum samples was investigated.	Afkhami et al., 2017
BoNT/A	Voltammetric, impedimetric	SPCE/AuND/CSNP/Glu/Ab/BSA/BoNT-A	• LOD was determined as 0.15 pg/mL. • Detection of BoNT/A was performed in milk and serum samples.	Sorouri et al., 2017

(Continued)

TABLE 15.1 (CONTINUED)
Some Electrochemical Biosensors Designed for the Detection of Chemical and Biological Agents

Target Agents	Sensor Type	Sensor Design	Results	Ref.
Bacillus anthracis spore	Impedimetric	SPGE/BAS-6R/*B. cereus*	• DNA-based aptasensor platform was designed. • *B. cereus* spores 14579 were detected with LOD of 3×10^3 CFU/mL. • The air sampling was performed.	Mazzaracchio et al., 2019
B. anthracis	Voltammetric, impedimetric	GCE/CoPC/CS/ Ab1/ BSA/*B. anthracis* Sap antigen/Au–Pd NPs@ BNNS/Ab$_2$	• Surface array protein (Sap) of *B. anthracis* was detected with the biosensor. • The bacteria can be detected in culture medium after 1 h of growth. • LOD was determined as 1 pg/mL.	Sharma et al., 2016
Brucella	Impedimetric, voltammetric	SPCE/RpG/Au/GO/*Brucella* Ab/BSA	• LOD was determined as 3.2×10^2 CFU/mL. • The biosensor was applied successfully to test *Brucella* in pasteurized milk samples.	Chen et al., 2022a
Brucella	Impedimetric voltammetric	SPGE/Cys-Au/4-MBA/	• Analytical parameters were evaluated with SWV and characterization studies were performed with EIS. • LOD was 5.12×102 cfu/mL. • *Brucella* were detected in fresh milk.	Chen et al., 2022b
Clostridium perfringens	Voltammetric impedimetric	MGCE/CMB/DNA walker/ BSA/HCR/MB	• CV, DPV, and EIS techniques were used. • LOD was determined as 1 CFU/g. • To evaluate the application of this bimodal DNA biosensor in meat samples, various meat samples (chicken, beef, mutton, duck, and pork) were analyzed by adding *C. perfringens* at known concentration.	Wang et al., 2022

(Continued)

TABLE 15.1 (CONTINUED)
Some Electrochemical Biosensors Designed for the Detection of Chemical and Biological Agents

Target Agents	Sensor Type	Sensor Design	Results	Ref.
Clostridium perfringens	Voltammetric, impedimetric	GCE/CH/CeO$_2$, ssDNA probe/tDNA	• CV and EIS techniques were used. • LOD was determined as 7.06×10^{-15} mol/L. • The fabricated biosensor was applied for the determination of *C. perfringens* DNA sequence in dairy products with RSD lower than 4.96%.	Qian et al., 2018
Salmonella Typhi	Voltammetric, impedimetric	SPCE/P-Cys@AuNP/CP/BSA/LIT/RP/ssDNA	• CV technique was used for electropolymerization of P-Cys@AuNP. • Analytical parameters were evaluated with DPV measurements. • LOD values for the fabricated biosensor were calculated as 6.8 $\times 10^{-25}$ mol/L in *S. Typhi* complementary linear target and 1 CFU/ml in a real *S. Typhi* sample. • Target DNA was spiked in the human blood, raw milk, egg, and poultry feces samples.	Bacchu et al., 2022
Salmonella spp.	Impedimetric, voltammetric	GE/Cys/Nisin/*Salmonella*	• Peptides in bacterial cell membranes were bound by electrostatic attraction with nisin. • LOD for *Salmonella* cells was determined as 1.5×10^1 CFU/mL. • Bacteria cell was added in food matrices milk sample and were analyzed.	Malvano et al., 2020
Burkholderia pseudomallei	Voltammetric, impedimetric	GE/P-ssDNA/Cas14a-SgRNA/PtPd@PCN-224	• CRISPR-based electrochemical biosensor was designed. • In the presence of target DNA, CRISPR/Cas14a was activated and P-ssDNA was divided. The reduction peak decreased due to the lack of assembly of PtPd@PCN-224 nanozymes. • LOD was calculated as 12.8 aM. • *B. pseudomallei* DNA in human serum can successfully determine.	Li et al., 2023

(Continued)

TABLE 15.1 (CONTINUED)
Some Electrochemical Biosensors Designed for the Detection of Chemical and Biological Agents

Target Agents	Sensor Type	Sensor Design	Results	Ref.
Cryptosporidium parvum	Voltammetric	NMI electrode/aptamer/ parasitic organism	• The microfluid aptasensor system was designed as POC. • DPV technique was used for measurements. • LOD values were determined as 5 oocysts/mL in buffer medium and 10 oocysts/mL in stool and tap water media. • Detection time was determined as 40 min in different sample matrices.	Siavash Moakhar et al., 2023
Cryptosporidium parvum	Impedimetric, voltammetric	GE/Protein G/Thiol/ anti-*Cryptosporidium* antibodies (IgG3+BSA)	• Scalable chip-based electrochemical biosensor was designed. • CV, EIS, and SWV electrochemical methods were used. • With EIS, LOD was determined as 20 oocysts/5 µL. • With the biosensor, early detection of bacteria in water samples can be realized.	Luka et al., 2022
Vibrio cholerae	Voltammetric	SPCE/AuNP/SiNP/GA/ capture probe/cDNA/ AQMS/ reporter probe/ AQMS	• A sandwich-type DNA hybridization strategy was used to capture the target bacterial DNA with using a double DNA probe. • LOD was determined as $1.25 \times 10{-}18$ M (i.e., $1.1513 \times 10{-}13$ µg/µL). • The DNA-based biosensor can have long-term stability of up to 55 days. • Various bacterial strains, environmental water samples, and vegetable samples were used for the evaluation of the developed DNA biosensor.	Futra et al., 2023
Cholera toxin subunit B (CTxB)	Impedimetric	SPCE/CCB/CTxB	• Cell-membrane based biosensor system was developed. • LOD for CTxB was determined as $\sim$11.46 nM. • It was determined that the biosensor remained stable for three weeks.	Kim et al., 2023

(Continued)

TABLE 15.1 (CONTINUED)

Some Electrochemical Biosensors Designed for the Detection of Chemical and Biological Agents

Target Agents	Sensor Type	Sensor Design	Results	Ref.
Hantavirus Araucaria	Voltammetric, impedimetric	Proto pasta/Hantavirus Ab/ BSA/Hantavirus	• Sample preparation time was determined as ~30 min. • CCB was assessed using EIS with the main components of diarrhea stools, including saccharides, steroids, acids, and various bacterial toxins. • A commercial 3D conductive filament consisting of carbon black and polylactic acid was used as an electrode system. • LOD for Hantavirus Araucaria nucleoprotein was determined as 22 µg/mL. • Immunosensor was applied with success for virus detection in 100x diluted human serum samples.	Martins et al., 2021
Hantavirus	Voltammetric, impedimetric	GCE/AuNPs/RGO/cDNA/ MCH/tDNA/ CuMOF-sDNA	• The biosensor design was based on metal-organic framework (MOF). • A sandwich-type DNA hybridization strategy was used to capture the target bacterial DNA using a double DNA probe. • LOD of tDNA was determined as 0.74×10^{-15} mol/L. • The biosensor was applied to detect real virus samples.	Yiwei et al., 2021
Sulfur Mustard	Voltammetric, chronoamperometric	Paper/Graphite ink/ CB- PBNPs/ChOx/SM or CEES	• Office paper-based electrode was designed for biosensor design. • With the biosensor, 2-chloroethyl ethyl sulfide (CEES) and bis(2-chloroethyl) sulfide (SM) detections were performed. • The H_2O_2 reduction current of ChOx enzyme was monitored chronoamperometrically. • LOD for the measurement of a sulfur mustard was determined as 0.9 mM.	Colozza et al., 2021a

(Continued)

TABLE 15.1 (CONTINUED)
Some Electrochemical Biosensors Designed for the Detection of Chemical and Biological Agents

Target Agents	Sensor Type	Sensor Design	Results	Ref.
Sulfur mustard	Voltammetric	SPE/oligonucleotide/SM/pAb-2F8/sAb/ALP	• The biosensor was designed for the detection of SM. • The biosensor was designed as enzyme-linked immunoassay. • LOD was determined as 12 µM.	Colozza et al., 2021b

GE: Gold electrode; CMCG: Carboxymethyl cashew gum; Ab: Antibody; BSA: Bovine serum albumin; SPCE: Screen printed carbon electrode; MNP: Magnetic nanoparticle; Ab-C: *E. coli* O157:H7 capture antibody; Ab-D: The *E. coli* O157:H7 detection antibody; HRB-Ab: Secondary HRP-labeled antibody; TMB: 3,3′,5,5′-tetramethylbenzidine liquid substrate; SPGE: Screen printed gold electrode; GO/Ppy: Graphene oxide/polypyrrole nanohybrids; Ab_1: Primary antibody (Ab_1); GO/Fe_3O_4/MB/Ab_2: Graphene oxide/nano-iron oxide/methylene blue/label the secondary antibody; AChE: Acetylcholinesterase; PoPD: Poly(o-phenylenediamine); ChOx: Choline oxidase; CS: Chitosan (CS); Glu: Glutaraldehyde; DenAu: Dendrimer gold; PTAA: Polymer polythiophene acetic acid; PANi: pH-sensitive polyaniline; OPH: Enzyme organophosphate hydrolase; AN: Acupuncture needle; rGO: Reduced graphene oxide; PMB: Poly(methylene blue); P[Bvim]Br: Poly- 1-vinyl-3-butylimidazolium bromide; MB: Methylene blue; Fc: Ferrocene; crRNA: Target-specific CRISPR RNA; GCE: Glassy carbon electrode; pDNA: Probe-DNA; tDNA: Target DNA; MPBA: 4-mercaptophenylboric acid; 3-APBA: 3-aminophenylboronic acid; CFGO: Conjugated carboxyl graphene oxide; CB: Conductive carbon black; GA: Glutaraldehyde; BSAT: BSA and Tween-20; ITO: Indium tin oxide; P-Mel: Poly-melamine; PGA: Poly(glutamic acid); DSS: Disuccinimidyl suberate; CP-DNA: Capture probe DNA; LT: Complementary linear targets; RP: Reporter probe; Au-Gr-Cs: Au nanoparticles/graphene-chitosan; MPA: 3-mercaptopropionic acid; AuNDs/CSNPs: Gold nanodendrites and chitosan nanoparticles; BAS-6R: *B. anthracis* spores-binding DNA aptamer; CoPC: Cobalt (II) phthalocyanine Ab1: Rabbit α-*B. anthracis* Sap antibodies; Au–Pd NPs@BNNSs: Gold–palladium nanoparticle@ poly (diallyldimethylammonium chloride) functionalized boron nitride nanosheets; Ab2: Mouse monoclonal anti-*B. anthracis* Sap antibodies; RpG/Au/GO: Recombinant protein G/gold nanoparticles/graphene oxide; 4-MBA: 4-mercaptobenzoic acid; MGCE: Magnetic glassy carbon electrode; CMBs: Carboxyl magnetic beads; HCR: Hybridization chain reaction; CeO2: Cerium oxide; tDNA: Target DNA; P-Cys: Poly cysteine; CP: Capture probe; LT: Complementary linear targets; RP: Reporter probe; P-ssDNA: Phosphorylated ssDNA; PtPd@PCN-224: PtPd nanoparticles functionalized porphyrin metal-organic framework nanoenzymes; NMIs: 3D gold nano-/microislands; AQMS: Anthraquinone-2-sulfonic acid monohydrate sodium salt; CCB: GM1-expressing (monosialo-tetra-hexosyl-ganglioside) Caco-2 cell membrane-coated biosensor; CuMOF: Cu-based metal-organic framework; CB/PBNPs: Carbon black and Prussian blue nanoparticles; ChOx: Choline oxidase; ALP: Alkaline phosphatase enzyme; pAb-2F8: Primary antibody 2F8; sAb: Secondary Ab (anti-mouse IgG).

is expressed by the potential applied between the working and reference electrodes. Thanks to the continuously measured current, precise and quantitative analytical information about the target analyte can be obtained. The magnitude of the current is proportional to the concentration of electroactive species present in the solution. Both cathodic (reducing) and anodic (oxidizing) reactions are recorded with the biosensor type. Amperometric measurement is widely used in biocatalytic and affinity sensors due to its simplicity and low limit of detection (LOD). Generally, these biosensors are more sensitive than potentiometric sensors. In addition, amperometric biosensors have comparable response times and energy ranges and are more suitable for mass production. (Ghindilis et al., 1998). The difference with this method as compared to the voltammetric biosensor is that there is no measurement potential. Amperometric sensors, which are very effective in various fields, are also used for the detection of chemical (Bu et al., 2020) and biological (Bazsefidpar et al., 2023; Chen et al., 2020; Melo et al., 2021) agents. In a study by Melo et. al (Melo et al., 2021), an antibody-based chronoamperometry biosensor was designed for the detection of *Salmonella*. For the design of a biosensor platform, carboxymethylated cashew gum (CMCG) film was electroplated onto a gold electrode surface and polyclonal anti-*Salmonella* antibodies were immobilized on the surface. The LOD of *Salmonella enterica* serovar Typhimurium was determined as 10 CFU/mL, and the detection time was 125 minutes. In addition, it was stated that the immunosensor could be used for the detection of *Salmonella* from ultra-high temperature milk without a pre-enrichment step. Another an antibody-based amperometric biosensor platform was developed for the analysis of *E. coli* O157:H7 (Bazsefidpar et al., 2023). In this study, a sandwich-based immunoassay platform was developed using magnetic nanoparticles (MNP). First, core-shell magnetic Fe_3O_4@Au nanoparticles were functionalized with an *E. coli* O157:H7 capture antibody (Ab-C). In the presence of *E. coli,* the bacteria were captured by the antibody. Then, the *E. coli* O157:H7 detection antibody (Ab-D) and the secondary HRP-labeled antibody (HRB-Ab) were bound to the surface of *E. coli* and a sandwich structure was formed. A screen-printed carbon electrode (SPCE) was used to fabricate the biosensor platform and the sandwich structure was immobilized on the electrode surface using a magnet. The enzymatic reaction was observed chronoamperometrically with the addition of 3,3′,5,5′-tetramethylbenzidine substrate (TMB). It was found that *E. coli* can be detected in a short period of one hour and in milk samples.

A potentiometric biosensor is based on measuring the potential difference between the working and reference electrodes. According to the International Union of Pure and Applied Chemistry (IUPAC), potentiometric biosensors have two important properties: (i) The biological component is an integral part of the sensor receptor that recognizes an analyte, (ii) the analytical signal generated by the biosensor is a potential (Koncki, 2007). The main advantages of these biosensors are their sensitivity and selectivity when using a highly stable, accurate reference electrode. They are easy to manufacture, inexpensive, can be used in a wide range of ion concentrations, and downsizing does not affect their performance. The use of potentiometric tools in the field of biosensors has opened many new doors for diagnosis and detection. There are a variety of sensors on the market that have been developed using potentiometric methods (Singh et al., 2021b). Compared to voltammetric and amperometric biosensors, potentiometric biosensors have the advantage of not being affected by electrode size. Thanks to this advantage, they can be miniaturized and tailor-made without sensitivity problems (Zdrachek and Bakker, 2021). According to these advantages, four-channel SPCE electrodes on polypropylene paper were fabricated for the simultaneous detection of two different bacteria (*Staphylococcus aureus* (SA) and *E. coli* O157:H7) (Xu et al., 2023). In this study, peptides were used as biorecognition molecules for bacterial recognition. The electrode surfaces were coated with dendrimer gold (DenAu) and conductive polymer polythiophene acetic acid (PTAA) to increase the effective area for peptide immobilization. LOD values for SA and *E. coli* were determined to be 38 and 241 CFU/mL, respectively. It was reported that simultaneous detection of milk samples was possible, and the biosensor had good specificity and reproducibility. Another study is a potentiometric wearable tattoo sensor developed for chemical warfare agent detection (Mishra et al., 2018). The nerve agent diisopropyl fluorophosphate (DFP) can be detected with the pH-sensitive flexible tattoo biosensor. In the development of the biosensor, tattoo paper was used as the electrode and the working electrode was formed with carbon ink. It was then coated with polyanaline (PANi) to make it pH sensitive. The surface was functionalized with organophosphate hydrolase (OPH) enzyme, to allow the degradation of DFP. Finally, the surface was coated with a PVA-acrylamide hydrogel. This coating allowed the distribution of DFP vapor on the surface. The biosensor system, based on the enzymatic hydrolysis of DFP, offered the possibility of wireless monitoring from a mobile device. It is a very promising system for analyzing chemical threats directly from the body.

Voltammetry belongs to a category of electroanalytical methods in which information about an analyte is obtained by changing a potential and then measuring the resulting current (Grieshaber et al., 2008). Voltammetric biosensors detect variations in both potential and current. As the potential is scanned over a wide range, the current and the result of the potential are measured. The current generated is directly proportional to the concentration of the target in the electrolyte (Ronkainen et al., 2010). Voltametric biosensors are widely used in sensor platforms because of their low cost, good selectivity, and high sensitivity. They have low noise, and these methods can be used to detect multiple species with different peak potentials in a single scan. In the biosensor, many techniques such as differential pulse voltammetry (DPV), cyclic voltammetry (CV), square wave voltammetry (SWV), and linear sweep voltammetry (LSV) are used for detection. The CV technique is one of the most used voltammetric methods. In the method, the current measured against the applied voltage, which changes at a constant rate, is measured between the specified start and end voltage values. This voltage is first applied in the positive or negative direction from the start potential to the end potential and then from the end potential to the start potential. The technique is used in electrochemistry to determine electrode reaction mechanisms, standard electron transfer rate constants, and diffusion coefficients (N. Aristov and A. Habekost, 2015). The CV technique provides information about the amount of oxidizable material in the working electrode. In addition, it can used to study the stability, repeatability, and reproducibility of the biosensor

(Serra, 2011). DPV is a sensitive electrochemical technique. It is commonly used for single or simultaneous analysis of compounds. In this technique, small amplitude potential pulses are applied to a linear ramp potential, and short pulses are in a staircase wave form (Deffo et al., 2024). A base potential value is chosen in DPV and applied to the electrode. The base potential is increased in equal steps between pulses. The current is measured before applying the pulse and the difference between the currents is calculated at the end of each pulse (Simões and Xavier, 2017). The main advantage of DPV is the low capacitive current, which leads to high sensitivity. Small step sizes in DPV also result in narrower voltammetric peaks. Therefore, DPV is often used to distinguish analytes with similar oxidation potentials (Venton and DiScenza, 2020). SWV is a large-amplitude differential technique in which a waveform consisting of a symmetric square wave superimposed on a base conductor potential is applied to the working electrode (Osteryoung and Osteryoung, 1985). Because of the minimal contribution of non-Faradaic currents, SWV can be used to achieve high sensitivity scanning. The use of differential current waveforms instead of reversal current waveforms and the significant time variation between potential reversal and current collection make this technique more sensitive than other electroanalytical techniques (Tolun and Altintas, 2023). More specifically, the timing of SWV is frequency dependent. As frequency increases, the time constant decreases and the measured Faradaic current increases, resulting in sharper and better-defined peaks in the voltammogram (Batista Deroco et al., 2020). This technique is one of the fastest and most sensitive voltammetry techniques when compared to chromatographic and spectroscopic techniques in terms of LOD value (Simões and Xavier, 2017). Another electrochemical method is impedimetric sensors, which can be used alone or in combination with voltammetric methods. It is an electrochemical biosensor system in which the change in impedance is used to detect analytes or biological entities (Singh et al., 2021b). The most common technique used in this method is electrochemical impedance spectroscopy (EIS). EIS is a sensitive indicator of a variety of chemical and physical properties. This method can be used to easily determine the properties of the electrode as well as the processes that occur at the interface of the electrode (Singh et al., 2021b). The bonding processes that occur at the electrode surface cause the electrical properties to change in the solution between the two electrodes. For the EIS technique to work in the frequency domain, it requires a circuit modeled as a solution that is a combination of these elements depending on the analytes present in the solution. When a voltage is applied to the circuit, the current is calculated using Ohm's law. This current flows through all elements of the system, such as the working electrode, the biological material, the solution, and the auxiliary electrode. The impedance value thus obtained represents the contributions of these elements. The EIS value of a biological substance is determined as a function of time or as a function of the concentration of the substance to be analyzed. It is used in the preparation of biosensors, in monitoring and analyzing the interactions of biomolecules, in determining the interaction between the bioreceptor and the analyte, and in surface characterizations.

In recent years, many studies have been conducted in which both voltammetric and impedimetric measurements have been widely used to detect biological

(Afkhami et al., 2017; Ali et al., 2023; Bacchu et al., 2022; Caratelli et al., 2021; Chen et al., 2022a, 2022b; Durmus et al., 2022; Futra et al., 2023; Hu et al., 2023; Kashefi-Kheyrabadi et al., 2023; Kim et al., 2023; Li et al., 2023; Liang et al., 2023, 2023; Luka et al., 2022; Malvano et al., 2020; Martins et al., 2021; Mazzaracchio et al., 2019; Qian et al., 2018; Sarwar et al., 2023; Sharma et al., 2016; Siavash Moakhar et al., 2023; Sorouri et al., 2017; Wang et al., 2022; Yang et al., 2023; Yin et al., 2023; Yiwei et al., 2021; Zhou et al., 2023) and chemical (Colozza et al., 2021a, 2021b) agents. With the COVID-19 pandemic, studies on electrochemical biosensing systems for the detection of the SARS-CoV-2 virus have intensified. Since specific detection of target molecules is very important, molecular imprinting technology has been used in sensor systems. For the detection of SARS-CoV-2 spike (S) protein, a miniature biosensor was designed with molecularly imprinted polymers (MIPs) (Yang et al., 2023). For the design of the miniaturized system, a small stainless steel acupuncture needle (AN) with good conductivity was used instead of an electrode. In the development of the biosensor, gold nanoparticles (AuNP) and graphene oxide (GO) were first deposited on the surface AN. Then an electroactive surface was formed by electropolymerization of polymethylene blue (PMB) and poly-1-vinyl-3-butylimidazolium bromide (P[Bvim]Br). The CV technique was used to achieve electropolymerization. After the S protein template of SARS-CoV-2 was applied to the surface, P[Bvim]Br was again deposited on the surface. After removal of the template, clinical serum samples were added to the surface. In the presence of spike proteins in the sample, a decrease in SWV peak currents was observed. It was said that the system, which has the advantage of being able to work with clinical samples and is a miniaturized system, has high reliability. In the electrochemical biosensor system, another popular approach is RNA detection by integrating CRISPR-Cas effectors. A similar approach for the detection of SARS-CoV-2 S and Orf1ab genes was adopted by Kheyrabadi et al (Kashefi-Kheyrabadi et al., 2023). An electrochemical biosensor powered by CRISPR/Cas13a was designed with a strategy that does not require nucleic acid amplification. In the system, gold electrodes were functionalized with methylene blue or ferrocene-labelled reporter RNA. Following, Cas13-target-specific CRISPR RNA (crRNA) and target RNA was added to the electrode surface. In the presence of target RNA, the cleavage capacity of Cas13a increases and leads to cleavage of ferrocene-labeled RNA, and a decrease in the SWV signal is observed. Similarly, in the absence of the target RNA, the cleavage activity of CAS13a is inhibited and the signal is not affected. It has been reported that it can be used as POC by integration with a portable potentiator, for further studies. A similar system is designed for the detection of *Staphylococcus aureus* (Huang et al., 2023). In this system, CRISPR/Cas12a was combined with saltatory rolling circle amplification (SRCA). The electrode surface was functionalized with methylene blue and ssDNA. The double-stranded DNAs obtained by SRCA were recognized by Cas12a /crRNAs. When Cas12a was activated in the presence of the target DNA, the ssDNA on the electrode surface began to be cleaved and a change in the peaks occurred. In this study, current changes were evaluated using the SWV technique, while CV and EIS techniques were used for electrode surface characterization.

15.3.2 OPTICAL BIOSENSORS

Optical systems are a type of biosensor defined as systems that convert information from photons into signals (Borisov and Wolfbeis, 2008). Visible or ultraviolet light is preferred by the sensor for analysis. The sensor essentially consists of a light source whose wavelength can be selected, a light detector, and a sensing material with which the light interacts with the target analyte (Askim and Suslick, 2017). Changes such as adsorption, fluorescence, refractive index, or light scattering, which occur as a result of the interaction between analyte and receptor are measured. Obtaining signals are related to target analyte concentrations (Garzón et al., 2019). There are two main types of optical biosensor systems. The first type shows a change in optical properties due to the intrinsic property of the bioreceptor molecule interacting with the target analyte, and this type of sensor system is referred to as label-free. In other types of sensors, biomolecules are covalently labeled with probes carrying an optical property in order to monitor the optical change (Borisov and Wolfbeis, 2008). This system is known as label-based and includes an additional step for labeling. Therefore, it is known that label-free systems are cheaper and simpler than label-based systems (Fan et al., 2008). Both types of optical biosensors are widely used because of their rapid results and portability, as well as their specificity and stability (Harshita et al., 2023). In recent years, they have been widely used to detect biological and chemical threats. Optical biosensor platforms for the detection of biological and chemical agents have been summarized in Table 15.2.

Optical systems in which changes in absorbance or reflected intensity of the complex formed between the analyte and the bioreceptor are evaluated are called colorimetric sensors. In these systems based on color changes, nanoparticles such as gold, magnetic, silver, and carbon nanotubes are widely used. There is no need for complex and expensive biodetection tools, because the color change can be detected even with the naked eye (Song et al., 2010). These features make these systems inexpensive, simple and practical, and advantageous over other methods (Song et al., 2011). In addition, these features mean they are widely used in the detection of biological (Chen et al., 2016; Cheng et al., 2019; Hu et al., 2023; Retout et al., 2022; Teengam et al., 2017; Ventura et al., 2020; Zhang et al., 2022, 2021) and chemical (Ahamed et al., 2023; Davidson et al., 2020) agents. Since 2019, the SARS-CoV-2 virus has taken its place in the group of biological agents that affect the whole world. Thus, studies on optical diagnostic systems have accelerated, especially the portability of these devices and the interpretation of results without the need for an expert are beneficial for SARS-CoV-2 diagnosis. In an optical platform developed on a colorimetric basis, AgNPs were used to observe the color change (Retout et al., 2022). In the system based on diffusion-limited aggregation, the AgNPs were coated with peptides. In the system designed for SARS-CoV-2 detection, the AgNPs were transformed into highly branched structures (AgFS) and color change occurred thanks to the repeated amino acids in the peptide molecule. This system can measure protease activity involved in SARS-CoV-2 and provide detection of SARS-CoV-2 from saliva or influenza-infected external breath condensate (EBC) samples. One of the main advantages of colorimetric test systems is their integration with smartphones.

TABLE 15.2
Some Optical Biosensors Designed for the Detection of Chemical and Biological Agents

Target Agents	Sensor type	Sensor Design	Results	Ref.
SARS-CoV-2	Colorimetric	Well plate/EBC or saliva sample-Peptide or Mpro-peptide/AgNP-BSPP	• A system based on AgNP and peptide was designed. • DLA structure formation was controlled by peptides. • Analysis from saliva or EBC media was provided. • LOD was determined as 0.5 nM.	Retout et al., 2022
SARS-CoV-2	Colorimetric	Plate/AuNP-Pep15/S-RBD/ AuNP-Pep12	• Sandwich-type bioassay was developed. • Two peptide molecules were used as bioreceptors. • S-RBD protein can rapidly detected from wastewater. (<30 min.) • LOD was determined as 0.01 nM.	Zhu and Zhou, 2022
SARS-CoV-2	Colorimetric	Cas12a/crRNA/RPA or RT-RPA/AuNP- ssDNA	• ORF1ab and N regions of SARS-CoV-2 genome were targets. • Analytical performance was increased due to RPA-coupled Cas12a.	Zhang et al., 2021
SARS-CoV-2	Colorimetric	AuNP/pAb/Sample	• Photochemical immobilization technique was used for functionalization of AuNP. • The color change can be visible to the naked-eye and color was changed from red to purple. • The virus can be detected from nasal and throat swabs.	Ventura et al., 2020
SARS-CoV-2	Colorimetric	Whatman paper/TE probe/ Urease/color indicator mixture/Viral RNA sample	• Viral RNA was detected with a paper-based test system called MARVE. • Origami paper was prepared with printer. • Color change was detected with smartphone.	Zhang et al., 2022
SARS-CoV-2	Colorimetric	Lateral flow test platforms (LFA) 1. T line: cocktail Ab; C line: anti-human IgG; conjugate ped: polymersome-pAb 2. T line: SARS-CoV-2 antigen; C line: anti-human IgG; conjugate ped: polymersome-pAb	• Two different LFA platforms were designed for detection of SARS-CoV-2 Ab and protein. • Polymerzom was used for colorimetric signal. • Cocktail antibody was purified from serum samples.	Ghorbanizamani et al., 2021

(Continued)

TABLE 15.2 (CONTINUED)
Some Optical Biosensors Designed for the Detection of Chemical and Biological Agents

Target Agents	Sensor type	Sensor Design	Results	Ref.
MERS-CoV, MTB, and HPV	Colorimetric	Paper/AgNP-acpcPNA/DNA target	• Paper-based multiplex DNA sensor was designed. • PNA probe and AgNPs were used in the platform design. • LOD values for MERS-CoV, MTB, and HPV were found to be 1.53, 1.27, and 1.03 nM, respectively.	Teengam et al., 2017
Bacillus anthracis	Colorimetric	UCNP-TPP/EBT/DPA	• In the presence of DPA, EBT was separated from the conjugate UCNP-TPP and the color changed from magenta to blue. • The analysis of DPA in human serum samples was provided. • LOD value was determined at 0.9 μM.	Cheng et al., 2019
E. coli O157:H7	Colorimetric	Plate/Pt@BSA/GA/E. coli Ab/ BSAT/E. coli/ HRP-Ab	• Sandwich-type colorimetric biosensor was developed. • In the presence of E. coli, the color changed to yellow. • LOD was determined as 900 cfu/mL.	Hu et al., 2023
Clostridium Botulinum	Colorimetric	Magnetic microparticle/ peptide fragment/AuNP/Cu^{2+}	• Botulinum neurotoxin that is produced by Clostridium Botulinum, was selected as target analyte. • In the presence of botulinum, peptide molecule was cleaved. • Cleaved peptide was bound with AuNP and agglomeration was induced. • LOD values were determined as 1 ng/mL (6.67 pM) by the naked eye, 0.25 ng/mL (1.67 pM) by spectral absorbance.	Chen et al., 2016
164 chemical warfare agent	Colorimetric	iSense	• 164 samples were successfully determined from a total of 174 independent samples. • Analysis time was 30 min. • 73 indicators were used for detection.	Davidson et al., 2020

(Continued)

TABLE 15.2 (CONTINUED)
Some Optical Biosensors Designed for the Detection of Chemical and Biological Agents

Target Agents	Sensor type	Sensor Design	Results	Ref.
Sarin gas mimic DCP	Colorimetric	BPDA/DCP	• The color of BPDA solution was yellow. • In the presence of DCP, color changed to fuchsia-pink color. • With the addition of TEA, the BPDA was separated, and the color turned yellow again. • The designed system was transferred to paper strips-based sensing. • LOD was determined as 44.5 nM.	Ahamed et al., 2023
SARS-CoV-2	Fluorescence	CdSe-ZnS QDs/Peptide/ Sample	• Highly sensitive B-cell epitopes predictions of SARS-CoV-2 were used for peptide synthesis. • LOD was determined as 100 pM.	Zheng et al., 2022
Clostridium botulinum	Fluorescence	Plate/PEG-Peptide/ superparamagnetic bead- anti-PEG Ab/BoNT/A LC/SAPE dye	• Analysis time was determined as 6 hours. • Carrot juice was used for real sample. • LOD value was determined as 0.5 nM (25 ng/mL).	Klisara et al., 2019
E. coli	Fluorescence	Tb-BTC/E. coli Ab/E. coli	• In the system, MOF-based biosensor was designed. • LOD value was determined as 3 cfu/mL. • Response time and analysis time were determined as 5 min and 20–25 min, respectively. • Fruit juice was used as real sample.	Gupta et al., 2020
E. coli	Fluorescence	MNP-aptamer/cDNA-UCNP/E. coli	• In pork meat, E. coli can be detected with the system. • In the presence of E. coli, aptamer was removed from cDNA and was bound with target bacteria. • LOD was determined as 10 cfu/mL.	Li et al., 2020

(Continued)

TABLE 15.2 (CONTINUED)
Some Optical Biosensors Designed for the Detection of Chemical and Biological Agents

Target Agents	Sensor type	Sensor Design	Results	Ref.
E. coli	Fluoressence	SNP-RB/*E. coli*	• The working principle of the test is related to the fluorescence quenching mechanism of SNP-RB. • Silica nanoparticles were modified with Rhodamine B. • Response time was determined as 15 min. • LOD was determined as 8 CFU/mL.	Jenie et al., 2021
Vibrio cholera	Fluorescence	CQD/Aptamer/*V. cholera*	• CQDs were used as fluorescence probe. • Water samples were used as real sample. • LOD was determined as 426 CFU/mL.	Karthikeyan et al., 2023
Tabun mimic DCNP	Fluorescence	IMP-Py or IMP-Py-OH/ DCNP	• Two probes were designed. • LOD value was determined as 16.9 nM for IMP-Py and 10.9 µM for IMP-Py-OH	Thakur et al., 2023
Bacillus anthracis	Colorimetric and fluorescent dual-modal platform	CdTe QDs/XO/DPA	• CdTe QDs were used and XO, metal indicator, was coordinated with Cd²⁺. • In the presence of DPA, XO and DPA were exchanged and the intensity of red fluorescence color intensity increased while the absorption was yellow due to the released XO. • The reaction time was reported to be 1 min and the LOD values were determined as 240 nM and 42 nM for colorimetric and fluorescence, respectively. • DPA levels in urine were successfully determined.	Cao et al., 2023
Bacillus anthracis	Colorimetric and fluorescent dual-mode platform	AgNP- citrate/Eu³⁺/DPA	• After the addition of DPA, the color of the solution changed from pink to yellow. • A red fluorescent color appeared after the addition of DPA. • The LOD values were determined to be 0.31 µM for the ratiometric colorimetric assay and 17 nM for the fluorescence assay.	Yin and Tong, 2021

(Continued)

TABLE 15.2 (CONTINUED)

Some Optical Biosensors Designed for the Detection of Chemical and Biological Agents

Target Agents	Sensor type	Sensor Design	Results	Ref.
Bacillus anthracis	Colorimetric and fluorescent dual-mode platform	Eu-CDs/EBT/DPA	• Dual ratiometric optical biosensor was designed with carbon dots. • DPA found by a biomarker in the bacterial spore layer was selected as the target analyte. • LOD values were calculated as 10.6 nM and 1.0 μM for ratio colorimetric assay and naked eye, respectively. • DPA can successfully be detected in human urine samples.	Zhou et al., 2021
Bacillus anthracis	Colorimetric and fluorescent dual-mode platform	TPE-Ts@Eu/GMP ICP/DPA	• A mechanism that causes a fluorescence change from blue to red has been used for DPA analysis. • ICP nanoparticles were used as fluorescence probe. • A multi-channel optical test kit integrated with a smartphone has been developed. • LOD value was determined as 100 nM with smartphone.	Huang et al., 2020
Bacillus anthracis	Colorimetric and fluorescent dual-mode platform	Tb^{3+}-AR/DPA	• Dual sensing probe was developed with Tb^{3+} and Ar. • Tb-Ar was given orange florescence on the UV lamp, while mauve was observed as color. • When Tb^{3+} is conjugated with DPA, the fluorescence color turns green, while colorimetrically, yellow is observed.	Xu et al., 2019
E. coli	Colorimetric and fluorescent dual-mode platform	Fluorescent measurement: L-Trp/*E. coli* Colorimetric measurement: L-Trp/*E. coli*/ DMABA	• L-Try was used as fluorescent indicator, while indol that is degradation product L-Trp and DMABA were used as colorimetric materials. • Rose-red color can be detected with smartphone. • Target bacteria was successfully determined from river water and milk sample. • LOD values for fluorimetry/colorimetry were determined as 23 and 2.5×10^4 CFU/mL, respectively.	Liu et al., 2022

(Continued)

TABLE 15.2 (CONTINUED)
Some Optical Biosensors Designed for the Detection of Chemical and Biological Agents

Target Agents	Sensor type	Sensor Design	Results	Ref.
Phosgene	Colorimetric and ratiometric fluorescent	TLC plate/AC-6ED/Phosgene	• High response time was more rapid than 30 min. • LOD was determined as 0.09 nM.	Liu et al., 2019
Cyanide	Colorimetric and fluorogenic sensor	Paper/C1 or C2 probe/CN$^-$	• Two benzothiazole (C1) and phenanthroimidazole (C2)-based probes were designed. • The system was integrated with a smartphone. • LOD values were calculated at 0.45 for C1 and 0.52 μM for C2.	Erdemir and Malkondu, 2020
Cyanide	Fluorometric and colorimetric sensor	HBT-Br/CN$^-$	• Intramolecular charge transfer and ESIPT processes were based in probe. • Thiazolium conjugated HBT-Br derivative was synthesized as probe. • Cyanide was determined from water samples and apricot seeds. • LOD values were calculated at 1.79 μM.	Erdemir and Malkondu, 2021
SARS-CoV-2	SERS	Gold silmeco substrates/peptide-PEG/spike sample	• Peptide was synthesized from the domain of ACE2 area for use as bioreceptor. • LOD value was determined as 300 nM.	Payne et al., 2021
Botulinum neurotoxins (BoNT) types A and B	SERS	AuNP- MGITC or NBA-PEG-Antibody/BoNT spike sample/Fe$_3$O$_4$-Magnetic bead-anti-BoNT mAb	• MGITC and NBA were used as label. • Detection time of the system was less than 2 h and sample volume was less than ELISA. • LOD values were determined as 5.7 ng/mL and 1.3 ng/mL for BoNT/A and BoNT/B, respectively.	Kim et al., 2019
Staphylococcus aureus	SERS	Fe$_3$O$_4$@Au core-shell NC-Apt/*S. aureus*	• The system has advantages such as simplicity, sensitivity, and short analysis time. • LOD was determined as 25 cfu/mL. • The aptasensor can be detected in milk sample.	Zhao et al., 2022

(Continued)

TABLE 15.2 (CONTINUED)

Some Optical Biosensors Designed for the Detection of Chemical and Biological Agents

Target Agents	Sensor type	Sensor Design	Results	Ref.
VX, tabun, and cyclosarin	SERS	Si nanopillar/Au/4-PAO	• 4-PAO was used as a probe. • The probe introduced selectivity to the gold. • LOD values were determined as 0.2 µM for tabun, 0.4 µM for VX.	Juhlin et al., 2020
VX and HD	SERS	Whatman quartz filters/Au	• It is a system that allows gas phase analysis and uses a portable Raman spectrometer. • LODs in solution were 1.1×10^{-8} M and 9.3×10^{-9} M for HD and VX, respectively. In gas phase, the measured LODs for HD and VX were 0.054µg/L(~8ppbV) and 0.008µg/L (~0.73ppbV), respectively.	Heleg-Shabtai et al., 2020
Salmonella typhimurium	Microfluidic colorimetric biosensor	Sample/Aptamer-PS-cysteamine/cDNA-MNP/AuNP	• In the presence of Salmonella, sample was bound with aptamer conjugate and AuNP was captured in the cysteamine terminal of the conjugate. Color change occurred. • On the absence of Salmonella, aptamer was conjugated with cDNA and AuNP didn't bind. So, the color change didn't occur. • Color change can be evaluated with smartphone imaging application and Vis spectroscopy. • Detection time was determined as 45 min. • LOD was determined as 6.0×10^{1} cfu/mL. • Salad sample was used as real sample.	Man et al., 2021
Salmonella spp.	Microfluidic colorimetric biosensor	MNP-Ab/*Salmonella*/MnO$_2$ NFs/TMB	• With the addition of TMB, the reaction was catalyzed by the MnO$_2$ NF nanomimetic enzyme and the color was turned yellow. • Detection time was determined as 45 min. • Chicken samples were used for detection of the target bacteria. • LOD was determined as 44 CFU/mL. • Color change can be evaluated with smartphone imaging application.	Xue et al., 2021

(Continued)

TABLE 15.2 (CONTINUED)

Some Optical Biosensors Designed for the Detection of Chemical and Biological Agents

Target Agents	Sensor type	Sensor Design	Results	Ref.
Salmonella spp.	Microfluid biosensor-fluorescence labeling	MNP-mAb/Salmonella/FMS-pAb	• A microfluidic system based on fluorescence labeling and processing with a smartphone was designed. • Magnetic nanoparticles were functionalized with antibodies to remove the target bacteria from apple juice. • FR150C was used for labeling. • LOD value was determined as 58 CFU/mL.	Wang et al., 2019
Vibrio cholerae	Microfluid biosensor	LF primer (ctxA gene)/pond water/cell lyse/dragon green polystyrene fluorescent nanoparticles	• Smartphone-based particle diffusometry was designed. • Loop-mediated isothermal amplification technique was used. • Detection time was determined as 35 min. • LOD was determined as 0.66 aM. • Waterborne pathogen was detected by the system.	Moehling et al., 2020
Salmonella typhimurium	3D fluidic chip	MNP-mAb/bacteria/Au@PtNCs/pAb/H_2O_2-TMB	• Peroxide-like enzymatic activity was based in the system. • Target bacteria can be detected within 1 hour. • LOD was determined as 17 CFU/mL.	Zheng et al., 2020
SARS-CoV-2	SPR	SPR gold surface/Polypeptide/SARS-CoV-2 recombinant S protein/etanolamine/BSA/Serum sample	• COVID-19 antibodies can be determined from dried blood spots, plasma, and serum samples. • Bioanalysis time was determined as 30 min. • The sensor system was portable.	Djaileb et al., 2021

(*Continued*)

TABLE 15.2 (CONTINUED)
Some Optical Biosensors Designed for the Detection of Chemical and Biological Agents

Target Agents	Sensor type	Sensor Design	Results	Ref.
Influenza	LSPR	CdZnSeS/ZnSeS QDs/peptide/AuNP/anti-HA antibody/Virus	• In the presence of influenza virus, fluorescence spectra of QDs was quenched. • LOD was determined as 17.02 fg/mL. • Serum was used for real samples.	Nasrin et al., 2020
HIV	FRET	GO/Pep-FAM/HIV-1 PR	• HIV-1 protease was detected in human serum. • The detection time was determined as 30 min. • LOD was determined as 1.18 ng/mL.	Zhang et al., 2018

EBC: Influenza-infected external breath condensate; Mpro: Protease involved in SARS-CoV-2; AgNP: Silver nanoparticle; BSPP: Bis(p-sulfonatophenyl)phenylphosphine; ORF1ab: Open reading frame 1ab; RT-RPA: Reverse transcription recombinase polymerase amplification; TE probe: DNA probe; pAb: Polyclonal antibody; acpcPNA: Pyrrolidinyl peptide nucleic acid; UCNPs: Upconversion nanoparticles; TPP: Sodium tripolyphosphate; DCP: Diethylchlorophosphate; BPDA: Benzoxazole-scaffold molecule; PEG: Polyethylene glycol; BoNT/A LC: Botulinum neurotoxin serotype A light chain; SAPE: Streptavidin-R-phycoerythrin; Tb-BTC: Terbium-based benzene-1,3,5 tricarboxylic acid; UCNP: Lanthanide upconversion nanoparticles; SNP-RB: Fluorescence silica nanoparticles; IMP-Py: 2-(pyren-1-yl)imidazo[1,2-a]pyridine; DCNP: Diethylcyanochlorophosphonate; XO: Xylenol orange; DPA: Dipicolinic acid; Eu-CDs: Carbon dots doped with europium; EBT: Eriochrome Black T; TPE-TS: water-soluble tetra(4-sulfophenyl)ethene; Eu: europium; GMP: Ligand guanine monophosphate; AR: Alizarin red; L-Trp: L-tryptophan; DMABA: 4-dimethylaminobenzaldehyde; AC-6ED: ethylenediamine at the 6th position of 1,2-anthracenecarboximide; NBA: Nile blue A; MGITC: Malachite green isothiocyanate; mAb: Monoclonal antibody; PAO: Pyridine (or pyridyl) amide oxime; cDNA – MNP: Complementary DNA - magnetic nanoparticle; MnO₂ NFs: Manganese dioxide nanoflowers; TMB: 3,3',5,5'-tetramethylbenzidine; FMS: Fluorescent microspheres; Au@PtNCs: gold@platinum nanocatalysts; GO: Graphene oxide; Pep-FAM: 5-carboxyfluorescein-labeled peptide; HIV-1 PR: HIV-1 protease;

In one study (Zhang et al., 2022), the color change that occurs in response to RNA sequencing of SARS-CoV-2 mutations was measured using a smartphone. This sensor system was termed the MARVE (for multiplexed, nucleic-acid-amplification-free, single-nucleotide-resolved viral evolution) technique and involves nucleotide resolution and visual detection of RNAs instead of nucleic acid amplification. This test system was made from the Whatman paper using the origami technique. A DNA probe called TE was used. In the presence of viral RNA in the throat swab sample, it was replaced by a strand of the DNA probe and the silver ion was released from the TE probe causing enzymatic degradation of the analyzed urea. In this case, the deterioration with pH change and color change can be observed with a smartphone. It was found that the analysis process takes 25–30 minutes with this system, while the contamination problem caused by amplification was eliminated and the cost was reduced. Colorimetric-based optical sensor systems can be designed to detect more than one agent. Especially for such systems, the design of user-friendly and low-cost products is very important. An example of such an optical sensor system are the paper-based devices by Teengam et. al (Teengam et al., 2017). A paper-based multiplex colorimetric sensor was developed for the detection of Middle East coronavirus (MERS-CoV), Mycobacterium tuberculosis (MTB), and human papillomavirus (HPV). The sensory surface was occurred with AgNPs and a pyrolidinic PNA probes (acpcPNA) for DNA detection of target diseases. This system, which allows simultaneous detection of three different targets with high selectivity, was designed for single use. The channels were created using the technique of printing on paper, and then the test system was completed by folding the paper using the origami technique. In the middle of the paper, the samples dropped from the sample application area pass through the created channels and reach the AgNP peptide part, resulting in color formation. Although the response time of the developed sensor is fast, the detection limits were determined as nM 1.53 for MERS-CoV, 1.27 for MTB, and 1.03 nM for HPV. The colorimetric platform was developed by Cheng et al (Cheng et al., 2019). The system can be used to detect spores of *Bacillus anthracis* bacteria, which are an anthrax biomarker. The nanoprobe was constructed using lanthanide-based upconversion nanoparticles (UCNPs) and Eriochrome Black T (EBT). As a working principle of the system, EBT was released from UCNPs in the presence of DPA (dipicolinic acid). The main disadvantage of these systems is that DPA binds the vacancies instead of DPA and EBT competing with each other, which is due to the very large surface area of the UCNPs. In this case, no color change was observed. Therefore, sodium tripolyphosphate (TPP) was used as a placeholder molecule before conjugation with the EBT. The color of the conjugate UCNPs-TPP/EBT was magenta and in the presence of DPA the color changed to blue due to the release of free EBT.

Fluorescence-based biosensors, which are based on the detection of light emitted from the target sample because of radiative excitation, are used particles with fluorescence properties such as florescence nanoparticles, quantum dots (QDs), and fluorophore (Celikbas et al., 2018). In this system, fluorescence-based nanoparticles are not only biocompatible but also have good photostability. Moreover, they bind specifically to the target and cause a very rare change in the optical properties of the materials as a result of binding (S. Wolfbeis, 2015). In recent years, the system has been widely used to detect biological (Gupta et al., 2020; Jenie et al., 2021; Karthikeyan et al., 2023; Klisara et al., 2019; Li et al., 2020; Zheng et al., 2022) and chemical (Thakur et al., 2023) threats. Detection of B-cell epitopes of SARS-CoV-2 was performed with the fluorescent biosensing system developed by Zheng et. al (Zheng et al., 2022). In this system, CdSe-ZnS QDs coated with a peptide probe were used. The conjugates were used to detect SARS-CoV-2 antibodies. The fluorescence intensity of the QDs/peptide probe was higher in the absence of target antibodies. The fluorescence intensity of the probe was lower when QDs/peptide were associated with spike RBD antibodies. Reaction time for the association of the QDs/peptide probe with antibodies was determined as 5 minutes and the LOD value of the system was 100 pM. In another study, a fluorescent-based cost-effective paper strip was designed for the detection of Tabun mimic diethylcyanochlorophosphonate (DCNP) (Thakur et al., 2023). In this study, two different and novel probes were synthesized as 2-(pyren-1-yl)imidazo[1,2-a]pyridine (IMP-Py) and (2-(pyren-1-yl)imidazo[1,2-a] pyridin-3-yl)methanol (IMP-Py-OH). The working performance of both probes was compared, and it was determined that IMP-Py was more selective and specific. In the presence of DCNP, the reason for this was explained by the fact that the increase in aggregation-induced emission enhancement (AIEE) and the increase in fluorescence intensity—due to the phenomenon of photoinduced electron transfer (PET) in IMP-Py—cannot be observed in IMP-Py-OH due to phosphorylation of the hydroxyl group. It has been reported that in-situ and rapid detection of DCNP was achieved using test strips as POC systems, which were prepared by coating Whatman papers with IMP -Py.

Dual-mode sensing, which includes both colorimetric and fluorescence sensing systems, is widely used in the detection of biological (Cao et al., 2023; Huang et al., 2020; Liu et al., 2022; Xu et al., 2019; Yin and Tong, 2021; Zhou et al., 2021) and chemical (Erdemir and Malkondu, 2021, 2020; Liu et al., 2019) agents. Dual-mode sensing is advantageous in terms of verifying the accuracy and reliability of both methods with each other (Yin and Tong, 2021). Anthrax caused by *Bacillus anthrax* bacteria can be detected with 2,6-pyridinedicarboxylic acid (DPA). The DPA biomarker is located in the layer of bacterial spores. In a study by Zhou et al. (Zhou et al., 2021), carbon dots doped with europium were synthesized by the thermal pyrolysis method. Subsequently, the Eu-CDs were complexed with Eriochrome Black T (EBT). Both probes (Eu-CDs and EBT-Eu@CDs) exhibit blue fluorescence. In the presence of DPA, DPA and EBT were displaced and the fluorescent color changed to red. It was reported that DPA analysis of urine samples could be possible. Not only can the color change be read by eye, but it can also be monitored with a smartphone application that can be used as a portable system. With a similar detection strategy, a multichannel responsive test kit was designed for the detection of anthrax (Huang et al., 2020). Infinite coordination polymer nanoparticles (ICPs) were used in the system for optical signal changes. $Eu(NO_3)_3 \cdot 6H_2O$ were coated with ligand guanine monophosphate (GMP) and water-soluble tetra(4-sulfophenyl)ethene (TPE-TS) and TPE-TS@Eu/GMP ICP nanoparticles were formed. The POC test system was prepared using Whatman filter paper. DPA solutions of different concentrations were mixed with ICP nanoparticles, and the obtained solutions were dropped on the test paper. The obtained color changes on the paper strips were analyzed by smartphone using an imagine analysis application. In the presence of *Bacillus anthrax* in the environment, the ICP nanoparticle began to emit red fluorescence due to the conjugation of DPA with Eu and emitted a blue color. The conversion from blue to red, which varies depending on the amount of DPA, was visible to the eye and could be measured within 10 minutes, using a smartphone. In addition, colorimetric and ratiometric florescence probes have been developed for the detection of chemical substances. A probe with fast reaction time, high selectivity, and lower LOD value was developed by Liu et al. for the detection of phosgene (Liu et al., 2019). In the study, AC-6ED was designed as a fluorescence probe. The base of the probe was 1,2-anthracenecarboximide and ethylenediamine was added at the sixth position. In the presence of phosgene, phosgene was bound with a terminal primary aliphatic amine. A five-membered carbonyl ring was formed by an intramolecular cyclization reaction. The color changed from red to yellow and was visible to the naked eye.

Another optical sensing platform for the detection of chemical (Heleg-Shabtai et al., 2020; Juhlin et al., 2020) and biological (Kim et al., 2019; Payne et al., 2021; Zhao et al., 2022) agents is surface enhanced Raman scattering (SERS). In these systems, measurement of SERS signals is essential and the signals are generated by exciting the surface with laser excitation by sending Raman signals to the target analyte (Lee et al., 2017). In a study for the detection of SARS-CoV-2 biological agent by Payne et al. (Payne et al., 2021), a peptide modified SERS sensor was developed. The surface of the SERS was determined using the gold Silmeco substrate and the peptide that was synthesized from the ACE2 domain. Detection of SARS-CoV-2 spike proteins with high selectivity was achieved using the SERS surface. The peptide molecule is supplemented with PEG, and this has been reported to result in the spike protein being recognized at lower concentrations. As the SARS-CoV-2 virus variants cause mutations in the RBD region, the peptides will continue to bind variants. Therefore, it has been reported that peptide molecules need to be retrained for variant-specific recognition. The use of antibodies as bioreceptors is also quite common. A SERS sensor was developed for Botulinum neurotoxins, one of the most widely used and high-risk biological agents in bioterrorism. Similarly, SERS biosensors are used for the detection of chemical substances. In recent years by Juhlin et al. developed a SERS sensor for the rapid detection of tabun, VX, and cyclosarin (Juhlin et al., 2020). In the study, a Si nanopillar coated with gold was used. As a capture probe, 4-pyridine amide oxime (4-PAO), also known as a nerve agent antidote, was used. Specific binding was determined as tabun > VX > cyclosarin.

Studies of microfluidic platform design aimed at miniaturizing optical sensor systems for use as POCs have become quite common in recent years. The sensing systems, likened to a miniature laboratory, are a compact system that manages solution mixing, separation, sample introduction, reaction, and signal generation (Duocastella et al., 2010). In addition to fast and sensitive detection, online recognition is possible with microfluidic systems integrated with smartphones. These systems, which rely on the use of fluid mechanics to combine and automate conventional devices on a single, small platform, are very difficult to design (Kulkarni et al., 2022). Despite these difficulties, many systems are designed for the detection of biological agents

(Man et al., 2021; Moehling et al., 2020; Wang et al., 2019; Xue et al., 2021; Zheng et al., 2020). In a microfluidic system developed against *Salmonella*, monoclonal antibodies and MNP were used to separate the bacteria from apple juice. Then the bacteria labeled with fluorescent microspheres were injected into the microfluidic system. A fluorescence-based microscopic system and smartphone-based detection were provided in the system. Thanks to real-time video processing, it is reported that this system can detect pathogens in food within two hours.

In addition to all these optical systems, many different platforms such as SPR (Djaileb et al., 2021; Nasrin et al., 2020) and FRET (Zhang et al., 2018) are being developed for biological and chemical agents. Although there are designs for SARS-CoV-2 with the pandemic that has spread across the world in recent years, many optical platforms developed for other chemical and biological pathogens will continue to take their place in the literature.

15.3.3   Pros and Cons of Electrochemical and Optical Biosensors

The impact of chemical and biological agents on human life is increasing day by day. Therefore, there is greater need for POC diagnostic platforms for the detection of these agents that have advantageous features, such as small sample volume, speed of detection, and low cost. Many optical and electrochemical platforms have been designed for this purpose in recent years. These biosensor systems reduce the amount of samples required to detect target substances and their results are at least as sensitive as those obtained with equipment used in a laboratory (Harshita et al., 2023). One of the main advantages that allows the widespread use of both biosensor systems for the detection of these agents is that both systems can be small, while the analysis surfaces can be targeted to a specific analyte. The main aim is to provide the user with the ability to analyze in the field by designing the systems in a portable format. In addition, most systems are integrated with smartphones. It is possible to use these systems outside the laboratory environment. With these biosensor systems, which act as a bridge between the laboratory environment and daily life, dependence on specialists decreases (Celikbas et al., 2019). All the pros and cons of optical and electrochemical biosensors are summarized in Table 15.3.

TABLE 15.3
Pros and Cons of Electrochemical and Optical Biosensors

	Pros	Cons
Optical biosensor	• Real-time detection • Possibility of detection with the naked-eye • Short response time • Design as POC	• Light sensitivity • Low reproducibility and stability • Vulnerability to user errors
Electrochemical biosensor	• Low LOD • Design as POC • Robustness	• Requirement of expert specialists • Interference effect in the sample matrix

15.4 CONCLUSIONS AND FUTURE PERSPECTIVES

In this chapter, electrochemical and optical biosensor systems as alternative analytical methods for the detection of biological and chemical agents have been extensively evaluated. In recent years, with the pandemic, biosensor studies have focused on the detection of SARS-CoV-2, but the experience gained here will guide systems being developed to detect chemical and biological agents that threaten the health of all living things. Instead of the traditionally used laboratory and expert-dependent detection methods, the portable designs of electrochemical and optical sensor systems are noteworthy. Although most studies have highlighted portable optical biosensor systems that provide naked eye detection, these systems are not used as a commercial product. The situation is similar with electrochemical sensors. In addition to portable electrochemical designs, integrations with smartphones stand out. Biosensors that can detect multiple analytes simultaneously and in wearable form continue to develop day by day. We can predict that commercialized products of both systems will increase in the future. Undoubtedly, with these platforms, it will be possible to detect biological and chemical agents that can affect life in all areas, in very small amounts, within seconds, and our lives will become easier.

REFERENCES

Afkhami, A., Hashemi, P., Bagheri, H., Salimian, J., Ahmadi, A., Madrakian, T., 2017. Impedimetric immunosensor for the label-free and direct detection of botulinum neurotoxin serotype A using Au nanoparticles/graphene-chitosan composite. *Biosensors and Bioelectronics*, Special Issue Selected papers from the 26th Anniversary World Congress on Biosensors (Part II) 93, 124–131. https://doi.org/10.1016/j.bios.2016.09.059

Ahamed, S., Sultana, T., Mahato, M., Tohora, N., Rahman, Z., Ghanta, S., Kumar Das, S., 2023. Fabrication of a re-usable benzoxazole-based colorimetric sensor for selective and sensitive recognition of sarin mimic, diethylchlorophosphate. *Microchemical Journal* 193, 108982. https://doi.org/10.1016/j.microc.2023.108982

Ali, M.R., Bacchu, M.S., Das, S., Akter, S., Rahman, M.M., Saad Aly, M.A., Khan, M.Z.H., 2023. Label free flexible electrochemical DNA biosensor for selective detection of Shigella flexneri in real food samples. *Talanta* 253, 123909. https://doi.org/10.1016/j.talanta.2022.123909

Aristov, N., Habekost, A., 2015. Cyclic voltammetry – A versatile electrochemical method investigating electron transfer processes. *World Journal of Chemical Education* 3, 115–119. https://doi.org/10.12691/wjce-3-5-2

Askim, J.R., Suslick, K.S., 2017. 8.04 – Colorimetric and fluorometric sensor arrays for molecular recognition, in: Atwood, J.L. (Ed.), *Comprehensive Supramolecular Chemistry II*. Elsevier, Oxford, pp. 37–88. https://doi.org/10.1016/B978-0-12-409547-2.12616-2

Bacchu, M.S., Ali, M.R., Das, S., Akter, S., Sakamoto, H., Suye, S.-I., Rahman, M.M., Campbell, K., Khan, M.Z.H., 2022. A DNA functionalized advanced electrochemical biosensor for identification of the foodborne pathogen Salmonella enterica serovar Typhi in real samples. *Analytica Chimica Acta* 1192, 339332. https://doi.org/10.1016/j.aca.2021.339332

Balkourani, G., Brouzgou, A., Archonti, M., Papandrianos, N., Song, S., Tsiakaras, P., 2021. Emerging materials for the electrochemical detection of COVID-19. *Journal of Electroanalytical Chemistry (Lausanne)* 893, 115289. https://doi.org/10.1016/j.jelechem.2021.115289

Batista Deroco, P., Giarola, J. de F., Wachholz Júnior, D., Arantes Lorga, G., Tatsuo Kubota, L., 2020. Chapter four – Paper-based electrochemical sensing devices, in: Merkoçi, A. (Ed.), *Comprehensive Analytical Chemistry, Paper Based Sensors*. Elsevier, pp. 91–137. https://doi.org/10.1016/bs.coac.2019.11.001

Bazsefidpar, S., Freitas, M., Pereira, C.R., Gutiérrez, G., Serrano-Pertierra, E., Nouws, H.P.A., Matos, M., Delerue-Matos, C., Blanco-López, M.C., 2023. Fe3O4@Au core–shell magnetic nanoparticles for the rapid analysis of E. coli O157:H7 in an electrochemical immunoassay. *Biosensors* 13, 567. https://doi.org/10.3390/bios13050567

Borisov, S.M., Wolfbeis, O.S., 2008. Optical biosensors. *Chemical Reviews* 108, 423–461. https://doi.org/10.1021/cr068105t

Bu, L., Guo, L., Xie, J., 2020. An in situ assay of nerve agents enabled by a self-assembled bienzymatic electrochemical biosensor. *New Journal of Chemistry* 44, 7460–7466. https://doi.org/10.1039/D0NJ00929F

Cao, Y., Gong, X., Li, L., Li, H., Zhang, X., Guo, D.-Y., Wang, F., Pan, Q., 2023. Xylenol orange-modified CdTe quantum dots as a fluorescent/colorimetric dual-modal probe for anthrax biomarker based on competitive coordination. *Talanta* 261, 124664. https://doi.org/10.1016/j.talanta.2023.124664

Caratelli, V., Fillo, S., D'Amore, N., Rossetto, O., Pirazzini, M., Moccia, M., Avitabile, C., Moscone, D., Lista, F., Arduini, F., 2021. Paper-based electrochemical peptide sensor for on-site detection of botulinum neurotoxin serotype A and C. *Biosensors and Bioelectronics* 183, 113210. https://doi.org/10.1016/j.bios.2021.113210

Carvalhais, C., Querido, M., Pereira, C.C., Santos, J., 2021. Biological risk assessment: A challenge for occupational safety and health practitioners during the COVID-19 (SARS-CoV-2) pandemic. *Work* 69, 3–13. https://doi.org/10.3233/WOR-205302

CDC., 2019. Centers for Disease Control and Prevention (CDC) | Bioterrorism Agents/ Diseases (by category) | Emergency Preparedness & Response [WWW Document]. https://emergency.cdc.gov/agent/agentlist-category.asp (accessed 9.26.23).

Celikbas, E., Balaban, S., Evran, S., Coskunol, H., Timur, S., 2019. A bottom-up approach for developing aptasensors for abused drugs: Biosensors in forensics. *Biosensors* 9, 118. https://doi.org/10.3390/bios9040118

Celikbas, E., Guler Celik, E., Timur, S., 2018. Paper-based analytical methods for smartphone sensing with functional nanoparticles: Bridges from smart surfaces to global health. *Analytical Chemistry* 90, 12325–12333. https://doi.org/10.1021/acs.analchem.8b03120

Chaubey, A., Malhotra, B.D., 2002. Mediated biosensors. *Biosensors and Bioelectronics* 17, 441–456. https://doi.org/10.1016/S0956-5663(01)00313-X

Chen, H., Cui, C., Ma, X., Yang, W., Zuo, Y., 2020. Amperometric biosensor for Brucella testing through molecular orientation technology in combination with signal amplification technology. *ChemElectroChem* 7, 2672–2679. https://doi.org/10.1002/celc.202000569

Chen, H., Liu, H., Cui, C., Zhang, W., Zuo, Y., 2022a. Recombinant protein G/Au nanoparticles/graphene oxide modified electrodes used as an electrochemical biosensor for Brucella Testing in milk. *Journal of Food Science and Technology* 59, 4653–4662. https://doi.org/10.1007/s13197-022-05544-8

Chen, H., Liu, H., Cui, C., Zhang, X., Yang, W., Zuo, Y., 2022b. Highly sensitive detection of Brucella in milk by cysteamine functionalized nanogold/4-Mercaptobenzoic acid electrochemical biosensor. *Food Measure* 16, 3501–3511. https://doi.org/10.1007/s11694-022-01428-9

Chen, S., Chu, L.T., Chen, T.-H., 2016. Colorimetric detection of active botulinum neurotoxin using Cu2+ mediated gold nanoparticles agglomeration. *Sensors and Actuators B: Chemical* 235, 563–567. https://doi.org/10.1016/j.snb.2016.05.118

Cheng, Z.-H., Liu, X., Zhang, S.-Q., Yang, T., Chen, M.-L., Wang, J.-H., 2019. Placeholder strategy with upconversion Nanoparticles–Eriochrome Black T conjugate for a colorimetric assay of an anthrax biomarker. *Analytical Chemistry* 91, 12094–12099. https://doi.org/10.1021/acs.analchem.9b03342

Colozza, N., Kehe, K., Popp, T., Steinritz, D., Moscone, D., Arduini, F., 2021a. Paper-based electrochemical sensor for on-site detection of the sulphur mustard. *Environmental Science and Pollution Research* 28, 25069–25080. https://doi.org/10.1007/s11356-018-2545-6

Colozza, N., Mazzaracchio, V., Kehe, K., Tsoutsoulopoulos, A., Schioppa, S., Fabiani, L., Steinritz, D., Moscone, D., Arduini, F., 2021b. Development of novel carbon black-based heterogeneous oligonucleotide-antibody assay for sulfur mustard detection. *Sensors and Actuators B: Chemical* 328, 129054. https://doi.org/10.1016/j.snb.2020.129054

Davidson, C.E., Dixon, M.M., Williams, B.R., Kilper, G.K., Lim, S.H., Martino, R.A., Rhodes, P., Hulet, M.S., Miles, R.W., Samuels, A.C., Emanuel, P.A., Miklos, A.E., 2020. Detection of chemical warfare agents by colorimetric sensor arrays. *ACS Sensors* 5, 1102–1109. https://doi.org/10.1021/acssensors.0c00042

Deffo, G., Nde Tene, T.F., Medonbou Dongmo, L., Zambou Jiokeng, S.L., Tonleu Temgoua, R.C., 2024. Differential pulse and square-wave voltammetry as sensitive methods for electroanalysis applications, in: Wandelt, K., Bussetti, G. (Eds.), *Encyclopedia of Solid-Liquid Interfaces* (First Edition). Elsevier, Oxford, pp. 409–417. https://doi.org/10.1016/B978-0-323-85669-0.00040-4

Djaileb, A., Jodaylami, M.H., Coutu, J., Ricard, P., Lamarre, M., Rochet, L., Cellier-Goetghebeur, S., Macaulay, D., Charron, B., Lavallée, É., Thibault, V., Stevenson, K., Forest, S., S. Live, L., Abonnenc, N., Guedon, A., Quessy, P., Lemay, J.-F., Farnós, O., Kamen, A., Stuible, M., Gervais, C., Durocher, Y., Cholette, F., Mesa, C., Kim, J., Cayer, M.-P., Grandmont, M.-J. de, Brouard, D., Trottier, S., Boudreau, D., Pelletier, J.N., Masson, J.-F., 2021. Cross-validation of ELISA and a portable surface plasmon resonance instrument for IgG antibody serology with SARS-CoV-2 positive individuals. *Analyst* 146, 4905–4917. https://doi.org/10.1039/D1AN00893E

Dryden, M.D.M., Wheeler, A.R., 2015. DStat: A versatile, open-source potentiostat for electroanalysis and integration. *PLoS One* 10, e0140349. https://doi.org/10.1371/journal.pone.0140349

Duocastella, M., Fernández-Pradas, J.M., Morenza, J.L., Zafra, D., Serra, P., 2010. Novel laser printing technique for miniaturized biosensors preparation. *Sensors and Actuators B: Chemical* 145, 596–600. https://doi.org/10.1016/j.snb.2009.11.055

Durmus, C., Balaban Hanoglu, S., Harmanci, D., Moulahoum, H., Tok, K., Ghorbanizamani, F., Sanli, S., Zihnioglu, F., Evran, S., Cicek, C., Sertoz, R., Arda, B., Goksel, T., Turhan, K., Timur, S., 2022. Indiscriminate SARS-CoV-2 multivariant detection using magnetic nanoparticle-based electrochemical immunosensing. *Talanta* 243, 123356. https://doi.org/10.1016/j.talanta.2022.123356

Erdemir, S., Malkondu, S., 2021. Visual and quantitative detection of CN– ion in aqueous media by an HBT-Br and thiazolium conjugated fluorometric and colorimetric probe: Real samples and useful applications. *Talanta* 221, 121639. https://doi.org/10.1016/j.talanta.2020.121639

Erdemir, S., Malkondu, S., 2020. On-site and low-cost detection of cyanide by simple colorimetric and fluorogenic sensors: Smartphone and test strip applications. *Talanta* 207, 120278. https://doi.org/10.1016/j.talanta.2019.120278

Fan, X., White, I.M., Shopova, S.I., Zhu, H., Suter, J.D., Sun, Y., 2008. Sensitive optical biosensors for unlabeled targets: A review. *Analytica Chimica Acta* 620, 8–26. https://doi.org/10.1016/j.aca.2008.05.022

Futra, D., Tan, L.L., Lee, S.Y., Lertanantawong, B., Heng, L.Y., 2023. An ultrasensitive Voltammetric Genosensor for the detection of bacteria vibrio cholerae in vegetable and environmental water samples. *Biosensors* 13, 616. https://doi.org/10.3390/bios13060616

Garzón, V., Pinacho, D.G., Bustos, R.-H., Garzón, G., Bustamante, S., 2019. Optical biosensors for therapeutic drug monitoring. *Biosensors* 9, 132. https://doi.org/10.3390/bios9040132

Ghindilis, A.L., Atanasov, P., Wilkins, M., Wilkins, E., 1998. Immunosensors: Electrochemical sensing and other engineering approaches. *Biosensors and Bioelectronics* 13, 113–131. https://doi.org/10.1016/S0956-5663(97)00031-6

Ghorbanizamani, F., Tok, K., Moulahoum, H., Harmanci, D., Hanoglu, S.B., Durmus, C., Zihnioglu, F., Evran, S., Cicek, C., Sertoz, R., Arda, B., Goksel, T., Turhan, K., Timur, S., 2021. Dye-loaded polymersome-based lateral flow assay: Rational design of a COVID-19 testing platform by repurposing SARS-CoV-2 antibody cocktail and antigens obtained from positive human samples. *ACS Sensors* 6, 2988–2997. https://doi.org/10.1021/acssensors.1c00854

Green, M.S., LeDuc, J., Cohen, D., Franz, D.R., 2019. Confronting the threat of bioterrorism: Realities, challenges, and defensive strategies. *The Lancet Infectious Diseases* 19, e2–e13. https://doi.org/10.1016/S1473-3099(18)30298-6

Grieshaber, D., MacKenzie, R., Vörös, J., Reimhult, E., 2008. Electrochemical biosensors – Sensor principles and architectures. *Sensors* 8, 1400–1458. https://doi.org/10.3390/s80314000

Gubala, V., Harris, L.F., Ricco, A.J., Tan, M.X., Williams, D.E., 2012. Point of care diagnostics: Status and future. *Analytical Chemistry* 84, 487–515. https://doi.org/10.1021/ac2030199

Gupta, A., Garg, M., Singh, S., Deep, A., Sharma, A.L., 2020. Highly sensitive optical detection of Escherichia coli using terbium-based metal–organic framework. *ACS Applied Materials & Interfaces* 12, 48198–48205. https://doi.org/10.1021/acsami.0c14312

Harshita, Wu, H.-F., Kailasa, S.K., 2023. Recent advances in nanomaterials-based optical sensors for detection of various biomarkers (inorganic species, organic and biomolecules). *Luminescence* 38, 954–998. https://doi.org/10.1002/bio.4353

Heleg-Shabtai, V., Sharabi, H., Zaltsman, A., Ron, I., Pevzner, A., 2020. Surface-enhanced Raman spectroscopy (SERS) for detection of VX and HD in the gas phase using a hand-held Raman spectrometer. *Analyst* 145, 6334–6341. https://doi.org/10.1039/D0AN01170C

Hu, C., Wei, G., Zhu, F., Wu, A., Luo, L., Shen, S., Jia, N., Zhang, J., 2023. Label-free electrochemical biosensing architecture based on a protein-decorated Pt@BSA nanocomposite for sensitive immunoassay of pathogenic Escherichia coli O157:H7. *ACS Sustainable Chemistry & Engineering* 11, 7894–7907. https://doi.org/10.1021/acssuschemeng.3c01241

Huang, C., Ma, R., Luo, Y., Shi, G., Deng, J., Zhou, T., 2020. Stimulus response of TPE-TS@Eu/GMP ICPs: Toward colorimetric sensing of an anthrax biomarker with double ratiometric fluorescence and its coffee ring test kit for point-of-use application. *Analytical Chemistry* 92, 12934–12942. https://doi.org/10.1021/acs.analchem.0c01570

Huang, L., Yuan, N., Guo, W., Zhang, Y., Zhang, W., 2023. An electrochemical biosensor for the highly sensitive detection of Staphylococcus aureus based on SRCA-CRISPR/Cas12a. *Talanta* 252, 123821. https://doi.org/10.1016/j.talanta.2022.123821

Jansson, D., Åstot, C., 2015. Analysis of paralytic shellfish toxins, potential chemical threat agents, in food using hydrophilic interaction liquid chromatography–mass spectrometry. *Journal of Chromatography A* 1417, 41–48. https://doi.org/10.1016/j.chroma.2015.09.029

Jenie, S.N.A., Kusumastuti, Y., Krismastuti, F.S.H., Untoro, Y.M., Dewi, R.T., Udin, L.Z., Artanti, N., 2021. Rapid fluorescence quenching detection of Escherichia coli using natural silica-based nanoparticles. *Sensors* 21, 881. https://doi.org/10.3390/s21030881

Juhlin, L., Mikaelsson, T., Hakonen, A., Schmidt, M.S., Rindzevicius, T., Boisen, A., Käll, M., Andersson, P.O., 2020. Selective surface-enhanced Raman scattering detection of Tabun, VX and Cyclosarin nerve agents using 4-pyridine amide oxime functionalized gold nanopillars. *Talanta* 211, 120721. https://doi.org/10.1016/j.talanta.2020.120721

Karthikeyan, M., Venkatasubbu, G.D., Rathinasabapathi, P., 2023. A label-free carbon dots-based fluorescent aptasensor for the detection of V. cholerae O139. *Diamond and Related Materials* 137, 110173. https://doi.org/10.1016/j.diamond.2023.110173

Kashefi-Kheyrabadi, L., Nguyen, H.V., Go, A., Lee, M.-H., 2023. Ultrasensitive and amplification-free detection of SARS-CoV-2 RNA using an electrochemical biosensor powered by CRISPR/Cas13a. *Bioelectrochemistry* 150, 108364. https://doi.org/10.1016/j.bioelechem.2023.108364

Kim, K., Choi, N., Jeon, J.H., Rhie, G., Choo, J., 2019. SERS-based immunoassays for the detection of botulinum toxins A and B using magnetic beads. *Sensors* 19, 4081. https://doi.org/10.3390/s19194081

Kim, Y., Lee, D., Seo, Y., Jung, H.G., Jang, J.W., Park, D., Kim, I., Kim, J., Lee, G., Hwang, K.S., Kim, S.-H., Lee, S.W., Lee, J.H., Yoon, D.S., 2023. Caco-2 cell-derived biomimetic electrochemical biosensor for cholera toxin detection. *Biosensors and Bioelectronics* 226, 115105. https://doi.org/10.1016/j.bios.2023.115105

Klisara, N., Peters, J., Haasnoot, W., Nielen, M.W.F., Palaniappan, A., Liedberg, B., 2019. Functional fluorescence assay of botulinum neurotoxin A in complex matrices using magnetic beads. *Sensors and Actuators B: Chemical* 281, 912–919. https://doi.org/10.1016/j.snb.2018.10.100

Koncki, R., 2007. Recent developments in potentiometric biosensors for biomedical analysis. *Analytica Chimica Acta* 599, 7–15. https://doi.org/10.1016/j.aca.2007.08.003

Kulkarni, M.B., Ayachit, N.H., Aminabhavi, T.M., 2022. Biosensors and microfluidic biosensors: From fabrication to application. *Biosensors* 12, 543. https://doi.org/10.3390/bios12070543

Lee, J., Takemura, K., Park, E.Y., 2017. Plasmonic nanomaterial-based optical biosensing platforms for virus detection. *Sensors* 17, 2332. https://doi.org/10.3390/s17102332

Li, C., Liu, C., Liu, R., Wang, Y., Li, A., Tian, S., Cheng, W., Ding, S., Li, W., Zhao, M., Xia, Q., 2023. A novel CRISPR/Cas14a-based electrochemical biosensor for ultrasensitive detection of Burkholderia pseudomallei with PtPd@PCN-224 nanoenzymes for signal amplification. *Biosensors and Bioelectronics* 225, 115098. https://doi.org/10.1016/j.bios.2023.115098

Li, H., Ahmad, W., Rong, Y., Chen, Q., Zuo, M., Ouyang, Q., Guo, Z., 2020. Designing an aptamer based magnetic and upconversion nanoparticles conjugated fluorescence sensor for screening Escherichia coli in food. *Food Control* 107, 106761. https://doi.org/10.1016/j.foodcont.2019.106761

Li, H., Liu, X., Li, L., Mu, X., Genov, R., Mason, A.J., 2017. CMOS electrochemical instrumentation for biosensor microsystems: A review. *Sensors* 17, 74. https://doi.org/10.3390/s17010074

Liang, Q., Huang, Y., Wang, M., Kuang, D., Yang, J., Yi, Y., Shi, H., Li, J., Yang, J., Li, G., 2023. An electrochemical biosensor for SARS-CoV-2 detection via its papain-like cysteine protease and the protease inhibitor screening. *Chemical Engineering Journal* 452, 139646. https://doi.org/10.1016/j.cej.2022.139646

Liu, J., Yu, Z., Chen, Q., Jia, L., 2022. L-Tryptophan assisted construction of fluorescent and colorimetric dual-channel biosensor for detection of live Escherichia coli. *Microchemical Journal* 174, 107085. https://doi.org/10.1016/j.microc.2021.107085

Liu, P., Liu, N., Liu, C., Jia, Y., Huang, L., Zhou, G., Li, C., Wang, S., 2019. A colorimetric and ratiometric fluorescent probe with ultralow detection limit and high selectivity for phosgene sensing. *Dyes and Pigments* 163, 489–495. https://doi.org/10.1016/j.dyepig.2018.12.031

Luka, G.S., Najjaran, H., Hoorfar, M., 2022. On-chip-based electrochemical biosensor for the sensitive and label-free detection of Cryptosporidium. *Scientific Reports* 12, 6957. https://doi.org/10.1038/s41598-022-10765-0

Malvano, F., Pilloton, R., Albanese, D., 2020. A novel impedimetric biosensor based on the antimicrobial activity of the peptide nisin for the detection of Salmonella spp. *Food Chemistry* 325, 126868. https://doi.org/10.1016/j.foodchem.2020.126868

Man, Y., Ban, M., Li, A., Jin, X., Du, Y., Pan, L., 2021. A microfluidic colorimetric biosensor for in-field detection of Salmonella in fresh-cut vegetables using thiolated polystyrene microspheres, hose-based microvalve and smartphone imaging APP. *Food Chemistry* 354, 129578. https://doi.org/10.1016/j.foodchem.2021.129578

Martins, G., Gogola, J.L., Budni, L.H., Janegitz, B.C., Marcolino-Junior, L.H., Bergamini, M.F., 2021. 3D-printed electrode as a new platform for electrochemical immunosensors for virus detection. *Analytica Chimica Acta* 1147, 30–37. https://doi.org/10.1016/j.aca.2020.12.014

Mazzaracchio, V., Neagu, D., Porchetta, A., Marcoccio, E., Pomponi, A., Faggioni, G., D'Amore, N., Notargiacomo, A., Pea, M., Moscone, D., Palleschi, G., Lista, F., Arduini, F., 2019. A label-free impedimetric aptasensor for the detection of Bacillus anthracis spore simulant. *Biosensors and Bioelectronics* 126, 640–646. https://doi.org/10.1016/j.bios.2018.11.017

Melo, A.M.A., Furtado, R.F., de Fatima Borges, M., Biswas, A., Cheng, H.N., Alves, C.R., 2021. Performance of an amperometric immunosensor assembled on carboxymethylated cashew gum for Salmonella detection. *Microchemical Journal* 167, 106268. https://doi.org/10.1016/j.microc.2021.106268

Mishra, R.K., Barfidokht, A., Karajic, A., Sempionatto, J.R., Wang, J., Wang, J., 2018. Wearable potentiometric tattoo biosensor for on-body detection of G-type nerve agents simulants. *Sensors and Actuators B: Chemical* 273, 966–972. https://doi.org/10.1016/j.snb.2018.07.001

Moehling, T.J., Lee, D.H., Henderson, M.E., McDonald, M.K., Tsang, P.H., Kaakeh, S., Kim, E.S., Wereley, S.T., Kinzer-Ursem, T.L., Clayton, K.N., Linnes, J.C., 2020. A smartphone-based particle diffusometry platform for sub-attomolar detection of Vibrio cholerae in environmental water. *Biosensors and Bioelectronics* 167, 112497. https://doi.org/10.1016/j.bios.2020.112497

Naresh, V., Lee, N., 2021. A review on biosensors and recent development of nanostructured materials-enabled biosensors. *Sensors* 21, 1109. https://doi.org/10.3390/s21041109

Nasrin, F., Chowdhury, A.D., Takemura, K., Kozaki, I., Honda, H., Adegoke, O., Park, E.Y., 2020. Fluorometric virus detection platform using quantum dots-gold nanocomposites optimizing the linker length variation. *Analytica Chimica Acta* 1109, 148–157. https://doi.org/10.1016/j.aca.2020.02.039

Osteryoung, J.G., Osteryoung, R.A., 1985. Square wave voltammetry. *Analytical Chemistry* 57, 101–110. https://doi.org/10.1021/ac00279a004

Payne, T.D., Klawa, S.J., Jian, T., Kim, S.H., Papanikolas, M.J., Freeman, R., Schultz, Z.D., 2021. Catching COVID: Engineering peptide-modified surface-enhanced Raman spectroscopy sensors for SARS-CoV-2. *ACS Sensors* 6, 3436–3444. https://doi.org/10.1021/acssensors.1c01344

Pohanka, M., Skládal, P., Kroèa, M., 2007. Biosensors for biological warfare agent detection (Review paper). *Defence Science Journal* 57, 185–193. https://doi.org/10.14429/dsj.57.1760

Qian, X., Qu, Q., Li, L., Ran, X., Zuo, L., Huang, R., Wang, Q., 2018. Ultrasensitive electrochemical detection of clostridium perfringens DNA based morphology-dependent DNA adsorption properties of CeO_2 nanorods in dairy products. *Sensors* 18, 1878. https://doi.org/10.3390/s18061878

Retout, M., Mantri, Y., Jin, Z., Zhou, J., Noël, G., Donovan, B., Yim, W., Jokerst, J.V., 2022. Peptide-induced fractal assembly of silver nanoparticles for visual detection of disease biomarkers. *ACS Nano* 16, 6165–6175. https://doi.org/10.1021/acsnano.1c11643

Ronkainen, N.J., Halsall, H.B., Heineman, W.R., 2010. Electrochemical biosensors. *Chemical Society Reviews* 39, 1747–1763. https://doi.org/10.1039/B714449K

Sarwar, S., Lin, M.-C., Amezaga, C., Wei, Z., Iyayi, E., Polk, H., Wang, R., Wang, H., Zhang, X., 2023. Ultrasensitive electrochemical biosensors based on zinc sulfide/graphene hybrid for rapid detection of SARS-CoV-2. *Advanced Composites and Hybrid Materials* 6, 49. https://doi.org/10.1007/s42114-023-00630-7

Saylan, Y., Akgönüllü, S., Denizli, A., 2020. Plasmonic sensors for monitoring biological and chemical threat agents. *Biosensors* 10, 142. https://doi.org/10.3390/bios10100142

Serra, P.A., 2011. *Biosensors for Health, Environment and Biosecurity.* BoD – Books on Demand.

Sharma, M.K., Narayanan, J., Pardasani, D., Srivastava, D.N., Upadhyay, S., Goel, A.K., 2016. Ultrasensitive electrochemical immunoassay for surface array protein, a Bacillus anthracis biomarker using Au–Pd nanocrystals loaded on boron-nitride nanosheets as catalytic labels. *Biosensors and Bioelectronics* 80, 442–449. https://doi.org/10.1016/j.bios.2016.02.008

Siavash Moakhar, R., Mahimkar, R., Khorrami Jahromi, A., Mahshid, S.S., del Real Mata, C., Lu, Y., Vasquez Camargo, F., Dixon, B., Gilleard, J., J Da Silva, A., Ndao, M., Mahshid, S., 2023. Aptamer-based electrochemical microfluidic biosensor for the detection of cryptosporidium parvum. *ACS Sensors* 8, 2149–2158. https://doi.org/10.1021/acsnano.2c01349

Simões, F.R., Xavier, M.G., 2017. 6 – Electrochemical sensors, in: Da Róz, A.L., Ferreira, M., de Lima Leite, F., Oliveira, O.N. (Eds.), *Nanoscience and Its Applications, Micro and Nano Technologies.* William Andrew Publishing, pp. 155–178. https://doi.org/10.1016/B978-0-323-49780-0.00006-5

Singh, A., Sharma, A., Ahmed, A., Sundramoorthy, A.K., Furukawa, H., Arya, S., Khosla, A., 2021a. Recent advances in electrochemical biosensors: Applications, challenges, and future scope. *Biosensors* 11, 336. https://doi.org/10.3390/bios11090336

Singh, A., Sharma, A., Ahmed, A., Sundramoorthy, A.K., Furukawa, H., Arya, S., Khosla, A., 2021b. Recent advances in electrochemical biosensors: Applications, challenges, and future scope. *Biosensors* 11, 336. https://doi.org/10.3390/bios11090336

Song, S., Qin, Y., He, Y., Huang, Q., Fan, C., Chen, H.-Y., 2010. Functional nanoprobes for ultrasensitive detection of biomolecules. *Chemical Society Reviews* 39, 4234–4243. https://doi.org/10.1039/C000682N

Song, Y., Wei, W., Qu, X., 2011. Colorimetric biosensing using smart materials. *Advanced Materials* 23, 4215–4236. https://doi.org/10.1002/adma.201101853

Sorouri, R., Bagheri, H., Afkhami, A., Salimian, J., 2017. Fabrication of a novel highly sensitive and selective immunosensor for botulinum neurotoxin serotype A based on an effective platform of electrosynthesized gold nanodendrites/chitosan nanoparticles. *Sensors* 17, 1074. https://doi.org/10.3390/s17051074

Strozier, E.D., Mooney, D.D., Friedenberg, D.A., Klupinski, T.P., Triplett, C.A., 2016. Use of comprehensive two-dimensional gas chromatography with time-of-flight mass spectrometric detection and random forest pattern recognition techniques for classifying chemical threat agents and detecting chemical attribution signatures. *Analytical Chemistry* 88, 7068–7075. https://doi.org/10.1021/acs.analchem.6b00725

Teengam, P., Siangproh, W., Tuantranont, A., Vilaivan, T., Chailapakul, O., Henry, C.S., 2017. Multiplex paper-based colorimetric DNA sensor using pyrrolidinyl peptide nucleic acid-induced AgNPs aggregation for detecting MERS-CoV, MTB, and HPV oligonucleotides. *Analytical Chemistry* 89, 5428–5435. https://doi.org/10.1021/acs.analchem.7b00255

Thakur, A., Chaudhran, P.A., Sharma, A., 2023. Simple and efficient PET and AIEE mechanism-based fluorescent probes for sensing Tabun mimic DCNP. *Analytica Chimica Acta* 1239, 340727. https://doi.org/10.1016/j.aca.2022.340727

Tolun, A., Altintas, Z., 2023. Chapter 16 – Chemical sensing of food phenolics and antioxidant capacity, in: Barhoum, A., Altintas, Z. (Eds.), *Advanced Sensor Technology*. Elsevier, pp. 593–646. https://doi.org/10.1016/B978-0-323-90222-9.00004-2

Upadhyayula, V.K.K., 2012. Functionalized gold nanoparticle supported sensory mechanisms applied in detection of chemical and biological threat agents: A review. *Analytica Chimica Acta* 715, 1–18. https://doi.org/10.1016/j.aca.2011.12.008

Venton, B.J., DiScenza, D.J., 2020. Chapter 3 – Voltammetry, in: Patel, B. (Ed.), *Electrochemistry for Bioanalysis*. Elsevier, pp. 27–50. https://doi.org/10.1016/B978-0-12-821203-5.00004-X

Ventura, B.D., Cennamo, M., Minopoli, A., Campanile, R., Censi, S.B., Terracciano, D., Portella, G., Velotta, R., 2020. Colorimetric test for fast detection of SARS-CoV-2 in nasal and throat swabs. *ACS Sensors* 5, 3043–3048. https://doi.org/10.1021/acssensors.0c01742

Wang, S., Zheng, L., Cai, G., Liu, N., Liao, M., Li, Y., Zhang, X., Lin, J., 2019. A microfluidic biosensor for online and sensitive detection of Salmonella typhimurium using fluorescence labeling and smartphone video processing. *Biosensors and Bioelectronics* 140, 111333. https://doi.org/10.1016/j.bios.2019.111333

Wang, W., Yuan, W., Wang, D., Mai, X., Wang, D., Zhu, Y., Liu, F., Sun, Z., 2022. Dual-mode sensor based on the synergy of magnetic separation and functionalized probes for the ultrasensitive detection of Clostridium perfringens. *RSC Advances* 12, 25744–25752. https://doi.org/10.1039/D2RA04344K

Wolfbeis, O.S., 2015. An overview of nanoparticles commonly used in fluorescent bioimaging. *Chemical Society Reviews* 44, 4743–4768. https://doi.org/10.1039/C4CS00392F

Xu, C., Cheng, S., Zhang, F., Wang, Q., 2023. Simultaneous detection of Staphylococcus aureus and E. coli O157:H7 using a self-calibrated potentiometric sensors array based on peptide recognition. *IEEE Sensors Journal* 23, 17399–17406. https://doi.org/10.1109/JSEN.2023.3285617

Xu, M., Huang, W., Lu, D., Huang, C., Deng, J., Zhou, T., 2019. Alizarin Red–Tb3+ complex as a ratiometric colorimetric and fluorescent dual probe for the smartphone-based detection of an anthrax biomarker. *Analytical Methods* 11, 4267–4273. https://doi.org/10.1039/C9AY01235D

Xue, L., Jin, N., Guo, R., Wang, S., Qi, W., Liu, Y., Li, Y., Lin, J., 2021. Microfluidic colorimetric biosensors based on MnO2 nanozymes and convergence–divergence spiral micromixers for rapid and sensitive detection of salmonella. *ACS Sensors* 6, 2883–2892. https://doi.org/10.1021/acssensors.1c00292

Yang, X., Yin, Z.-Z., Zheng, G., Zhou, M., Zhang, H., Li, J., Cai, W., Kong, Y., 2023. Molecularly imprinted miniature electrochemical biosensor for SARS-CoV-2 spike protein based on Au nanoparticles and reduced graphene oxide modified acupuncture needle. *Bioelectrochemistry* 151, 108375. https://doi.org/10.1016/j.bioelechem.2023.108375

Yin, S., Tong, C., 2021. Europium(III)-modified silver nanoparticles as ratiometric colorimetric and fluorescent dual-mode probes for selective detection of dipicolinic acid in bacterial spores and lake waters. *ACS Applied Nano Materials* 4, 5469–5477. https://doi.org/10.1021/acsanm.1c00838

Yin, Z.-Z., Liu, Z., Zhou, M., Yang, X., Zheng, G., Zhang, H., Kong, Y., 2023. A surface molecularly imprinted electrochemical biosensor for the detection of SARS-CoV-2 spike protein by using Cu7S4-Au as built-in probe. *Bioelectrochemistry* 152, 108462. https://doi.org/10.1016/j.bioelechem.2023.108462

Yiwei, X., Yahui, L., Weilong, T., Jiyong, S., Xiaobo, Z., Wen, Z., Xinai, Z., Yanxiao, L., Changqiang, Z., Lele, A., Hong, L., Tingting, S., 2021. Electrochemical determination of hantavirus using gold nanoparticle-modified graphene as an electrode material and Cu-based metal-organic framework assisted signal generation. *Microchim Acta* 188, 112. https://doi.org/10.1007/s00604-021-04769-2

Yu, L., Peel, G.K., Cheema, F.H., Lawrence, W.S., Bukreyeva, N., Jinks, C.W., Peel, J.E., Peterson, J.W., Paessler, S., Hourani, M., Ren, Z., 2020. Catching and killing of airborne SARS-CoV-2 to control spread of COVID-19 by a heated air disinfection system. *Materials Today Physics* 15, 100249. https://doi.org/10.1016/j.mtphys.2020.100249

Yue, G., Su, S., Li, N., Shuai, M., Lai, X., Astruc, D., Zhao, P., 2016. Gold nanoparticles as sensors in the colorimetric and fluorescence detection of chemical warfare agents. *Coordination Chemistry Reviews* 311, 75–84. https://doi.org/10.1016/j.ccr.2015.11.009

Zdrachek, E., Bakker, E., 2021. Potentiometric sensing. *Analytical Chemistry* 93, 72–102. https://doi.org/10.1021/acs.analchem.0c04249

Zhang, T., Deng, R., Wang, Y., Wu, C., Zhang, K., Wang, C., Gong, N., Ledesma-Amaro, R., Teng, X., Yang, C., Xue, T., Zhang, Y., Hu, Y., He, Q., Li, W., Li, J., 2022. A paper-based assay for the colorimetric detection of SARS-CoV-2 variants at single-nucleotide resolution. *Nature Biomedical Engineering* 6, 957–967. https://doi.org/10.1038/s41551-022-00907-0

Zhang, W.S., Pan, J., Li, F., Zhu, M., Xu, M., Zhu, H., Yu, Y., Su, G., 2021. Reverse Transcription recombinase polymerase amplification coupled with CRISPR-Cas12a for facile and highly sensitive colorimetric SARS-CoV-2 detection. *Analytical Chemistry* 93, 4126–4133. https://doi.org/10.1021/acs.analchem.1c00013

Zhang, Y., Chen, X., Roozbahani, G.M., Guan, X., 2018. Graphene oxide-based biosensing platform for rapid and sensitive detection of HIV-1 protease. *Analytical and Bioanalytical Chemistry* 410, 6177–6185. https://doi.org/10.1007/s00216-018-1224-2

Zhao, W., Zhang, D., Zhou, T., Huang, J., Wang, Y., Li, B., Chen, L., Yang, J., Liu, Y., 2022. Aptamer-conjugated magnetic Fe3O4@Au core-shell multifunctional nanoprobe: A three-in-one aptasensor for selective capture, sensitive SERS detection and efficient near-infrared light triggered photothermal therapy of Staphylococcus aureus. *Sensors and Actuators B: Chemical* 350, 130879. https://doi.org/10.1016/j.snb.2021.130879

Zheng, L., Cai, G., Qi, W., Wang, S., Wang, M., Lin, J., 2020. Optical biosensor for rapid detection of salmonella typhimurium based on porous Gold@Platinum nanocatalysts and a 3D fluidic chip. *ACS Sensors* 5, 65–72. https://doi.org/10.1021/acssensors.9b01472

Zheng, Y., Song, K., Cai, K., Liu, L., Tang, D., Long, W., Zhai, B., Chen, J., Tao, Y., Zhao, Y., Liang, S., Huang, Q., Liu, Q., Zhang, Q., Chen, Y., Liu, Y., Li, H., Wang, P., Lan, K., Liu, H., Xu, K., 2022. B-cell-epitope-based fluorescent quantum dot biosensors for SARS-CoV-2 enable highly sensitive COVID-19 antibody detection. *Viruses* 14, 1031. https://doi.org/10.3390/v14051031

Zhou, Q., Fang, Y., Li, J., Hong, D., Zhu, P., Chen, S., Tan, K., 2021. A design strategy of dual-ratiomentric optical probe based on europium-doped carbon dots for colorimetric and fluorescent visual detection of anthrax biomarker. *Talanta* 222, 121548. https://doi.org/10.1016/j.talanta.2020.121548

Zhou, Y., Li, Z., Huang, J., Wu, Y., Mao, X., Tan, Y., Liu, H., Ma, D., Li, X., Wang, X., 2023. Development of a phage-based electrochemical biosensor for detection of Escherichia coli O157: H7 GXEC-N07. *Bioelectrochemistry* 150, 108345. https://doi.org/10.1016/j.bioelechem.2022.108345

Zhu, Q., Zhou, X., 2022. A colorimetric sandwich-type bioassay for SARS-CoV-2 using a hACE2-based affinity peptide pair. *Journal of Hazardous Materials* 425, 127923. https://doi.org/10.1016/j.jhazmat.2021.127923

Index

For Product Safety Concerns and Information please contact our EU
representative GPSR@taylorandfrancis.com
Taylor & Francis Verlag GmbH, Kaufingerstraße 24, 80331 München, Germany

www.ingramcontent.com/pod-product-compliance
Lightning Source LLC
Chambersburg PA
CBHW060806220326
41598CB00022B/2553